MOLECULAR IMMUNOLOGY OF MYCOTIC AND ACTINOMYCOTIC INFECTIONS

MOLECULAR IMMUNOLOGY OF MYCOTIC AND ACTINOMYCOTIC INFECTIONS

ERROL REISS, Ph.D.
Division of Mycotic Diseases
Center for Infectious Diseases
Centers for Disease Control
Public Health Service
U.S. Department of Health and Human Services
Atlanta

ELSEVIER
New York • Amsterdam • London

Elsevier Science Publishing Co., Inc.
52 Vanderbilt Avenue, New York, New York 10017

Sole distributors outside the United States and Canada:
Elsevier Science Publishers B.V.
P.O. Box 211, 1000 AE Amsterdam, The Netherlands

© 1986 by Elsevier Science Publishing Co., Inc.

This book has been registered with the Copyright Clearance Center, Inc. For further information, please contact the Copyright Clearance Center, Salem, Massachusetts.

Library of Congress Cataloging in Publication Data

Reiss, Errol.
  Molecular immunology of mycotic and actinomycotic infections.

    Includes index.
    1. Mycoses—Immunological aspects. 2. Actinomycosis—Immunological aspects. I. Title. [DNLM: 1. Actinomycosis—immunology. 2. Mycoses—immunology. WC 450 R378m]
  QR201.M98R45   1986   616.9'6907'9   86-8969
  ISBN 0-444-01039-4

Current printing (last digit):
10 9 8 7 6 5 4 3 2 1

Manufactured in the United States of America

Contents

PREFACE xv
ACKNOWLEDGMENTS xvii

Chapter 1 Introduction 1

Chapter 2 Molecular Organization
of the Fungal Cell Wall 5
  2.1 INTRODUCTION 5
  2.2 OVERVIEW OF PEPTIDOMANNANS 6
    2.2.1 Mural 7
    2.2.2 Capsular 9
    2.2.3 Secreted 10
    2.2.4 Peptidomannan Structure 10
  2.3 GLUCANS 11
  2.4 CHITIN 14
    2.4.1 Chitin Detection in Fungi 15
    2.4.2 Molds 16
    2.4.3 Yeasts 17
    2.4.4 Pathogenic Genera 19
  2.5 FUNCTIONS OF THE CELL WALL 20
  2.6 PREPARATION AND CRITERIA FOR PURITY 22
    2.6.1 Isolation and Purification 23
    2.6.2 Summary of Criteria for Wall Preparation and Purity 24
      2.6.2a Antigenic Structure 24
      2.6.2b Compositional Analysis 24
      2.6.2c Outer Layers and Periplasmic Enzymes 25
      2.6.2d Summary 25

2.7 PROTEIN IN THE CELL WALL ... 26
2.8 SUMMARY ... 29

Chapter 3 *Blastomyces dermatitidis* ... 41

3.1 INTRODUCTION ... 41
3.2 BLASTOMYCIN ... 41
3.3 CELL WALL EXTRACTS ... 42
3.4 CYTOPLASMIC SOLUBLES ... 45
3.5 HUMORAL RESPONSE ... 46
    3.5.1 Complement Fixation and Immunodiffusion ... 46
    3.5.2 A-Factor ... 46
3.6 PHAGOCYTE INTERACTIONS ... 47
3.7 CELL-MEDIATED IMMUNITY ... 48
    3.7.1 Human Infections ... 48
    3.7.2 Experimental Murine Blastomycosis ... 48
3.8 SUMMARY ... 50

Chapter 4 *Coccidioides immitis* ... 53

4.1 INTRODUCTION ... 53
4.2 ANTIGENIC STRUCTURE ... 54
    4.2.1 Coccidioidin ... 54
    4.2.2 Wall Antigens ... 55
    4.2.3 Spherulin ... 60
4.3 HUMORAL RESPONSES ... 61
    4.3.1 Tube Precipitin ... 61
    4.3.2 Complement-Fixing Antibodies ... 61
    4.3.3 Immunoglobulin E ... 62
4.4 CELL-MEDIATED IMMUNITY ... 63
    4.4.1 Cutaneous Hypersensitivity and Anergy ... 63
    4.4.2 Non-Antigen-Specific Responses ... 65
        4.4.2a Erythema Nodosum ... 65
        4.4.2b Mixed Lymphocyte, Phytohemagglutinin, and E-Rosette Responses ... 65
    4.4.3 Antigen-Specific Responses ... 66
        4.4.3a Blastogenesis and MIF Production ... 66
        4.4.3b Serum Blocking Activity and Soluble Immune Complexes ... 67
        4.4.3c Antigens Evoking Cellular Responses ... 68
    4.4.4 Summary ... 69
4.5 IMMUNOTHERAPY WITH TRANSFER FACTOR ... 69
4.6 EXPERIMENTAL MURINE COCCIDIOIDOMYCOSIS ... 70
    4.6.1 Spherule Vaccine ... 70
    4.6.2 Protective Cellular Mechanisms ... 71
4.7 SUMMARY ... 72

Chapter 5 *Histoplasma capsulatum* — 77

- 5.1 INTRODUCTION — 77
- 5.2 CELL WALL — 78
  - 5.2.1 Serotypes — 79
  - 5.2.2 Chemotypes — 79
  - 5.2.3 Mannan — 80
- 5.3 HISTOPLASMIN — 82
  - 5.3.1 H and M Factors — 82
    - 5.3.1a *Purification of H and M Antigens of Histoplasmin* — 83
    - 5.3.1b *Polyacrylamide Gel Electrophoresis and Monoclonal Antibodies* — 85
- 5.4 HUMORAL RESPONSES — 85
  - 5.4.1 Complement Fixation and Precipitins — 85
  - 5.4.2 Radioimmunoassay and Enzyme Immunoassay — 88
- 5.5 DELAYED CUTANEOUS HYPERSENSITIVITY — 90
- 5.6 INTERACTION WITH MACROPHAGES — 91
- 5.7 IMMUNOMODULATION — 93
  - 5.7.1 Human Infections — 93
  - 5.7.2 Presumed Ocular Histoplasmosis Syndrome — 95
  - 5.7.3 Murine Infections — 95
  - 5.7.4 Ribosomal Vaccines — 96
- 5.8 SUMMARY — 97

Chapter 6 *Paracoccidioides brasiliensis* — 103

- 6.1 INTRODUCTION — 103
- 6.2 CELL WALL — 103
  - 6.2.1 α-1,3-Glucan — 105
  - 6.2.2 Galactomannan — 105
- 6.3 SECRETED ANTIGENS — 106
  - 6.3.1 Factor "E" — 106
  - 6.3.2 Paracoccidioidin — 108
- 6.4 CELL-MEDIATED RESPONSES — 108
  - 6.4.1 Delayed Cutaneous Hypersensitivity — 109
    - 6.4.1a *Specific* — 109
    - 6.4.1b *General* — 109
  - 6.4.2 In Vitro Cellular Responses — 109
    - 6.4.2a *Blastogenesis* — 109
    - 6.4.2b *Leukocyte Inhibitory Factor* — 110
    - 6.4.2c *E-Rosettes* — 110
    - 6.4.2d *Gammopathy* — 110
    - 6.4.2e *Killing by Granulocytes* — 110
    - 6.4.2f *Immunotherapy* — 111
- 6.5 SUMMARY — 111

Chapter 7 Sporothrix schenckii — 115
- 7.1 INTRODUCTION — 115
- 7.2 CELL WALL — 116
  - 7.2.1 Peptido-L-rhamno-D-mannan — 116
- 7.3 DELAYED CUTANEOUS HYPERSENSITIVITY — 120
  - 7.3.1 Sporotrichin and Alternatives — 120
  - 7.3.2 Skin Test Surveys — 122
- 7.4 IMMUNODIAGNOSIS — 123
  - 7.4.1 Agglutinins and Precipitins — 123
- 7.5 INTERACTION WITH PHAGOCYTES — 124
  - 7.5.1 Asteroid Bodies — 124
- 7.6 EXPERIMENTAL INFECTIONS AND IMMUNITY — 125
- 7.7 SUMMARY — 126

Chapter 8 Aspergillus fumigatus — 129
- 8.1 INTRODUCTION — 129
- 8.2 POLYSACCHARIDE ANTIGENS — 131
  - 8.2.1 Galactomannan — 132
    - 8.2.1a Antigenemia — 133
  - 8.2.2 C-Substance — 137
  - 8.2.3 β-1,3-Glucan — 137
- 8.3 PROTEIN ANTIGENS — 137
  - 8.3.1 Secretion During Growth and Autolysis — 138
  - 8.3.2 Immunoenzyme Analysis — 138
  - 8.3.3 Mycelial Extracts — 139
  - 8.3.4 Purified Antigens — 142
  - 8.3.5 Summary of Secreted and Cell Extract Proteins — 143
- 8.4 ALLERGIC BRONCHOPULMONARY ASPERGILLOSIS — 143
- 8.5 HUMORAL IMMUNITY AND IMMUNOPATHOLOGY — 144
  - 8.5.1 Immunoglobulin E — 144
  - 8.5.2 Immune Complexes — 146
  - 8.5.3 Precipitins and Primary Binding Assays — 146
  - 8.5.4 Crossreactivity Among Aspergillus Species Antigens — 149
- 8.6 CELL-MEDIATED IMMUNITY — 149
- 8.7 SUMMARY — 150

Chapter 9 Dermatophytes — 157
- 9.1 INTRODUCTION — 157
- 9.2 CELL WALL — 158
- 9.3 GALACTOMANNAN POLYSACCHARIDES AND GLYCOPEPTIDES — 159
  - 9.3.1 Isolation — 159
  - 9.3.2 Purification — 160
  - 9.3.3 Structure — 161

9.3.4 C-Substance	163
9.3.5 Crossreactivity and Taxonomic Value	164
9.3.6 Humoral Responses	164
9.4 GLUCANS	165
9.5 PROTEINASES	166
9.6 CHRONIC DERMATOPHYTOSIS	167
9.6.1 Immunodeficits	168
9.6.2 Soluble Suppressor Substances	168
9.6.3 IgE-Mediated Suppression	169
9.6.4 Tolerogenic Potential of *Trichophyton rubrum*	170
9.6.5 Variation in Antigenic Potency	170
9.6.6 Ratio of T-Helper to T-Suppressor Cells	171
9.7 SUMMARY	171

Chapter 10 Zygomycetes — 177

10.1 INTRODUCTION	177
10.2 CELL WALL	177
10.2.1 Mucoran	178
10.2.2 Peptido-L-fuco-D-mannan	179
10.3 RHINOCEREBRAL ZYGOMYCOSIS	183
10.4 *RHIZOPUS* INHIBITORY FACTOR	184
10.5 NEUTROPHIL CIDAL MECHANISMS	185
10.6 EXPERIMENTAL ZYGOMYCOSIS	185
10.7 SUMMARY	187

Chapter 11 Candida albicans — 191

11.1 INTRODUCTION	191
11.2 CELL WALL ARCHITECTURE	193
11.2.1 Protein	193
11.2.2 Mannan	193
11.2.3 Glucan	194
11.2.4 Chitin	194
11.2.5 Summary of Ultrastructure	195
11.3 MANNAN IMMUNOCHEMISTRY AND IMMUNOLOGY	196
11.3.1 General Plan	196
11.3.2 Serotypes	197
11.3.3 Preparation	199
11.3.4 Methylation-Fragmentation	200
11.3.5 Acetolysis	202
11.3.6 *O*-Phosphonomannan	205
11.3.7 Exo-α-Mannan-Hydrolase	208
11.3.8 Mannan as an Immunomodulator	210
11.3.8a Regulation of Blastogenesis	210
11.3.8b Complement Activation	211
11.3.9 Mannanemia	212

11.4	Glucan	218
	11.4.1 Chemistry	218
	11.4.2 Role in Infection	218
11.5	PROTEIN ANTIGENS	219
	11.5.1 Acid Phosphatase	221
	11.5.2 Acidic Carboxyl Proteinase	223
	11.5.3 Major Cytoplasmic Proteins	225
	11.5.4 Membrane-Bound Antigens	227
11.6	IMMUNE RESPONSES TO INFECTION: THYMIC AND HUMORAL	227
	11.6.1 Murine Infections	227
	11.6.2 Chronic Mucocutaneous Candidiasis	230
	11.6.2a Classification and Immune Defects	231
	11.6.2b Mechanisms of Blocking Blastogenesis	233
	11.6.2c Hyper-IgE-Recurrent Infection Syndrome	234
	11.6.2d Lesions of CMC	234
	11.6.2e Immunotherapy	235
11.7	PHAGOCYTIC CANDIDACIDAL MECHANISMS	236
11.8	SUMMARY	239

Chapter 12 *Cryptococcus neoformans* 251

12.1	INTRODUCTION	251
12.2	THE CAPSULE AS AN AGGRESSIN	252
	12.2.1 Physical Barrier	252
	12.2.2 Persistence in Body Fluids and Tissues	252
	12.2.3 Rodent Virulence	253
	12.2.4 Tolerance	253
	12.2.5 Antiphagocytic Nature	255
12.3	CAPSULAR AND CELL WALL HETEROGLYCANS OF *CRYPTOCOCCUS* AND *TREMELLA* SPECIES	256
	12.3.1 Structure of Glucuronoxylomannan	257
	12.3.2 Structural Features of the Galactoxylomannan	260
	12.3.3 Serologic Heterogeneity in the *Cryptococcus neoformans* Capsule	261
12.4	OPSONINS	264
	12.4.1 Genetic Control of Humoral Immunity	265
12.5	PROTEIN ANTIGENS	265
	12.5.1 Cryptococcin	266
	12.5.2 Homogenate Fractions	267
12.6	HOST DEFENSES	268
	12.6.1 Acquired Immunity	270
	12.6.2 Cell-Mediated Cytotoxicity	270
	12.6.3 T-Lymphocyte Functions	271
	12.6.3a Mice	271
	12.6.3b Humans	273
12.7	SUMMARY	274

Chapter 13 Nocardiae — 281

- 13.1 INTRODUCTION — 281
- 13.2 MURAL COVALENT SKELETON — 282
  - 13.2.1 Nocardomycolic Acids — 283
  - 13.2.2 Arabinogalactan and Arabinomannan — 283
  - 13.2.3 Peptidoglycan — 285
- 13.3 ANTIGENIC COMPLEXES — 285
  - 13.3.1 Serologic Analysis — 285
    - 13.3.1a *Agglutinins* — 287
    - 13.3.1b *Complement Fixation* — 287
    - 13.3.1c *Immunodiffusion* — 287
    - 13.3.1d *Serotypes* — 289
    - 13.3.1e *Mycobacterial Reference Precipitins* — 290
  - 13.3.2 Delayed Hypersensitivity — 290
- 13.4 FRACTIONATION OF NOCARDIAL EXTRACTS — 291
  - 13.4.1 *Nocardia*-Active Polypeptides — 291
  - 13.4.2 Superoxide Dismutase — 293
  - 13.4.3 Enzyme-Linked Immunoelectro Transfer Blot — 293
  - 13.4.4 Polysaccharides — 294
  - 13.4.5 Ribosomal Proteins — 295
  - 13.4.6 Glycolipid — 296
- 13.5 INTERACTION WITH MACROPHAGES — 296
- 13.6 EXPERIMENTAL MURINE NOCARDIOSIS — 297
  - 13.6.1 L-Forms — 298
- 13.7 NOCARDIAL IMMUNOSTIMULANTS — 300
  - 13.7.1 B-Lymphocyte Mitogen — 301
- 13.8 SUMMARY — 303

Chapter 14 Microaerophilic Actinomycetes — 311

- 14.1 INTRODUCTION — 311
  - 14.1.1 Actinomycetes as Agents of Periodontal Disease — 311
- 14.2 POLYSACCHARIDES — 312
  - 14.2.1 Levan — 312
  - 14.2.2 6-Deoxy-L-talose Polymer — 313
  - 14.2.3 Viscous Exocellular *N*-Acetylglucosamine Polymer — 313
- 14.3 FIMBRIAE AND VIRULENCE ASSOCIATION — 314
- 14.4 B-CELL MITOGEN OF *ACTINOMYCES VISCOSUS* — 315
  - 14.4.1 Cellular Requirements for Activation — 315
  - 14.4.2 Blastogenesis Correlates with Periodontitis — 317
- 14.5 SUMMARY — 318

Chapter 15 Thermophilic Actinomycetes — 321

- 15.1 INTRODUCTION — 321
- 15.2 ANTIGENIC STRUCTURE AND PRECIPITINS — 322
  - 15.2.1 Precipitin Score, C Region Antigens — 322

	15.2.2	Double Dialysis Technique	322
	15.2.3	Glycoprotein "a"	323
	15.2.4	Glycopeptide "1"	324
	15.2.5	Two-Dimensional Immunoelectrophoresis, Isoelectric Focusing, and Enzyme Immunoassays	325
	15.2.6	Summary of Serology	328
15.3	PROTEINASES AND COMPLEMENT ACTIVATION		329
15.4	IMMUNOPATHOLOGY		330
	15.4.1	Humans	330
	15.4.2	Animals	332
	15.4.3	Model for Immunopathogenesis	332
15.5	SUMMARY		333

Chapter 16 Review of Innate and Humoral Immunity — 337

16.1	INTRODUCTION		337
	16.1.1	Serum Fungistasis	338
	16.1.2	Neutrophil Fungicidal Properties	338
	16.1.3	Natural Killer Cells	340
16.2	OVERVIEW OF MECHANISMS REGULATING HUMORAL RESPONSES		340
	16.2.1	T-Dependent Responses	341
	16.2.2	T-Independent Responses	343
	16.2.3	T-Suppression	343
16.3	FUNGAL POLYSACCHARIDE ANTIGENS		344
	16.3.1	Mannans	344
	16.3.2	Galactomannans	346
	16.3.3	Mannanemia	348
16.4	PROTEIN ANTIGENS		349
	16.4.1	Species-Specific Secreted Factors	349
	16.4.2	Secreted Proteinases	350
	16.4.3	Extracts of Mycelial and Yeast Forms	351
	16.4.4	Wall-Bound	351
	16.4.5	Membrane-Bound	352
16.5	ENZYME IMMUNOASSAYS		352
16.6	MONOCLONAL ANTIBODIES		358
16.7	ROLE OF HUMORAL IMMUNITY IN INFECTION		362
	16.7.1	Polyclonal B-Cell Activation	362
	16.7.2	Immune Complex Formation	362
	16.7.3	Protective Role for Humoral Immunity	363
	16.7.4	No Measurable Protective Effect	364
	16.7.5	Pathogenetic Antibodies	364

Chapter 17 Review of Cell-Mediated Immunity — 373

17.1	INTRODUCTION		373
	17.1.1	T-Cell–B-Cell–Macrophage Collaboration	373

17.2	LYMPHOKINES	375
17.3	T-CELL–T-CELL INTERACTIONS	376
17.4	T-CELL EFFECTOR MECHANISMS	376
17.5	GENERATION OF T-SUPPRESSOR CELLS AND SOLUBLE SUPPRESSOR SUBSTANCES	376
	17.5.1 Antigen-Specific Suppression	377
	17.5.2 Modulation of Suppression by Immune Complexes	378
	17.5.3 General Suppression	378
	17.5.4 Suppression by Anti-Idiotype Antibodies	379
	17.5.5 Contrasuppression	379
17.6	CELLULAR BASIS OF DELAYED HYPERSENSITIVITY	380
	17.6.1 Organization of the Hypersensitivity Granuloma and the Associated Immunopathology	381
	17.6.2 Neutral Proteinases and Lysosomal Hydrolases	381
	17.6.3 Fibrosis	381
17.7	ANERGY OF INFECTION	382
	17.7.1 Primary Systemic Mycoses	382
	17.7.2 Chronic Dermatophyte and *Candida* Species Infections	383
17.8	PROTECTIVE FUNCTIONS OF CELL-MEDIATED IMMUNITY	384
	17.8.1 Experimental Mycoses	384
	17.8.2 Human Mycoses	385
17.9	ANTIGENIC STIMULANTS OF CELL-MEDIATED IMMUNITY	392

EPILOGUE   401

INDEX   409

Preface

This book describes the interactions of antigens of the yeasts, molds, and actinomycetes with the immune systems of the human and animal hosts. In the relatively short history of this subject, the key antigens have been enumerated, and important observations have been made about the major humoral and cell-mediated responses that occur in these infections. It is hoped that this text will stimulate further research on mycotic pathogens by unifying knowledge about the structure and biological functions of clinically important antigens in the framework of the well-regulated and of the deranged immune responses.

The teaching goal of the text is to provide an immunochemical perspective cutting across generic boundaries of the mycotic pathogens. It is intended for basic and clinical investigators as well as for students entering the field from other disciplines. Clinical correlations of the antigen-antibody and cell-mediated reactions are highlighted because it is only through the corrective of clinical relevance that gains in basic studies can be translated into control and prevention of the mycoses.

One could ask why it is necessary to devote much research to a group of diseases that on a global scale are not great scourges of humanity and, indeed, are collectively paid scant attention in many general texts. The most compelling answer to me is that no less attention is capable of unlocking the mysteries of why fungi are pathogenic and what strategies have been evolved to restrict these agents to their saprophytic or commensal niches. Medical mycology is moving from the descriptive to the analytical mode, out of isolation and into the scientific mainstream. Once persuaded of the need for an analytical approach, the rationale becomes clear for using methods of the highest resolving power for structural analysis of mycotic antigens.

However, a survey of the immunological literature involving mycotic agents gives the impression that "the organism is the antigen." Careful and complex manipulations of lymphocyte and leukocyte populations are often done only to use whole killed organisms or unpurified antigens as the antigenic stimulus. To an extent this reflects the unavailability of more purified antigens but may, regrettably, represent an idée fixe point of view. The intact viable organism is the most natural antigenic stimulus, but the use of killed fungi or unpurified extracts to monitor immune responses in vitro has a major pitfall. The response being measured is a composite of several individual reactions. Important antigens may be masked or missing because they were not induced by the conditions under which the fungus was cultivated in vitro. The text will cite examples of these eventualities.

The purification and characterization of mycotic antigens are worthwhile goals but cannot by themselves lead to profound insight into pathogenesis. Essential information is needed linking the structure of the antigenic molecule to its biological function in the metabolism of the fungus or actinomycete. The possibility exists that the antigen may be a determinant of pathogenicity. Biochemists mainly have preferred to use nonpathogens as models for the cell wall structure and for characterization of secreted proteins. In this way, useful knowledge has been gained that applies by extension to mycotic pathogens, although processes can be overlooked that the pathogens have evolved to secure their peculiar status. Despite considerable success in understanding the cell wall structure of a model yeast, *Saccharomyces cerevisiae*, or a model mold, *Neurospora crassa*, we are not closer to an understanding of how or whether the cell wall constituents of *Histoplasma capsulatum* contribute to its survival within macrophages and granulomas.

The value of biochemical characterization of antigens is limited without instruction from cellular and molecular immunology. Several phenomena important in the immune response to mycotic infections have been defined but the mechanisms underlying these phenomena have not yielded to critical experimentation.

Acknowledgments

It is fortunate to have this opportunity to express appreciation to my mentors whose collective advice and encouragement provided the stimulus for this book. Professor Walter John Nickerson guided my graduate studies. His critical insight into scientific problems, along with that of the other members of the Waksman Institute of Microbiology, Rutgers University, created an atmosphere for research that still endures. Dr. Herbert F. Hasenclever of the Medical Mycology Section at the National Institutes of Health made me aware of the immunochemistry of *Candida albicans,* a topic that is a frequent preoccupation.

The Division of Mycotic Diseases at the Centers for Disease Control is my scientific home. Dr. Libero Ajello, the founder and director of these laboratories, has created a unique scientific resource. The scientists who visit here from all corners of the world have expanded my scientific and cultural horizons, and have become my colleagues and friends. Dr. Leo Kaufman, chief of the Immunology Branch, has helped me by developing a harmonious environment for research. His insights into clinical mycology have helped to focus our joint projects.

Without teaching experience, I would not have had the confidence to write this book. For giving me the opportunity to teach immunology, I am grateful to Dr. Ahmed Abdelal, chairman of the Department of Biology, Georgia State University, and Dr. Judith Lumb of the Department of Biology, Atlanta University. Dr. Judith E. Domer deserves my deepest gratitude for her critical review of the manuscript.

I have been fortunate to have several bright and stimulating collaborators in research but none so steadfast as my friend Dr. Robert Cherniak. I could not have written this book without the knowledge imparted to me by these scientists.

I am grateful for my association with John Lawrence, senior editor of Elsevier whose bright spark of life was snuffed in his prime.

It has been a distinct pleasure and great good luck to have Cindy Knudsen as editorial assistant for the whole text.

My wife Cheryl and children, Brendan and Merryl, sustained me in this endeavor. I hope they will forgive me for the large blocks of time taken away from family life in order to write.

MOLECULAR IMMUNOLOGY OF MYCOTIC AND ACTINOMYCOTIC INFECTIONS

Chapter 1

Introduction

Throughout its evolution the human species has adapted to life on our moldy earth. In some respects the activities of fungi have given us cause to celebrate our existence with leavened bread, wine, and aged cheese. The antibiotics that fungi and actinomycetes use to survive among competing soil microflora have been redirected to give us lifesaving "magic bullets." However, humanity has suffered much from the molds that afflict edible plants. Thus, fungi and actinomycetes constitute a multitude of harmless saprophytes, many plant pathogens, several beneficent species, and a few that have somehow become adapted as human pathogens.

Some of the pathogenic yeasts, molds, and actinomycetes lead an inconspicuous existence as endogenous commensals on the skin and mucosae of warm-blooded animals and humans. Others complete their life cycle in the soil independent of humans and only by accidental inhalation or traumatic implantation do they evade the host defenses and cause disease.

In the historical development of medical mycology the causative agents and clinical features of mycoses have been described. It is natural that interest has gradually turned towards an exploration of the molecules these microbes produce that evoke a host response. Moreover, humans have evolved a hierarchy of innate and inducible responses that allow us to prevail over the mycoses most of the time.

This text will encompass the antigenic structures of the major mycotic agents of disease and their interactions with the immune systems of their human and animal hosts. The major thrust is to explain the macromolecular structures of these microbes drawing attention to similarities and differences that are reflected in antigenic crossreactivity or specificity. Even though these microbes are a diverse group, some generalities will emerge that may help to unify our knowledge.

There is no precedent for this text; therefore, its organization follows no previously set pattern. The agents of disease are discussed individually and the chapters are divided into groups. Five chapters on the primary, systemic, dimorphic fungi are given first, followed by three chapters on filamentous fungi. Next, two chapters on the yeastlike fungi are presented. The actinomycetes occupy three chapters. The key findings are summarized at the end of each chapter to give a unity of tone to the volume and as an aid to students. United by their common property as simple, eukaryotic microbes, these fungi and the diseases they cause are remarkably different. The actinomycetes, because of their rudimentary, filamentous structure are allied with the fungi for historical reasons; certainly they are true bacteria that also cause a variety of diseases. Their continued study by mycologists encourages an understanding of the larger microbial world.

Although the cell biology of fungi is outside the scope of the text, a chapter on the fungal cell wall precedes the other chapters. It is included because the cell wall is the surface in direct contact with the host and is a barrier to the destruction of the invading fungi. Cell wall composition and the resulting antigenic mosaic are major points of difference between bacteria and fungi. The themes of Chapter 16, Innate and Humoral Immunity, and Chapter 17, Cell-Mediated Immunity, allow a review of knowledge across genus and species lines. The chapter devoted to humoral immunity is also an opportunity to address the topics of enzyme immunoassays and monoclonal antibodies. A selection of key mycotic antigens is proposed as potential biological standards. The T-cell-mediated immune responses of the host are a finely tuned orchestra of up- and down-regulatory mechanisms. These mechanisms are imperfectly understood, but recent pioneering research in this area has enabled the formulation of hypothetical composites of the immune responses included in the text. Within each chapter the material is arranged along parallel lines. The antigenic mosaic is described in terms of the cell wall, culture supernatants, cell extracts and subcellular compartments. When they are known, the immunochemical properties of purified factors are discussed. A recurrent theme is the prevalence of mannose-based polysaccharides, mannans, and secreted proteins, many of which are proteinases. Both of these classes of molecules are antigens.

Homoglycans and heteroglycans containing mannose from different genera frequently contain sequences that are responsible for crossreactions. But one hastens to add that epitopes directing species and serotype specificity also exist in mannans. Mannans must be separated from proteins in order to independently assess their respective contributions to antigenicity. The protein antigens are more apt to be species-specific and may function as aggressins as, for example, the acidic carboxyl proteinase of *Candida albicans*. Polysaccharides may also be aggressins, eg, the viscous, acidic, glucuronoxylomannan capsule of *Cryptococcus neoformans*.

Although ample justification exists for a compendium of methods for the preparation of mycotic antigens, this volume does not address that subject directly. Yet inclusion of criteria for purity of antigens has been necessary, frequently expressing these criteria in terms of the sequence of purification steps. This is done to prompt the reader to question, in a given study, the type of antigen used and its relation, if any, to those used previously. By so doing, emphasis is placed on the need for greater standardization. Methods of preparing antigens for chemical analysis may not be appropriate for maximum preservation of antigenic determinants. Discussion of separations based on size, charge, affinity, and immunoaffinity should enable the reader to identify the state of the art in each subject area.

The humoral limb of the immune system is approached first, in each chapter, in terms of those serologic tests that are useful benchmarks in the diagnosis and monitoring of disease activity. This topic brings up the burgeoning development of sensitive primary binding assays for the detection of class-specific antibodies and antigenemia. Monoclonal antibodies are being implemented in some of these immunoassays. Another aspect of humoral immunity covered is the functions of complement in the host responses to fungi.

Assays of humoral immunity are considered for their advantages in monitoring antigen purifications, especially two-dimensional crossed "rocket" immunoelectrophoresis and the enzyme-linked immunoelectrotransfer blot assays. Here, too, the early impact of monoclonal antibodies as probes and immunoadsorbents is addressed. Humoral responses that have been implicated as contributing to immunopathology are explored, particularly the occurrence of immunoglobulin E and of soluble complexes.

After identifying the key antigens and humoral responses, the stage is set, in each chapter, to introduce the critical cell-mediated immune responses in the host defense against fungi. This limb of the immune system is presented on different levels. Delayed cutaneous hypersensitivity is presented in relation to its occurrence in the community among healthy exposed persons, in acute infection, and through a discussion of the conditions for its replacement by the anergy of infection.

Attention is drawn to the in vitro correlates of cell-mediated immunity and to the antigens most capable of evoking those responses. Data derived from human infections and from experimentally infected animals are treated separately. Investigations with athymic nude mice serve as models to assess the effect on mycotic infection of deleting T-cell-mediated immunity.

Insights gained from clinical studies of human inborn immunodeficits are reviewed. These accidental experiments of nature that result in chronic dermatophytic and *Candida* species infections, among others, tell much about the requirements for a successful host defense. When data are available, comparisons are made between the protective immune effector functions and those, like granuloma formation, that contribute to immunopathology.

Important studies have begun to investigate in vitro cell–cell interactions in the immune response to mycotic antigens and deserve special attention. These more rigorous studies suggest models of the cooperative responses to some fungi and actinomycetes. Such models are included in order to provoke the questions that, when tested experimentally, will unravel the mechanisms of the host–fungus and host–actinomycete interactions.

Chapter 2

Molecular Organization of the Fungal Cell Wall

2.1 INTRODUCTION

The protoplasts of filamentous and yeastlike fungi are encased in a rigid, laminated, largely polysaccharide protective covering that reinforces the cylindrical or ellipsoid cell shape. The cell wall is plasticized at the tips of growing hyphae and at the radius of maximum curvature in yeasts that undergo multipolar budding. Thus fungi demonstrate the phenomenon of apical growth. Changes in the cell shape are often reflected in changes in the cell wall composition. The fundamental observations of (i) apical growth, (ii) the wall as a determinant of shape, and (iii) the corollary that morphogenesis is reflected in the mural composition were introduced by Walter J. Nickerson (1963).

The purpose of this chapter is to introduce the structure of the wall as a surface that the host recognizes as a foreign antigenic complex. If the wall is sufficiently porous, it can be penetrated by the oxidative and nonoxidative fungicidal mechanisms, which then act on the plasmalemma. On the other hand, the host phagocytes may have to grapple with the wall itself, in an attempt to degrade the slimy outer coat and the inner fibrillar and more chemically inert mural layers. We are still far from understanding the host–fungus interaction. Fundamental knowledge of the molecular organization of the cell wall can help to attain this goal.

Reviews of various aspects of the fungal cell wall appear with increasing frequency. The important subject of mannan has been addressed both structurally (Ballou, 1976) and in terms of its genetics (Ballou, 1980). Frequent revisions of our concept of the antigenic structure of yeast mannans have been necessary (Okubo et al, 1980; Suzuki, 1981). The diversity of mannose-based heteroglycans is known in general (Gander, 1974), with special reference to the pathogenic fungi (San Blas, 1982), and in the intricate complexity of a model lipopeptidophosphogalactomannan (Gander et al, 1980; Gander and Fang, 1980). The structure and localization of yeast glucans and chitin are becoming well-understood (Bacon, 1981; Marchessault and Deslandes, 1980; Cabib and Shematek, 1981). Chitin is particularly interesting because of its role in conferring strength, assisting in orderly cell division (Berkeley, 1980; Jolles et al, 1974; Hunsley and Gooday, 1974; Hunsley and Kay, 1976), and as a potential target for specific drugs (Gooday, 1977). The glycanases have been classified and their action patterns discussed in relation to enzymatic cell wall degradation (Phaff, 1977). A compendium of the analyses of carbohydrate and amino acid composition of fungal cell walls has reached a considerable size (Bartnicki-Garcia and Lippman, 1982). More general monographs have been directed towards cell wall biosynthesis (Farkas, 1979), morphogenesis (Stewart and Rogers, 1978), and composition (Rosenberger, 1976).

The fundamental differences between bacterial and fungal cell walls are important in understanding the phagocyte–fungus interaction, in developing chemotherapeutic strategy, and of more direct concern in delineating the antigenic mosaic. Fungi lack peptidoglycan (muramic acid-N-acetylglucosaminyl polymer cross-linked by peptide bridges), glycerol or ribitol teichoic acids, and the lipopolysaccharide endotoxin classes of macromolecules. As a result, fungi are resistant to antibacterial agents like penicillin that interdict bacterial cell wall synthesis. In place of these wall structures, fungi have outer layers of readily soluble peptidomannans embedded in a matrix of α- and β-glucans. Structural rigidity is increased by crystalline chitin fibrils arrayed as rings or discs in the bud scars of yeast forms and as a thin microcrystalline sleeve, interwoven with glucan fibrils, that covers the lateral walls of mycelial forms. Because of the importance of chitin in providing a channel for the nucleus to migrate into the daughter cell during the budding process, chitin antagonists like polyoxin D can immobilize or kill fungi. Not all genera of zoopathogenic fungi have an identical macromolecular construction, and the taxonomic considerations are discussed below.

2.2 OVERVIEW OF PEPTIDOMANNANS

Peptidomannans are a family of mannose-based polymers that are the readily soluble outer layers of the cell wall. They are either homopolymers of mannose or heteroglycans with α-D-mannan backbones. Peptidomannans are the major mural antigens that determine serotypes.

Evidence for the surface location of mannan rests on several structural probes illustrated best by the research on *Candida albicans*. Agglutinins produced by immunization of rabbits with heat-killed blastoconidia of *C albicans* are considered to be specific for mannan because it is the major heat-stable surface antigen. After adsorption of anti-*C albicans* serotype A with cells of serotype B, agglutinins specific for type A are retained. These determinants have been shown to reside in mannan (Summers et al, 1964). The lectin concanavalin A (Con A) has specificity for sugar termini with the 3,4,6-arabino-α-D-glycopyranosyl structure (So and Goldstein, 1968), especially α-D-mannose residues. *C albicans* is agglutinated by Con A, and this reaction is reversed by methyl-α-D-mannoside (Cassone et al, 1978). Phagocytosed *C albicans* can be differentiated from free blastoconidia by staining of the latter with fluorescein-Con A (Richardson et al, 1982). The cell surface peptidorhamnomannan of *Sporothrix schenckii* yeast forms and mycelial forms binds Con A, as visualized with horseradish peroxidase-diaminobenzidine (Travassos and Lloyd, 1980). In an analogous lectin-binding experiment, the galactomannan of *Schizosaccharomyces pombe* is located on the surface by scanning electron microscopy of cells stained with the colloidal gold-labeled lectin of *Bandeiraea simplicifolia*, having specificity for galactopyranosyl residues (Horisberger and Rosset, 1977). Although it is potentially a less specific probe, the periodic acid–thiosemicarbazide–silver stain permits ultrastructural contrast between mural layers susceptible to periodate oxidation and those which are not. Periodate-sensitive polysaccharides have vicinal unsubstituted hydroxyl groups.

The outermost mural layer of *C albicans* reacts strongly, as predicted from the susceptibility of the 1,2- and 1,6-mannosyl linkages in mannan (Poulain et al, 1978) (Figure 2.1). Serologic crossreactions occur to a moderate extent among pathogenic fungi, but in many genera there are substituents that confer a degree of specificity. Three general plans for fungal mannan can be recognized.

2.2.1 Mural

Most mannans are embedded in the glucan-chitin wall matrix and are major structural elements. In this type, α-1,6-linked mannan backbones predominate with side chains two to five sugar residues in length. Structural analyses by Ballou's group on the mannans of *Saccharomyces cerevisiae* have resulted in the structure proposed in Figure 2.2 that serves as a model for the mannan of *Candida* species. Sequences of α-1,6-mannan are joined by glycopeptide bonds to structural protein. The mannan is organized into an inner core, an outer chain, and base-labile oligomannosides. Antigenic specificity is determined by the outer chain region, consisting of three or four repeating blocks of linear α-1,6-mannan substituted with oligomannosides having a degree of polymerization (dp) of 1 to 4 (Figure 2.3). *Candida albicans* mannan has longer side chains in which mannopentaose and mannohexaose are encountered. The predominant linkage in the side chains is α-1,2 with α-1,3 resi-

Figure 2.1. Blastoconidia of *Candida albicans* cultured for 48 hours on rice agar with Tween nonionic detergent at 24 °C showing the reaction to periodic acid–thiosemicarbazide-silver stain. Electrondense granules indicate mannan. Mannan is seen sloughing off the outer wall and as a narrow zone near the plasmalemma, the presumed site of outer chain synthesis by mannose transferases (Poulain et al, 1978). *Source:* Reprinted with permission of the publisher Annales de Microbiologie de l'Institut Pasteur.

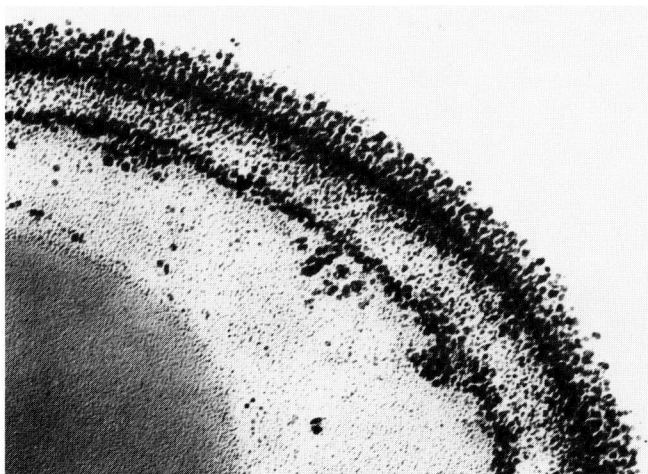

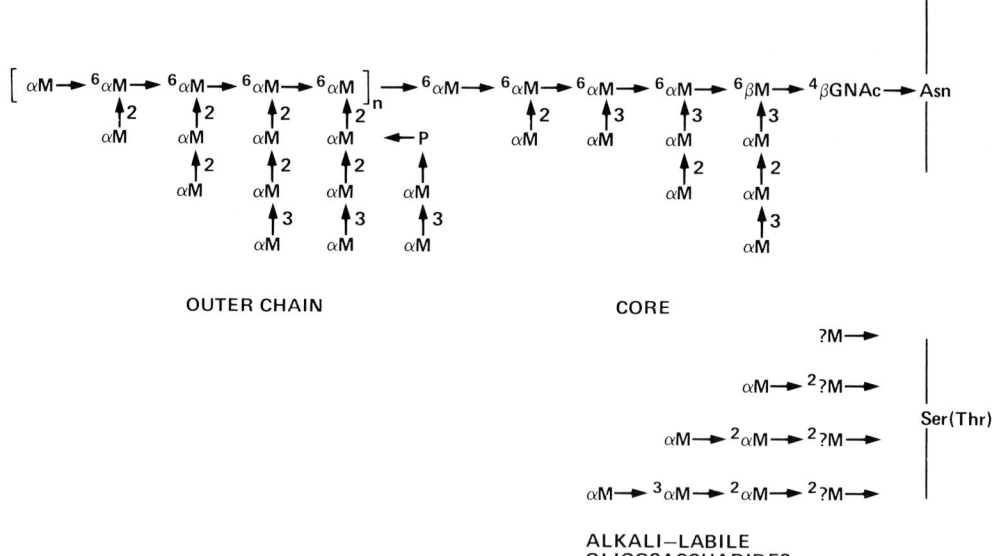

Figure 2.2. Proposed repeat unit structure of the carbohydrate chains linked to protein in *Saccharomyces cerevisiae* X2180 mannoprotein. Four types of mannose units are linked to serine (ser) and threonine (thr) whereas a more complex polysaccharide chain is attached to asparagine (asn). The subscript in the outer chain is 3 to 5 (Ballou 1980). The inner core unit is synthesized via polyisoprenol carrier lipid intermediates. The dolichol phosphate mannose derivatives are assembled intracellularly in the golgi apparatus. The process is similar to that involved in the synthesis of mammalian glycoproteins. The inner core linked to protein is then translocated outside the plasmalemma where the outer chain region is assembled by stepwise addition of guanosine-5'-diphosphate (GDP) mannose units by at least ten mannosyl transferases of different specificities (Farkas, 1979). *Source:* Reprinted with permission of the publisher from Fungal Polysaccharides. P. A. Sandford and K. Matsuda, (eds.), copyright 1980, American Chemical Society.

dues disposed as nonreducing termini. Other substituents have been described in the outer chain region-phosphodiester-mannose and *N*-acetylglucosamine (GlcNAc) (Ballou, 1976). The inner core region connects the α-1,6-mannan to protein via an *N,N'*-diacetylchitobiosyl-asparagine glycopeptide bond. Yeast mannan shares this initiation sequence with mammalian glycoproteins (Walborg, 1978). Base-labile oligomannosides are joined to the structural protein via hydroxyamino acid ester bonds between mannose residues and serine or threonine of the peptide moiety. The observed molecular weight of mannans depends on the method of preparation. The use of Fehling solution results in some alkaline degradation and a molecular weight of mannans of about 40,000 da, whereas complexation with borate-cetyltrimethylammonium bromide (CTAB) preserves a higher molecular weight species, 133,000 da (Nakajima and Ballou, 1974). Mannan prepared by the Peat et al (1961) method is serologically active and the serotype specificity of *C albicans* mannan is preserved (Summers et al, 1964), but it is not immunogenic in depot injections with complete Freund adjuvant (Lehmann and Reiss, 1980). Antimannan IgG can be induced more easily by injection of soluble peptidoglucomannan, which is extracted with cold alkali from isolated cell walls (Lehmann and Reiss, 1980; Reiss et al, 1974). The higher antibody-precipitating activity of baker's yeast mannan prepared by the CTAB method as compared with that obtained by Fehling precipitation has been attributed to β-1,3-mannobiosyl and mannosyl phosphodiesters (Okubo et al, 1981). The mannan isolated with Fehling solution, in which acid is used to dissociate the Cu^{++}-mannan complex, is therefore a degradation product that has lost alkali-labile and acid-labile oligomannosyl residues. The same interpretation may hold for the mannan of *C albicans*, which is more highly phosphorylated than baker's yeast (Hasenclever and McAtee, 1977).

2.2.2 Capsular

This type of mannan is restricted to a single genus of the pathogenic fungi, *Cryptococcus*, representing the genetically compatible A and D serotypes (*C neoformans* var *neoformans*) and the B and C serotypes (*C neoformans* var *gattii*) (Kwon-Chung et al, 1982). The serotypes are based on agglutinin-adsorption (Evans and Kessel, 1950; Wilson et al, 1968) and by immunofluorescent cross-staining and adsorptions (Kaplan et al, 1981). There has been concern about the serotyping of *C neoformans*, especially of serotypes A and D. Bennett's group has serotyped some clinical isolates as "AD" and Bhattacharjee et al, (1981) reported the isolation of D clones from a culture originally thought to represent only A. Serotypic determinants are considered to reside in the capsular polysaccharide, but thus far structural analyses have not yielded definitive information about type-specific epitopes. Capsule diameters vary, depending on the isolate, from barely detectable to greater than 4 μm. The capsular polysaccharide is a viscous glucuronoxylomannan of high molecular weight, measured by gel permeation chromatography as 7×10^5 to 8×10^5 da (Bhattacharjee et al, 1978, 1980). The heteroglycan consists of linear α-1,3-mannan backbone singly substituted with β-1,2-glucuronosyl or β-1,2-xylosyl residues. The ratio of xylose : mannose : glucuronic acid varies according to the serotype (see Table 12.1). In serotypes B and C the mannose residues in the backbone are all trisubstituted and some are tetrasubstituted with xylosyl and glucuronosyl at O-2, O-4 (Bhattacharjee et al, 1979). Serotypes A and D have some unbranched mannosyl residues and no tetrasubstituted ones. A significant O-acetyl content occurs in these mannans that contributes to their antigenicity (Cherniak et al, 1980; Goren and Middlebrook, 1967). The sites of the O-acetylation and other structural features are

Figure 2.3. Immunodominant side chains in saprophytic yeast mannans. The curved lines indicate the antibody binding sites inferred from hapten inhibition studies. Mannans of *Candida albicans* are more complex (Ballou 1980; also see Chapter 11). *Source:* Reprinted with permission of the publisher from Fungal Polysaccharides, P. A. Sandford and K. Matsuda, (eds.), copyright 1980, American Chemical Society.

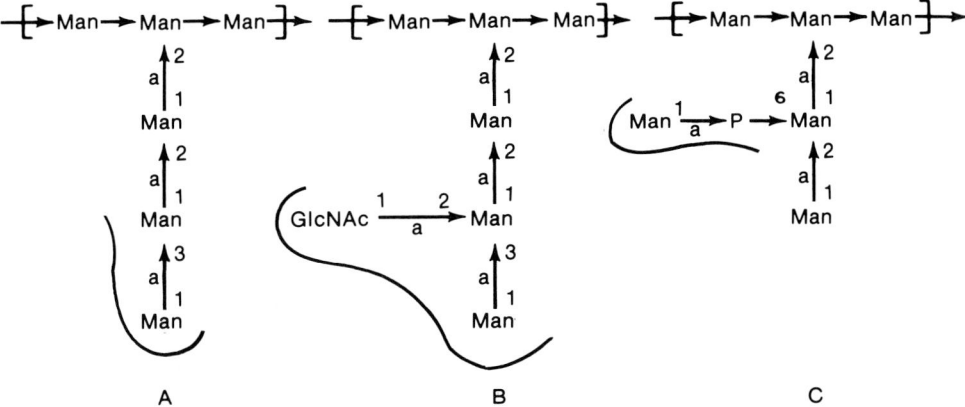

discussed in Chapter 12. The presence of galactose in preparations of the major viscous acidic glucuronoxylomannan is an artifact resulting from contamination with another secreted complex of heteroglycans, galactoxylomannan (Cherniak et al, 1982).

2.2.3 Secreted

Some mannans appear to occur in equilibrium between the plasma membrane and the extracellular milieu. Such a polysaccharide is not considered a major structural component of the wall. An example of this type is the peptidophosphogalactomannan (PPGM) of *Penicillium charlesii* (Gander et al, 1980). The galactoxylomannan of *C neoformans* may also belong in this category (Cherniak et al, 1982). The PPGM secreted into the growth medium is at first of high molecular weight (68,000 da) and has a high galactose content (69.9%). Later, as the culture ages, the weight of the PPGM declines by two thirds and only 12% galactose is present. The decline is due to the action of exo-β-galactofuranosidase on the polymer. The heteroglycan is composed of linear 1,6-linked mannose residues with linear 1,2-mannosyl side chains and major side chains of 1,4-linked galactofuranosides. The proposed functions for this mannan are: (i) as the carbohydrate portion of a glycoenzyme that fastens the enzyme to the cell wall; (ii) the carbohydrate component is amphipathic, containing ethanolamine, and it reacts with charged groups in the plasma membrane thereby anchoring an enzyme in the region between the plasma membrane and the cell wall (Gander et al, 1980); (iii) the carbohydrate serves as a safe conduct pass by protecting mural enzymes from premature proteolysis while they travel from the ribosome to the plasmalemma (Gander et al, 1980). This protection may continue once the enzyme is situated in the periplasmic space.

These polymers are reminiscent of the lightly and heavily galactosylated mannans that have been characterized from *Trichophyton mentagrophytes* (Bishop et al, 1965, 1966). The precise structural role of phosphodiester-linked mannan is incompletely understood. There is some evidence that this type of linkage contributes to the structural integrity of segments of mannans or serves as a bridge between mannan and glucan. Interchain acid-labile phosphodiesters link building blocks of mannan in the peptidophosphogalactomannan of *Exophiala (Cladosporium) werneckii* (Lloyd, 1972). The phosphodiesters in the mannoprotein that acts as a sexual agglutinin in *Hansenula wingei* are cleaved by a mild acid hydrolysis that liberates both glucose and mannose, implying that these hexoses are attached as glycosyl phosphate units (Hashimoto et al, 1980). During this treatment phosphate remains bound to mannan yielding mannose-6-phosphate. Other evidence for the participation of phosphodiester linkages in joining mannan to glucan is considered below in Chapter 11.

2.2.4 Peptidomannan Structure

The peptidomannan macromolecules are comprised of branched mannose-based polymers that are the readily soluble outer coating of the fungal cell wall. They are the principal, if not the only, cell wall glycans capable of producing an antigenic stimulus. In the main chain, they are linked to structural protein via glycosylamine or hydroxyamino acid ester linkages. The occurrence of a small (about 10%) protein component influences the way in which the antigens are processed, so it is not clear if, like pure polysaccharides, the induction of antibodies to them is independent of

T-lymphocyte help. A notable exception is the major viscous, acidic glucuronoxylomannan of *Cryptococcus neoformans*, which appears to be a pure polysaccharide. Specificity in peptidomannans resides in the short or single substituents disposed along the linear mannan sequences (Figure 2.3). The major linkage in the linear portions is α-1,6 although α-1,3 linkages occur in cryptococcal polysaccharides. If, as in the case of *Candida* species, the mannan is a homopolymer, specificity is provided by the glycosidic bond arrangements in the oligomannosides of the outer chain region. Otherwise, substituents of xylose, rhamnose, fucose, 3-O-methyl mannose and, perhaps, glucosamine, will be discussed in relation to their occurrence in individual genera. Galactose is a constituent of mannans from several genera including all the primary systemic fungi, the dermatophytes, and the apergilli. The role of galactosyl residues as a crossreactive determinant seems major, but more fine structural analysis will be required to see how this substituent could contribute to specificity. Noncarbohydrate determinants, particularly O-acetyl and O-phosphonomannan, also affect antigenic specificity.

Two properties of mannans that aid in their purification are the formation of Cu^{++} complexes with Fehling solution, and affinity for the lectin concanavalin A. Both of these properties are dependent on the presence of nonreducing terminal mannosyl residues. Heavily galactosylated mannans, for example galactomannan (GM) II of the dermatophytes, phosphogalactomannan of *Exophiala werneckii*, and a galactoxylomannan of *C neoformans* neither bind to Con A nor form Cu^{++} complexes.

Hot alkali should be discouraged as a means of extracting mannans from fungi if the goal is maximum preservation of antigenic determinants. Alkaline degradation is a peeling reaction releasing sugars from reducing termini. Under more drastic alkaline conditions fragmentation occurs as, for example, the conversion of glucose to D,L-lactate by intramolecular oxidation–reduction (Whistler and BeMiller, 1958). Dilute alkali can effect de-O-acetylation and β-elimination of hydroxyamino acid ester glycopeptide bonds, whereas boiling in 1M alkali will destroy relatively stable glycosylamine bonds.

The structure and function of mannan antigens will be considered in the context of each mycotic agent. Finally in Chapter 16 (Review of Humoral Immunity) information about the various mannans will be integrated.

2.3 GLUCANS

Glucans are a major constituent of fungi. Among the zoopathogenic species, the glucans are α-1,3-, β-1,3-, or β-1,6-glucosyl polymers. To the extent that they occur as linear sequences, the 1,3-glucans associate to form fibrils that increase the strength of the wall. However, none of these glucans is completely unbranched and as the percentage of branching increases, opportunities for helical formations are reduced. The importance of glucan in the maintenance of cell shape is illustrated by the lytic effect of the glucan synthesis inhibitor echinocandin B on growing cultures of *S cerevisiae* (McCammon and Parks, 1982).

A close association of glucan and chitin as an insoluble matrix exists in the mycelial forms (Stagg and Feather, 1977; Molano et al, 1980). Linkages between insoluble glucan and chitin are known in the basidiomycete *Schizophyllum commune* (Sietsma and Wessels, 1979, 1981). The glucan became soluble after digestion of the associated chitin with chitinase from *Streptomyces tatsumaensis*. The residue after chitinase digestion was enriched in lysine and citrulline accounting for 80% of the total amino acid content. These same amino acids were detected after the glucan-chitin matrix was digested with β-1,3-glucanase. Bridges between glucan chains and chitin contain lysine and citrulline connected to the chitin moiety by glycosylamine bonds. Much of the structural analysis of glucans has been accomplished with the saprophytes, and these must serve as models until more definitive work on the pathogens is available. Cabib et al (1982) proposed a model for the ultrastructural arrangement of glucans in the cell wall of *Saccharomyces cerevisiae*. Beneath the peripheral mannan layer is a globular zone of β-1,6-glucan containing a minority of β-1,3 linkages. This layer is superimposed upon the major structural and fibrillar β-1,3-glucan that is the mesh work for maintenance of cell shape.

Glucan cannot be released from whole *S cerevisiae* with hot dilute alkali because alkali-resistant β-1,6-glucan is a superficial layer that acts as a filter to block passage of the β-1,3-glucan. Preparation of isolated cell walls is thus justified not only to account for the mural constituents, but is required for the extraction of β-1,3-glucan. A three-stage extraction is used for β-glucan preparation. Isolated walls are extracted with hot dilute alkali to solubilize the major β-1,3-glucan, then the alkali-resistant residue is treated with hot 0.5 M acetic acid to liberate mainly the β-1,6-glucan that is a minor component of the wall dry weight (Bacon, 1981; Cabib and Shematek, 1981). Exposure of the acid-resistant residue to alkali results in a further release of β-1,3-glucan. The β-1,3-glucan of *S cerevisiae* contains 3% to 4% of branching through C-6 (Bacon, 1981). This point is important for understanding the extent to which this glucan can associate into fibrils. Known standards of linear β-1,3-glucan, ie, paramylon of euglenophytes, are fibrillar and the long unbranched β-1,3 sequences associate in a hydrated triple helix (Marchessault and Deslandes, 1980). X-ray diffraction patterns very similar to the one observed with paramylon are found with acid-treated cell walls of *S cerevisiae* and of the basidiomycete *Armillaria mellea* (Jelsma and Kreger, 1975). In spite of the branching that occurs in *S cerevisiae*, glucan microfibrils are distributed over the surface of yeast walls in (i) shadowed wall fragments, (ii) next to the plasmalemma in freeze-etched samples, (iii) and during regeneration of protoplasts (Bacon, 1981). The sharpening of X-ray diffraction patterns obtained after hot dilute acid treatment is explained by the unmasking of β-1,3-glucan fibrils by solubilization of β-1,6-glucan, but more important, acid treatment cleaves branching points by hydrolysis, resulting in freedom of β-1,3-glucan chains to crystallize (Jelsma and Kreger, 1975). Glucan fibrils in native walls are thin, densely woven, and have a low order of crystallinity (Bacon, 1981).

- *Saccharomyces cerevisiae* and *Candida albicans*. Glucan accounts for 47% of the dry weight of *C albicans* yeast form cell walls (Chattaway et al, 1968). Structural analysis of glucan solubilized from whole *C albicans* serotype B yeast with boiling dilute alkali indicated a majority of β-1,6 linkages (67.4 mol%) and 6 mol% 1,3 linkages (Yu et al, 1967). In light of subsequent work on *S cerevisiae* mentioned above, the choice of whole yeast cells as the starting material, instead of isolated walls, probably inhibited release of β-1,3-glucan. Further investigations of *C albicans* glucans will thus be necessary, using either the three-stage extraction method and/ or cell walls. Three distinct glucans were characterized from *S cerevisiae* according to solubility properties, molecular size, and linkage-sequence analysis: alkali-insoluble β-1,3-glucan; alkali-insoluble–acetic acid-soluble, highly branched β-1,6-glucan, and alkali-soluble, water-insoluble β-1,3-glucan (Fleet and Manners, 1976). The difference in alkali solubility of the two β-1,3-glucans is related to the presence in the soluble glucan of 8% to 12% β-1,6 linkages occurring in repeat sequences of at least three such bonds, and as branch points (Fleet and Manners, 1976). The β-1,6-glucan has a lower molecular weight than alkali soluble β-1,3-glucan. Linkages between glucan and protein or between mannan and glucan in the intact wall are suspected but not proven. Soluble glucan-mannan complexes have been extracted from defatted C albicans cell walls with dilute alkali, but further characterization and linkage-sequence analysis of these fractions have not appeared (Kessler and Nickerson, 1959; Reiss et al, 1974). The occurrence in the intact yeast wall of linkages between glucan and protein or between mannan and glucan is suspected but not proven.

A stable mannan-Alcian blue precipitate is formed in S cerevisiae involving only a small portion of bulk mannan that has a high mannose:phosphate molar ratio and approximately equivalent amounts of glucose and mannose (Friis and Ottolenghi, 1970). Further details are presented below in Chapter 11. Glucan–mannan complexes are suspected in cell walls of another yeast Pichia polymorpha (Villa et al, 1980). In that instance, the alkali-soluble polysaccharides were treated with Fehling solution to precipitate mannan. Intrinsic radiolabeling showed radioactivity associated with both mannose and glucose in the Fehling precipitate.

- *Aspergillus niger.* A crystalline glucan, nigeran, composed of alternating α-1,3 and α-1,4 linkages is extracted from A niger with hot water (Bobbitt and Nordin, 1978). This polymer does not occur in Aspergillus fumigatus. Under conditions of N-depletion the nigeran content can exceed 30% of the cell dry weight of Aspergillus awamori (Bobbitt and Nordin, 1982). It accumulates in the outer layers of the cell wall and is secreted into the medium complexed with 2% to 5% protein.

A highly dextrorotatory α-1,3-glucan that is alkali-soluble and acid-precipitable is isolated from A niger (Johnston, 1965). About 54% of the cell walls of A niger is not soluble in alkali and consists of a chitin-associated β-1,3-glucan (Stagg and Feather, 1973). The alkali insoluble wall fraction contained 52.4% glucan and 29.8% chitin. This fraction was solubilized in dimethylsulfoxide after nitrous acid treatment. Nitrous acid does not deaminate acetylated amino groups, so that sufficient depolymerization occurred to enhance solubility without destroying chitin. The methylation gas chromatography–mass spectrometry (GLC–MS) analysis of this fraction showed that the major glucosyl linkage was 1,3 with a minority (up to 15%) of 1,4 linkages.

- Dimorphic systemic fungi. Two types of glucans are encountered in walls of *Histoplasma capsulatum.* A β-1,3-glucan has been detected on the basis of enzymatic digestion by β-1,3-glucanase purified from snail (Helix pomatia) digestive juice (Kanetsuna et al, 1974) or from Cladosporium resinae (Reiss, 1977). About 22% of the glucans in H capsulatum G-184B yeast form cell walls (chemotype 2) and 52% of the alkali-resistant residues were solubilized by β-1,3-glucanase. Chemotype 2 of H capsulatum contains large amounts of alkali-soluble α-1,3-glucan. This polymer is absent in cell walls of chemotype 1 (Pine and Boone, 1968; Domer, 1971; Reiss, 1977; San Blas et al, 1978). The α-1,3-glucan is considered to occupy an exterior position in the wall as short thick fibrils; once these are removed by alkali extraction, a network of fine fibrils remains. These are considered to be composed of chitin (Kanetsuna et al, 1974), but a role for a fibrillar β-1,3-glucan is conceivable in view of its occurrence as fibrils in baker's yeast walls. Structural analysis of the β-1,3-glucan of H capsulatum remains to be determined.

The main glucosyl polymer of *Paracoccidioides brasiliensis* yeast form cell walls is an α-1,3-glucan that can account for 45% of the mural dry weight (Carbonell et al, 1970; Kanetsuna and Carbonell, 1970; Kanetsuna et al, 1972; San Blas and San Blas, 1977). The α-1,3-glucan was characterized by its specific rotation, $[\alpha]_D^{25} = +233°$ (1M NaOH) and susceptibility to α-1,3-glucanase of *Trichoderma viride*. It is resistant to β-1,3-glucanases present in digestive juice of *Helix pomatia* and that from *Basidiomycete QM806*. Methylation-fragmentation GLC–MS showed the major linkage was 1,3 with a small number of branch points through C-6. The large α-1,3-glucan component of yeast forms contributes to a wall thickness of 200 to 600 nm compared with 80 to 150 nm thick wall of mycelial forms. Yeast form cell walls contain about 4% β-glucan.

When the temperature of growth is shifted from 37 °C to 20 °C, conversion to the mycelial form of *P brasiliensis* occurs and α-1,3-glucan synthesis is repressed. There is a corresponding increase in long, thin, interwoven β-1,3-glucan fibrils. Ultrastructural evidence has shown that α-1,3-glucan occurs as short thick fibrils in an outer wall layer and covered by a thin coat of galactomannan. The α-1,3-glucan is solubilized with alkali and precipitates upon acidification. Mutants deficient in α-1,3-glucan can convert to the yeast form but have diminished mouse virulence (San Blas and San Blas, 1977) and human polymorphonuclear neutrophils more readily kill α-1,3-glucan-deficient mutants (Goihman-Yahr et al, 1980). Thus, α-1,3-glucan is itself a determinant of virulence or is genetically linked to a virulence factor.

- Dermatophytes. Glycogenlike glucans with allergenic potential have been characterized from surface mats of *Trichophyton rubrum* (How et al, 1972, 1973). Enzymic digestions with pullulanase (specific for α-1,6-glucosyl flanking α-1,4-glucosyl residues), α-amylase, and β-amylase, established that at least 80% of the glucan consists of linear sequences of α-1,4-glucosyl units joined by α-1,6-branch points.

2.4 CHITIN

Chitin exists as insoluble sheets or fibrils composed of long unbranched chains of poly-β-1,4-N-acetylglucosamine (GlcNAc). Chitin is more inert and rigid than cellulose because the acetamido group at C-2 limits rotation around the glycosidic bond (Winterburn, 1974). The rigidity of chitin chains is further enhanced by intramolecular H-bonding through O'-3 and O'-5 of neighboring GlcNAc residues and by intermolecular H-bonding with the amide groups. Chitin chains stack on top of each other by hydrophobic contacts. Stacks of chains may be packed in the antiparallel α-chitin form, which probably results from chain folding, or in the parallel β-chitin form (Winterburn, 1974). The crystallinity of deproteinized chitin is reflected in characteristic X-ray diffraction patterns. When α-chitin from crabshell (*Cancer magister*) was used as a standard, the X-ray diffraction pattern agreed with chitin from *Mucor rouxii* (Ruiz-Herrera and Bartnicki-Garcia, 1974). This finding is consistent with other studies of fungal cell walls (Bartnicki-Garcia, 1968; Potgieter and Alexander, 1965). Chitin is a major constituent of the cell walls of filamentous fungi (Bartnicki-Garcia and Lipmann, 1982; Berkeley, 1981) (see Table 2.1). At the other extreme is *Saccharomyces cerevisiae* where the chitin content is 1% of the mural dry weight (Cabib and Shematek, 1981). Despite being a minor component, chitin has a crucial role in the cell division of *S cerevisiae* (Bowers et al, 1974). The yeast forms of the primary systemic fungal pathogens have a high chitin content comparable to that of filamentous fungi.

Table 2.1. Chitin in the Cell Walls of Some Pathogenic Fungi

Fungi	Chitin, Percent of Dry Weight of Cell Wall		
	Yeast Form	Mycelial Form	Reference
Candida albicans	1.7	7.6	Chattaway et al, 1968
Sporothrix schenckii	7.0	7.0	Previato et al, 1979
Phialophora verrucosa	na	5.9	Szaniszlo et al, 1972
Fonsecaea pedrosoi	na	5.7	Szaniszlo et al, 1972
Cladosporium carrionii	na	7.6	Szaniszlo et al, 1972
Trichophyton mentagrophytes	na	30.4	Noguchi et al, 1971
Aspergillus fumigatus	na	22.0	Hearn and Mackenzie, 1979
Coccidioides immitis			
Silveira	32.5 spherules	nd	Collins et al, 1976
M11	28.0 spherules	33.0	Wheat et al, 1977
Histoplasma capsulatum			
Chem 1 SWA	37.6	20.1	Domer, 1971
Chem 2 G184B	11.5	25.8	Kanetsuna et al, 1974
Blastomyces dermatitidis			
F, Mc	32.1	24.9–30.4	Domer, 1971
Ga-1, KL-1	22.7, 34.9	nd	Cox and Best, 1972
BD 64	48.0	13.9	Kanetsuna et al, 1969, 1971
Paracoccidioides brasiliensis 7193	32.2	18.0	Kanetsuna et al, 1969
Pb 73	29.8	5.9	San Blas and San Blas, 1977

Source: Reprinted from Reiss (1985) with permission of the publisher, Marcel Dekker, Inc.
Abbreviations: na, not applicable; nd, not determined.

2.4.1 Chitin Detection in Fungi

X-ray diffraction patterns consistent with α-chitin have been detected in several genera of fungi and have been thoroughly studied in fibrillar cell wall residues of *Neurospora crassa* after cycles of alkali and acid extraction (Hunsley and Kay, 1970; Potgieter and Alexander, 1965). Isolated fungal cell walls demonstrate X-ray diffraction bands with the same d values as purified crustacean chitin (Troy and Koffler, 1969). The diffuse pattern obtained with walls is sharpened after digestion with glucanases. Microbial chitinases and N-acetylglucosaminidases from *Streptomyces griseus* (Berger and Reynolds, 1958; Berkeley, 1981) and *Serratia marcescens* (Berkeley, 1981; Monreal and Reese, 1969) act on chitin-containing fungal cell walls, releasing chito-oligosaccharides and GlcNAc (Domer, 1971; Hunsley and Burnett, 1970; Reiss, 1977; Skujins et al, 1965). Molano et al (1980) localized chitin in the wall of *S cerevisiae* with fluorescein-labeled chitinase. The lectin wheat germ agglutinin, having specificity for GlcNAc, is also a suitable probe for chitin. Colloidal gold- and fluorescein-labeled wheat germ agglutinin bound to chitin in baker's yeast walls (Horisberger and Rosset, 1976; Molano et al, 1980). A significant portion of ^{14}C-glucosamine taken up by *S cerevisiae* is deposited in the cell wall (Molano et al, 1980) and is released from cell ghosts by chitinase. Cell-free synthesis of chitin microfibrils was reported with extracts of *Mucor rouxii* (Ruiz-Herrera and Bartnicki-Garcia, 1974) and chitin synthetase is known to occur in many genera of fungi (Farkas, 1979).

2.4.2 Molds

The wall at the hyphal tips exemplified in *Neurospora crassa* has an inner layer of chitin fibrils covered by a thin, amorphous layer (Figure 2.4). The growing point and the hyphal septa are where chitin synthetase is most active, as shown by autoradiography with tritiated GlcNAc-labeled hyphae (Gooday, 1977; Leiva et al, 1982). The apical dome shows bright fluorescence when stained with the optical brightener Calcofluor White M2R, which is also considered a probe for chitin (Hunsley and Kay, 1976). As growth occurs, the wall increases in thickness and the microcrystalline sleeve of chitin is embedded with glucan and proteins, giving a laminated reinforced resin quality and high tensile strength to the intact wall (Hunsley and Kay, 1976; Winterburn, 1974). The location of chitin in *N crassa* walls was studied

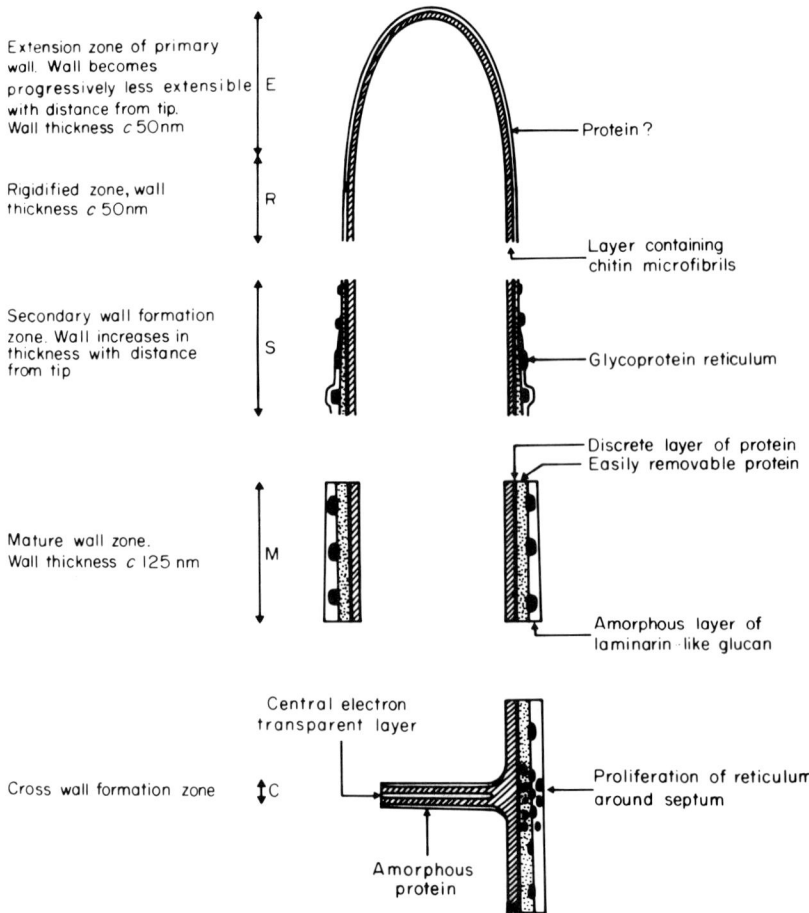

Figure 2.4. Cell wall at the growing tip of a hypha of *Neurospora crassa* showing primary and secondary wall layers (Trinci 1978). *Source:* Reprinted with permission of the publisher, Blackwell Scientific Publications, Ltd.

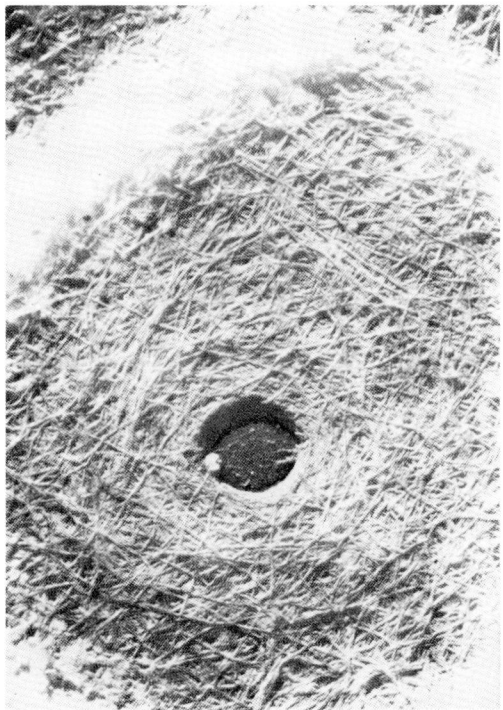

Figure 2.5. Septum from 5-day-old culture of *Neurospora crassa*. Hyphae were mechanically disrupted and septa were subjected to alkali, permanganate oxidation, and acid. Septal microfibrils shown resisted this treatment but underwent rapid dissolution after chitinase digestion (Hunsley and Gooday, 1974). *Source:* Reprinted with permission of the authors and publisher, Springer Verlag, Inc., N.Y.

after either chitinase digestion or cycles of acid and alkali extractions (Mahadevan and Tatum, 1967). A chitinase released only GlcNAc from the walls, but digestion proceeded to a limited extent. The additional presence of β-1,3-glucanase potentiated the attack of chitinase. These findings are evidence that chitin occurs as a skeletal core protected by an external glucan layer. Cycles of acidic and alkaline extraction reveal a fibrillar chitin layer distributed uniformly along the lateral walls of the hyphae. The crosswalls seem especially resistant to chitinase unless steps are taken to dissolve glucan-peptide in alkali, whereupon rings of chitin-rich undigested crosswalls are released (Figure 2.5).

Septa of *N crassa* have been isolated by mechanical disruption of the cells (Hunsley and Gooday, 1974). The inner layer of the hyphal wall, known to contain chitin, is contiguous with the septum. An outer amorphous layer of the septum is removed with pronase, revealing microfibrils that are oriented tangentially and, to an extent, radially. The fibrillar structure resists drastic chemical treatment with alkali, permanganate, and acid, but is readily dissolved by chitinase, not by β-1,3-glucanase. Autoradiography of the incorporation of tritiated GlcNAc shows that silver grains occur in a highly ordered distribution, consistent with the microfibrillar arrangement. These observations agree with the hypothesis that septa arise as a rim of chitin that develops centripetally, leaving a small central pore.

2.4.3 Yeasts

Chitin is a minor component of the *S cerevisiae* wall comprising less than 1% of the mural dry weight (Cabib and Bowers, 1971). Greater than 90% of this chitin forms the primary septum between the mother cell and bud, providing a channel through which the nucleus can pass (Cabib and Shematek, 1981). The chitin ring that forms around the neck of the mother cell develops further by centripetal growth as a disk of chitin to give a crosswall that is the primary septum of the bud scar (Molano et al, 1980). Chitin in the bud scars of *S cerevisiae* was investigated by electron microscopy before and after glucanase or chitinase treatment. The chitin content of cell ghosts made by cyclic alkaline and acetic acid extractions was enriched to 14.5% compared with 0.8% in cell walls (Cabib and Bowers, 1971). The ghosts consisted of bud scars held together by a thin matrix that retained the ellipsoid cell shape. When ghosts were digested with glucanase, most of the matrix was solubilized releasing intact bud scars. Treatment with chitinase, on the other hand, left the ghosts intact but the thick ridges of the bud scars disappeared.

Wheat germ agglutinin (WGA) is a probe for β-1,4-GlcNAc polymers that has been used to identify chitin in *S cerevisiae*. Horisberger and Rosset (1976) prepared bud scars by extraction of cell walls with β-1,3-glucanase and showed that colloidal gold-labeled WGA bound to rings that delimit the bud scars. The binding of fluorescein-WGA to yeast cell ghosts was localized in the bud scars but also occurred to a weaker extent along the whole contour of the wall. Chitinase-digested ghosts retained considerable fluorescence in the bud scars and this was removed either by repeated digestion with the same enzyme or with β-1,6-glucanase (Molano et al, 1980). The proof that the fluorescent brightener, Calcofluor white M2R, is a specific probe for chitin in yeasts is that it is localized exclusively in the birth and bud scars, and not when glucan is being synthesized in regenerating protoplasts under conditions wherein chitin synthesis is blocked by cycloheximide (Elorza et al, 1983). Calcofluor interrupts the cell cycle in *C albicans* but chitin continues to accumulate. The synthesis of chitin is not blocked by disrupting hydrogen-bonding sites in the nascent polymer but the self-assembly of crystalline microfibrils is inhibited. *S cerevisiae* cells were grown with ^{14}C-GlcNAc under conditions where the label was incorporated into chitin and not into other wall polymers. Enzymolysis of the corresponding radiolabeled cell ghosts showed that a minority (about 8%) of the chitin is in the lateral walls and the remainder is in the bud scars. Chitin not present in bud scars forms a thin layer uniformly distributed over the lateral walls in *S cerevisiae* and also in *C albicans* (Tronchin et al, 1981). This layer can be demonstrated in chemically produced cell wall ghosts or in pronase-digested thin sections of whole yeast cells. The reaction of labeled WGA with chitin in the lateral walls was ablated by chitinase digestion. A mutually supportive role for β-1,6-glucan and chitin in the inner layers of the *S cerevisiae* cell wall was proposed because digestion with β-1,6-glucanase from *Bacillus circulans* could also remove ^{14}C-chitin from cell wall ghosts (Molano et al, 1980). The occurrence of a thin, uniform chitin layer in *S cerevisiae* may represent the remnant of an evolutionary link between yeasts and the filamentous fungi. In the latter, chitin exists not only in the septa but also as a substantial microcrystalline sleeve for the growing hyphae.

The critical role that chitin plays in yeast morphogenesis is illustrated by the effect of the inhibitor of chitin synthesis, polyoxin D, a structural analog of UDP-GlcNAc (Gooday, 1977). Budding occurs and the daughter cell is fully grown when the cell cycle is interrupted just before separation. Extrusion of the cytoplasm occurs at the neck between the two cells giving an exploded pair configuration (Bowers et al, 1974). This is the effect of failure to form the primary septum because of inhibition of chitin synthesis.

Nutritional conditions can also exert strong influence on the chitin content. For example, cultivation of the yeast *Rhodotorula glutinis* in an unbuffered medium causes the pH to drop to 2. Under those conditions the chitin-hexosamine content of the walls was 18% with 71.9% of that being chitin, in contrast to growth at a constant pH of 5.5 where the hexosamine content was 8.7% and only 18.7% of that was chitin (Berthe et al, 1981)

2.4.4 Pathogenic Genera

The chitin content of some principal genera of zoopathogenic fungi are shown in Table 2.1. These estimates are based either on acid hydrolyzates of cell walls or on chitinase digests. Rarely has more than one isolate of each species been compared in a single study (except see Domer, 1971). These data should thus be regarded as pending confirmation by other, definitive studies. The chitin content of the tissue form of the systemic dimorphic fungi is approximately one third of the mural dry weight.

Two yeast form chemotypes are known in *H capsulatum* that differ markedly in their cell wall composition. Yeast forms of chemotype 1 have a high chitin content ranging from 31% to 53.7% of the dry weight (Reiss, 1977; San Blas et al, 1978). Chemotype 2 walls have a reduced amount of chitin (12.3% to 20.1%) (Domer, 1971; San Blas et al, 1978), and instead there is a major amount of α-1,3-glucan that is absent in walls of chemotype 1. The chitin content of yeast form walls is higher than that of mycelial walls. Chitinases from two sources differed in the ability to digest yeast form walls of chemotype 1 (Davis et al, 1977). The chitinase–β-1,3-glucanase complex of *Streptomyces* sp released 25% to 38% of the cell wall as GlcNAc and a minor amount of glucose, but the chitinase-chitobiase of *Serratia marcescens*, which does not coproduce a glucanase, only hydrolyzed 12% to 15% of the cell wall to GlcNAc. This finding argues for an occluding effect or linkage between β-1,3-glucan and chitin in the *H capsulatum* cell wall.

The wall of the mature spherule form of *Coccidioides immitis* is highly susceptible to chitinase of *Streptomyces antibioticus* containing β-1,3-glucanase as a minor component (Collins et al, 1976). Dissolution of the wall occurs, leaving a resistant residue of less than 10% of the mass of the mature spherule. Glucose and GlcNAc, equivalent to 19% and 32.5% of the mural dry weight, respectively, are liberated in the process. Egg white lysozyme has a significant but limited ability to release GlcNAc from spherule walls, opening the question of whether lysozyme in polymorphonuclear neutrophils may contribute to host defense against *C immitis*. Spherules incubated in buffer alone undergo moderate autolysis within 48 hours, releasing glucose, but not GlcNAc, into the supernate. The effects of chitinase-glucanase complex on the ultrastructure of immature spherules was assessed (Hector and Pappagianis, 1982). This chitinase complex removed an inner wall layer along the entire perimeter of the lateral wall and a contiguous layer within the cleavage planes. The further effect of α-1,3-glucanase from *Bacillus circulans WL-12* was to partially or completely lyse the remaining outer mural layer.

Differences in the chitin content of the yeast form cell walls from two isolates of *B dermatitidis* have been noted (Cox and Best, 1972). The avirulent *Ga-1* isolate has a chitin content of 24%, whereas 43% was detected in walls of the *KL-1* isolate. The trypsin-digested and alkali-insoluble portion of the virulent-strain cell walls evoked tissue necrosis in mice (Cox et al, 1974). The intact walls of the virulent strain, in addition, could promote granuloma formation, an effect related to the occurrence of an alkali-soluble antigen. Chitin as a factor that makes killing by polymorphonuclear neutrophils, or macrophages, more difficult seems a tempting possibility, but one for which no proof is available. Reasoning that in *P brasiliensis* the large α-1,3-glucan content of the yeast form cell wall is not essential for dimorphism, because α-glucan deficient mutants still are able to convert to the yeast form, Kanetsuna (1981) investigated the role of chitin in the maintenance of cell shape. Yeast and mycelial forms were treated sequentially with dilute alkali, β-1,3-glucanase and pronase. Yeast forms retained their spherical shape but mycelial forms disintegrated. The explanation for this is that the higher chitin content and the random orientation of microfibrils in yeast forms improve structural rigidity whereas a linear orientation of chitin microfibrils and a lower chitin content exists in the mycelial form.

Hen egg white lysozyme (EC 3.2.1.17) and human leukocyte lysozyme are effective muramidases with only weak chitinase activity. They attack chitopentaoses resulting in 54% digestion in one hour, rising to 88% digestion after 16 hours, but colloidal chitin is a relatively poor substrate (Jolles et al, 1974). Colloidal crustacean chitin is probably not a suitable substrate to judge the efficacy of lysozyme against fungi, particularly in view of the lytic effect of lysozyme on dividing cells of *C albicans* (Kamaya, 1970). Although chitin is a minor constituent by weight, its critical role in providing a channel for nuclear migration during budding suggests that even minor interruption of chitin synthesis at this stage may cause a sufficient derangement to arrest cell division or even lyse the yeast. Further investigation of lysozyme interactions with growing fungi is indicated.

2.5 FUNCTIONS OF THE CELL WALL

The primary role of the cell walls of fungi is protective. The wall provides a supporting superstructure for the growing protoplast and assists in orderly cell division. The abundance of murolytic enzymes secreted by bacteria, actinomycetes, and fungi suggests the existence of antagonism among competing soil microflora (Phaff, 1977). It has been pointed out over again that outer wall layers act as a filter to block aqueous extraction of more deeply situated glucans and, moreover, that the fibrillar β-glucan–chitin matrix in the inner-wall layer limits the activity of chitinase. The same protective devices that fungi have evolved for survival in the soil, serve them well in the host–fungus interaction. Primary systemic, dimorphic fungal pathogens, perhaps because of the high chitin content in their walls, resist lysis in the host and evoke a granulomatous reaction. This is end-stage, cell-mediated immunity, the classic form of hypersensitivity of infection. A vivid illustration of persistence in tissue is the multiple, calcified foci in both lung fields that are the hallmarks of healed acute histoplasmosis resulting from exposure to heavily spore-laden dusts (Schwarz, 1981). *Histoplasma capsulatum* yeast forms survive and multiply within macrophages. The oxidative and nonoxidative fungicidal mechanisms appear insufficient for killing the yeast forms, or else the fungus can turn off the lysosomal process (Domer and Moser, 1980; Howard, 1973). Some researchers hold that the cell wall is a highly porous structure and does not present a barrier to lysosomal factors which produce lethal hits in the fungal plasmalemma. For reasons that are unclear, lethal hits do not occur efficiently for yeast forms phagocytosed by macrophages. Lysis of the *H capsulatum* cell wall also is inefficient, perhaps because the correct assortment of enzymes is not present. Indirect evidence of this is the limited success with which lysozyme can degrade chitin in the spherules of *C immitis* (Collins et al, 1976). In the extreme where a zoopathogen like *C neoformans* has evolved a viscous, acidic, polysaccharide capsule, opportunities are increased to evade phagocytosis, penetrate the blood-brain barrier, and establish meningitis. Although *C neoformans* may be the gastronomic delight of a soil ameba (Bunting et al, 1979), in experimental murine models the capsule has been characterized as antiphagocytic and tolerogenic (Kozel, 1977; Murphy and Cozad, 1972).

The description of phagocytic lysosomal enzymes capable of acting on fungal cell walls is an active area of research. Pseudohyphal forms of *C albicans*, too large for endophagocytosis, are subject to damage by polymorphonuclear neutrophils. Adherence is followed by the discharge of granules, loss of integrity of cytoplasmic organelles, and eventual dissolution of the cell wall (Diamond et al, 1978; Diamond and Krzesicki, 1978; Diamond and Haudenschild, 1980).

Another aspect of protection is the locus that the wall provides for periplasmic enzymes. Acid phosphatase has been identified in seven pathogenic *Candida* species, in *C neoformans*, and *Torulopsis glabrata* (Odds and Hierholzer, 1973). In *C albicans* this enzyme is a mannoprotein with a hexose:protein ratio of 7:1. N-Acetylglucosaminidase was identified in a superficial wall layer of *C albicans* (Pugh and Cawson, 1978). Acidic carboxyl proteinase is a glycoprotein secreted by most *C albicans* isolates (Rüchel, 1981). Although its location has not been cytochemically confirmed, it probably occupies a periplasmic niche. Gander and Fang (1980) have proposed that peptidophosphogalactomannan of *Penicillium charlesii* functions to protect mural enzymes from proteolytic cleavage during transport between the ribosome and the wall, and secondly, to anchor enzymes in the periplasmic space. These speculations may have broad implications for the pathogenic genera in which galactomannans are well-represented.

Other physicochemical properties of the fungal cell wall have implications for protection. Mannose-based glycans occur almost universally in the outer mural layers. As readily soluble polymers, these mannans probably exist in a hydrated form and serve as a gel in the cell microenvironment giving some protection against desiccation. Charged wall polymers have the potential to act as ion exchangers. Glucuronosyl groups in cryptococcal polysaccharides can bind cations, and the chitosan of *Mucor* species can act as an anion exchanger (Bartnicki-Garcia and Nickerson, 1962).

The fungal cell wall is sometimes regarded as a highly porous structure that does not pose a serious barrier to the oxidative and cationic protein-mediated fungicidal mechanisms of granulocytes. The porosity of *C albicans* cell walls was surveyed with polyethylene glycols as model compounds (Cope, 1980). When the molecular weight of the solute exceeded 600 da, corresponding to an Einstein–Stokes radius of 0.8 nm, the polymer was effectively excluded by the cell wall.

Adherence to buccal and vaginal epithelia cells is displayed by *C albicans* and to a much lesser extent by other *Candida* species (King et al, 1980). Adherence is inhibited by digestion of yeast with proteinase, suggesting that the outer coat protein or glycoprotein of the cell wall has a functional role in mediating adherence, which may explain how *C albicans* has become well-adapted to life on the mucosae of warm-blooded animals and humans.

The wall should also be considered as the source of antigen most accessible to the host. The weight of evidence favors peptidomannans as the principal mural antigens. The proteins occurring in the periplasmic space and those being secreted into the external milieu are also capable of providing potent antigenic stimuli, as shown in the instances of acid phosphatase and acidic carboxyl proteinase of *C albicans* (Macdonald and Odds, 1980; Odds and Hierholzer, 1973) and the H and M factors of histoplasmin (Domer and Moser, 1980; Pine, 1977).

Antigen processing involving T-lymphocyte–B-lymphocyte–macrophage collaboration is the major circuit of specific immune recognition. Many examples of mural and other mycotic antigens that evoke these responses are discussed in the following chapters and collated in Chapter 16. Other less specific forms of recognition serve as accessory host defense mechanisms especially in the latent period of the primary immune response.

The direct activation of complement was first described using zymosan, an alkali-extracted baker's yeast (Holan et al, 1980). The coating of fungi with the C3 component of complement covers antiphagocytic sites on the cell surface and provides a ligand between the target cell and C3 receptors on phagocytes. Other fungal polysaccharides are known as alternative pathway activators including the cell wall of *C neoformans* (Laxalt and Kozel, 1979) and mannan of *C albicans* (Kind et al, 1972). Alternative pathway activation is also evoked by whole cells of *Rhizopus oryzae* and other zygomycetes leading to production of factors chemotactic for polymorphonuclear neutrophils (Marx et al, 1982).

Acute phase reactants like C-reactive protein also have affinity in plasma for fungal polysaccharides. This observation was made by Longbottom and Pepys (1964), who described C-substance activity in polysaccharides of *Aspergillus fumigatus*. Peptidomannan from *Epidermophyton floccosum* was characterized as containing phosphorylcholine, and was precipitated by human serum containing C-reactive protein (Baldo et al, 1977).

The functions of the cell wall can be summarized: (i) maintenance of shape and orderly cell division; (ii) resistance to lysis by competing microflora and host phagocytes; (iii) as a locus for periplasmic enzymes; (iv) prevention of desiccation, action as a filter and ion exchanger; (v) mediation of adherence; (vi) immune recognition at the level of antibody responses, direct complement activation, and C-reactive protein.

2.6 PREPARATION AND CRITERIA FOR PURITY

Cell walls are typically prepared by the mechanical disruption of cells and the removal of cytoplasmic membranes by repeated washing. A consideration in manipulating batches of pathogenic fungi for cell wall preparation is killing them before processing. Killing can be accomplished by immersion in 0.2% neutral, buffered formaldehyde. After a suitable time standing in the cold (about 48 hours) a sample should be centrifuged and checked for sterility on enriched medium.

The phenomenon of cell wall autohydrolysis has been described. Washed cells of *S pombe* incubated in buffer release proteins, β-1,3-glucose oligosaccharides, and soluble glycans (Phaff, 1977). Endogenous glucanases have also been characterized in *H capsulatum* (Davis et al, 1977). The culture fluid from 5-day-old *Aspergillus nidulans* contained enzymes that could partially digest the alkali-resistant fraction of the *A nidulans* cell wall (Zonneveld, 1971). Inhibition of glucanases is desirable to prevent autohydrolysis during the multiple washings used in cell wall purification.

Glucono-δ-lactone is a potent inhibitor of glucanase reactions. Some glucanases, ie, zymolyase, are rapidly inactivated at 60 °C (Phaff, 1977). Boiling cell walls for brief periods (about 5 minutes) has been used to inactivate glucanases (Fleet and Phaff, 1974), but controls to determine the presence of heat-labile determinants would be necessary if the antigenic structure of the walls is to be investigated.

2.6.1 Isolation and Purification

Yeast form cells can be directly disrupted, but mycelial mats or the balls of mycelia that form in liquid shake cultures may need pretreatment to break up large aggregates. Pretreatment is done conveniently in a Waring blender, or Virtis homogenizer. Experience in this laboratory favors the use of the Braun MSK cell homogenizer as the method of choice for disrupting fungi to produce cell walls. Less efficient methods of fungal cell disruption such as sonication, French pressure cell, or Ribi cell fractionator also have a place, especially if the goal is to strip mural outer layers without substantial cytoplasmic contamination. The Ribi fractionator stripped the outer layer of the *C immitis* endospores, after which the remaining inner layer was disrupted with the Braun cell homogenizer (Cole, 1982). Under different conditions, the Ribi fractionator disrupted *C immitis* spherules releasing intact endospores that were separated from large pieces of spherule wall by differential centrifugation (Wheat et al, 1977). The MSE microblender has been used effectively to prepare detached septa of *N crassa* (Hunsley and Gooday, 1974) and to macerate *A fumigatus* mycelia before Braun homogenization (Hearn and Mackenzie, 1979). Cell walls prepared by disrupting baker's yeast in a Braun cell homogenizer, followed by multiple washings in buffer, still contained trapped membrane vesicles (Robertson et al, 1980). These could be removed by repeated treatments with a Lourdes multi-mix and washings.

The conditions for successful homogenization require close attention to the duration and to temperature controls (Orenstein, 1971). A preferred method for preparation of fungal cell walls consists of disruption with glass ballotini in the Braun cell homogenizer and multiple washings in a nonionic detergent. Details for a general method are given elsewhere (Reiss, 1985).

Further purification of isolated cell walls — the Gram reaction and phase contrast microscopy are used to determine the extent of breakage. Intact cells are Gram-positive and refractile, and lose these properties once they are broken. Electron microscopy of negatively stained or cross-sectioned walls will indicate the relative absence of gross debris and the extent of fragmentation. Sedimentation of a whole cell homogenate, or partially purified walls, through a density gradient removes intact cells and provides a more uniform wall fraction. A 9% to 21% dextran gradient was used for baker's yeast walls (Nurminen et al, 1970) or a discontinuous 0.4 to 1.6M sucrose gradient aided in purifying walls of *Trichophyton mentagrophytes* (Noguchi et al, 1971). Cell walls accumulate at the interface between the two sucrose solutions.

Digestion of walls with enzymes is another means of purifying them of cytoplasmic contamination. Walls of *B dermatitidis* were digested with trypsin (Cox and Best, 1972). Trypsin and pepsin reduced the mural weight of *H capsulatum* walls 27% to 69%, depending on the isolate (Pine and Boone, 1968). Both trypsin and chymotrypsin were used on mycelial walls of *N crassa* (Potgieter and Alexander, 1965). A sequence of proteinase treatments with papain, nagarse, and trypsin was applied to *T mentagrophytes* walls (Noguchi et al, 1971; 1975). The effects of proteinase treatments among different studies are difficult to compare because of variation in the choice of enzymes and conditions for their use. Criteria for the proteinase digestions can only be suggested. Measurements should be made of the release of ninhydrin-positive material so that the extent of digestion can be determined. Antimicrobial agents should be added to inhibit contamination. A combination of cycloheximide and chloramphenicol has been adequate for this purpose (Reiss, 1977). The removal of extraneous nucleic acid with RNAse is recommended by Taylor and Cameron (1973) who advise that tests to exclude the presence of nucleic acid should rely on measurement of nucleotide bases rather than on pentoses, because the latter may be part of the mural heteroglycans.

Reports of the lipid content of the walls in excess of 8% to 10% should be regarded critically to ensure that the purification procedure rigorously excludes cytoplasmic contamination. If the objective is antigenic analysis, organic solvents should be avoided for defatting in favor of nonionic detergents. The third stage of the lipid extraction procedure that consists of (i) ethanol-diethyl ether, (ii) chloroform, (iii) acidified ethanol-ether (Bartnicki-Garcia and Nickerson, 1962) is risky for the preservation of acid-labile antigenic determinants such as phosphodiester-linked mannan.

2.6.2 Summary of Criteria for Wall Preparation and Purity

Selection of a method to prepare cell wall fractions is obviously influenced by the objective of the investigation: *antigenic structure,* where avoidance of denaturing conditions is necessary for maximum preservation of antigenic determinants, or *composition analysis* of the cell wall matrix, where exclusion of all noncovalent lipid and protein requires more rigorous purifications involving denaturing conditions. Complete breakage of fungal cells is usually necessary. The isolation of *outer mural layers* and *periplasmic enzymes* is only indirectly concerned with the cell wall composition. Rupture of the cells is an obvious disadvantage leading to cytoplasmic contamination. Activation of autolysis may aid in the release of outer mural layers and enzymes. The following guides are intended to accomplish these different objectives.

2.6.2a Antigenic Structure

Nonviability of fungal cells is necessary for safety considerations.

Provide for inactivation of autolytic enzymes.

Break cells, but avoid excessive fragmentation.

Remove intact cells and subcellular organelles by differential centrifugation. This step most likely requires a dense osmoticum or a density gradient.

Removal of lipid and protein with a nonionic detergent.

Divide walls into aliquots and extract one portion with a proteinase of relatively narrow specificity, ie, trypsin. Digestion with an RNAse is optional.

Monitor the purification by release of A_{260nm} or A_{280nm} material and by the total N content of the walls.

2.6.2b Compositional Analysis

The steps taken are the same as in the previous paragraph except that lipid and noncovalent protein are removed with sodium dodecyl sulfate; and proteinases of broader specificity may be used on aliquots of the walls. Digestion with RNAse is desirable. Monitoring of purification should measure the removal of nucleic acid and the amino acid content of walls at different stages of purification.

2.6.2c *Outer Layers and Periplasmic Enzymes*

The choice of agents for killing cells is limited because formaldehyde may fix and inactivate periplasmic enzymes. Activation of autolytic enzymes is a desirable means of releasing mural outer layers and increasing the access to enzymes. Mild sonication provides the maximum tolerable force to remove outer layers while avoiding fungal cell breakage. Reducing compounds, especially dithiothreitol, and chelating agents like EDTA can loosen the mural outer layers with the aid of added proteinase (Torres-Bauza and Riggsby, 1980) or even without proteinase.

The objectives of recovering readily soluble outer mural layers and periplasmic enzymes and studying the antigenic structure and chemical composition of cell walls could conceivably both be met by applying the above methods in series. To do so, first soluble outer layers and mural enzymes must be isolated that would otherwise be lost during the cell disruption and multiple washing steps. Next, disruption of the cells and removal of cytoplasm would be accomplished with nonionic detergents. Finally, a portion of the wall preparation would be exposed to stronger detergents and enzymes to remove traces of noncovalent polymers.

2.6.2d *Summary*

Absence of intact cells indicated by phase contrast microscopy and staining reactions.

The retention of cell shape and the absence of cytoplasmic debris is verified by electron microscopy.

Total nitrogen and amino acid nitrogen are reduced to constant levels, usually less than 15% protein in the wall. This may require the use of a proteinase like trypsin.

The absence or reduction of lipids to constant amounts less than 10% of the dry weight. This can be ascertained by extraction of a sample with chloroform-methanol and subsequent thin-layer chromatography.

The absence of nucleic acids, measured as nucleotide bases instead of as pentoses.

Supplementary criteria increase confidence in the purity of the cell wall preparation.

If walls are prepared by methods that do not lead to enzyme inactivation, then cytoplasmic contamination may be assayed by the presence of the membrane marker enzyme adenosine triphosphatase (Abrams et al, 1974).

The infrared spectra of fungal cell walls provide a means of inferring the presence of cellulose, β-1,3-glucan and chitin (Michell and Scurfield, 1967) that can also aid in determining cell wall purity (Rehacek et al, 1969). The infrared spectra of isolated cell walls show a marked decrease in absorption at most wavelengths during purification especially at an 8.1 μm absorption band corresponding to $P = O$ of nucleic acids and polyphosphate.

The purity of baker's yeast cell walls was assessed at various stages by two-dimensional polyacrylamide gel electrophoresis (PAGE) consisting of isoelectric focusing followed by sodium dodecyl sulfate-PAGE (Robertson et al, 1980). Approximately 170 proteins were present in crude cell walls. After several washings the number of proteins was reduced to 107 and they were weaker in intensity except for component 16w that was the only wall-associated protein that copurified with the wall. It was a glycoprotein with a pI of 5.0 and a M_r of 25,000 da. Two-dimensional PAGE in conjunction with electron microscopy is a useful way of assessing the occurrence of proteins firmly bound to the cell wall and those that are there due to contamination with cytoplasm.

The carbon 13 nuclear magnetic resonance (^{13}C-NMR) spectra of intact cells and cell walls are markedly different. This observation was adapted to monitor cell walls of *Penicillium ochro chloron* during purification (Matsunaga et al, 1981). The disappearance of resonances attributable to triglycerides and mannitol occurred during purification but the polysaccharide resonances persisted. The major polysaccharide resonances were due to β-galactofuranosides present in a galactan. Chitin and β-1,3-glucan because of their highly ordered rigid structures do not yield resonances.

2.7 PROTEIN IN THE CELL WALL

The sources of protein in the cell wall vary. Certainly some of it may be cytoplasmic in origin resulting from insufficient purification (Phaff, 1977). Walls of fungi subjected to extensive washings still contain 10% to 15% protein which is considered to be the structural component (Rosenberger, 1976). Harsh disruption and washing procedures may remove readily soluble mural outer layers with accidental loss of native protein. Reports of either high protein content or none at all (Troy and Koffler, 1969) are open to question. Model studies on baker's yeast (Ballou, 1976) have emphasized the peptido or protein content of mannan that is linked to the polysaccharide via N,N'-diacetylchitobiosyl glycosylamine bonds to asparagine, and by alkali-labile hydroxyamino acid ester linkages to mannosyl reducing termini. Recent studies indicate the occurrence in *Schizosaccharomyces pombe* of glycopeptide bonds between β-1,3-glucan ("R"-glucan) and chitin, in which the amino acids involved are lysine and citrulline (Sietsma and Wessels, 1979).

The chitin–β-glucan mural core of fungal cell walls can be isolated by a sequence of vigorous chemical treatments consisting of partial solubilization of walls with 1M KOH, dimethylsulfoxide, and 40% KOH. Next, selective depolymerization of deacetylated chitin with nitrous acid releases additional β-glucan. This was taken as a criterion for the linkage of chitin to glucan. After establishing this linkage as a glycopeptide bond in *S commune*, Sietsma and Wessels (1981) went on to show similar behavior in the walls of *S cerevisiae, N crassa, A nidulans,* and *Coprinus cinereus*. This broad similarity across generic lines increases the probability that glycopeptide bonds between chitin and β-1,3-glucan are a general phenomenon among fungi. Other glycopeptide bonds will probably be discovered, for example, those that may link glucan to mannan.

Proteins associated with cell wall preparations have received attention in *B dermatitidis, A fumigatus,* and *C immitis* (Collins et al, 1977). Cell walls of both dimorphic states of *B dermatitidis* were prepared by disruption in the Braun cell homogenizer; then the walls were washed in buffer and defatted by the three-stage extraction procedure mentioned above (Bartnicki-Garcia and Nickerson, 1962). The walls were then partially solubilized with 1M alkali and the extract was partitioned by adding ammonium sulfate. Two proteins were located in the supernate and also in the precipitated material by SDS-PAGE; each of the four proteins had distinctive mobility (Roy and Landau, 1972). Cell walls of *B dermatitidis* yeast forms were digested with trypsin and exposed to 1M alkali (Cox and Larsh, 1974a, 1974b). The soluble portion was dialyzed, precipitating an α-1,3-glucan. The protein content of the alkali-soluble, water soluble (ASWS) fraction was 44% to 70% of the total weight. SDS-PAGE revealed the presence of three to four proteins of between 30,000 to 50,000 da. The ASWS fraction is antigenically potent.

Cell walls were produced from 3-day-old mycelia of *A fumigatus* (Hearn and Mackenzie, 1979; 1981; Wilson and Hearn, 1982). The walls were prepared by shaking the mycelia with glass ballotini and multiple washings with hot water. Next the detergent, triton X-100, was used to release five or more proteins that were found to be serologically active against sera from aspergillosis patients. The antigens were resolved by two-dimensional crossed rocket immunoelectrophoresis, and their immunologic activity was ablated by pronase digestion or by placing Con A or wheat germ agglutinin in the intermediate gel. These observations are evidence for the glycoprotein nature of the triton-extracted mural antigens of *A fumigatus*.

The antigens of *C immitis* responsible for the tube precipitin (TP) and the complement fixation (F) reactions were found in walls isolated from spherule forms. The TP antigen is heat stable but the F factor is heat-labile and is probably a protein. Walls of mature spherules incubated with lysozyme or chitinase released both of the antigens into the supernate. However, immature spherules appear to lack the F factor (Collins et al, 1977).

Our impression of the fungal cell wall as a dynamic structure is strengthened by its accessory function as a locus for enzymatic activity. Dense exudates over the cell surface of the invasive mycelial form of *C albicans* have been observed in scrapings from human oral lesions and from infected chorioallantoic membranes of chick embryos (Rajasingham and Cawson, 1982). These exudates seem to be involved in assisting the organism to invade the host. The hyphal tips have a bulging appearance resembling the appressoria of plant pathogens.

Beyond the structural and other wall-associated proteins are the mural enzymes, or glycoenzymes that reside in the periplasmic space or are secreted (Table 2.2). These enzymes are important from the viewpoint of antigenic structure, because they potentially provide an antigenic stimulus in the host. As attention is focused on the host–fungus interaction, factors mediating adherence are becoming known, such as the protein microfibrils that provide the means of adherence of *C albicans* to vaginal epithelia (King et al, 1980).

Two major categories of mural enzymes are the proteinases and β-1,3-glucanases. In later chapters the evidence that these proteinases promote pathogenicity will be summarized. The β-1,3-glucanases have the potential to act as autolysins but their activity on autologous cell walls is weak and the conditions for activation of these enzymes in vivo are not known. Other factors like the M-antigen of histoplasmin are included, but evidence for its enzymic role is slight. The M-factor is a heat-labile protein secreted by young mycelial forms and by the yeast forms in human infections. It is not produced by other pathogenic fungi or by morphologically similar saprophytes. Different, even mutually exclusive approaches, are needed to study these sources of protein in the cell wall. Mild sonication may suffice to shear off outer wall layers containing surface microfibrils. The detection of mural enzymes first requires that the yeast or mold is grown under the correct induction conditions. The proteinase of *C albicans* is induced by serum albumin (Remold et al, 1968), and acid phosphatase is repressed by high levels of inorganic phosphate (Arnold, 1972).

Table 2.2. Cell Wall Composition of the Mycelial, Germ Tube, and Blastoconidial Forms of *Candida albicans*, as Percent of the Wall Dry Weight

Isolate:	Unspecified		IM806	ATCC 10261		ATCC 26555		67	13
	Mycelia	Blastoconidia	Blastoconidia	Blastoconidia	Germ tubes	Mycelia	Blastoconidia	Mycelia	Blastoconidia
Medium	100% ox serum	Yeast extract peptone-glucose	Chemically defined	Chemically defined	Chemically defined	Chemically defined, Lee et al, 1975[c]		Chemically defined: 1 ng/ml biotin; 10 ng/ml biotin	
Age of culture, hr.	18	18	48	22	18[a]	22	22	48	48
Method of wall preparation	French Pressure Cell		Waring blender glass beads	Virtis homogenizer, glass beads		Incorporation of U-^{14}C-glucose[d]		Braun homogenizer, glass beads	
Criteria for purity	Orcinol test for pentose, light, electron microscopy, methylene blue exclusion		Light, electron microscopy	Light, electron microscopy		na	na	Microscopy, nucleotide content measured	
Defatted	yes	yes	yes	no	no	no	no	yes	yes
Glucose	23.2	29.5	56.5	58.0	55.0	74.0	71.0	30.9	21.7
Mannose	18.0	15.2	22.9	21.0	20.0	4.0	19.0	27.7	33.3
N-acetylglucosamine	7.6	1.7	1.4	0.6	2.7	21.0	9.0	2.6	1.6
Protein	19.8	35.4	5.2	3.8	5.2	nd	nd	26.5	28.3
Phosphate	1.4	2.4	nd	nd	nd	nd	nd	2.4	2.8
Recovery	75.5	84.8	87.1	83.4	82.9	99.0[e]	99.0[e]	90.1	87.7
Reference	Chattaway et al, 1968[b]		Kessler and Nickerson, 1959	Sullivan et al, 1983		Elorza et al, 1983		Yamaguchi, 1974	

Abbreviations: na, not applicable; nd, not determined.
[a] 18 hour blastoconidia starved by aeration for 24 hours. Germ tubes induced on N-acetylglucosamine, L-glutamine.
[b] Data recalculated from original report.
[c] See reference and Table in Chapter 11.
[d] Walls not isolated, see text.
[e] Recovery based on total incorporation of radionuclide.

Some enzymes are directly assayable in the intact cells including acid phosphatase and the β-1,4-xylanase of *Cryptococcus albidus* (Notario et al, 1979). Sometimes "permeabilizing" fungal cells by exposure to toluene or ethyl acetate (Arnold, 1972) improves the conditions for enzyme assays by activating the autolytic enzyme system. Trapping of mural enzymes behind disulfide-bridged wall components has been proposed for the invertase of baker's yeast (Kidby and Davies, 1970). In that event, sulfhydryl reagents like dithiothreitol aid in the release of invertase and acid phosphatase (Arnold, 1972; Chattaway et al, 1974).

2.8 SUMMARY

The cell wall analyses of individual mycotic agents are placed within their respective chapters so that a more complete picture of the antigenic mosaic could be realized. Yet inclusion of an example of the types of studies that have appeared is necessary in order to illustrate the progress made, and the unresolved difficulties that remain. The major focus of those who have attempted cell wall analyses of *C albicans* is the wall as the immediate effector of morphogenesis. This research spans 25 years but disparities still exist that cannot be reconciled with the existing data (Table 2.3). Comparisons among studies are complicated by uncontrolled variables, including the isolate used, medium for growth, methods for (i) cell disruption, (ii) cell wall purifications, (iii) criteria for purity, and (iv) the number of wall components that were analyzed. Despite these differences, some general conclusions can be drawn from consistent results among these studies (Table 2.3). The chitin content of the cell wall increases during the transition from blastoconidia through the germ tube stage and into the mycelial form. This increase was most carefully studied by Sullivan et al (1983) who incorporated U-^{14}C-glucose into cells, then identified that portion which was converted to N-acetylglucosamine in hydrolyzates of wall residues resistant to acid- and alkali-extractions.

A similar radioisotopic tracer approach was accomplished by Elorza et al (1983), but their data diverge markedly from other studies so that further confirmation is desirable. In that study walls were not isolated, rather the composition of mural glycans was determined by extracting *C albicans* cells with three cycles consisting of

Table 2.3. Mural Enzymes of Zoopathogenic Fungi and Related Genera

Enzymes	Species	Reference
Mural location proven		
Acidic carboxyl proteinase	*Candida albicans*	Remold et al., 1968
Acid phosphatase	*Candida albicans*	Odds and Hierholzer, 1973 Chattaway et al, 1974
β-1,4-xylanase	*Cryptococcus albidus* var *aerius*	Notario et al, 1979
Mural location suspected		
Chymotrypsin-like proteinase	*Aspergillus fumigatus*	Tran Van Ky et al, 1969
H and M proteins of histoplasmins[a]	*Histoplasma capsulatum*	Pine, 1977
Keratinase II and III	*Trichophyton mentagrophytes*	Yu et al, 1971

Source: Reprinted from Reiss (1985) with permission from the publisher, Marcel Dekker, Inc.

[a] Catalase activity has been shown to be associated with the M but not the H protein, but it is uncertain whether the enzyme is the M antigen or closely associated with it (D. H. Howard, personal communication).

boiling 2M NaOH, 0.5M acetic acid, and 2M NaOH. The mannan fraction was determined as the precipitate with Fehling solution. Acid-soluble glucan was determined, and most likely was enriched in β-1,6 linkages. The third cycle of alkali released β-1,3-glucan leaving behind a residue that was determined to contain only glucosamine. A disadvantage of this approach is that estimations of cell wall protein, lipid, and phosphate were not possible. The conditions for extracting mannan may have led to underestimation resulting from alkaline degradation.

The various studies in Table 2.3 are also in general agreement about the mannan content of cell walls which appears to be about 20.2%, and this component is not noticeably influenced by morphogenesis. A notable exception is the fivefold reduction in the mannan content of mycelial forms of *C albicans* (Elorza et al, 1983). No agreement exists yet on the total glucan component and the scanty data on the assortment of the glucan into its constituent types. This aspect requires a more systematic, analytic approach. Sullivan et al (1983) estimated the alkali-soluble glucan at between 5% to 8% in both stationary phase blastoconidia and germ tubes, whereas acid-soluble glucan accounted for 24% to 27% of the wall dry weight irrespective of the stage or form of growth. Similarly, the acid-insoluble glucan varied over a narrow range of between 22% to 26% of the wall dry weight.

Protein in the cell wall is the most elusive and paradoxical component probably because it reflects directly on the state of purity of the walls, and serves to underscore the need for more objective criteria for wall purity. There is no agreement on the best method to estimate wall proteins. The typical approach (Kessler and Nickerson, 1959; Chattaway et al, 1968) was to determine the total cell wall nitrogen and then to convert this to protein nitrogen by subtracting the nitrogen contributed by N-acetylglucosamine. Sullivan et al (1983) preferred the coomassie blue dye-binding assay. This is somewhat surprising because dye-binding is usually reserved for readily soluble proteins, and even then the response factor varies according to the protein being measured. For example, bovine serum albumin Cohn fraction V, traditionally used as a standard for protein determinations, yields a very high response factor in the coomassie blue assay and its use leads to underestimation of other proteins. The response factor of the glycoprotein and other proteins of the *C albicans* wall are unknown. For that reason, use of more than one method for estimating cell wall protein would seem advisable.

The estimation of protein in the cell wall was a problematic procedure in the analysis by Yamaguchi (1974). *C albicans* were disrupted in the Braun homogenizer, and washed by centrifugation 30 times. Protein was extracted from the walls by boiling them in 1M KOH and was estimated in the alkaline extract with folin-phenol. The result was that protein constituted about 25% of the wall dry weight of either morphogenetic form. The large disparities among these analyses leave open the question of what constitutes "cell wall protein."

An alternative method is simply to extract cell wall proteins with heat and sodium dodecyl sulfate solution for analysis by PAGE (Chaffin and Stocco, 1983). This method affords a more critical enumeration of individual proteins, which may provide benchmarks to follow during purifications. However, the structural mannoprotein of the wall, because of its high glycosidic component, does not enter the gel.

The cell walls of fungi pathogenic for humans and animals are a legitimate subject for further investigations because the functions of the wall are important for host–fungus interactions. The mural outer layers, consisting of readily soluble mannans, are the principal surface antigens. Superficial protein microfibrils of *C albicans* that are lost during work-up in conventionally prepared walls supply an important function in mediating adherence to the mucosae. The wall provides a microenvironment for periplasmic enzymes that also have an antigenic role, a role that is probably more dependent than peptidomannans on T-lymphocyte modulation. The fibrillar and more chemically inert mural matrix composed of glucans and chitin is a physical barrier to host phagocytes, but one whose significance has not been clarified. Much circumstantial evidence implies that in histoplasmosis the residual cell walls cannot be cleared and that the ensuing granulomatous reaction contributes to immunopathology.

Some stages of wall synthesis can be blocked by antibiotics, for example, the inhibition of chitin synthesis by polyoxin. Fungi are vulnerable at the sites of budding or at the hyphal apex, but the selective interruption of wall synthesis as a rational means of chemotherapy has not been realized.

Walls of mechanically disrupted cells are purified by differential centrifugation in detergent solutions. Standard criteria for purity, such as removal of extraneous lipid and protein, have been supplemented by the novel use of infrared spectra, two-dimensional PAGE, and ^{13}C-NMR as devices for monitoring purification. The selection of a method of purification is influenced by the sometimes incompatible goals of cell wall analysis. Thus, the preparation of superficial layers of the wall and isolation of periplasmic enzymes should avoid cell disruption. Antigenic analysis of the mural glycans requires disruption and exposure to detergents, but should minimize exposure to extremes of pH and heat. Analysis of the chemical composition is carried out with walls purified with stronger detergents and proteinases.

Mannans of fungi can be classified as mural, capsular or secreted. The first category is the most diverse, whereas true capsular mannan is restricted in the pathogenic fungi to the glucuronoxylomannan of *Cryptococcus neoformans*. Actively secreted mannans are not well recognized in the pathogenic genera, but the galactoxylomannan of *C neoformans* may belong to this type. The general plan of mannans is a comblike structure composed of linear α-linked backbones with short oligosaccharides bearing antigenic determinants. Antigenic specificity resides in terminal nonreducing residues of mannose, galactose, fucose, xylose, rhamnose, glucuronic acid, or hexosamine. Noncarbohydrate determinants such as O-acetyl and O-phosphonomannan also contribute to antigenicity. Glycopeptide linkages in fungal cell walls occur as base-labile hydroxyamino acid esters and relatively more stable asparaginyl-glycosylamine bonds to GlcNAc. Bridges between β-1,3-glucan chains containing lysine and citrulline are connected to the chitin moiety, although in the pathogenic genera proof of this linkage is lacking.

Galactomannans are common antigens, widely distributed in the pathogenic fungal genera, in which acid-labile galactofuranosyl residues predominate. Galactomannans vary in the extent of galactosyl substitution and in their ability to precipitate with Fehling solution. The occurrence of galactomannans in the primary systemic dimorphic genera hinders the attainment of diagnostic specificity, because of their crossreactivity and persistence in culture supernates and cytoplasmic extracts.

Comparative analysis of the walls of the pathogenic fungi is still at an early stage despite much solid pioneering work. The gaps in knowledge can be addressed by adopting more objective criteria for cell wall purity, more complete analysis of the constituents under optimal conditions for hydrolysis of labile monosaccharides, and a comparison of several isolates grown under readily reproducible conditions. The library of existing cell wall analyses should be increased with further definitive studies. For example, the chemotypes of a larger number of *H capsulatum* isolates need to be determined so that measurements of the effect of wall composition on pathogenicity can be made.

In the present state of knowledge we have a general model of the organization of the fungal cell wall (Figure 2.6). Superficial layers of protein microfibrils extend from the periphery of the wall and are loosely attached to readily soluble peptidomannans. The mannans are embedded in a fibrillar glucan–chitin matrix. α-1,3-Glucans are present in the primary systemic pathogens and are enriched in the yeast forms. These short, thick fibrils are exterior to an interwoven network of fine fibrils of β-1,3-glucan and chitin. *Candida albicans* is a special case because it contains no α-1,3-glucan and the chitin content is low. Chitin is localized in the bud scars of this

Figure 2.6. Hypothetical model for the molecular organization of the cell wall of *Candida albicans* is based on that proposed for *Saccharomyces cerevisiae* by Kidby and Davies, 1970. The original plan is altered here to account for the increased amount of chitin in the wall of *C albicans* which is concentrated in bud scars and distributed as a thin lateral layer (Tronchin et al, 1981). Linkages between major wall polymers are elusive and those shown are based on fragmentary data. Peptides containing lysine and citrulline are enriched in the fibrillar cell wall residues containing glucan and chitin from baker's yeast and other fungi (Sietsma and Wessels, 1979). Enzymatic digestion of the glucan releases GlcNAc joined to lysine and/or citrulline. The major structural and, depending on the extent of branching, fibrillar cell wall polymer of baker's yeast and *C albicans* is β-1,3-glucan. In its pure form this polymer associates into a triple helix (Marchessault and Deslandes, 1980). A superficial nonfibrillar layer of β-1,6-glucan is postulated here to be linked to outer chain mannan via phosphodiesters. Friis and Ottolenghi, (1970) first provided data of such mannan glucan bonds, but they have not been confirmed. Mannan itself is shown here organized in the inner core-outer chain model linked to structural protein by chitobiosyl-asparagine bonds (Ballou, 1980). Mannan protrudes into the external milieu. Chelating and sulfhydryl reagents potentiate the protoplasting effects of mixed glucanases (Torres-Bauza and Riggsby, 1980). In order to justify these effects interchain disulfide bridges are shown as in the Kidby and Davies model. Negative charges on phosphate esters could be chelated by cations but no formal proof exists for this effect. Periplasmic enzymes, shown as "e", such as acid phosphatase and secreted acidic carboxyl proteinases may also reside in the periplasmic space.

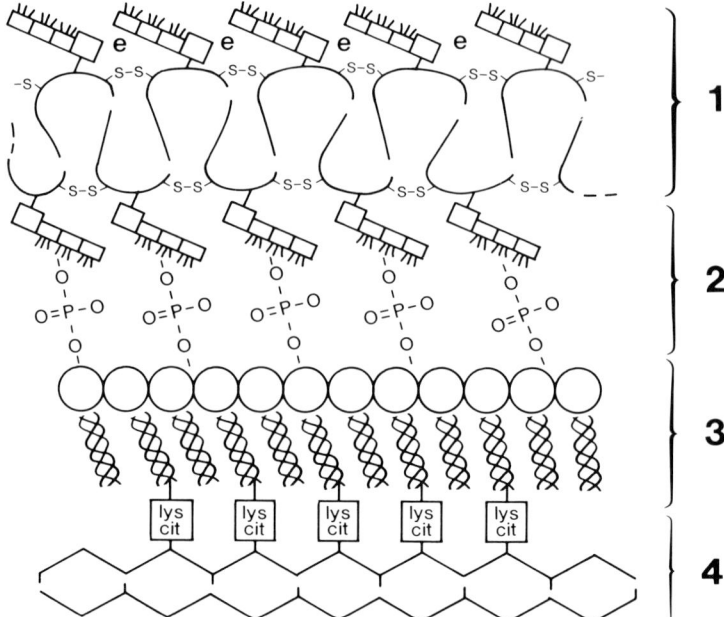

yeastlike fungus, and only a minor amount is distributed along the lateral walls. The major structural component of *C albicans* walls is β-1,3-glucan. Data obtained from fungi with a higher chitin content shows evidence for covalent glycopeptide linkages between β-1,3-glucan and chitin. In the mycelial forms of fungi, chitin is concentrated in the hyphal septa and is also abundant as a microcrystalline sleeve for the growing hypha. The wall is an important locus for the enzymes and glycoenzymes present in the periplasmic space. Some, like acid phosphatase, are assayable in the intact cell, whereas others, like the acidic carboxyl proteinase of *C albicans*, are secreted into the growth medium.

REFERENCES

Abrams A, Baron C, Schnebli HP, 1974. The isolation of bacterial membrane ATPase and nectin. Methods Enzymol 32:428–439.

Arnold, WN, 1972. The structure of the yeast cell wall- solubilization of a marker enzyme, β-fructofuranosidase, by the autolytic enzyme system. J Biol Chem 247:1161–1169.

Bacon JSD, 1981. Nature and disposition of polysaccharides within the cell envelope. pp. 85–96. In Yeast Cell Envelopes, Biochemistry, Biophysics and Ultrastructure. Arnold WN, ed. Boca Raton FL: CRC Press.

Baldo BA, Fletcher TC, Pepys J, 1977. Isolation of a peptido-polysaccharide from the dermatophyte *Epidermophyton floccosum* and a study of its reaction with human C-reactive protein and a mouse anti-phosphorylcholine myeloma serum. Immunology 32:831–842.

Ballou C, 1976. Structure and biosynthesis of the mannan component of the yeast cell envelope. Adv Microbial Physiol 14:93–158.

Ballou CE, 1980. Genetics of yeast mannoprotein biosynthesis. pp. 1–14. In Fungal Polysaccharides. Sandford PA, Matsuda K, eds. Washington, DC: American Chemical Society.

Bartnicki-Garcia S, 1968. Cell wall chemistry, morphogenesis and taxonomy of fungi. Annu Rev Microbiol 22:87–108.

Bartnicki-Garcia S, Lippman E, 1982. Fungal cell wall composition. pp. 229–252. In CRC Handbook of Microbiology. Laskin AI, and Lechevalier HA, eds. 2nd Edition. Vol. IV. Boca Raton, FL: CRC Press.

Bartnicki-Garcia S, Nickerson WJ, 1962. Isolation, composition and structure of cell walls of filamentous and yeastlike forms of *Mucor rouxii*. Biochim Biophys Acta 58:102–119.

Berger LR, Reynolds DM, 1958. The chitinase system of a strain of *Streptomyces griseus*. Biochim Biophys Acta 29:522–534.

Berkeley RCW, 1980. Chitin, chitosan and their degradative enzymes. pp. 205–236. In Microbial Polysaccharides and Polysaccharases. Berkeley RCW, Gooday GN, Ellwood DC, eds. London: Academic Press.

Berthe MC, Charpentier C, Lematre J, Bonaly R, 1981. Glucosamine and chitin accumulation in cell walls of the yeast *Rhodotorula glutinis* CBS 3044. Influence of culture and conditions. Biochem Biophys Res Comm 100:1504–1514.

Bhattacharjee AK, Kwon-Chung KJ, Glaudemans CPJ, 1978. On the structure of the capsular polysaccharide from *Cryptococcus neoformans* serotype C. Immunochem 15:673–679.

Bhattacharjee AK, Kwon-Chung KJ, Glaudemans CPJ, 1979. On the structure of the capsular polysaccharide from *Cryptococcus neoformans* serotype C-II. Mol Immunol 16:531–532.

Bhattacharjee AK, Kwon-Chung KJ, Glaudemans CPJ, 1980. Structural studies on the major capsular polysaccharide from *Cryptococcus bacillisporus* serotype B. Carbohydr Res 82:103–111.

Bhattacharjee AK, Kwon-Chung KJ, Glaudemans CPJ, 1981. Capsular polysaccharides from a parent strain and from a possible mutant strain of *Cryptococcus neoformans* serotype A. Carbohydr Res 95:237–247.

Bishop CT, Perry MB, Blank F, Cooper FP, 1965. The water-soluble polysaccharides of dermatophytes IV. Galactomannans I from *Trichophyton granulosum, Trichophyton interdigitale, Microsporum quinckeanum, Trichophyton rubrum,* and *Trichophyton schoenleinii.* Can J Chem 43:30–39.

Bishop CT, Perry MB, Blank F, 1966. The water-soluble polysaccharides of dermatophytes V. Galactomannans II from *Trichophyton granulosum, Trichophyton interdigitale, Microsporum quinckeanum, Trichophyton rubrum,* and *Trichophyton schonleinii.* Can J Chem 44:2291–2298.

Bishop CT, Perry MB, Hulyalkar RK, Blank F, 1966. The water-soluble polysaccharides of dermatophytes. VI. Glucans from *Trichophyton granulosum, Trichophyton interdigitale, Microsporum quinckeanum, Trichophyton rubrum* and *Trichophyton schoenleinii.* Can J Chem 44:2299–2303.

Bobbitt TF, Nordin JH, 1978. Hyphal nigeran as a potential phylogenetic marker for *Aspergillus* and *Penicillium* species. Mycologia 70:1201–1211.

Bobbitt TF, Nordin JH, 1982. Production and composition of an extracellular nigeran-protein complex isolated from cultures of *Aspergillus awamori.* J Bacteriol 150:365–376.

Bowers B, Levin G, Cabib E, 1974. Effect of polyoxin D on chitin synthesis and septum formation in *Saccharomyces cerevisiae.* J Bacteriol 119:564–575.

Bunting LA, Neilson JB, Bulmer GS, 1979. *Cryptococcus neoformans:* gastronomic delight of a soil ameba. Sabouraudia 17:225–232.

Cabib E, Bowers B, 1971. Chitin and yeast budding. Localization of chitin in yeast bud scars. J Biol Chem 246:152–159.

Cabib E, Roberts R, Bowers B, 1982. Synthesis of the yeast cell wall and its regulation. Annu Rev Biochem 51:763–793.

Cabib E, Shematek EM, 1981. Structural polysaccharides of plants and fungi: comparative and morphogenetic aspects. pp. 52–90. In Biology of Carbohydrates. Ginsburg V, Robbins P, eds. Vol. 1. New York: John Wiley.

Carbonell LM, Kanetsuna F, Gil F, 1970. Chemical morphology of glucan and chitin in the cell wall of the yeast phase of *Paracoccidioides brasiliensis.* J Bacteriol 101:636–642.

Cassone A, Mattia E, Boldrini L, 1978. Agglutination of blastospores of *Candida albicans* by concanavalin A and its relationship with the distribution of mannan polymers and the ultrastructure of the cell wall. J Gen Microbiol. 105:263–273.

Chaffin WL, Stocco DM, 1983. Cell wall proteins of *Candida albicans.* Can J Microbiol 29:1438–1444.

Chattaway FW, Holmes MR, Barlow AJE, 1968. Cell wall composition of the mycelial and blastospore forms of *Candida albicans.* J Gen Microbiol 51:367–376.

Chattaway FW, Shenolikar S, Barlow AJE, 1974. The release of acid phosphatase and polysaccharide and protein containing components from the surface of dimorphic forms of *Candida albicans* by treatment with dithiothreitol. J Gen Microbiol 83:423–425.

Cherniak R, Reiss E, Slodki ME, Plattner RD, Blumer SO, 1980. Structure and antigenic activity of the capsular polysaccharide from *Cryptococcus neoformans* serotype A. Mol Immunol 17:1025–1032.

Cherniak R, Reiss E, Turner SH, 1982. A galactoxylomannan antigen of *Cryptococcus neoformans* serotype A. Carbohydr Res 103:239–250.

Cole GT, 1982. Cell wall structure and chemistry. 13th Inter Congress Microbiol, session 77: Boston. unpublished.

Collins MS, Pappagianis D, Lee J, 1976. Enzymatic solubilization of precipitin and complement-fixing antigen from endospores, spherules and spherule fraction of *Coccidioides immitis.* pp. 429–444. In Coccidioidomycosis — Proc Third Inter Symp - Tucson. Ajello L, ed. New York: Stratton.

Cope JE, 1980. The porosity of the cell wall of *Candida albicans.* J Gen Microbiol 119:253–255.

Cox RA, Best GK, 1972. Cell wall composition of two strains of *Blastomyces dermatitidis* exhibiting differences in virulence for mice. Infect Immun 5:449–453.

Cox RA, Larsh HW, 1974. Yeast and mycelial phase antigens of *Blastomyces dermatitidis:* comparison using disc gel electrophoresis. Infect Immun 10:48–53.

Cox RA, Mills LR, Best GK, Denton JF, 1974. Histological reactions to cell walls of an avirulent and a virulent strain of *Blastomyces dermatitidis*. J Infect Dis 129:179–186.

Davis TE, Domer JE, Yu-teh L, 1977. Cell wall studies on *Histoplasma capsulatum* and *Blastomyces dermatitidis* using autologous and heterologous enzymes. Infect Immun 15:978–987.

Diamond RD, Haudenschild CC, 1981. Monocyte-mediated serum-independent damage to hyphal and pseudohyphal forms of *Candida albicans* in vitro. J Clin Invest 67:173–182.

Diamond RD, Krzesicki R, 1978. Mechanisms of attachment of neutrophils to *Candida albicans* pseudohyphae in the absence of serum and of subsequent damage to pseudohyphae by microbicidal processes of neutrophils in vitro. J Clin Invest 61:360–369.

Diamond RD, Krzesicki R, Jao W, 1978. Damage to pseudohyphal forms of *Candida albicans* by neutrophils in the absence of serum in vitro. J Clin Invest 61:349–359.

Domer JE, 1971. Monosaccharide and chitin content of cell walls of *Histoplasma capsulatum* and *Blastomyces dermatitidis*. J Bacteriol 107:870–877.

Domer JE, Moser SA, 1980. Histoplasmosis—a review. Rev Med Vet Mycol 15:159–182.

Elorza MV, Rico H, Gozalbo D, Sentandreu R, 1983. Cell wall composition and protoplast regeneration in *Candida albicans*. Antonie van Leeuwenhoek J 49:457–469.

Elorza MV, Rico H, Sentandreu R, 1983. Calcofluor white alters the assembly of chitin fibrils in *Saccharomyces cerevisiae* and *Candida albicans* cells. J Gen Microbiol 129:1577–1582.

Evans EE, Kessel JF, 1950. The antigenic composition of *Cryptococcus neoformans* II. Studies with the capsular polysaccharide. J Immunol 67:109–114.

Farkas V, 1979. Biosynthesis of cell walls of fungi. Microbiol Rev 43:117–144.

Fleet GH, Manners DJ, 1976. Isolation and composition of an alkali-soluble glucan from the cell walls of *Saccharomyces cerevisiae*. J Gen Microbiol 94:180–192.

Fleet GH, Phaff HJ, 1974. Glucanases in *Schizosaccharomyces*- isolation and properties of the cell wall-associated beta(1 → 3) glucanases. J Biol Chem 249:1717–1728.

Friis J, Ottolenghi P, 1970. The genetically determined binding of Alcian blue by a minor fraction of yeast cell walls. Comp Rend Trav Lab Carlsberg. 37:327–341.

Gander JE, 1974. Fungal cell wall glycoproteins and polysaccharides. Annu Rev Microbiol 28:103–119.

Gander JE, Beachy J, Unkefer CJ, Tonn SJ, 1980. Toward understanding the structure, biosynthesis, and function of a membrane-bound fungal glycopeptide—Structural studies. pp. 49–79. In Fungal Polysaccharides. Sandford PA, Matsuda K, eds. Washington, DC: American Chemical Society.

Gander JE, Fang F, 1980. Toward understanding the structure, biosynthesis, and function of a membrane-bound fungal glycopeptide—Biosynthetic studies. pp. 35–48. In Fungal Polysaccharides. Sandford PA, Mutsuda K, eds. Washington, DC: American Chemical Society.

Goihman-Yahr M, Essenfeld-Yahr E, de Albornoz MC, Yarzabal L, et al, 1980. Defect of in vitro digestive ability of polymorphonuclear leukocytes in paracoccidioidomycosis. Infect Immun 28:557–566.

Gooday GW, 1977. Biosynthesis of the fungal cell wall—Mechanisms and implications. J Gen Microbiol 99:1–11.

Goren MB, Middlebrook GM, 1967. Protein conjugates of polysaccharide from *Cryptococcus neoformans*. J Immunol 98:901–913.

Gorin PA, Spencer JFT, 1968. Galactomannans of *Trichosporon fermentans* and other yeasts: proton magnetic resonance and chemical studies. Can J Chem 46:2299–2304.

Hasenclever HF, McAtee FJ, 1977. Antigenic relationships of *Candida albicans, Saccharomyces telluris* and *Saccharomyces cerevisiae*. pp. 126–137. In Host Parasite Relationships in Systemic Mycoses. Beemer AM, Ben-David A, Klingberg MA, Kuttin ES, eds. Basel: S. Karger.

Hashimoto C, Cohen RE, Ballou CE, 1980. Characterization of phosphorylated oligomannosides from *Hansenula* wingei mannoprotein. Biochem 19:5932–5938.

Hearn VM, Mackenzie DR, 1979. The preparation and chemical composition of fractions from *Aspergillus fumigatus* wall and protoplasts possessing antigenic activity. J Gen Microbiol 112:35–44.

Hector RF, Pappagianis D, 1982. Enzymatic degradation of the walls of spherules of *Coccidioides immitis*. Exper Mycol 6:136–152.

Holan Z, Beran K, Miler I, 1980. Preparation of zymosan from yeast cell walls. Folia Microbiol 25:501–504.

Horisberger M, Rosset J, 1976. Localization of wheat germ agglutinin receptor sites on yeast cells by scanning electron microscopy. Experientia 32:998–1000.

How MJ, Withnall MT, Cruickshank CND, 1972. Allergenic glucans from dermatophytes. Part 1. Isolation, purification, and biological properties. Carbohyd Res 25:341–353.

How MJ, Withnall MT, Somers PJ, 1973. Allergenic glucans from dermatophytes. Part II. Enzymic degradation. Carbohyd Res 26:21–31.

Howard DH, 1975. The role of phagocytic mechanisms in defense against *Histoplasma capsulatum*. pp. 50–59. In Mycoses, Scientific Publication no. 304. Washington, DC: Pan American Health Organization.

Hunsley D, Burnett JH, 1970. The ultrastructural architecture of the walls of some hyphal fungi. J Gen Microbiol 62:203–218.

Hunsley D, Gooday GW, 1974. The structure and development of septa in *Neurospora crassa*. Protoplasma 82:125–146.

Hunsley D, Kay D, 1976. Wall structure of *Neurospora* hyphal apex—immunofluorescent localization of wall surface antigens. J. Gen Microbiol 95:233–248.

Jolles P, Bernier I, Berthou J, Charlemagne D, et al, 1974. From lysozymes to chitinases: structural, kinetic and crystallographic studies. pp. 31–54. In Lysozyme, Osserman EF, Canfield RE, Beychok S, eds. New York and London: Academic Press.

Jelsma J, Kreger DR, 1975. Ultrastructural observation of the $(1 \rightarrow 3)$-β-D-glucan of fungal cell walls. Carbohyd Res 43:200–203.

Johnston IR, 1965. The partial acid hydrolysis of a highly dextrorotatory fragment of the cell wall of *Aspergillus niger*. Biochem J 96:659–664.

Kamaya T, 1970. Lytic action of lysozyme on *Candida albicans*. Mycopathologia 42:197–207.

Kanetsuna F, 1981. Ultrastructural studies on the dimorphism of *Paracoccidioides brasiliensis, Blastomyces dermatitidis,* and *Histoplasma capsulatum*. Sabouraudia 19:275–286.

Kanetsuna F, Carbonell LM, 1970. Cell wall glucans of the yeast and mycelial forms of *Paracoccidioides brasiliensis*. J Bacteriol 101:675–680.

Kanetsuna F, Carbonell LM, 1971. Cell wall composition of the yeastlike and mycelial forms of *Blastomyces dermatitidis*. J Bacteriol 106:946–948.

Kanetsuna F, Carbonell LM, Azuma I, Yamamura Y, 1972. Biochemical studies on the thermal dimorphism of *Paracoccidioides brasiliensis*. J Bacteriol 110:208–218.

Kanetsuna F, Carbonell LM, Gil F, Azuma I, 1974. Chemical and ultrastructural studies on the cell walls of the yeastlike and mycelial forms of *Histoplasma capsulatum*. Mycopath et Mycologia Appl 54:1–13.

Kanetsuna F, Carbonell LM, Moreno RE, Rodriguez J, 1969. Cell wall composition of the yeast and mycelial forms of *Paracoccidioides brasiliensis*. J Bacteriol 97:1036–1041.

Kaplan W, Bragg SL, Crane S, Ahearn DG, 1981. Serotyping *Cryptococcus neoformans* by immunofluorescence. J Clin Microbiol 14:313–317.

Kessler G, Nickerson WJ, 1959. Glucomannan-protein complexes from the cell walls of yeasts. J Biol Chem 234:2281–2285.

Kidby DK, Davies R, 1970. Invertase and disulphide bridges in the yeast wall. J Gen Microbiol 61:327–333.

Kind LS, Kaushal PK, Drury P, 1972. Fatal anaphylaxis-like reaction induced by yeast mannans in nonsensitized mice. Infect Immun 5:180–182.

King RD, Lee JC, Morris AL, 1980. Adherence of *Candida albicans* and other *Candida* species to mucosal epithelial cells. Infect Immun 27:667–674.

Kozel TR, 1977. Nonencapsulated variant of *Cryptococcus neoformans* II. Surface receptors for cryptococcal polysaccharide and their role in inhibition of phagocytosis by polysaccharide. Infect Immun 16:99–106.

Kwon-Chung KJ, Polacheck I, Bennett JE, 1982. Improved diagnostic medium for separation of *Cryptococcus neoformans* var *neoformans* (serotypes A and D) and *Cryptococcus neoformans* var *gattii* (serotypes B and C). J Clin Microbiol 15:535–537.

Laxalt KA, Kozel, TR, 1979. Chemotaxigenesis and activation of the alternative complement pathway by encapsulated and nonencapsulated *Cryptococcus neoformans*. Infect Immun 26:435–440.

Lehmann PF, Reiss E, 1980. Detection of *Candida albicans* mannan by immunodiffusion, counterimmunoelectrophoresis and enzyme-linked immunoassay. Mycopathol 70:83–88.

Leiva S, Gonzalez J, Olea N, Pincheira G, 1982. Radioautographic detection of macromolecules involved in synthesis of *Neurospora crassa* cell wall. Cell Molec Biol 28:159–165.

Lloyd KO, 1972. Molecular organization of a covalent peptido-phospho-polysaccharide complex from the yeast form of *Cladosporium werneckii*. Biochemistry 11:3884–3890.

Longbottom JL, Pepys J, 1964. Pulmonary aspergillosis: diagnostic and immunological significance of antigens and C-substance in *Aspergillus fumigatus*. J Pathol Bacteriol 88:141–151.

Macdonald F, Odds FC, 1980. Inducible proteinase of *Candida albicans* in diagnostic serology and in the pathogenesis of systemic candidosis. J Med Microbiol 13:423–435.

Mahadevan PR, Tatum EL, 1967. Localization of structural polymers in the cell wall of *Neurospora crassa*. J Cell Biol 35:295–302.

Marchessault RH, Deslandes Y, 1980. Texture and crystal structure of fungal polysaccharides. pp. 221–250. In Fungal Polysaccharides. Sandford PA, Matsuda K, eds. American Chemical Society Symposium series no. 126. Washington, DC: American Chemical Society.

Matsunaga T, Okubo A, Fukami M, Yamazaki S, Toda S, 1981. Identification of β-galactofuranosyl residues and their rapid internal motion in the *Penicillium ochro-chloron* cell wall probed by ^{13}C NMR. Biochem Biophys Res Commun 102:524–530.

Marx RS, Forsyth KR, Hentz SK, 1982. *Mucorales* species activation of serum leukotactic factor. Infect Immun 38:1217–1222.

McCammon MT, Parks LW, 1982. Enrichment for auxotrophic mutants in *Saccharomyces cerevisiae* using the cell wall inhibitor, echinocandin B. Molec Gen Genetics 186:295–297.

Michel AJ, Scurfield G, 1967. Composition of extracted fungal cell walls as indicated by infrared spectroscopy. Arch Biochem Biophys 120:628–637.

Molano J, Bowers B, Cabib E, 1980. Distribution of chitin in the yeast cell wall. An ultrastructural and chemical study. J Cell Biol 85:199–212.

Monreal J, Reese ET, 1969. The chitinase of *Serratia marcescens*. Can J Microbiol 15:689–696.

Murphy JW, Cozad, GC, 1972. Immunological unresponsiveness induced by cryptococcal capsular polysaccharide assayed by the hemolytic plaque technique. Infect Immun 5:896–901.

Nakajima T, Ballou CE, 1974. Characterization of the carbohydrate fragments obtained from *Saccharomyces cerevisiae* mannan by alkaline degradation. J Biol Chem 249:7679–7684.

Nickerson WJ, 1963. Symposium on biochemical basis of morphogenesis in fungi IV: molecular basis of form in yeasts. Bacteriol Rev 27:305–324.

Noguchi T, Banno Y, Watanabe T, Nozawa Y, Ito Y, 1975. Carbohydrate composition of the isolated cell walls of dermatophytes. Mycopathol 55:71–76.

Noguchi T, Kitazima Y, Nozawa Y, Ito Y, 1971. Isolation, composition and structure of cell walls of *Trichophyton mentagrophytes*. Arch Biochem Biophys 146:506–512.

Notario V, Villa TG, Villanueva JR, 1979. Cell wall associated 1,4-beta-D-xylanase in *Cryptococcus albidus* var *aerius*: in situ characterization of the activity. J Gen Microbiol 114:415–422.

Nurminen T, Oura E, Suomalainen H, 1970. The enzymic composition of the isolated cell wall and plasma membrane of baker's yeast. Biochem J 116:61–69.

Odds FC, Hierholzer JC, 1973. Purification and properties of a glycoprotein acid phosphatase from *Candida albicans*. J Bacteriol 114:257–266.

Okubo Y, Ichikawa T, Suzuki S, 1980. Immunochemistry of *Candida albicans* mannan. pp. 95–112. In Fungal Polysaccharides. Sandford PA, Matsuda K, eds. Washington, DC: American Chemical Society.

Okubo Y, Shibata N, Ichikawa T, Chaki S, Suzuki S, 1981. Immunochemical study on baker's yeast mannan prepared by fractional precipitation with cetyltrimethylammonium bromide. Arch Biochem Biophys 212:204–215.

Orenstein NS, 1971. Cell wall molecular architecture in baker's yeast. PhD Dissertation. Rutgers University, New Brunswick, NJ.

Peat S, Whelan WJ, Edwards TE, 1961. Polysaccharides of baker's yeast. Part IV. Mannan. J Chem Soc (London) 1:29–34.

Phaff HJ, 1977. Enzymatic yeast cell wall degradation. pp. 244–282. In Food Proteins: Improvement through Chemical and Enzymatic Analysis. Feeney RE, ed. American Chemical Society, Advances in chemistry series no. 160. Washington, DC: American Chemical Society.

Pine L, 1977. Histoplasma antigens: their production, purification, and uses. pp. 138–168. In Proc 21st OHOLO Biological Conference Ma'alot, Israel, Contributions Microbiol and Immunol, Vol 3. Beemer AM, Ben-David A, Klingberg MA, Kuttin ES, eds. Basel: S. Karger.

Pine L, Boone CJ, 1968. Cell wall composition and serological activity of *Histoplasma capsulatum* serotypes and related species. J Bacteriol 96:789–798.

Potgieter HJ, Alexander M, 1965. Polysaccharide components of *Neurospora crassa* hyphal walls. Can J Microbiol 11:122–125.

Poulain D, Tronchin G, Dubremetz JF, Biguet J, 1978. Ultrastructure of the cell wall of *Candida albicans* blastospores: study of its constituitive layers by use of a cytochemical technique revealing polysaccharides. Ann Microbiol (Institut Pasteur) 129:141–153.

Previato JO, Gorin PAJ, Travassos LR, 1979. Cell wall composition in different cell types of the dimorphic species *Sporothrix schenckii.* Exp Mycol 3:83–91.

Pugh D, Cawson RA, 1978. The surface layer of *Candida albicans.* Microbios 23:19–23.

Rajasingham KC, Cawson RA, 1982. Ultrastructural identification of extracellular material and appressoria in *Candida albicans.* Cytobios 35:77–83.

Rehacek J, Beran J, Bicik V, 1969. Disintegration of microorganisms and preparation of yeast cell walls in a new type disintegrator. Appl Environ Microbiol 17:462-466.

Reiss E, 1977. Serial enzymatic hydrolysis of cell walls of two serotypes of yeast-form *Histoplasma capsulatum* with α (1 → 3)-glucanase, β (1 → 3)-glucanase, pronase, and chitinase. Infect Immun 16:181–188.

Reiss E, 1985. Cell wall composition. pp. 57–102. In Fungi Pathogenic for Humans and Animals, Part B II. Howard DH and Howard LF, eds. New York: Marcel Dekker.

Reiss E, Stone SH, Hasenclever HF, 1974. Serological and cellular immune activity of peptidoglucomannan fractions of *Candida albicans* cell walls. Infect Immun 9:881–890.

Remold H, Fasold H, Staib F, 1968. Purification and characterization of a proteolytic enzyme from *Candida albicans.* Biochim Biophys Acta 167:399–406.

Richardson MD, Kearns MJ, Smith H, 1982. Differentiation of extracellular from ingested *Candida albicans* blastospores in phagocytosis tests by staining with fluorescein-labelled concanavalin A. J Immunol Methods 52:241–244.

Robertson AJ, Gerlach JH, Rank GH, Fowke LC. 1980. Yeast cell wall, membrane, and soluble marker polypeptides identified by comparative two-dimensional electrophoresis. Can J Biochem 58:565–572.

Rosenberger RF, 1976. The cell wall. pp. 328–344. In The Filamentous Fungi Vol 2. Biosynthesis and Metabolism. Smith JE, Berry DR, eds. New York: John Wiley and Sons.

Roy I, Landau JW, 1972. Protein constituents of cell walls of the dimorphic phases of *Blastomyces dermatitidis.* Can J Microbiol 18:473–478.

Rüchel R, 1981. Properties of a purified proteinase from the yeast *Candida albicans.* Biochim Biophys Acta 659:99–113.

Ruiz-Herrera J, Bartnicki-Garcia S, 1974. Synthesis of cell wall microfibrils in vitro by a "soluble" chitin synthetase from *Mucor rouxii.* Science 186:357–359.

San Blas G, Ordaz D, Yegres, FJ, 1978. *Histoplasma capsulatum*: chemical variability of the yeast cell wall. Sabouraudia 16:276–284.

San Blas G, 1982. The cell wall of fungal human pathogens: its possible role in host–parasite relationships—a review. Mycopathol 79:159–184.

San Blas G, San Blas F, 1977. *Paracoccidioides brasiliensis*: cell wall structure and virulence. Mycopathol 62:77–86.

Schwarz J, 1981. Histoplasmosis. New York: Praeger, 472 p.

Sietsma JH, Wessels JGH, 1979. Evidence for covalent linkages between chitin and β-glucan in a fungal wall. J Gen Microbiol 114:99–108.

Sietsma JH, Wessels JGH, 1981. Solubility of $(1 \rightarrow 3)$-β-D/$(1 \rightarrow 6)$-β-D-glucan in fungal walls: importance of presumed linkage between glucan and chitin. J Gen Microbiol 125:209–212.

Skujins JJ, Potgieter HJ, Alexander M, 1965. Dissolution of fungal cell walls by a streptomycete chitinase and β-1,3-glucanase. Arch Biochem Biophys 111:358–364.

So LL, Goldstein IJ, 1968. Protein-carbohydrate interaction. XIII. The interaction of concanavalin A with α-mannans from a variety of microorganisms. J Biol Chem 243:2003–2007.

Staff CM, Feather MS, 1973. The characterization of a chitin-associated D-glucan from the cell walls of *Aspergillus niger*. Biochim Biophys Acta 320:64–72.

Stewart PR, Rogers PJ, 1978. Fungal dimorphism: a particular expression of cell wall morphogenesis. pp. 164–196. In The Filamentous Fungi. vol 3. Smith JE, Berry DR, eds. New York: John Wiley.

Sullivan PA, Yin CY, Molloy C, Templeton MD, Shepherd MG, 1983. An analysis of the metabolism and cell wall composition of *Candida albicans* during germ tube formation. Can J Microbiol 29:1514–1525.

Summers DF, Grollman AP, Hasenclever, HF, 1964. Polysaccharide antigens of *Candida* cell wall. J Immunol 92:491–499.

Suzuki S, 1981. Antigenic determinants. pp. 85–96. In Yeast Cell Envelopes: Biochemistry, Biophysics, and Ultrastructure—Vol I. Arnold WN, ed. Boca Raton, FL: CRC Press.

Szaniszlo PJ, Cooper BH, Voges HS, 1972. Chemical compositions of the hyphal walls of three chromomycosis agents. Sabouraudia 10:94–102.

Taylor IFP, Cameron DS, 1973. Preparation and quantitative analysis of fungal cell walls. Strategy and tactics. Annu Rev Microbiol 27:243–260.

Torres-Bauza LJ, Riggsby WS, 1980. Protoplasts from yeast and mycelial forms of *Candida albicans*. J Gen Microbiol 119:341–349.

Tran Van Ky P, Torck C, Vaucelle T, Floc'h F, 1969. Etude comparee sur immunoelectrophoregramme des enzymes de l-extrait antigenique d'*Aspergillus fumigatus* reveles par des serums experimentaux et des serums de malades atteints d'aspergillose. Sabouraudia 7:73–84.

Travassos LR, Lloyd KO, 1980. *Sporothrix schenckii* and related species of *Ceratocystis*. Microbiol Rev.44:683–721.

Trinci APJ, 1978. Wall and hyphal growth. Sci Prog Oxford 65:75–99.

Tronchin G, Poulain D, Herbaut J, Biguet J, 1981. Localization of chitin in the cell wall of *Candida albicans* by means of wheat germ agglutinin. Fluorescence and ultrastructural studies. Eur J Cell Biol 26:121–128.

Troy FA, Koffler H, 1969. The chemistry and molecular architecture of the cell walls of *Penicillium chrysogenum*. J Biol Chem 244:5563–5576.

Villa TG, Notario V, Villaneuva JR, 1980. Chemical and Enzymic analyses of Pichia polymorpha cell walls. Can J Microbiol 26:169–174.

Walborg EF Jr, 1978. Current concepts of glycoprotein structure. pp. 1–20. In Glycoproteins and Glycolipids in Disease Processes: ACS Symposium Series 80. Walborg EF Jr, ed. Washington, DC: American Chemical Society.

Wheat RW, Tritschler C, Conant NF, Lowe EP, 1977. Comparison of *Coccidioides immitis* arthrospore, mycelium and spherule cell walls and influence of growth medium on mycelial cell wall composition. Infect Immun 17:91–97.

Whistler RL, BeMiller JN, 1958. Alkaline degradation of polysaccharides. Adv Carbohydr Chem 13:289–329.

Wilson DE, Bennett JE, Bailey JW, 1968. Serologic grouping of *Cryptococcus neoformans*. Proc Soc Exp Biol Med 127:820–823.

Wilson EV, Hearn VM, 1982. A comparison of surface and cytoplasmic antigens of *Aspergillus fumigatus* in an enzyme-linked immunosorbent assay (ELISA). Mykosen 25:653–661.

Winterburn PJ, 1974. Polysaccharide structure and function. pp. 307–342. In Companion to Biochemistry. Bull AT, Lagnado JR, Thomas JO, Tipton KF, eds. London: Longman.

Yamaguchi H, 1974. Effect of biotin insufficiency on composition and ultrastructure of cell wall of *Candida albicans* in relation to its mycelial morphogenesis. J Gen Appl Microbiol 20:217–228.

Yu RJ, Bishop CT, Cooper FP, Blank F, Hasenclever HF, 1967. Glucans from *Candida albicans* (serotype B) and from *Candida parapsilosis*. Can J Chem 45:2264–2267.

Yu RJ, Harmon SR, Grappel SF, Blank F, 1971. Two cell-bound keratinases of *Trichophyton mentagrophytes*. J Invest Dermatol 56:27–32.

Zonneveld BJM, 1971. Biochemical analysis of the cell wall of *Aspergillus nidulans*. Biochim Biophys Acta 249:506–514.

Chapter 3

Blastomyces dermatitidis

3.1 INTRODUCTION

Blastomyces dermatitidis is a dimorphic fungus that has a filamentous saprophyte existence and grows as a yeast form in the human or animal host. Its ecological niche is unknown but presumed to be in the soil. Infection by *B dermatitidis* is found in the central USA, extending in the north around the Great Lakes and into Manitoba Province in Canada. The endemic zone also extends into the southeastern states. The world-wide endemic areas include Africa and northern South America. Exposure to *B dermatitidis* is the result of inhalation of conidia. In the majority of persons exposed to the organism, no active disease follows. Those unable to contain the infective dose may develop an acute pneumonia, or the pulmonary phase may be inapparent. The pneumonia may be self-limited or disseminated, especially to the skin. In the absence of an acute pulmonary phase, the onset may be insidious, leading to chronic pulmonary or disseminated infection. The skin lesions are at first closed nodules, but become abscesses with purulent exudates. In untreated individuals, slow progressive verrucous lesions cover large areas over a period of years. New skin lesions frequently are accompanied by either osseous or articular involvement. The histologic picture is characterized first by a polymorphonuclear neutrophil (PMN) exudate that is gradually replaced, or coexists with granulomas. These rarely caseate, unlike the granulomas of histoplasmosis (Chandler et al, 1980; Sarosi and Davies, 1979; Witorsch and Utz, 1968).

3.2 BLASTOMYCIN

Supernatants of old mold form cultures of *B dermatitidis*, termed blastomycins, are prepared by a method analogous to that of histoplasmin, but unlike the latter product, blastomycins do not produce delayed cutaneous hypersensitivity in a significant proportion of cases and frequently show crossreactions in histoplasmosis patients. Forty percent to 50% or fewer patients display positive delayed cutaneous hypersensitivity reactions to blastomycin (Cox and Larsh, 1974a). In a series of culture-proven, self-limited blastomycoses, blastomycin skin tests were positive in four out of nine patients (Recht et al, 1979). Witorsch and Utz (1968) reported that all

25 active blastomycosis patients were flatly unresponsive to blastomycin. Some of the observed inactivity of blastomycin is not a failure of antigenic potency but is attributable to the cutaneous anergy common in disseminated mycoses, including blastomycosis. Smith (1949) described a subset of blastomycosis patients with poor prognosis who had cutaneous anergy to blastomycin coupled with high levels of complement-fixing antibodies. A lack of both sensitivity and specificity of blastomycin is also well-documented in sensitized guinea pigs (Cox and Larsh, 1974a). Commercial blastomycin evoked positive delayed cutaneous hypersensitivity in 37% of guinea pigs sensitized with Merthiolate-killed *B dermatitidis* yeast forms in complete Freund adjuvant, and in 40% of animals sensitized with killed *H capsulatum*. An individually prepared blastomycin was more sensitive but this increased activity was accompanied by a corresponding rise in crossreactions among *H capsulatum*-sensitized animals. The histologic findings of blastomycin skin test sites at 24 hours show a mononuclear infiltrate compatible with type IV delayed cutaneous hypersensitivity. The specificity index—defined as reactors among homologously sensitized animals divided by the number of reactors in animals sensitized with the heterologous fungus—of blastomycin in sensitized guinea pigs was less than one, whereas the specificity index for histoplasmin was 7.2 (Cox and Larsh, 1974a). Unpurified blastomycin is an antigenic complex containing a significant amount of nonspecific sensitin masking a small amount of specific factor(s). Efforts to improve the potency and specificity of blastomycin have been made. In one such study (Lancaster and Sprouse, 1976), isotachophoresis resolved nine ultraviolet-absorbing fractions. Some of these proteins were unreactive when purified, others crossreacted with *H capsulatum*, and one, fraction "F," was specific for *B dermatitidis*. The purified protein evoked a smaller delayed cutaneous hypersensitivity reaction at 48 hours than did blastomycin, but was reactive in all guinea pigs tested that were sensitized with *B dermatitidis* and was not reactive with any *H capsulatum*-sensitized animals (Cox and Larsh, 1974a). Commercial and individually prepared blastomycins have been analyzed by polyacrylamide gel electrophoresis. These extracts contain relatively few proteins and their number varies from four to six or more. There is also variation in the R_f values of components among different lots. The majority of proteins in blastomycins display positive periodic acid-Schiff (PAS) reactions indicating that they are glycoproteins (Cox and Larsh, 1974b).

3.3 CELL WALL EXTRACTS

Both alkali-soluble and insoluble glucans from yeast and mycelial forms of *B dermatitidis* were exposed to the mixed glycosidases of snail digestive juice or the exo-β-1,3-glucanase of *Basidiomycete QM806* (Kanetsuna and Carbonell, 1971). Snail digestive juice contains glucanases, mannanase, and chitinase, but has low proteolytic activity. It entirely lacks α-1,3-glucanase (Phaff, 1977). The alkali-soluble–acid-precipitable glucan of yeast forms was attacked to only a minor extent (5%) by snail digestive juice, increasing the dextrorotation of the product. About half of the alkali-resistant cell wall of the yeast form was susceptible to snail digestive juice, but the analogous wall fraction from the mycelial forms was more susceptible, 89%. The basidiomycete's exo-β-1,3-glucanase completely removed the glucan from the alkali-resistant cell wall of mycelial forms, but only a minor portion was recovered as glucose. This is probably because this enzyme can bypass β-1,6-branch points, yielding gentiobiose (Phaff, 1977). About 27% of the mycelial form wall is an alkali-soluble–acid-precipitable, highly dextrorotatory glucan. The β-1,3-glucan present in the alkali-insoluble fraction was made accessible by first digesting chitin with commercial chitinase. After this treatment, the insoluble glucan became soluble in dilute alkali. This result provides indirect evidence that chitin and β-1,3-glucan are linked covalently, an idea that has been confirmed more recently (Sietsma and Wessels, 1979) in another fungus. The results of chemical and enzymatic analysis agree with the proposal that the glycans of the yeast form wall of *B dermatitidis* are a major α-1,3-glucan comprising nearly 95% of the polymeric glucose, and the remaining portion is a β-1,3-glucan. Chitin comprises about 37% of the yeast form wall. In the mycelial forms, the α-1,3-glucan and chitin contents are reduced with a corresponding increase in β-1,3-glucan. Galactomannan is also increased in mycelial form walls. Two wall compositional analyses of the same isolate of *B dermatitidis* have been reported (Kanetsuna et al, 1969; Kanetsuna and Carbonell, 1971: Table 3.1). The more recent analysis appears to give a more accurate picture, because the protein and chitin contents are in closer alignment with what is expected in the walls of the systemic dimorphic fungi. In contrast to *Paracoccidioides brasiliensis*, α-1,3-glucan is more easily demonstrated in the cell walls of *B dermatitidis* mycelial forms. The alkali-soluble protein fraction of *B dermatitidis* walls has been discussed above.

Proteins associated with cell walls of both dimorphic states of *B dermatitidis* were prepared by disruption of the cells in the Braun cell homogenizer. Walls then were washed in a buffer and defatted by a three-stage extraction procedure (Roy and Landau, 1972). The walls were partially solubilized with 1M alkali and the extract was partitioned by adding ammonium sulfate. Two proteins were located by sodium dodecyl sulfate-polyacrylamide gel electrophoresis (SDS-PAGE); each of the four proteins had distinctive mobility.

An alkali-soluble–water-soluble fraction of *B dermatitidis* yeast form cell walls (B-ASWS) is a source of antigens for assaying delayed cutaneous hypersensitivity and its in vitro correlates (Cox and Larsh, 1974a, 1974b; Lancaster and Sprouse, 1976b; Deighton et al, 1977; Hall et al, 1978). The need for such a preparation grew out of the shortcomings of blastomycin, namely its lack of both sensitivity and specificity (Witorsch and Utz, 1968; Levin, 1970). The B-ASWS is prepared from the cell wall fraction of mechanically disrupted yeast forms. Walls are washed in water, digested with trypsin, and then exposed to 1M NaOH for 3 hours at 25 °C. The

Table 3.1. Cell Wall Composition of Yeast and Mycelial Forms of *Blastomyces dermatitidis* BD64

	Cell Wall Dry Weight (percent)	
Component	Yeast Form	Mycelial Form
Total neutral sugar	36.2^b–47.1^a	43.5^a–51.0^b
N-acetylglucosamine	37^b–48^a	13.9^a–22.8^b
Protein (as amino acid)	7.1^a–7.8^b	10.9^b–26.8^a
Lipid	5.5^a	9.3^a
Phosphate	0.12^b–0.14^a	0.08^b–0.36^a

[a] Values calculated from Kanetsuna et al., 1969.
[b] Values calculated from Kanetsuna and Carbonell, 1971; glucose, 39.2%; galactose 3.9%; mannose 7.8%.

protein content of B-ASWS varies from lot to lot with the same culture (ATCC 26199) from 44% to 70% (Cox and Larsh, 1974b; Deighton et al, 1977). A carbohydrate component comprises 31.3% of the antigenic complex and contains mannose, glucose, and galactose (Cox and Larsh, 1974b). When analyzed by SDS-PAGE, this antigenic complex contains three to four proteins between 30,000 to 50,000 da and a PAS-positive component that does not correspond to a protein. This pattern of proteins in B-ASWS is reproducible for similarly prepared products of four different isolates, but differs from protein profiles of blastomycin and cytoplasmic solubles (Cox and Larsh, 1974b).

Unlike blastomycin, which has a low specificity index, B-ASWS has potency and specificity comparable to that achieved with histoplasmin in eliciting delayed cutaneous hypersensitivity in guinea pigs sensitized with Merthiolate-killed yeast forms (Cox and Larsh, 1974a). Histologic study of cutaneous sites in *B dermatitidis*-sensitized guinea pigs after B-ASWS skin tests revealed an early 4 hour PMN response changing at 24 hours to a more intense, predominantly mononuclear response that persisted on a reduced level at 48 hours.

The B-ASWS is active in driving peritoneal exudate cells from *B dermatitidis*-infected guinea pigs to produce migration inhibitory factor (MIF) (Deighton et al, 1977). Some MIF activity was generated in peritoneal exudate cells from *H capsulatum*-infected animals stimulated in vitro by B-ASWS, but this was below 20% inhibition, corresponding to the often-used threshold for specific inhibition of macrophage migration. Blastomycin, by comparison, was a weaker source of antigen, evoking borderline inhibition of migration and toxicity at higher concentrations. Lymphocytes from axillary nodes draining the site of subcutaneous infection with *B dermatitidis* were stimulated in vitro with B-ASWS resulting in a mean blastogenic index of 2.5. These lymph node cells were flatly unresponsive to blastomycin. Higher blastogenic indices of 6 to 7 were elicited in vitro by B-ASWS stimulation of peripheral blood lymphocytes from *B dermatitidis*-infected guinea pigs (Hall et al, 1978). This reaction was reasonably specific, the lymphocytes from *H capsulatum*-sensitized animals producing only low level blastogenesis when cultured with B-ASWS.

B-ASWS was fractionated by preparative discontinuous PAGE into four ultraviolet-absorbing peaks. Peak 4, a minor component by weight, had an increased specific activity to evoke delayed cutaneous hypersensitivity compared with unfractionated B-ASWS or to blastomycin, and was free of crossreactivity in guinea pigs sensitized with *H capsulatum* (Lancaster and Sprouse, 1976b). The remaining three fractions were antigenic, but crossreacted in *H capsulatum*-sensitized animals, as did unfractionated B-ASWS. Thus the genus-specific antigen resided in one component. The number of proteins detected in B-ASWS, four, accords with the analytic PAGE results (Cox and Larsh, 1974b). Isoelectric focusing of proteins in B-ASWS (Hall et al, 1978) yielded two proteins (pI 4.01, 4.69) that provided smaller blastogenic indices and reduced cutaneous reactivity than intact B-ASWS. The genus-specificity of isofocused proteins was not reported (Hall et al, 1978).

The cutaneous reactivity of B-ASWS was considerably improved by ultrafiltration. The material retained on an Amicon PM10 membrane provided more consistent activity among batches, reducing the standard deviation in the number of positive reactors (Cox and Larsh, 1974b). The reduction probably occurs because of removal of lower molecular weight polypeptides or glycopeptides that are not active antigens.

The conclusion that can be drawn from these studies is that a two stage procedure is desirable to obtain a potent, specific antigen from the B-ASWS complex. The ultrafiltrate from an Amicon PM30 membrane (nominal weight limit for proteins of 30,000 da) retained by PM10 membrane is then fractionated by disc electrophoresis to purify a specific antigen. A comparison of this factor with blastomycin and with the "factor A" is needed before a more coherent picture of *B dermatitidis* antigenic structure emerges (Table 3.2).

The biologic activity of *B dermatitidis* yeast form cell walls is dissociable into a granulomagenic fraction–alkali-soluble portion and a necrosis-inducing fraction which is the alkali-resistant wall residue (Cox et al, 1974). The alkali-soluble portion of walls from the virulent kL-1 isolate induced granulomas 72 hours after injection into normal mice whereas that obtained from avirulent GA-1 walls did not. B-ASWS is known to evoke a mixed type III and type IV delayed cutaneous hypersensitivity reaction in sensitized guinea pigs, but the granulomatous response to the alkali extract of cell walls in mice does not depend on prior sensitization. These findings are not easily reconciled with the presumed role of the chitin-glucan insoluble wall matrix as an inducer of granulomas, because of its resistance to digestion by host glycosidases.

3.4 CYTOPLASMIC SOLUBLES

The cytoplasmic soluble antigens of *B dermatitidis* yeast forms capable of evoking delayed cutaneous hypersensitivity partitioned between 10,000 and 50,000 da limit ultrafiltration membranes. The cutaneously active fractions from ultrafiltrates of cytoplasmic solubles and from blastomycins contained one to four proteins including a glycoprotein with identical mobility in discontinuous PAGE. The cytoplasm of yeast forms has been deemphasized as a source of antigens because only the B-ASWS fraction of cell walls was shown to be genus-specific (Cox and Larsh, 1974b).

Table 3.2. Properties of Two *Blastomyces dermatitidis* Antigens: B-ASWS and Factor A

Alkali-Soluble Water-Soluble Fraction of Cell Walls (B-ASWS)[a]	
Preparation	Isolated cell walls extracted with trypsin, then 1M NaOH. Ultrafiltration over PM30 membrane.
Properties	Contains 3 to 4 proteins (polyacrylamide gel electrophoresis). Genus specificity resides in one protein. Molecular weight 30,000 to 35,000 da.
	Elicits positive Arthus and delayed cutaneous hypersensitivity in guinea pigs.
	Evokes MIF production and in vitro lymphocyte proliferation.
	Low degree of crossreaction in *H capsulatum*-sensitized guinea pigs.
Factor A[b]	
Preparation	Culture supernatant. Chromatographed on DEAE-cellulose. The antigen has a strong negative charge.
Properties	Resistant to trypsin and other proteinases, activity may be labile to boiling. Specific for *B dermatitidis* in a variety of humoral tests: precipitin, complement fixation, EIA.

Sources:
[a] Cox & Larsh, 1974; Deighton et al, 1977; Hall et al, 1978.
[b] Green et al, 1980.

Pertinent to the search for antigenic activity in B dermatitidis yeast forms is the finding (Salvin, 1952) that saline washings of whole killed yeast forms contained a toxin lethal for mice when injected with heat-killed tubercle bacilli. None of the control group mice receiving only tubercle bacilli and saline died.

3.5 HUMORAL RESPONSE

3.5.1 Complement Fixation and Immunodiffusion

The ability of serology to provide a useful diagnostic adjunct in blastomycosis is underrated. Antigens present in homogenates and culture filtrates of B dermatitidis yeast forms have been characterized with panels of blastomycotic and heterologous case sera (Kaufman et al, 1973). The homogenate of yeast forms is more satisfactory for complement-fixation tests than that of mycelial forms, but for the immunodiffusion test, the filtrate of yeast form cultures was the best source of antigens. The immunodiffusion test was more sensitive than complement-fixation in detecting disease in 79% of 113 blastomycosis cases compared with the 57% that reacted to the complement-fixation test. The combined application of both tests raised the sensitivity level to 88%. Crossreactions with sera from histoplasmosis and coccidioidomycosis patients seriously impaired the specificity of the complement-fixation test with B dermatitidis yeast form homogenate antigens. The operation of a positive reference antiserum in the immunodiffusion test ruled out immunoprecipitin arcs that did not form lines of identity with B dermatitidis antigens A and B. Antibodies to A occurred alone or with those against factor B, but the latter reaction was not observed by itself. The antibodies usually persisted for more than 2 months after chemotherapy was completed and radiography proved clearing of symptoms.

3.5.2 A-Factor

The A-antigen identified in yeast form culture filtrates as specific for B dermatitidis was purified by diethylaminoethyl- (DEAE) cellulose chromatography with a discontinuous salt gradient (Green et al, 1980). The antigen is tightly bound requiring 0.6M NaCl for elution. Maximal yields of factor A were obtained by resuspending washed yeast forms in buffered saline for 2 weeks. During that time the antigen was leached or secreted into the buffer. The specific activity of A-antigen increased fivefold after chromatography and the protein : carbohydrate ratio increased from one to ten. Activity, judged by "rocket" immunoelectrophoresis, was unaffected by exposure to trypsin or other proteinases but was labile to boiling (Table 3.2). Purified A-antigen is stable for several months at 5 °C in phosphate buffer. The serologic activity of purified factor A was compared in a panel of 27 blastomycosis patients by immunodiffusion and complement-fixation tests and by indirect enzyme immunoassay (EIA). One immunoprecipitin arc was observed in the immunodiffusion reaction between purified factor A and reference antiserum. Forty-eight percent of the panel were reactive in immunodiffusion. The specificity of the complement-fixation test was increased markedly with purified factor A as antigen but crossreactions were observed with sera from some histoplasmosis patients suggesting that impurities remain in the A-antigen. The increased specificity in complement-fixation was accompanied by a decline in sensitivity, indicating that in infection with B dermatitidis complement-fixing antibodies are also directed against other factors possibly ones that *B dermatitidis* shares with other systemic fungi. This decline in sensitivity was overcome by an indirect EIA that detected 93% of the blastomycosis test panel. The adaptation of purified factor A to indirect EIA achieved increased sensitivity and a simplified format that should be encouraged.

Monoclonal antibodies to factor A have been produced from fusions of spleen cells of mice immunized with purified A antigen. Mice primed with 3 μg of antigen and boosted twice were fused with the SP2/0 nonsecreting plasmacytoma cell line and gave rise to 25 hybridomas reactive with the A-antigen in indirect EIA (Green et al, 1982). Six clones were propagated as ascites tumors in mice: five of the six fixed complement. Four clones were analyzed for their IgG subclass; three were IgG_{2a} and one was IgG_1. Complete specificity of the monoclonal antibodies was observed in the complement-fixation test, with no crossreactions encountered when *H capsulatum* yeast forms, histoplasmin, or *C immitis* antigens were tested. These monoclonal reagents should lead the way toward eliminating the uncertainty about the value of serology in blastomycosis (Sarosi and Davies, 1979).

3.6 PHAGOCYTE INTERACTIONS

The ability of *B dermatitidis* to replicate in vivo varies depending on the isolate (Brummer et al, 1981) from completely avirulent *GA-1* to isolate *ATCC 26199* that caused fatal infection in mice after inoculation with only 10 colony-forming units (Harvey et al, 1978). When mature, the yeast forms are too large for endophagocytosis but despite this, adherence to murine macrophages is sufficient to inhibit in vitro replication of avirulent and virulence-attenuated isolates. Experiments with a virulent isolate showed that although adherence to macrophages apparently occurs, inhibition of in vitro replication does not last beyond 24 hours. Macrophages stimulated by thioglycollate were better able to inhibit replication of *B dermatitidis* yeast forms than those from immunized mice.

A cytophilic inhibitor of PMN chemotaxis was detected in sera from five blastomycosis patients (Repine et al, 1978). Zymosan-activated autologous serum was the attractant in Boyden chamber experiments. Locomotion of PMN from blastomycosis patients was significantly reduced compared with that from patients with other mycoses or from those of normal persons. When blastomycosis patients' PMN were washed and incubated with zymosan-activated normal sera, locomotion was increased to normal values. Blastomycotic sera also reduced chemotaxis of PMN provided by normal persons or those with other mycoses. Favorable response to therapy was accompanied by the disappearance of the serum inhibitor. The inhibitor is heat-stable (56 °C, 30 minutes) and cytophilic but can be removed from cells by repeated washing. This substance has not been characterized and neither *B dermatitidis* antigen, antibodies, or immune complexes have been ruled out.

Human polymorphonuclear neutrophils have potent cidal activity against *H capsulatum* yeast forms, yet the latter survive and multiply in macrophages (see Chapter 5). The opposite situation has been observed in the interaction between human phagocytes and yeast forms of *B dermatitidis* (Brummer and Stevens, 1982). Human PMN fail to kill *Blastomyces* yeast forms after 2 hours in vitro and stimulate their replication after 24 hours. Monocytes, on the other hand, partially inhibit *Blastomyces* yeast form cell division. The monocytes aggregate and trap yeast forms in small clumps.

Prompt and efficient chemotaxis and aggregation of human PMN with *B dermatitidis* yeast forms is known to occur (Sixby et al, 1979). These events are accompanied by chemiluminescence, indicating triggering of the oxidative burst, but only about one third of the yeast cells were killed over a range of effector:target ratios. The different proportions of yeast forms killed in these two studies may be related to the virulence of the strains that were used.

3.7 CELL-MEDIATED IMMUNITY

3.7.1 Human Infections

Findings of cutaneous anergy and flat blastogenic responses to antigens and mitogens correlate with a poor prognosis in histoplasmosis, coccidioidomycosis, and paracoccidioidomycosis. Attempts to probe these reponses in blastomycosis have been frustrated by the lack, until recently, of methods to produce *B dermatitidis*-specific antigens, and the difficulty in assembling a large enough series of patients to test. Of 28 blastomycosis patients, all were flatly negative in intradermal tests with blastomycin (Witorsch and Utz, 1968). Responses of lymphocytes from two relapsed pulmonary blastomycosis patients showed positive blastogenic indices and leukocyte-MIF production evoked by a homogenate of *B dermatitidis* yeast forms (Sohnle et al, 1980) and normal mitogen induced proliferation. This yeast form antigen also stimulated lymphocytes from an individual serving as a control without blastomycosis who had a strong delayed cutaneous hypersensitivity reaction to histoplasmin, probably denoting a lack of antigenic specificity. An outbreak of blastomycosis among four families on a hunting trip provided an opportunity to monitor the change, with time, of blastomycin-elicited cutaneous hypersensitivity and lymphocyte blastogenesis (Sarosi and King, 1977). All 18 infected persons remained healthy at the end of the 3-year observation period. Those 16 giving initial positive delayed cutaneous hypersensitivity reactors, an uncommonly high percentage of positive reactors, declined to five reactors (Recht et al, 1979); of 12 whose lymphocytes were tested during the acute phase and had blastogenic indices greater than three, only seven had blastogenic indices greater than three at the conclusion of the study. The high incidence of positive reactions in the acute stage of infection was unexpected because blastomycin is considered a poor antigen. The reduction in the size and extent of delayed cutaneous hypersensitivity and blast transformation over the 3 year period posed another problem because hypersensitivity of infection is considered to be a durable state. Whether other, more potent, *B dermatitidis* antigens are required to monitor patients or whether diminution of the skin test accompanies recovery from blastomycosis remains to be determined. Periodic monitoring of cutaneous and lymphocyte responses to *B dermatitidis*-specific antigens has not yet been reported. Until such monitoring is done the nature of immune deficits that precede or result from *B dermatitidis* infection remains an open question.

3.7.2 Experimental Murine Blastomycosis

Spencer and Cozad (1973) showed that subcutaneous injection of C57B1/6J mice with low doses of live *B dermatitidis* yeast forms produced a local infection that resolved, conferring specific delayed footpad swelling responsiveness that peaked at 15 days postinjection coinciding with maximal resistance to intraperitoneal challenge with one half of a 30 day LD_{50} dose of live yeast forms.

Immunization of outbred ICI mice with French pressure cell-killed yeast forms produced significant protection against intravenous challenge with 10^5 conidia, but no protection could be demonstrated when immunized mice were challenged intravenously with 10^5 viable yeast forms (Landay et al, 1972). In contrast, the resistance of C57B1/6J mice to challenge with live *B dermatitidis* was increased by immunization with Merthiolate killed yeast forms (Cozad and Chang, 1980). The highest index of resistance occurred 18 days after immunization and coincided with the peak delayed footpad swelling, with killed yeast forms as antigen, suggesting that maturation of cell-mediated immunity is an important aspect of resistance. The predominant target organ was the lung even when the yeast forms were injected intravenously. The importance of the lung as target organ in the murine blastomycosis model was also emphasized by Harvey et al (1978).

Differences in susceptibility to *B dermatitidis* infection were observed among inbred strains of mice (Morozumi et al, 1981). When challenged by the respiratory route DBA/1J were most resistant, C3H/HeJ the most susceptible, and BALB/c intermediate in susceptibility. These differences could not be explained by the H-2 haplotype or referred to complement deficiency. When the intraperitoneal route was used the susceptibility of these two mouse strains was reversed. No difference was detected in the ability of peritoneal macrophages from subcutaneously infected mice of either DBA/1 or C3H/HeJ strain to inhibit in vivo replication of *B dermatitidis* yeast forms.

Normal BALB/cByJ mice were infected sublethally with *B dermatitidis* yeast forms and the development of responses was studied to (i) the B-ASWS cell wall extract, (ii) a urea-induced lysate of yeast forms, and (iii) a yeast form culture supernate (Morozumi et al, 1982). Humoral responses (IgG) to the culture supernate measured by indirect EIA were evident 1 to 2 weeks after infection and rose to a plateau at 4 weeks. Proliferative responses of regional lymph node cells showed significant tritiated thymidine uptake after 4 weeks, but spleen cell cultures were responsive 1 or 2 weeks earlier. These responses also reached a plateau by the fourth week. The urea-induced lysate antigen evoked higher in vitro and delayed cutaneous hypersensitivity responses than B-ASWS. Peritoneal exudate cells removed from mice 1 week after infection inhibited the replication of *B dermatitidis* yeast forms.

The subcutaneous priming infection confers significant protection against a lethal respiratory challenge 4 weeks later. No deaths occurred in the immunized group whereas 160 colony forming units achieved a 30-day $LD_{80\%}$ in unprotected mice. The results of this series of experiments are somewhat predictable and do not give new insight into which limb of the immune response is critical for host defense or what fractions of *B dermatitidis* could substitute for a live primary infection in conferring immunity. The ability of activated peritoneal exudate cells to successfully inhibit the replication of *B dermatitidis* is distinct from the well-described intracellular survival of *H capsulatum* in macrophages. *Blastomyces dermatitidis*, because of its large size, is primarily extracellular. The responsiveness of lymphocytes to *B dermatitidis* soluble antigen is in contrast to the anergy of spleen cell cultures to *H capsulatum* (see Chapter 5). The localization of infection with *B dermatitidis* to the subcutaneous site may avoid the systemic anergy of infection.

Isolated cell walls of avirulent *B dermatitidis* GA1 induced necrosis but failed to induce granulomas in guinea pigs, in contrast to the granulomagenic cell walls of the virulent tester strain (Cox et al, 1974).

Other products of *B dermatitidis* yeast forms of potential significance as aggressins or antigens have been reported. A saline extract of acetone-dried *B dermatitidis* yeast forms was lethal for mice within 48 hours after injection (Salvin, 1952). Glycosidase activities were detected in culture filtrates of *B dermatitidis* yeast forms including β-1,3-glucanase, α-mannosidase and N-acetyl-β-glucosaminidase (Davis et al, 1977). The relationship of toxin or glycosidase activity to blastomycosis is unknown.

3.8 SUMMARY

Blastomycin has a low cutaneous reactivity and high crossreactivity occurs in histoplasmosis patients. Increased reactivity can be achieved by preparative isotachophoresis. Alkaline extracts of *B dermatitidis* cell walls are soluble antigens that have increased specificity for detecting delayed cutaneous hypersensitivity. This "ASWS" antigenic complex contains four proteins. Variation among batches is reduced and specific activity is increased by ultrafiltration that removes inert low molecular weight polypeptides. The alkali-soluble wall fraction appears to evoke granulomas in normal animals whereas the alkali-resistant fraction evokes local necrosis. Cytoplasmic soluble antigens have not been emphasized as a source of antigens because only the ASWS is genus-specific. Filtrates of yeast form cultures contain a protein antigen, termed A-factor, that participates in a precipitin reaction in agar gel, forming an arc that is diagnostic for blastomycosis. Factor A is an acidic trypsin-resistant glycoprotein. Although monoclonal antibodies have been produced that fix complement to factor A, the complement-fixing antibodies in human infections are directed against other antigens. The combination of immunodiffusion reactions for the A arc and complement-fixing antibodies against a homogenate of yeast forms detect the disease in a large majority of patients. A further increase in diagnostic sensitivity may result from the use of the A-factor in an indirect EIA for antibodies. Investigations of the antigenic structure are needed to find the relationship, if any, between factor A and those antigens present in ASWS.

Human PMN fail to kill *B dermatitidis* yeast forms but monocytes from the blood aggregate around and trap them, partially inhibiting yeast form replication. A cytophilic inhibitor of PMN chemotaxis has been described in the sera of some blastomycosis patients. The failure of PMN to kill yeast forms is significant in view of the accumulation of these phagocytes that is a hallmark of blastomycotic lesions.

Assessment of cellular response has been hindered by the lack of potency in blastomycin. Where positive delayed cutaneous hypersensitivity reactions were encountered these tended to wane over a follow up period despite the fact that the hypersensitivity of infection is typically a durable state.

REFERENCES

Brummer E, Morozumi PA, Philpott DE, Stevens DA, 1981. Virulence of fungi: correlation of virulence of *Blastomyces dermatitidis* in vivo with escape from macrophage inhibition of replication in vitro. Infect Immun 32:864–871.

Brummer E, Stevens DA, 1982. Opposite effects of human monocytes, macrophages and polymorphonuclear leukocytes on replication of *Blastomyces dermatitidis* in vitro. Infect Immun 36:297–303.

Chandler FW, Kaplan W, Ajello L, 1980. A Colour Atlas and Textbook of Histopathology of Mycotic Diseases 333 pp. London: Wolfe Medical Publications.

Cox RA, Larsh HW, 1974a. Isolation of skin-test active preparations from yeast-phase cells of *Blastomyces dermatitidis*. Infect Immun 10:42–47.

Cox RA, Larsh HW, 1974b. Yeast and mycelial phase antigens of *Blastomyces dermatitidis*: comparison using disc gel electrophoresis. Infect Immun 10:48–53.

Cox RA, Mills LR, Best GK, Denton JF, 1974. Histological reactions to cell walls of an avirulent and a virulent strain of *Blastomyces dermatitidis*. J Infect Dis 129:179–186.

Cozad GC, Chang CT, 1980. Cell-mediated immunoprotection in blastomycosis. Infect Immun 28:398–403.

Davis TE Jr, Domer JE, Li Y-T, 1977. Cell wall studies of *Histoplasma capsulatum* and *Blastomyces dermatitidis* using autologous and heterologous enzymes. Infect Immun 15:978–987.

Deighton F, Cox RA, Hall NK, Larsh HW, 1977. In vivo and in vitro cell mediated immune responses to a cell wall antigen of *Blastomyces dermatitidis*. Infect Immun 15:429–435.

Green JH, Harrell WK, Johnson JE, Benson R, 1980. Isolation of an antigen from *Blastomyces dermatitidis* that is specific for the diagnosis of blastomycosis. Curr Microbiol 4:293–296.

Green JH, Harrell WK, Aloisio C, 1982. The preparation of monoclonal antibodies for use as control reagents in the serological diagnosis of blastomycosis. Abst Ann Mtg Amer Soc Microbiol (Atlanta) No F67:337.

Hall NK, Deighton F, Larsh HW, 1978. Use of an alkali-soluble water-soluble extract of *Blastomyces dermatitidis* yeast-phase cell walls and isoelectrically focused components in peripheral lymphocyte transformations. Infect Immun 19:411–415.

Harvey RP, Schmid ES, Carrington CC, Stevens DA, 1978. Mouse model of pulmonary blastomycosis: utility, simplicity, and quantitative parameters. Amer Rev Respir Dis 117:695–703.

Kanetsuna F, Carbonell LM, 1971. Cell wall composition of the yeastlike and mycelial forms of *Blastomyces dermatitidis*. J Bacteriol 106:946–948.

Kanetsuna F, Carbonell LM, Moreno RE, Rodriguez J, 1969. Cell wall composition of the yeast and mycelial forms of *Paracoccidioides brasiliensis*. J Bacteriol 97:1036–1041.

Kaufman L, McLaughlin DW, Clark MJ, Blumer S, 1973. Specific immunodiffusion test for blastomycosis. Appl Environ Microbiol 26:244–247.

Lancaster MV, Sprouse RF, 1976. Preparative isotachophoretic separation of skin test antigens from blastomycin purified derivative. Infect Immun 13:758–762.

Lancaster MV, Sprouse RF, 1976b. Isolation of a purified skin test antigen from *Blastomyces dermatitidis* yeast phase cell wall. Infect Immun 14:623–625.

Landay ME, Hotchi M, Soares N, 1972. Effect of prior vaccination on experimental blastomycosis. Mycopath Mycol Appl 46:61–64.

Levin S, 1970. The fungal skin test as a diagnostic hindrance. J Infect Dis 122:343–345.

Morozumi PA, Brummer E, Stevens DA, 1981. Strain differences in resistance to infection that are uniquely dependent on the route of challenge: studies in murine pulmonary blastomycosis. Infect Immun 34:623–625.

Morozumi PA, Brummer E, Stevens DA, 1982. Protection against pulmonary blastomycosis: correlation with cellular and humoral immunity in mice following subcutaneous non-lethal infection. Infect Immun 37:670–678.

Phaff HJ, 1977. Enzymatic yeast cell wall degradation. pp. 244–282. In Food Proteins: Improvement Through Chemical and Enzymatic Analysis. American Chemical Society Advances in Chemistry series no 160. Feeney RE, ed. Washington, DC: American Chemical Society.

Recht LD, Phillips JP, Eckman MR, Sarosi GA, 1979. Self-limited blastomycosis, a report of 13 cases. Amer Rev Respir Dis 120:1109–1120.

Repine JE, Clawson CC, Rasp FL Jr, Sarosi GA, Hoidal JR, 1978. Defective neutrophil locomotion in human blastomycosis: evidence for a serum inhibitor. Amer Rev Respir Dis 118:325–334.

Roy I, Landau JW, 1972. Protein constituents of cell walls of the dimorphic phases of *Blastomyces dermatitidis*. Can J Microbiol 18:473–478.

Salvin SB, 1952. Endotoxin in pathogenic fungi. J Immunol 69:89–99.

Sarosi GA, Davies SF, 1979. Blastomycosis. State of the art. Amer Rev Respir Dis 120:911–938.

Sarosi GA, King RA, 1977. Apparent diminution of the blastomycin skin test; follow-up of an epidemic of blastomycosis. Amer Rev Respir Dis 116:785–788.

Sietsma JH, Wessels JGH, 1979. Evidence for covalent linkages between chitin and β-glucan in a fungal wall. J Gen Microbiol 114:99–108.

Sixby JW, Fields BT, Sun CN, Clark RA, Nolan CM, 1979. Interactions between human granulocytes and *Blastomyces dermatitidis*. Infect Immun 23:41–44.

Smith DT, 1949. Immunologic types of blastomycosis: a report on 40 cases. Ann Internal Med 31:463–469.

Sohnle P, Varkey B, Rose H, 1980. Lymphocyte function in relapsed pulmonary blastomycosis. J Allergy Clin Immunol 65:276–280.

Spencer HP, Cozad GC, 1973. Role of delayed hypersensitivity in blastomycosis in mice. Infect Immun 7:329–334.

Witorsch P, Utz JP, 1968. North American blastomycosis. A study of 40 patients. Medicine (Balt.) 47:169–200.

Chapter 4

Coccidioides immitis

4.1 INTRODUCTION

Coccidioides immitis is a soil-dwelling fungus found in scattered sites in the southwestern USA, especially the San Joaquin Valley, California, and overlapping the Mexican border. Other local endemic zones occur in Mexico, Central, and South America. The areas where *C immitis* is found are semi-arid, with a short intense rainy season, the Lower Sonoron Life Zone (Drutz and Catanzaro, 1978). Positive skin tests or actual disease due to *C immitis* occur in some tropical and woodland regions but these cases probably represent (i) the limits of its range, (ii) exposure to contaminated fomites, or (iii) exposure of persons remote from the endemic zone to arthroconidia borne on dust clouds carried into the atmosphere by windstorms (Pappagianis and Einstein, 1978). The fungus is dimorphic, producing filaments in the soil that fragment into the easily dispersed arthroconidia. The tissue form consists of spherules (20 to 200 μm) that divide by internal cleavage and then rupture, releasing endospores which grow and propagate the spherule form.

Approximately 60% of persons infected with *C immitis* are asymptomatic, with a positive skin test to coccidioidin as the only enduring evidence of exposure. Of the 40% who develop symptoms, most experience a benign influenza-like disease. About 5% to 10% are left with a pulmonary cavity or with granulomas. Symptoms of primary pulmonary coccidioidomycosis generally clear in 2 to 3 weeks. Pneumonia requiring hospitalization may be acute, persistent, or chronic. A small fraction of those with pneumonia have miliary lesions. Otherwise, single or multiple nodules evolve in many cases of coccidioidal pneumonia. The liquid centers of such nodules contain viable organisms that can persist for months or years. Gradual clearing of nodules or calcification may occur. Pulmonary disease involving cavitation is also found, and is sometimes discovered on routine chest radiographs. Cavities may close spontaneously over many months. Extrapulmonary dissemination occurs in less than 1% of cases, with osseous, articular, meningeal, and cutaneous forms being well-recognized. Renal and ocular coccidioidomycosis have also been described. Detailed information on many aspects of coccidioidomycosis is found in Stevens (1980); histopathology is presented in depth in Chandler et al (1980), and the mycology of *C immitis* is explicated in McGinnis (1980) and Rippon (1982).

4.2 ANTIGENIC STRUCTURE

4.2.1 Coccidioidin

The only coccidioidin (CDN) of the mycelial form of *C immitis* licensed for human skin tests by the Food and Drug Administration is the particle-free culture filtrate obtained after 2 months' static culture of the fungus on synthetic asparagine–glycerol–salts medium (Smith et al, 1948). The fungus is inactivated or killed in 1:10,000 Merthiolate and filtered off. The original CDN was a pool of culture filtrates from ten isolates, but the current accepted standard, lot 64D2.5, is a pool of products from 65 isolates. There is no hard evidence that these many isolates are necessary, although strain differences in potency have been detected in guinea pig skin tests (Chaparas et al, 1974). CDN prepared this way is remarkably heat stable; autoclaving for 30 minutes at 15 psi, or standing in flowing steam for 1 hour resulted in a 20% loss of activity. The general intradermal dose of CDN for healthy subjects is 0.1 ml of a 1:100 dilution. Patients with coccidioidomycosis should first receive a 1:1000 dilution in case of hyperreactivity and erythema nodosum.

The solubilization of a large portion of the cell contents from young mycelia of *C immitis* by autolysis was first reported by Pappagianis et al (1961). This rapid production of CDN afforded other advantages over its preparation from old culture filtrates in that resuspension of washed mycelia in distilled water removed potentially toxic staling products from the medium, and the use of young mycelia favored preservation of antigenic determinants. In the original method (Pappagianis et al, 1961) the *Silveira* isolate was grown for 3 days at 34 °C in glucose-yeast extract shake culture, then washed in distilled water and autolyzed in toluene-saturated water (3% toluene vol/vol). Autolysis of old mycelia or of young mycelia in $CHCl_3$ or Merthiolate did not produce active products. Approximately 60% to 70% of the cell weight was solubilized; of this only 10% was nondialyzable. Most of the nondialyzable fraction was a mannose-based polysaccharide.

A variation of the toluene-autolysis procedure of preparing CDN (Huppert and Bailey, 1965a) used the dialyzable portion of yeast extract in the growth medium and pooled the young mycelia of 24 *C immitis* isolates prior to autolysis. Dialysis of CDN against a membrane of the conventional 10,000 to 12,000 molecular weight limit cannot be recommended because the dialyzable constituents may contain some of the cutaneous and in vitro generating activity. This is disputed by Deresinski et al (1974) who compared dialyzed CDN with spherulin to study lymphocyte blastogenesis in coccidioidomycosis patients. Dialysis of CDN did not result in any apparent loss of activity. This may be because the CDN used was a toluene-autolyzate of young mycelia and antigen was present in high molecular weight form. Definitive studies comparing the dialyzed CDN and dialyzate produced from both types of CDN (culture filtrate and autolyzate) have not been done in dose-response fashion. This type of experiment is a necessary step towards standardization of reagents.

Although standardization on the basis of biologic activity has been, and will continue to be, the prime focus, a moderate effort has been made at chemical characterization of *C immitis* antigens. CDN prepared by toluene autolysis was concentrated by ultrafiltration over a 2000 da limit membrane. Then two-dimensional "rocket" electrophoresis was carried out using sera from a burro immunized with CDN in complete Freund adjuvant (CFA) for up to 8 months (Huppert et al, 1978; 1979) (Figure 4.1). Twenty-six antigenic components were separated. Of these 16 were unique to CDN and were not found in spherulin. Unexpectedly, this same

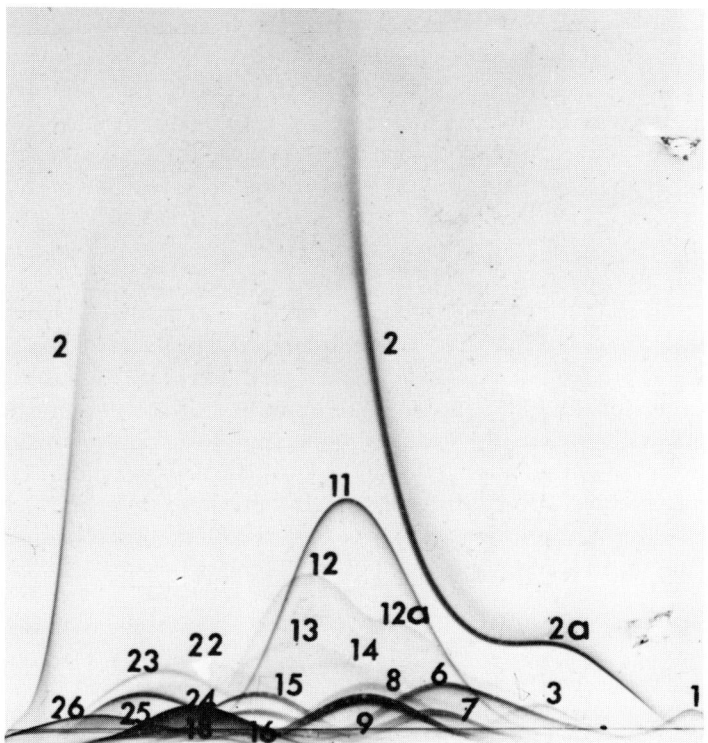

Figure 4.1. Two-dimensional crossed "rocket" immunoelectrophoresis of coccidioidin against burro anticoccidioidin-IgG. Well with antigen is at the cathodal end of the plate. Factors 2 and 11 are abundant and specific for *Coccidioides immitis* (Huppert et al, 1978, 1979). *Source:* Reprinted with permission of the publisher, American Society for Microbiology.

serum recognized 16 antigens in spherulin (SPH) that were not detected in the homologous CDN–anti-CDN reaction. Factors designated 2 and 11 in CDN were abundant. The same anti-CDN IgG preparation recognized six antigens held in common with histoplasmin, in *Histoplasma* – alkali-soluble – water-soluble extract of cell walls (H-ASWS), and with five from blastomycin, and *Blastomyces*-ASWS. These factors had a low rate of migration, possibly indicative of a large carbohydrate component. The extraction of the antigens from crude cell walls after trypsinization and alkali extraction suggests that these common factors are wall-associated.

4.2.2 Wall Antigens

The loosely adherent electron-dense outer wall layer of arthroconidia is a remnant of the parent hypha and may function to inhibit phagocytosis (Huppert et al, 1982). Under phase transformation conditions (20% CO_2 on Converse medium) enlargement and multiple nuclear replications occur and then cease. Next, segmentation of the protoplasm occurs by invagination of the electron-lucent inner wall. New wall synthesis is rapid and segmentation proceeds. New endospores of the "tissue" form are uninucleate, held in clusters surrounded by a hyaline membrane. After release from the ruptured spherule, the endospores are held together by fine fibrils (Figures 4.2, 4.3) (Huppert et al, 1982). The wall of newly released endospores is derived from the segmentation apparatus, but the cell acquires an electron-dense outer wall as maturation proceeds. Cell walls of *C immitis* were extracted with organic solvents, sodium dodecyl sulfate, and pronase before analysis (Table 4.1). Spherules have a chitin content almost two times higher than mycelial walls. This finding is consistent with results in other systemic dimorphic pathogens where the chitin content of the tissue form is higher than the saprophytic form (Table 2.1). The use of pronase to remove the protein not covalently linked to the cell wall matrix does not seem advisable because the specificity of that enzyme is too broad to exclude attack on covalent structural glycoprotein.

The cell wall composition of *C immitis* arthroconidia was studied because of the importance of these conidia as the infectious propagule (Cole et al, 1983). The key to understanding the approach taken is that different methods of cell disruption selectively removed layers of the cell wall. Less harsh methods, such as sonication and passage through a French pressure cell, do not cause significant disruption (< 10%). However, they do remove the outer cell wall layer of the arthroconidia. This layer has a distinct ultrastructural appearance consisting of protuberances and sleeve-like terminal extensions that are remnants of the natural disarticulation that gives rise to arthroconidia. This layer confers hydrophobicity to the arthroconidia. The inner

Figure 4.2. Freeze-fracture preparation of *Coccidioides immitis* spherules in infected mouse lung. One spherule contains hundreds of endospores. The spherule wall is thick and endospores retain a layer of outer material from the segmentation planes (Drutz and Huppert, 1983). *Source:* Reprinted from the Journal of Infectious Diseases with permission of the publisher. Copyright 1983, The University of Chicago Press.

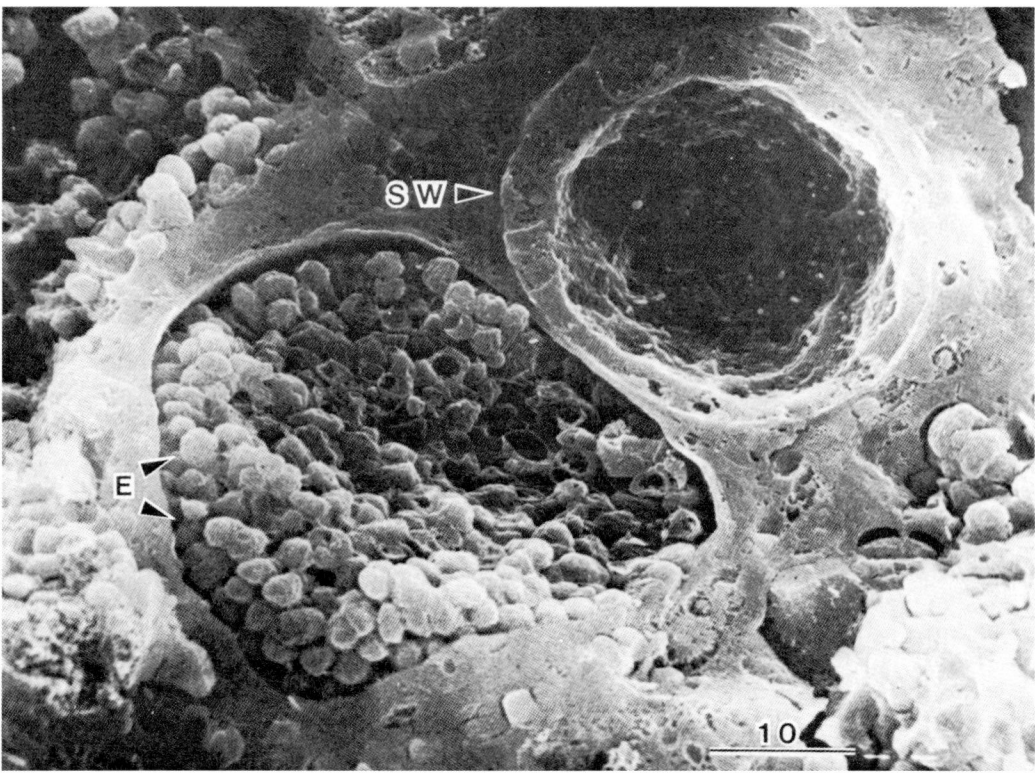

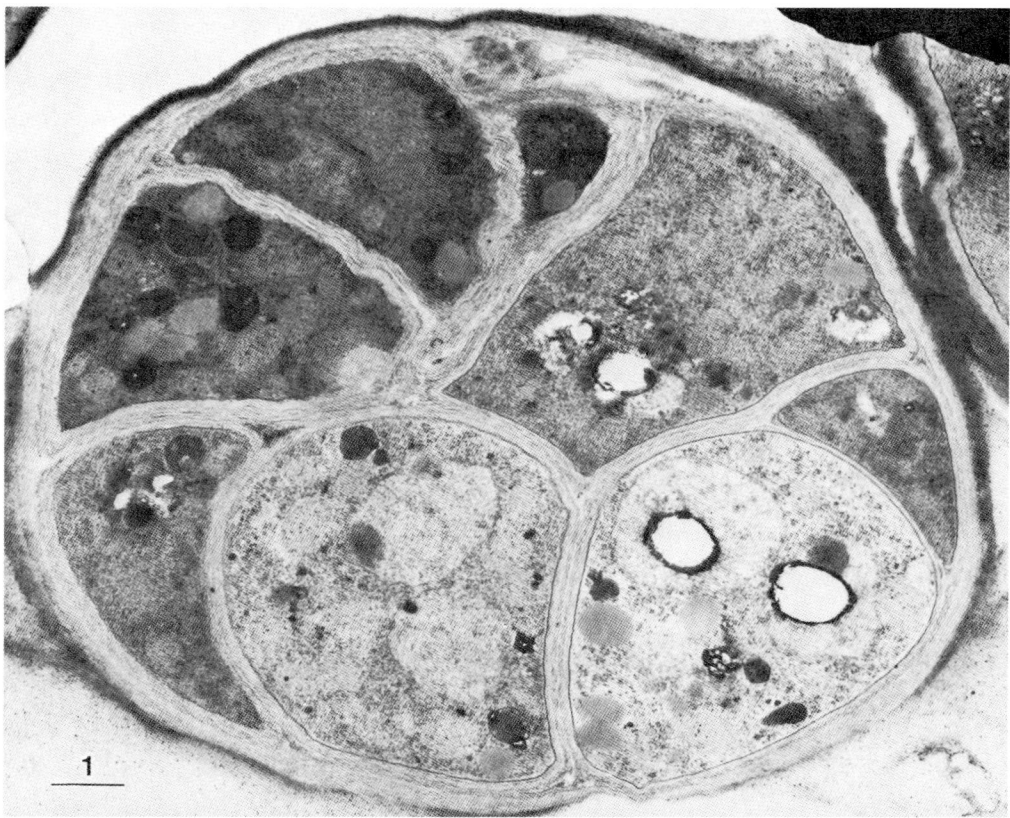

Figure 4.3. Young *Coccidioides immitis* spherule in early stages of cytoplasmic segmentation. The walls of segments are laminated (Drutz and Huppert, 1983). *Source:* Reprinted from the Journal of Infectious Diseases with permission of the publisher. Copyright 1983, The University of Chicago Press.

surface of the outer cell wall layer revealed by surface fractures is comprised of small clusters of "rodlet fascicles." After it is liberated from the conidia, the outer wall layer can be fractionated by ultracentrifugation into an 81,500 g residue and a supernatant fraction. The inner wall layer is freed from cytoplasm by disruption with ballotini in the Braun cell homogenizer, and multiple washings.

Table 4.1. Composition of *Coccidioides immitis* M11 Cell Walls

Component[a]	Spherules[b]	Mycelia[b]
Glucose	66	55
Mannose	27	42
3-O-methyl mannose	8	2
Galactose	trace	3
N-acetylglucosamine[c]	28	33
Protein (amino acids)[c]	9	18

Source: Reprinted from Reiss (1985) with permission of the publisher, Marcel Dekker, Inc.

[a] Sugars recorded as percent of neutral sugar content.
[b] Spherules grown in Converse medium; mycelial cells grown in mannitol-Converse medium.
[c] N-acetylglucosamine and amino acids determined on sodium dodecyl sulfate-treated, pronase digested, defatted cell wall preparations. Figures for N-acetylglucosamine and protein are percent dry weight of cells.

The compositional analysis of the inner and outer wall layers of *C immitis* arthroconidia is shown in Table 4.2. Notice that the outer wall layer is the source of most of the protein and lipid, whereas the inner wall layers contain all of the 3-*O*-methyl mannose and chitin. Functionally, then, the outer wall layer acts as a hydrophobic and antiphagocytic shield. Antigens held in common with coccidioidin were demonstrated in conidial wall fractions by tandem two-dimensional crossed "rocket" immunoelectrophoresis in comparison with the coccidioidin reference system. Antigen 11 was found in the phosphate buffer saline wash from the outer cell walls. The inner conidial cell wall was extracted with 1M NaOH in a manner identical to that used to prepare the "alkali-soluble–water-soluble" antigenic complex of Cox et al (1981). This processing liberated antigen "2," with respect to the coccidioidin reference system. This antigen may contain the 3-*O*-methylmannose determinant.

Coccidioidin has been compared with an alkali-soluble, water soluble extract of cell walls of the mycelial form of *C immitis* for the ability to evoke delayed cutaneous hypersensitivity and lymphocyte blastogenesis (Ward et al, 1975; Cox et al, 1977). Cell walls prepared by disruption of mycelial forms in a Braun homogenizer were digested with trypsin, and then extracted with 1N NaOH at 25 °C. The supernate was dialyzed versus distilled water and then clarified by centrifugation. This method of preparation does not define the extent of wall breakage nor the criteria for purity of cell walls with respect to lipid and nucleic acid content, nor is there explicit monitoring of the extent of trypsin digestion. The extract is not defined in terms of its nucleic acid, total nitrogen, amino nitrogen, or the kinds of sugars present. Most likely the extract is a heterogeneous mixture of polysaccharides and alkali-denatured protein. In guinea pigs injected with formalin-killed mycelia and mice receiving 1000 arthroconidia, more of the guinea pigs responded to the ASWS antigen in skin tests than to CDN. Is it possible that the reaction to ASWS is due to the presence of polysaccharide which induces an Arthus reaction that merges confusingly with delayed cutaneous hypersensitivity? This possibility should be considered because formalin-killed cells have altered or inactivated protein determinants and are poor

Table 4.2. Chemical Composition of Arthroconidial Wall Fractions of *Coccidioides immitis* C634

	Conidial Wall Fraction (percent of dry weight)		
	Outer		
	I	II	Inner
Total neutral carbohydrate	12.0	31.6	32.3
Mannose[a]	64.6	30.3	45.8
Glucose[a]	22.5	58.7	35.1
Galactose[a]	11.5	7.9	7.1
3-*O*-methylmannose[a]	1.4	3.1	12.0
Hexosamine	1.7	0.6	21.9
Peptides	49.8	28.5	27.4
Lipids			
readily extracted	15.4	7.4	4.5
bound	9.7	5.2	12.9

Source: Reprinted with permission from Cole et al. (1983) and Academic Press, Inc.
[a] As % of total neutral carbohydrate.

immunogens for anti-protein immunoglobulin. It is typical of delayed cutaneous hypersensitivity measurements with fungi that histologic examination of the cellular infiltrate was not analyzed.

The report that alkaline extracts of cell walls of *C immitis* could evoke delayed cutaneous hypersensitivity responses prompted investigation of the carbohydrate composition of *C immitis* extracts (Wheat and Scheer, 1977; Wheat et al, 1978). Previously 3-*O*-methylmannose was detected in CDN toluene autolyzate (Anderson et al, 1971). Mycelia were killed and dehydrated with acetone, defatted with chloroform–methanol, and extracted with perchloric acid to remove DNA (Wheat and Scheer, 1977). After these treatments, the insoluble fraction was washed in water, then extracted with 1N NaOH at room temperature. The alkali-soluble fraction was neutralized and clarified. When the alkali extract was analyzed for constituent sugars by gas-liquid chromatography, no 3-*O*-methylmannose was found. Instead, glucose was predominant (46% of total carbohydrate), mannose and galactose were also present. Where was the 3-*O*-methylmannose? Extraction of defatted mycelia with hot aqueous phenol (44% phenol, 68 °C) resulted in extraction of a 3-*O*-methylmannose-containing polymer into the phenol phase. This solubility characteristic implies a glycoprotein (Westphal procedure). The phenol phase then was dialyzed and the water-soluble portion was tested for delayed cutaneous hypersensitivity in guinea pigs infected 8 weeks previously with 500 arthrospores. This 3-*O*-methylmannose-containing fraction was active in detecting cutaneous hypersensitivity but less so than CDN. The phenol-extracted material was 68% protein, 15% neutral sugar. The carbohydrate component contained 3-*O*-methylmannose (20%), mannose (74%), with traces of galactose and glucose. The ability of the 3-*O*-methylmannose-containing fraction to evoke delayed cutaneous hypersensitivity was due to its linkage to a protein, acting as a hapten–protein conjugate. Further support for this view depends on chromatographic purifications which are yet to be done. Alkaline extraction of *C immitis* mechanically disrupted cell walls results in a mixture of glucan and galactomannan, whereas hot aqueous phenol selectively removes a 3-*O*-methylmannose polymer from defatted mycelia, possibly as a glycoprotein.

Hot aqueous phenol extracts of defatted *C immitis* mycelial walls contain small amounts (0.1% of cell weight) of phenol-soluble–water-soluble heteromannans (Wheat et al, 1983). These elute in the void volume of Sephacryl S-200 columns and are heterogeneous on Sepharose CL-4B. These latter fractions contain mannans that are 4.5% to 5.9% 3-*O*-methylated. Galactose and glucose are also present. When cell walls are first cleaned of proteins by extraction with sodium dodecyl sulfate, they no longer contain phenol-soluble–water-soluble material. Alkali-soluble–water-soluble extracts of trypsinized cell walls contain mannan with a substantial 3-*O*-methylated component (up to 22% of the total hexose). Sodium dodecyl sulfate- (SDS) extracted mycelial walls subsequently exposed to dilute alkali release 12.4% of the total carbohydrate as 3-*O*-methylmannose.

3-*O*-Methylated heteromannans are thus present in the mycelial wall but are also loosely associated with the wall as a glycoprotein or lipoglycoprotein. The high molecular weight phenol-soluble 3-*O*-methyl-heteromannan preparation reacts in immunodiffusion against anti-*C immitis* spherules and forms a precipitin line that spurs over the line containing the alkali-soluble–water-soluble antigenic complex. When the phenol-soluble 3-*O*-methylated heteromannan is treated with alkali the additional determinant is lost suggesting that epitopes other thn 3-O-methylmannose contribute to antigenicity.

4.2.3 Spherulin

In theory, the antigens derived from the tissue form of a dimorphic fungus should more closely resemble the antigenic mosaic recognized by the host. The ability to induce the spherule-endospore phase in vitro depended on the presence of a detergent, tamol N, and on elevated pCO_2 (Converse and Besemer, 1959). The supernates of spherule cultures grown on this synthetic medium displayed cutaneous toxicity, which was overcome by resuspending spherules in distilled water and allowing autolysis to occur for up to 40 days at 34 °C. The particulates then were removed by centrifugation, and filter sterilization (Levine et al, 1969; Levine et al, 1977). Difficulty was encountered in inducing morphogenesis to a synchronized and purely spherule phase in vitro, requiring many serial transfers. Spherules had to be recovered in mature form, but before endospore release, prior to resuspending them in water for autolysis.

The optimal dose for human skin tests was 1.4 to 2.8 µg on a dry weight basis; guinea pigs and mice required higher doses to achieve positive responses. Forty times greater specific activity in human skin tests is claimed for spherulin compared with CDN (Levine et al, 1977). This greater activity may be due, at least in part, to staling products in the growth medium of CDN (Levine et al, 1977). The specific activity of dialyzed toluene-lysate-CDN was not tested.

Spherulin (SPH) has been suggested as a more effective substitute for CDN as an elicitor of delayed cutaneous hypersensitivity and its in vitro correlates. A comparison of those antigens in humans are discussed below. A study in guinea pigs (Chaparas et al, 1974) sensitized with either killed mycelia, killed spherules, or CDN in adjuvant showed that CDN and SPH elicited equivalent areas of induration over a sixteenfold range in concentrations. This relationship held for animals sensitized with each of three isolates of C immitis and tested with antigens extracted from the homologous strains. In order to estimate the difference in specific activity on a weight basis, the nondialyzable solids were recovered from CDNs. When this was done spherulin appeared to be 17 times more active. However the authors took note of earlier observations (Anderson et al, 1971; Ibrahim and Pappagianis, 1973) that the dialyzable constituents of CDN contain skin reactive or anergy-inducing substances, which may account for much of the reactivity. The main conclusions drawn from this study were that the three groups of animals were either high or low responders depending on the isolate used to sensitize them, but that SPH and CDN were about equivalent in detecting delayed cutaneous hypersensitivity.

The effects of storage temperature and dialysis on the delayed cutaneous hypersensitivity-eliciting ability of SPH prepared from C immitis 46 were determined (Levine and Scalarone, 1975). Lyophilization, freezing at −30 °C, refrigeration, or storage at ambient temperature, 22 °C, for 10 months resulted in equivalent retention of biologic activity. After dialysis the retentate contained 53.4% of the starting material, of this 11.1% was protein, and 83.3% was polysaccharide. Delayed cutaneous hypersensitivity activity was high in the retentate fraction tested in guinea pigs, but the dialyzate fraction was poorly reactive even at 25 times the dose of the retentate. Delayed cutaneous hypersensitivity activity of SPH was stable to trypsin, but was heat-labile (60 °C, 30 minutes) suggesting that at least one protein component must not be denatured in order to retain full activity. No unequivocal examples exist of nitrogen-free polysaccharides capable of eliciting delayed cutaneous hypersensitivity, and their presence in fractions enriched in activity is either coincidental or indicative of glycoprotein.

4.3 HUMORAL RESPONSES

4.3.1 Tube Precipitin

The tube precipitin (TP) reaction is the first laboratory sign of a primary immune response to *C immitis*. The reaction is present in about 90% of patients by the third week of illness. At that time less than one third have produced complement-fixing antibodies. The TP antibodies slowly decline and after 5 months, fewer than 25% of the patients' sera react in that test, but the incidence of positive complement-fixation rises to 85% (Pappagianis et al, 1965). The TP antigen is heat-stable (60 °C, 30 minutes) and probably is a cell surface carbohydrate present in CDN. The antibodies participating in this reaction are of the IgM class. This classification was made by two-step immunoelectrophoresis in which a patient's serum was first electrophoresed, then diffused against either anti-IgM or anti-IgG (Sawaki et al, 1966). Slides then were washed and a second diffusion was made against ^{131}I-labeled CDN. The resulting autoradiograms showed that sera displaying the TP reaction and no complement-fixing activity had arcs of the IgM class. Where only complement-fixing activity was present, arcs identified with IgG were observed.

The TP reaction conforms to the idea that agglutinating IgM antibodies herald the onset of infection. The reaction first observed as an almost gelatinous precipitate formed over 2 to 3 days can be observed using latex particles coated with heat-inactivated CDN (Huppert and Bailey, 1965b; Huppert et al, 1968). In order of decreasing sensitivity these three tests measuring the same heat-stable antigen are latex agglutination, immunodiffusion version of TP, and TP. The latex agglutination test is subject to false-positive results and is therefore considered presumptive. Precipitins are rarely found in cerebrospinal fluid, yet the latex agglutination test shows many reactions—these are false-positives (Pappagianis et al, 1976). When cerebrospinal fluid did contain precipitins these were directed against the heat-labile F antigen discussed below.

4.3.2 Complement-Fixing Antibodies

The reaction between IgG and a heat-labile (60 °C, 30 minutes) component of CDN rises in direct proportion to disease severity and according to the evaluation of Smith et al (1956) titers of 1:16 or greater create apprehension about extrapulmonary dissemination. This assumption has been corroborated (Sawaki et al, 1966; Kaufman and Reiss, 1986). The reaction can also be estimated by a quantitative version of the immunodiffusion test (Huppert and Bailey, 1965a). The immunodiffusion test first relied on the observation that CDN produced by the toluene-induced lysis of young mycelia (Pappagianis et al, 1961) and diafiltered over a parlodion membrane (Huppert and Bailey, 1965a) retained only the "CF," or more recently termed "F" antigen. In side-by-side comparison with the complement-fixation test the results using the immunodiffusion version detecting the F antigen agreed in 98% of the 319 sera.

Favorable response to antimycotic treatment results in a decreased concentration of this antibody. The polyclonal gammopathy (Cox and Arnold, 1979) observed in patients with coccidioidomycosis is probably due to elevated levels of the complement-fixing antibodies. Although no role for the complement-fixing antibodies has been found in neutralizing the growth of spherules it is reasonable that opsonization of spherules results in some complement-mediated damage to them.

The latex agglutination test to detect the heat-stable antigen, first associated with the tube precipitin reaction, in combination with the immunodiffusion test to measure the F antigen, correctly identified 93% of the 133 case sera (Huppert et al, 1968). This compared favorably with the combination of the slower and more tedious tube precipitin and complement-fixation tests. However, the latter two tests are still considered confirmatory in the laboratory diagnosis of coccidioidomycosis.

Radioimmunoassay (RIA) with coccidioidin as the solid-phase adsorbed antigen afforded more accurate quantitation of the initial antibody titer than the complement-fixation test and provided better documentation of serial improvement especially when the antibody concentration was low (Catanzaro and Flatauer, 1983). However, the two tests correlated poorly ($r = 0.46$), possibly because the antigens used were not identical.

4.3.3 Immunoglobulin E

Elevated total and antigen-specific IgE occurs in disseminated coccidioidomycosis that is proportional to disease severity and is not the result of a preexisting atopic condition (Cox and Arnold, 1979; Cox et al, 1982). (Figure 4.4) These observations are of interest because general increases in this immunoglobulin class are associated with defects in cell-mediated immunity such as Wiskott–Aldrich and DiGeorge syndromes. Rats and mice treated with sublethal irradiation or antithymocyte serum before or after injection with dinitrophenylated *Ascaris* extract produce elevated IgE (Ishizaka, 1976). Speculation that these diseases have a failure of T-cell regulation in common is tempting, but this idea is probably an oversimplification since in Wiskott–Aldrich disease the hypercatabolism of immunoglobulins tends to affect

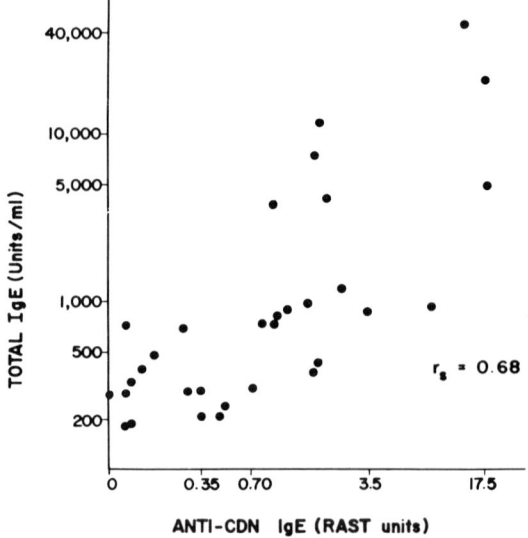

Figure 4.4. Correlation between total serum IgE concentrations and anticoccidioidin-specific IgG in 32 coccidioidomycosis patients with serum IgE 185 U/mL or greater. The correlation between total serum IgE and *C immitis*-specific IgE was statistically significant (Cox et al, 1982) (RAST, radioallergosorbent test). *Source:* Reprinted with permission of the authors and the publisher, American Society for Microbiology.

the levels of those immunoglobulin classes with long half-lives, thus accentuating IgE which has a half-life of 2 to 3 days (Ishizaka, 1976). Twenty-three of 59 patients (39%) with disseminated coccidioidomycosis had elevated total IgE levels (> 185 U/ml) (Cox et al, 1982). A lower percentage (27%) of those with the pulmonary form of the disease had increased IgE. In two series of patients the highest levels, approximately 45,000 U/ml, occurred in the intractable cutaneous form of *C immitis* infection (Cox and Arnold, 1979; Cox et al, 1982). As with total IgE, *C immitis* specific IgE correlated with disease severity when any of three antigens, (coccidioidin, spherulin, or the ASWS-mycelial-form antigen) were bound to CNBr-cellulose discs as the solid phase for the radioallergosorbent test (RAST). Coinciding with elevated *C immitis* specific IgE, positive RAST scores for common allergens such as Bermuda grass were also encountered. The IgE levels had a low but significant correlation with the concentration of complement-fixing antibodies except in two subjects whose elevated IgE persisted for 5 years after clinical recovery. With these notable exceptions, the defective regulation of IgE is most likely acquired because IgE typically declines upon recovery.

The explanation for hyper-IgE production in coccidioidomycosis is not understood. This disease results in diminished cell-mediated immune responses, even anergy. Faulty modulation of IgE, in the opinion of Cox et al (1982) is a failure of T-suppressor cell subset to control the *C immitis* specific and general IgE production. Proof is needed to support this hypothesis. *Coccidioides immitis* antigens, like those of helminths, (Ishizaka, 1976) may also be allergenic and stimulants of this immunoglobulin class.

4.4 CELL-MEDIATED IMMUNITY

4.4.1 Cutaneous Hypersensitivity and Anergy

The delayed cutaneous hypersensitivty reaction to CDN is a classical reaction of type IV in Gell and Coombs's scheme, the hypersensitivity of infection (Gell et al, 1975). The persistence of the reaction is a reminder of a pool of antigen-specific T-lymphocytes. Cutaneous hypersensitivity is present in healthy immune individuals in the endemic zones and sensitization is a function of duration of residence. The acquisition of delayed cutaneous hypersensitivity is a sign of normal cell-mediated immunity and its loss during pulmonary or systemic *C immitis* infection indicates declining immunocompetence. A primary immune response after exposure to dusts containing *C immitis* arthroconidia is accompanied by development of delayed cutaneous hypersensitivity within the first week of illness 84% of the time (Drutz and Catanzaro, 1978). Disseminated coccidioidomycosis in about half of the cases results in cutaneous anergy to CDN and, in some instances, poor cutaneous responses to common microbial antigens.

The antigens currently used for skin tests are CDN and SPH and the composition of these antigenic complexes is discussed in section 4.2. The incidence of skin test positive reactors in regions endemic for *C immitis* was studied in populations of factory workers and volunteers at a local health clinic in the San Joaquin Valley, California (Stevens et al, 1974a). In a second endemic area, the state of Sonora, Mexico, soldiers, Yaqui Indians, and prenursing students were skin-tested (Levine et al, 1973). The California series included 298 persons, 16.4% of whom reacted to 2.8 μg of spherulin, 12.4% reacted to the 1:100 dilution of CDN. Of those living in the area for less than 1 year, 3.6% reacted, the rate increasing to 14.5% for 1 to 5 year residents. Greater than 5 years residence was associated with a 19.5% reaction rate to SPH and 15.7% to CDN. In this study CDN and SPH were about equivalent in detecting delayed cutaneous hypersensitivity in healthy persons in the endemic area. The Mexican series tested 243 persons, 64% of whom reacted to spherulin and 41% to CDN. About one third of the spherulin reactors did not respond to CDN and the former reagent had superior sensitivity. The reactor rate for the group with less than 1 year's residence was 31.3% to SPH, and 12.5% to CDN increasing in 2 to 10 year residents to 62.1% and 37.5% respectively. The reason for the higher degree of sensitization in the Mexican population may relate to the greater exposure to *C immitis* of rural persons.

Asymptomatic pulmonary infection results in strong delayed cutaneous hypersensitivity to coccidioidin. Persons with unstable chest films, the "valley fever" chest X-ray pattern, which resolves without hospitalization or antimycotic therapy, are in this category. Twelve such children had positive delayed cutaneous hypersensitivity to coccidioidin (Cox and Vivas, 1977). When active pulmonary coccidioidomycosis with progressive fibrocaseation occurs, *C immitis*-specific cutaneous anergy is noticed. Twelve patients in this category were skin-tested and six had delayed hypersensitivity. Another series of 18 pulmonary coccidioidomycosis patients (Catanzaro, 1977) showed seven as positive reactors to coccidioidin.

Patients with extrapulmonary dissemination are *more* prone to anergy. None of the four patients tested by Cox and Vivas (1977) reacted to coccidioidin, and 23 of 28, or 82%, tested by Catanzaro (1977) failed to react. This anergy implies a sequestration or tolerance of antigen-specific T-lymphocytes. Opelz and Scheer (1975) measured coccidioidin dermal sensitivity in 78 disseminated cases and reported 87% were reactive. This much higher rate of reactors suggests the clinical status of the patients at the time of the skin test (whether acute, chronic, or convalescent) should be specified.

Successful amphotericin B treatment of disseminated coccidioidomycosis and remission for more than a year restored delayed cutaneous hypersensitivity in four of five patients (Cox and Vivas, 1977), implying that the cutaneous anergy is an acquired and reversible defect. The report of Levine et al (1977) is particularly interesting because of the claim that spherulin is better able to detect delayed cutaneous hypersensitivity in patients with disseminated coccidioidomycosis. Twenty-three such patients were skin-tested and 17 reacted to spherulin, only nine to coccidioidin. This points out the desirability of skin-testing patients with spherulin before concluding that anergy exists.

Cutaneous anergy to coccidioidin was induced in guinea pigs after daily injections of large amounts of coccidioidin (nine doses of 130 mg) (Ibrahim and Pappagianis, 1973). The anergy was shortlived and responsiveness returned in 3 days. Anergic animals also exhibited an inhibition of migration inhibitory factor production to coccidioidin but not to the control antigen, tuberculin. Complement-fixing antibodies were present at the time of anergy. All anergy-inducing activity resided in a low molecular weight dialyzable portion of coccidioidin.

The lack of crossreactivity of spherulin in skin tests was surveyed in Colombian soldiers residing in an area endemic for *Histoplasma capsulatum* but not for *C immitis*. The rate of reactors of the 265 soldiers tested with coccidioidin was less than 2%, and less than 1% for spherulin. At ten times the normal dose, the positive reactors were about 7% for either of the two antigens (Levine et al, 1977). Persons who reside in the *H capsulatum* endemic zones have positive histoplasmin skin tests, and rarely show crossreactions to coccidioidin. On the other hand, residents of the areas where *C immitis* is found, who have positive delayed cutaneous hypersensitivity to coccidioidin, *frequently* show cutaneous crossreactions with histoplasmin (Huppert, 1970).

The coincidence of high complement-fixing antibody titers and cutaneous anergy to coccidioidin prompted Rea et al (1979) to study the ability of coccidioidomycosis patients to become sensitized to dinitrochlorobenzene (DNCB) as a functional test for the induction of delayed hypersensitivity. Of 15 patients with severe pulmonary disease, 14 responded normally to DNCB. In the group with disseminated disease, nine of the patients whose complement-fixation titers were lower than 1:8 had normal DNCB responses, but 41 patients with complement-fixation titers above 1:16 had a significant reduction of sensitization. The occurrence of positive delayed cutaneous hypersensitivity to coccidioidin in 39% of those with disseminated coccidioidomycosis did not correlate with the ability to be sensitized to DNCB. The basis for reduced contact sensitization in disseminated coccidioidomycosis is unknown, but it could be due to a disruption of lymphocyte traffic resulting from the trapping of *C immitis* in the lymph nodes. The question of how high levels of complement-fixing antibodies may suppress delayed hypersensitivity responses has not been explained, but immune complexes can interfere with blastogenic responses (Cox et al, 1982).

4.4.2 Non-Antigen-Specific Responses

4.4.2a Erythema Nodosum

Some skin tests with coccidioidin that cause severe local reactions, also precipitate lesions over the anterior tibial or gluteal regions that continue for a few days, a reaction known as erythema nodosum. The eruptions do not occur in old reactors, prompting Smith et al (1948) to speculate that the effect is seen only in still active, unstable infection. One view is that the lesions are hyperactive type IV delayed cutaneous hypersensitivity but no biopsy evidence has been presented in support of this idea. Catanzaro (personal communication) found that blastogenesis to coccidioidin in the peripheral blood of lymphocytes of patients with active erythema nodosum is not elevated. The possibility exists that a type III or Arthus reaction is involved. The vesicular eruptions on the hands and legs that are the "ids" of dermatophytosis may be analogous to erythema multiforme of acute *C immitis* infection. The presence of immune complexes, C3, or polymorphonuclear neutrophil infiltration has not been reported.

4.4.2b Mixed Lymphocyte, Phytohemagglutinin, and E-Rosette Responses

Mixed lymphocyte reactions were lower in coccidioidomycosis patients than in healthy donors (Cox and Vivas, 1977), but the total number of sheep erythrocyte E-rosetting lymphocytes from these patients was normal. Responses to the predominantly T-dependent mitogen, phytohemagglutinin (PHA), were used as an index of immunosuppression in coccidioidomycosis (Catanzaro et al, 1975; Catanzaro, 1981). Depressed PHA responses were restricted to active cases with multifocal dissemination. Addition of prostaglandin synthesis inhibitors indomethacin or R020-5720 restored mitogen responsiveness, as did removal of an adherent leukocyte population on a nylon wool column. These results are consistent with the presence of a suppressor cell population in the most seriously ill coccidioidomycosis patients. This suppression is removed by inhibition of prostaglandins. The phenotype of the suppressor cell has not been characterized. Other evidence favors a subclass of patients with disseminated coccidioidomycosis who have a general T-cell anergy. Four of eight patients with disseminated disease not only were anergic to coccidioidin, but also to streptokinase–streptodornase, trichophytin, or mumps (Catanzaro et al, 1975), and as these patients responded favorably to therapy, PHA responses and skin tests to these and to other common microbial antigens also improved. A state of *C immitis* antigen-specific tolerance is expressed in the peripheral blood lymphocytes of some coccidioidomycosis patients, and in an even smaller group a more general and profound T-cell anergy occurs.

4.4.3 Antigen-Specific Responses

4.4.3a Blastogenesis and MIF Production

The usual response of persons in the endemic area to *C immitis* exposure is conversion to positive delayed cutaneous hypersensitivity with no ensuing disease. These healthy immune persons are the source of active lymphocyte cultures when stimulated in vitro with *C immitis* antigens (Deresinski et al, 1974). The mean blastogenic index elicited in persons to a variety of antigens was for C-ASWS, 22.1; CDN (toluene autolyzate), 25.6; CDN (culture filtrate), 6.2; and spherulin 19.4 (Cox et al, 1977). In healthy skin test-negative persons all antigens were slightly stimulatory, with blastogenic indices of 3.8 or less. In another series of healthy, immune persons and those with asymptomatic pulmonary exposure to *C immitis* (Cox and Vivas, 1977) C-ASWS and CDN (toluene autolyzate) elicited closely similar blastogenic indices in the range of 10.3 to 13.9. Lymphocytes from this same category of skin test-positive normal subjects were capable of being stimulated by CDN to produce MIF (Catanzaro et al, 1975; Cox and Vivas, 1977). Using the C-ASWS antigen complex as a probe, it was shown (Cox et al, 1976) that school children in the Texas endemic zone who had unstable chest films and those with calcified pulmonary residuals, were equally competent to produce MIF and to undergo blastogenesis. These children had positive delayed cutaneous hypersensitivity to CDN and negative PPD reactions.

Compared to the activity of lymphocytes from healthy skin test-positive persons or those with asymptomatic pulmonary exposure, active pulmonary coccidioidomycosis can lead to diminished blastogenesis to *C immitis* antigens. Of 18 patients with pulmonary coccidioidomycosis, 31% were negative in CDN skin tests and good agreement with in vitro tests was seen in 38% of these patients who had a flat blastogenic index to CDN and also a reduction in PHA mitogenesis (Catanzaro, 1977). Before that, agreement among delayed cutaneous hypersensitivity to CDN, blastogenic indices, and MIF production was found in ten pulmonary cases (Catanzaro et al, 1975). Cox and Vivas (1977) noted sharp reductions in blastogenic indices among patients with active pulmonary coccidioidomycosis.

The number of cutaneously anergic patients increases in disseminated coccidioidomycosis. According to one study of 36 disseminated cases, 62% were unresponsive to a 1:100 dilution of CDN (Catanzaro, 1977). An uncoupling of delayed cutaneous hypersensitivity and blastogenic indices to CDN was observed because lymphocytes from only 28% of these patients failed to transform in vitro. Put another way, blastogenesis is a more sensitive indicator of fluctuations in cell-mediated immunity than is the skin test. There was close correlation of suppression of delayed cutaneous hypersensitivity, MIF production, and blastogenic indices greater than 2, in eight disseminated cases (Catanzaro et al, 1975). This correlation does not hold true for the size of the skin reaction in relation to the magnitude of the blastogenic index to CDN (Opelz and Scheer, 1975).

In two other series of disseminated coccidioidomycosis patients the blastogenic index evoked by various *C immitis* antigens was depressed compared with healthy immune persons (Cox et al, 1977). The C-ASWS induced the most significant levels of blastogenesis. In the second series (Cox and Vivas, 1977) blastogenesis evoked by either CDN or C-ASWS was most suppressed in disseminated cases (blastogenic index range: 2.3 to 2.5). Migration inhibitory factor production was negligible in six coccidioidomycosis patients, two of whom had positive delayed cutaneous hypersensitivity to CDN.

The observation that recovery from *C immitis* infection usually results in a return of delayed cutaneous hypersensitivity (Smith et al, 1948; Cox and Vivas, 1977) is borne out by in vitro tests of cell-mediated immunity (Cox and Vivas, 1977; Catanzaro, 1977). The difference is that the magnitude of the blastogenic index is usually below that of persons with asymptomatic pulmonary exposure, indicating a prolonged, partial, antigen-specific T-cell deficit in the peripheral lymphocyte population. Twelve persons with more than a year's remission of coccidioidomycosis made gains in blastogenic indices but were less responsive than normal immune persons (Catanzaro et al, 1975). Later, (Catanzaro, 1977) 18 patients were studied in both pulmonary and disseminated categories who were initially unreactive to CDN but their ability to make MIF was gradually restored after 8 months' therapy. Those not capable of mounting a blastogenic response in the first 5 months did so over the next 10 months. This did not insure a favorable outcome as four patients who converted to positive delayed cutaneous hypersensitivity later experienced a reactivation of *C immitis* disease.

In another series (Cox and Vivas, 1977), four of five patients who had recovered from disseminated coccidioidomycosis and remained in remission for more than 1 year reexpressed delayed cutaneous hypersensitivity to CDN. Their blastogenic indices and MIF production rose too, but fell short of levels present in normal immune persons.

4.4.3b Serum Blocking Activity and Soluble Immune Complexes
Blocking activity was demonstrated in the plasma of one disseminated case that could suppress blastogenic responses of autologous and some allogeneic lymphocytes (Opelz and Scheer, 1975). The patient's complement-fixing antibody level of greater than 1 : 50,000 implicated immunoglobulin or circulating immune complexes as the blocking factor.

Five disseminated coccidioidomycosis patients were anergic to spherulin in both cutaneous and in vitro lymphocyte transformation assays (Harvey and Stevens, 1981). The in vitro proliferative response was augmented by replacement of autologous serum with normal AB serum. Thus no intrinsic cellular defect was observed in these patients. Patients with positive delayed cutaneous hypersensitivity did not show a potentiation in the blastogenic index when autologous serum was exchanged with AB serum.

Immune complexes occurred in 22 of 55 patients with disseminated coccidioidomycosis and their incidence increased with disease activity, measured as the extent of organ involvement (Cox et al, 1982). The presence of soluble immune complexes correlates with total IgG which is in turn related to the complement-fixing antibody titer to CDN. However, no direct relation exists between the concentration of immune complexes measured by ^{125}I-Clq radioimmunoprecipitation and complement-fixing antibodies (Cox et al, 1982). This paradoxical situation could result from the more rapid clearance of complexes formed at high immunoglobulin concentrations. *C immitis* antigen has been localized in the immune complexes (Yoshinoya et al, 1980) and their size was estimated at between 7s and 19s. Clinical effects of these immune complexes are suspected but unproven. These effects include the erythema nodosum and arthritis (desert rheumatism) that occurs in a minority of patients at an early stage of the illness. Evidence is accumulating that immune complexes are the serum factors blocking blastogenic responses to CDN and contributing to anergy in coccidioidomycosis (Harvey and Stevens, 1980; Cox et al, 1982). The mechanism for blocking activity may be the interaction of the Fc portion of the complexed antibodies with Fc receptors on lymphocytes.

Antigenemia can be detected in active coccidioidomycosis (Weiner, 1983) coexisting with complement-fixing antibody. The necessity to dissociate serum-antigen complexes at acidic pH (pH 2.7) and heat (97 °C, 20 minutes) infers that the circulating antigen occurs in the form of soluble immune complexes. The antigen that circulates shares determinants with a heat-stable, periodate-labile factor in coccidioidin. The method of detecting antigen is a competitive binding radioimmunoassay in which the radiolabeled polysaccharide-protein complex competes with antigen in dissociated patients' serum for a limited number of anti-*C immitis* IgG molecules. Instead of the Farr type of RIA that Weiner has used previously to detect antigenemia in aspergillosis and candidiasis, in the *C immitis* assay staphylococcal protein A was used as the solid phase adsorbent. The polysaccharide-protein complex had an apparent molecular weight of 200,000 to 230,000 da (Sephacryl S-300 column), migrated as a broad band in immunoelectrophoresis, and gave a diffuse zone of migration in disc polyacrylamide gel electrophoresis. That the circulating antigen shares these same physicochemical properties has not been proven as yet, but Weiner suggests that the antigen that circulates may be similar to the heat-stable factor in coccidioidin that participates in the "TP" reaction (see 4.3.1).

4.4.3c Antigens Evoking Cellular Responses

As previously mentioned, side-by-side comparison of four different *C immitis* antigens in one study (Cox et al, 1977) showed the alkali-soluble cell wall antigen, C-ASWS, to be superior, even including spherulin. However, the practice of dialyzing coccidioidin may interfere with its effectiveness in stimulating lymphocytes by removal of antigenic polypeptides. If the reason for dialysis is to remove Merthiolate, then a membrane of low molecular weight limit, ie, 2,000 da, can be used. In view of the claims of greater sensitivity of spherulin as an indicator of delayed cutaneous hypersensitivity (Stevens et al, 1974), inclusion of this antigenic complex when testing human in vitro lymphocyte responses is advisable until enough parallel studies show whether spherulin, coccidioidin, or the C-ASWS is a more useful product, or if a battery of antigens offers additional benefit (Catanzaro et al, 1976; Cox et al, 1977).

The efficacy of various antigenic preparations of *C immitis* to evoke delayed cutaneous hypersensitivity and its in vitro correlates was compared in infected guinea pigs (Cox et al, 1981). Of the four antigens tested, the toluene-induced mycelial lysate, coccidioidin-TS, was clearly superior in cutaneous tests. In the direct MIF assay, no such superiority was evident: the coccidioidin and alkali extracts of mycelial and spherule walls had a comparable ability to evoke MIF. Spherulin was stimulatory to the migration of normal peritoneal exudates and was toxic above 50 µg/ml. The alkali extract of spherule walls (ASWS-spherules) was, weight-for-weight, a better antigen for stimulating specific lymphocyte blastogenesis. Previously, coccidioidin-TS was found to contain 26 immunoprecipitating factors, spherulin, 12, and the ASWS of mycelial forms only four. The ASWS extracts, although containing fewer antigenic components, can evoke the full spectrum of in vitro cell-mediated immune responses and seem only to be deficient in complement-fixing activity.

4.4.4 Summary

What value do the in vitro tests offer that is not simply defined by skin tests? The greatest concern develops when a coccidioidomycosis patient with active disease loses delayed cutaneous hypersensitivity and when the complement-fixing antibody level rises above 1:16. The in vitro tests can monitor fluctuations in cell-mediated immunity below the threshold for positive delayed cutaneous hypersensitivity, they can measure cell-mediated immune parameters that may be uncoupled from cutaneous anergy, and they may signal improvement before delayed cutaneous hypersensitivity returns. A reason for measuring in vitro cell-mediated immunity is to monitor the response to immunotherapy, and to the immunoadjuvant effects of amphotericin B. At present, immunotherapy consists of transfer factor, discussed below. Immunoprophylaxis in the form of spherule vaccine is being implemented and the stimulation of in vitro cell-mediated immunity may provide an acceptable means of measuring the response to vaccination.

The mechanistic problems of the immune response to *C immitis* can best be studied by in vitro cell-mediated assays. The potential for circulating immune complexes to block lymphocyte blastogenesis has been described by Cox et al, 1982, and plasma blocking factors are known from the work of others (Opelz and Scheer, 1975; Harvey and Stevens, 1981). The nature of *C immitis* antigen-specific unresponsiveness may be the result of plasma blocking factors, or of a tolerant state induced by suppressor T-cells, or to a lack of T-helper cells. Further explication is needed of the general T-cell anergy in coccidioidomycosis that is suggested by poor responses to phytohemagglutinin and to contact sensitins.

The pattern of in vitro cell-mediated immune responses in this disease provide sensitive probes of the acquired and partially reversible T-cell deficit that occurs in the pulmonary and disseminated forms. Intensive studies of refractory cases have shown evidence for preexisting immunoregulatory defects in patients unable to respond to chemo- or immunotherapy (Stevens et al, 1974b).

4.5 IMMUNOTHERAPY WITH TRANSFER FACTOR

The experience of transfer factor (TF) in coccidioidomycosis is greater than with any other acquired immunodeficiency (Catanzaro et al, 1976). This is because the disease is relatively common and some patients fail to improve after amphotericin B therapy. TF has been considered for patients with a poor biologic response to *C immitis* in terms of chronic or recurring infections over a period of years. The accepted dose is the dialyzable lysate of 5×10^8 to 1×10^9 lymphocytes from human donors with strong delayed cutaneous hypersensitivity and in vitro cell-mediated immunity to *C immitis* (blastogenic index to coccidioidin >2.5; and a MIF response $>20\%$ inhibition, Catanzaro and Spitler, 1976). Clinical improvement observed with TF is assumed, but not proven, to be due to the transfer of antigen-specific factors. The course of TF treatment is usually over a 6-month period, with doses injected subcutaneously at least once a month for a cumulative mean of ten doses. One series of 49 TF-treated patients included 25 pulmonary, 11 disseminated, and 13 with coccidioidal meningitis. The immunologic responses to TF were: conversion of the CDN skin test in 32 patients, blastogenic indices to CDN greater than 2 stimulated in 27, and MIF produced in response to CDN in 24. Thus, T-cell responses to CDN were reconstituted in the majority of patients receiving TF. The immune status of the remaining patients was that eight had positive delayed cutaneous hypersensitivity to CDN and 12 had positive blastogenic indices to CDN before TF treatment. Two difficulties in interpretation worthy of mention are that some patients convert to positive delayed cutaneous hypersensitivity without TF, and that $CDN_{\text{toluene autolyzate}}$ is to some extent mitogenic in that blastogenic indices greater than 2.5 are stimulated even in skin test-negative normal persons. The close temporal relationship between TF injection and conversion of the skin test, usually within a 10-day period, strongly suggests cause and effect.

Clinical improvement was observed in 30 patients treated with TF, 19 failing to respond. In 12 patients improvement closely followed TF injection. Improvements were seen in clearing of pulmonary infiltrates, sputa becoming free of *C immitis*, toleration of exercise, healing of draining abscesses, reduction in fever, and halt in progress of lesions (Catanzaro and Spitler, 1976; Catanzaro et al, 1976; Graybill et al, 1973). Clinical improvement occurred in seven patients with extensive pulmonary or disseminated coccidioidomycosis after TF (Catanzaro et al, 1976). In four of them, improvement was transient, which was also reflected in fluctuations in lymphocyte function tests. When in vitro cell-mediated immunity assays reverted to anergy, more doses of TF were required to reconvert them. The impression of these studies was that the most clinical improvement occurred in patients whose in vitro responses became and remained positive. The occurrence of anergy to polyclonal activators of T-lymphocytes such as PHA, gives strong evidence of the generally immunosuppressed state of some coccidioidomycosis patients (Catanzaro et al, 1974; Catanzaro and Spitler, 1976; Catanzaro et al, 1976). In some of these patients, blastogenic responses to PHA did not increase after TF and chemotherapy. In the majority, who were flatly unresponsive, or whose blastogenic index to PHA was less than ten at the start of therapy, gains were made gradually over a period of months to levels above ten, and in some instances, to over 100.

If the chemotherapeutic choices for this disease remain limited in effectiveness, there will be more incentive to continue immunotherapy and give impetus to the double blind trials that are needed to properly evaluate TF.

4.6 EXPERIMENTAL MURINE COCCIDIOIDOMYCOSIS

4.6.1 Spherule Vaccine

Early mouse protection experiments as reviewed by Levine et al (1977) found that sublethal infection conferred strong resistance, but that formalin-killed mycelia or arthroconidia vaccines were poor substitutes. The availability of an in vitro technique to culture spherule forms (Converse and Besemer, 1959) provided the means to a formalin-killed spherule-endospore vaccine that protected mice against 150 LD_{50} *C immitis* doses (compared with protection against 14 LD_{50} doses for the mycelial vaccine). Later, the importance of the isolate selected for vaccine production was appreciated (Levine et al, 1977). Vaccine from strain *Silveira* protected mice from arthoconidia of a variety of isolates, whereas vaccines made from other strains afforded narrow protection restricted to the homologous isolate (Levine et al, 1977). For optimal immunogenicity, spherules harvested prior to rupture and killed with 0.5% formalin were injected in three intramuscular doses of 0.7 mg given 1 week apart. The intravenous route was not successful (Levine et al, 1977).

Mice immunized with formalin-killed spherule vaccine are resistant to challenge with lethal arthroconidia doses. Moderate success has more recently been achieved by substituting soluble *C immitis* spherule or mycelial extracts for the formalin-killed spherule vaccine. A 1 mg dose of phosphate-buffer saline extract of *C immitis* spherule walls administered in alum adjuvant protected mice against intranasal challenge with 1000 arthroconidia (Pappagianis et al, 1979). The ASWS preparation from *C immitis* mycelia (see section 4.2.2) at a 1 mg dose in complete Freund's adjuvant also gave significant protection against intranasal challenge (Lecara et al, 1983).

Trials of spherule vaccine in humans are underway and are expected to take years to complete. Interim reports on the vaccine trials are lacking because of the need to conduct the study using encoded experimental and placebo groups.

4.6.2 Protective Cellular Mechanisms

Adoptive transfer experiments strongly support a role for cell-mediated immunity in protection. Some direct evidence supports this claim, but clearly up to now no protection can be ascribed to antibody-mediated immunity. Athymic nu/nu mice did not survive a challenge with 50 arthroconidia compared with survival after 21 days of all euthymic heterozygotes challenged with 350 arthroconidia. Normal mice were not protected from a challenge dose of arthroconidia by passive transfer of precipitin-positive immune serum (Kong et al, 1963). More recently, coating the infective dose of arthroconidia with serum from immune mice was also ineffective as a means of protection (Beaman et al, 1979). These experiments did not characterize the antibodies being transferred either qualitatively or quantitatively, so that further extension of this line of research to include monoclonal antibodies is warranted. Transfer of spleen cells from immune donor mice conferred protection, and the number of lymphocytes required to effect protection was reduced by using the T-cell enriched effluent from a nylon wool column (Beaman et al, 1979). Mice immunized with formalin-killed spherules by an intramuscular route were protected and showed no detectable complement-fixing antibodies in contrast to mice immunized intravenously that were not protected and in which complement-fixing antibodies were observed.

T-Lymphocytes from immune mice cannot directly kill *C immitis* endospores, but rather act indirectly through normal or immune macrophages (Beaman et al, 1981). Endospores are rapidly phagocytosed by macrophages but are not killed in the absence of immune lyt 1^+ 2^- lymphocytes because endospores and arthroconidia inhibit the fusion of phagosomes and lysosomes. The blockage is overcome by the presence of lymphocytes from mice immunized with formaldehyde-killed spherules of supernates from lyt 1^+ 2^- lymphocytes stimulated by spherules (Beaman et al, 1983). The lymphocytes that proliferate activate the macrophages and stimulate phago-lysosome fusion, but no role has yet been ascribed to oxygen intermediates in the killing process.

The available data support the views that (i) T-cell-mediated functions are critical in the development of protective immunity to *C immitis* and (ii) that no protective role is known for complement-fixing antibodies.

4.7 SUMMARY

Early evidence of the humoral response in coccidioidomycosis is provided by IgM antibodies reactive with a heat-stable coccidioidin factor in the tube precipitin or latex agglutination test. Then complement-fixing antibodies to heat-labile factors in coccidioidin appear and rise in proportion to disease severity (Figure 4.5). Elevated total IgE is observed in a significant minority of disseminated coccidioidomycosis cases and to lesser extent in the pulmonary form of the disease.

Primary immune responses to *C immitis* result in positive delayed cutaneous hypersensitivity within the first week that illness becomes apparent. Lymphocytes from healthy skin test-positive persons in the endemic area undergo brisk blastogenic and MIF responses to *C immitis* antigens. *C immitis*-specific anergy appears in the active pulmonary phase and becomes more profound in extrapulmonary dissemination (Figue 4.5). Some patients, who are anergic to coccidioidin will, nevertheless, react to spherulin.

Disseminated coccidioidomycosis patients with elevated complement-fixing antibodies have a reduced ability to be sensitized by contact sensitins such as DNCB. Lymphocytes from these patients are low responders in the mixed lymphocyte reaction and to phytomitogens. A subclass with disseminated disease have general T-cell anergy to common microbial antigens. Delayed cutaneous hypersensitivity to coccidioidin can be uncoupled from blastogenic responses so that lymphocytes from cutaneously anergic patients may still respond in vitro to *C immitis* antigens. Measurement of antigen-elicited blastogenic and MIF responses is justified because they are more sensitive indicators of the return of immunocompetence during recovery than skin tests. They also allow assessments of (i) serum factors blocking blastogenesis and (ii) the response to spherule vaccine.

Figure 4.5. Immune status as a function of disease progress in human coccidioidomycosis, R. A. Cox, personal communication. C-F, concentration of complement-fixing antibodies; CMI, cell-mediated immune parameters, specifically delayed cutaneous hypersensitivity, lymphocyte blastogenesis, leukocyte migration inhibitory factor. Printed with permission from Dr. R. A. Cox.

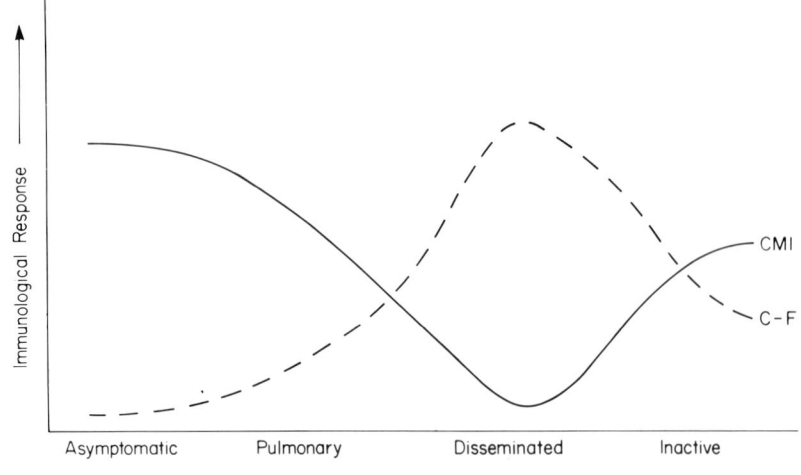

Serum factors in anergic patients that block in vitro lymphocyte responses to coccidioidin have been characterized as soluble immune complexes containing *C immitis* antigens. Clinical symptoms, including erythema nodosum and arthritis associated with acute coccidioidomycosis, may result from accumulations of soluble immune complexes.

Coccidioidin produced by the toluene-induced lysis of mycelial forms is superior to other antigens in evoking delayed cutaneous hypersensitivity in guinea pigs, but the only type of coccidioidin approved for human use is the filtrate from 2-month-old cultures representing a pool of many isolates. The toluene-induced coccidioidin, the alkali extract of mycelial cell walls (C-ASWS), or of spherules, are the most potent elicitors of lymphocyte blastogenesis. Toluene-induced coccidioidin contains 26 immunoprecipitating factors including large amounts of specific factors 2 and 11. Spherulin produces twelve precipitin arcs and the C-ASWS of mycelial forms only four. Six of the coccidioidin antigens crossreact with histoplasmin factors and five with *Blastomyces dermatitidis* preparations. The C-ASWS evokes a full spectrum of immune responses except for antibody-mediated complement activation. Spherulin is the supernate of autolyzed spherules cultivated in vitro. It contains factors labile to inactivation at 60 °C. Another antigenic preparation of *C immitis* is extracted by exposure of defatted mycelia to warm aqueous phenol and is of interest because it contains an unusual sugar, 3-O-methyl mannose, that may confer antigenic specificity.

Dialyzable human leukocyte extract, transfer factor, stimulates conversion of anergy to positive delayed-type hypersensitivity and improves the in vitro proliferative and MIF responses of lymphocytes from anergic patients with disseminated coccidioidomycosis. Clinical improvement has been observed after transfer factor therapy especially when cell-mediated responses become and remain positive. Even when patients were initially unresponsive to transfer factor and chemotherapy, gradual gains occurred over a period of months.

Vaccination of mice with formalin-killed spherules of the *Silveira* strain protects them from lethal challenge and T-enriched lymphocytes can adoptively transfer this immunity. Endospores are ingested by murine macrophages, but they inhibit phago-lysosome fusion. The inhibition is overcome by the presence of immune T-cells.

A partially reversible specific T-cell deficit occurs in pulmonary and disseminated coccidioidomycosis. The deficit is reflected in cutaneous anergy and depressed in vitro MIF and blastogenic responses to coccidioidin. The number of E-rosetting lymphocytes in the peripheral blood remains normal, implying a lack, sequestration, or suppression of *Coccidioides*-specific lymphocytes. Accompanying this anergy is a derangement in the control of antibody synthesis resulting in high levels of complement-fixing antibodies, and an elevated serum IgE that is partly attributed to anti-coccidioidin IgE. Evidence exists that patients with disseminated disease who are refractory to chemotherapy have a preexisting immunoregulatory defect, and this may be related to the HLA-A9 haplotype (Scheer et al, 1973).

T-lymphocyte functions are critical to successful host defense against *C immitis*. The success of spherule vaccine is correlated with its ability to evoke delayed hypersensitivity and not complement-fixing antibodies. Favorable clinical responses to chemotherapy or transfer factor also are accompanied by restoration of cutaneous hypersensitivity, MIF, and blastogenic responses, but to levels below those of healthy exposed persons.

Humoral responses, in the form of complement-fixing antibodies, have no obvious protective effect, and to the extent that they participate in soluble immune complexes, they probably (i) contribute to anergy by blocking blastogenic and MIF responses and (ii) accumulate in the joints and skin.

REFERENCES

Anderson KL, Wheat RW, Conant NF, 1971. Fractionation and composition studies of skin test active components of sensitins from *Coccidioides immitis*. Appl Microbiol 22:294–299.

Beaman L, Benjamini E, Pappagianis D, 1981. Role of lymphocytes in macrophage-induced killing of *Coccidioides immitis* in vitro. Infect Immun 34:347–353.

Beaman L, Benjamini E, Pappagianis D, 1983. Activation of macrophages by lymphokines: enhancement of phagosome–lysosome fusion and killing of *Coccidioides immitis*. Infect Immun 39:1201–1207.

Beaman LV, Pappagianis D, Benjamini E, 1979. Mechanisms of resistance to infection with *Coccidioides immitis* in mice. Infect Immun 23:681–685.

Catanzaro A, 1981. Suppressor cells in coccidioidomycosis. Cell Immunol 64:235–245.

Catanzaro A, 1977. Development of immunologic and clinical staging for immunotherapy. pp 325–334. In Coccidioidomycosis. Proc Third Intern Symp, Tucson. Ajello L, ed. New York: Stratton.

Catanzaro A, Flatauer F, 1983. Detection of serum antibodies in coccidioidomycosis by solid-phase radioimmunoassay. J Infect Dis 147:32–39.

Catanzaro A, Spitler L, Moser KM, 1974. Immunotherapy of coccidioidomycosis. J Clin Invest 54:690–701.

Catanzaro A, Spitler LE, Moser KM, 1975. Cellular immune responses in coccidioidomycosis. Cell Immunol 15:360–371.

Catanzaro A, Spitler L, Moser KM, 1976. Clinical and immunologic results of transfer factor therapy in coccidioidomycosis. pp 477–494. In Transfer Factor, Basic Properties and Clinical Applications. Ascher MS, Gottlieb AA, Kirkpatrick CH, eds. New York: Academic Press.

Catanzaro A, Spitler L, 1976. Transfer factor in diseases of the lung. pp 519–548. In Immunologic and Infectious Reactions in the Lung. Kirkpatrick CH, Reynolds HY, eds. New York: Marcel Dekker.

Chandler FW, Kaplan W, Ajello L, 1980. A Colour Atlas and Textbook of Histopathology of Mycotic Diseases. 333 pp. London: Wolfe Medical Publications.

Chaparas SD, Levine HB, Pappagianis D, Scalarone G, 1974. Comparison of skin-reactivity of spherule and mycelial coccidioidins produced by different strains of *Coccidioides immitis*. Proc Soc Exp Biol Med 145:806–810.

Cole GT, Pope LM, Huppert M, Sun SH, Starr P, 1983. Ultrastructure and composition of conidial wall fractions of *Coccidioides immitis*. Exper Mycol 7:297–318.

Converse JL, Besemer AR, 1959. Nutrition of the parasitic phase of *Coccidioides immitis* in a chemically defined liquid medium. J Bacteriol 78:231–239.

Cox RA, Arnold DR, 1979. Immunoglobulin E in coccidioidomycosis. J Immunol 123:194–200.

Cox RA, Baker BS, Stevens DA, 1982. Specificity of immunoglobulin E in coccidioidomycosis and correlation with disease involvement. Infect Immun 37:609–616.

Cox RA, Brummer E, Lecara G, 1977. In vitro lymphocyte responses of coccidioidin skin-test positive and negative persons to coccidioidin, spherulin, and a *Coccidioides* cell wall antigen. Infect Immun 15:751–755.

Cox RA, Mead CG, Pavey EF, 1981. Comparisons of mycelia- and spherule-derived antigens in cellular immune assays of *Coccidioides immitis* infected guinea pigs. Infect Immun 31:687–692.

Cox RA, Vivas JR, 1977. The spectrum of in vivo and in vitro cell-mediated immune responses in coccidioidomycosis. Cell Immunol 31:130–141.

Cox RA, Vivas JR, Gross A, Lecara G, Miller E, Brummer E, 1976. In vivo and in vitro cell mediated responses in coccidioidomycosis. I. Immunological responses of persons with primary asymptomatic infections. Amer Rev Respir Dis 114:937–943.

Deresinski SC, Levine HB, Stevens DA, 1974. Soluble antigens of mycelia and spherules in the in vitro detection of immunity to *Coccidioides immitis*. Infect Immun 10:700–704.

Drutz DJ, Catanzaro A, 1978. State of the art: coccidioidomycosis—part I. Amer Rev Respir Dis 117:559–585.

Drutz DJ, Huppert M, 1983. Coccidioidomycosis: factors affecting the host-parasite interaction. J Infect Dis 147:372–390.

Gell PGH, Coombs RRA, Lachmann PJ, 1975. Clinical Aspects of Immunology. pp 774–778. London: Blackwell Scientific Publications.

Graybill JR, Silva J Jr, Alford RH, Thor DE, 1973. Immunological and clinical improvement of progressive coccidioidomycosis following administration of transfer factor. Cell Immunol 8:120–135.

Harvey RP, Stevens DA, 1981. In vitro assays of cellular immunity in progressive coccidioidomycosis. Amer Rev Respir Dis 123:665–669.

Huppert M, 1970. Standardization of immunological reagents. pp. 243–252. In Proc First Intern Symp on Mycoses. PAHO Publ No 205. Washington, DC: Pan American Health Organization.

Huppert M, Adler JP, Rice EH, Sun SH, 1979. Common antigens among systemic disease fungi analyzed by two dimensional immunoelectrophoresis. Infect Immun 23:479–485.

Huppert M, Bailey JW, 1965a. The use of immunodiffusion tests in coccidioidomycosis I. The accuracy and reproducibility of the immunodiffusion test which correlates with complement fixation. Amer J Clin Path 44:364–368.

Huppert M, Bailey JW, 1965b. The use of immunodiffusion tests in coccidioidomycosis II. An immunodiffusion test as a substitute for the tube precipitin test. Amer J Clin Path 35:369–373.

Huppert M, Peterson ET, Sun SH, Chitjian PA, Derrevere WJ, 1968. Evaluation of a latex particle agglutination test for coccidioidomycosis. Amer J Clin Path 49:96–102.

Huppert M, Spratt NS, Vukovich KR, Sun SH, Rice EH, 1978. Antigenic analysis of coccidioidin and spherulin determined by two dimensional immunoelectrophoresis. Infect Immun 20:541–551.

Huppert M, Sun SH, Harrison JL, 1982. Morphogenesis throughout saprobic and parasitic cycles of *Coccidioides immitis*. Mycopathol 22:107–122.

Ibrahim AB, Pappagianis D, 1973. Experimental induction of anergy of coccidioidin by antigens of *Coccidioides immitis*. Infect Immun 7:786–794.

Ishizaka K, 1976. Cellular events in the IgE antibody response. Adv Immunol 23:1–75.

Kaufman L, Reiss E, 1986. Serodiagnosis of fungal diseases. pp 446–466. In Manual of Clinical Immunology, 3rd edition, Rose NR, Friedman H, eds, Washington, DC: American Society for Microbiology.

Kong YM, Levine HB, Smith CE, 1963. Immunogenic properties of undisrupted and disrupted spherules of *Coccidioides immitis* in mice. Sabouraudia 2:131–142.

Lecara G, Cox RA, Simpson RB, 1983. *Coccidioides immitis* vaccine: potential of an alkali-soluble, water-soluble cell wall antigen. Infect Immun 39:473–475.

Levine HB, Cobb JM, Scalarone GM, 1969. Spherule coccidioidin in delayed dermal sensitivity reactions of experimental animals. Sabouraudia 7:20–32.

Levine HB, Gonzalez-Ochoa A, TenEyck DR, 1973. Dermal sensitivity to *Coccidioides immitis*. Amer Rev Respir Dis 107:379–386.

Levine HB, Scalarone GM, 1975. Properties of spherulin, a skin test reagent in coccidioidomycosis. pp. 101–110. In Proc Third Inter Conf Mycoses. PAHO Publ No 304. Washington, DC: Pan American Health Org.

Levine HB, Scalarone GM, Chaparas SD, 1977. Preparation of fungal antigens and vaccines: studies on *Coccidioides immitis* and *Histoplasma capsulatum*. pp. 106–125. In Host Parasite Relationships in Systemic Mycoses. Beemer AM, et al, ed. Proc 21st OHOLO Biol Conf Ma'alot, Israel. Basel: S. Karger.

McGinnis MR, 1980. Laboratory Handbook of Medical Mycology. 661pp. New York: Academic Press.

Opelz G, Scheer MI, 1975. Cutaneous sensitivity and in vitro responsiveness of lymphocytes in patients with disseminated coccidioidomycosis. J Infect Dis 132:250–255.

Pappagianis D, Einstein H, 1978. Tempest from Tehachapi takes toll or *Coccidioides* conveyed aloft and afar. Western J Med 129:527–530.

Pappagianis D, Hector R, Levine HB, Collins MS, 1979. Immunization of mice against coccidioidomycosis with a subcellular vaccine. Infect Immun 25:440–445.

Pappagianis D, Krasnow I, Beall S, 1976. False-positive reactions of cerebrospinal fluid and diluted sera with the coccidioidal latex-agglutination test. Amer J Clin Path 66:916–921.

Pappagianis D, Lindsey NJ, Smith CE, Saito MG, 1965. Antibodies in human coccidioidomycosis: immunoelectrophoretic properties. Proc Soc Exptl Biol Med 118:118–122.

Pappagianis D, Smith CE, Kobayashi GS, Saito MB, 1961. Studies of antigens from young mycelia of *Coccidioides immitis*. J Infect Dis 108:35–44.

Rea TH, Johnson R, Einstein H, Levan NE, 1979. Dinitrochlorobenzene responsivity: difference between patients with severe pulmonary coccidioidomycosis and patients with disseminated coccidioidomycosis. J Infect Dis 139:353–356.

Rippon JW, 1982. Medical Mycology, The Pathogenic Fungi and Pathogenic Actinomycetes, 2nd edition. 842 pp. Philadelphia, PA: W. B. Saunders, Co.

Sawaki Y, Huppert M, Bailey JW, Yagi Y, 1966. Patterns of human antibody reactions in coccidioidomycosis. J Bacteriol 91:422–427.

Scheer MG, Opelz G, Terasaki P, Hewitt W, 1973. The association of disseminated coccidioidomycosis and histocompatibility type. Prog and Abstr 13th Interscience Conf Antimicrob Agents Chemother No. 157.

Smith CE, Saito MT, Simons SS, 1956. Patterns of 39,500 serologic tests in coccidioidomycosis. JAMA 160:546–552.

Smith CE, Whiting EG, Baker EE, Rosenberger HG, Beard RR, Saito MT, 1948. The use of coccidioidin. Amer Rev Tuberc 57:330–360.

Stevens DA, 1980. Coccidioidomycosis—a Text. 279 pp. New York: Plenum Medical Book Co.

Stevens DA, Levine HB, TenEyck DR, 1974a. Dermal sensitivity to different doses of spherulin and coccidioidin. Chest 65:530–533.

Stevens DA, Pappagianis D, Marinkovich VA, Wadell TF, 1974b. Immunotherapy in recurrent coccidioidomycosis. Cell Immunol 12:37–48.

Ward ER Jr, Cox RA, Schmitt JA Jr, Huppert M, Sun SH, 1975. Delayed-type hypersensitivity responses to a cell wall fraction of the mycelial phase of *Coccidioides immitis*. Infect Immun 12:1093–1097.

Weiner MH, 1983. Antigenemia detected in human coccidioidomycosis. J Clin Microbiol 18:136–142.

Wheat RW, Chung KSS, Ornellas EP, Scheer ER, 1978. Extraction of skin-test activity from *Coccidioides immitis* mycelia by water, perchloric acid and aqueous phenol extraction. Infect Immun 19:152–159.

Wheat RW, Scheer ER, 1977. Cell walls of *Coccidioides immitis*-neutral sugars of aqueous alkaline extract polymers. Infect Immun 15:340–341.

Wheat RW, Woodruff WW III, Haltiwanger RS, 1983. Occurrence of antigenic (species-specific?) partially 3-O-methylated heteromannans in cell wall and soluble cellular (nonwall) components of *Coccidioides immitis* mycelia. Infect Immun 41:728–734.

Yoshinoya S, Cox RA, Pope RM, 1980. Immune complexes in coccidioidomycosis. Detection and characterization. J Clin Invest 66:655–663.

Chapter 5

Histoplasma capsulatum

5.1 INTRODUCTION

Histoplasma capsulatum is a soil-dwelling filamentous fungus that produces two types of asexual spores: microconidia and round tuberculate macroconidia. *Histoplasma capsulatum* displays temperature-sensitive dimorphism and converts to a yeast form at 37 °C. The tissue form is a budding yeast ($2-3 \times 3-4$ μm) with a characteristic narrow neck between the mother and daughter cells. *Histoplasma capsulatum* is heterothallic and compatible (+) and (−) mating types unite to form the ascomycetous perfect stage, *Ajellomyces capsulata*. Skin test surveys and isolations of the fungus from soils near bird roosting sites and bat habitats show that the areas of highest endemicity in the USA are the Mississippi and Ohio river valleys. Scattered pockets of low level colonization of the soil occur along the St. Lawrence river in the north and the Rio Grande on the Mexican border. *H capsulatum* is found in Mexico, Central and South America. The *duboisii* variety of *H capsulatum* occurs in equatorial Africa. The reference works on histoplasmosis include Mandell et al (1979) on human medicine, Schwarz (1981) on pathology, Ajello (1971) on epidemiology, McGinnis (1980) on the mycology, and Rippon (1982) on further background information.

Persons who have had subclinical primary pulmonary or disseminated histoplasmosis commonly have residual fibrocaseous pulmonary nodules. The smaller lesions clear or eventually calcify, but larger lesions with central necrosis may persist for years. Such lesions contain many yeast forms in a dormant but probably viable state. Overwhelming parasitism of alveolar macrophages is characteristic of acute dissemination. The failure to eradicate these yeast forms from pulmonary foci leads to granulomas with infiltration of mononuclear cells, a classic example of the hypersensitivity of infection referred to in the Gell and Coombs scheme (Gell et al, 1975) as type IV delayed hypersensitivity, and is the mature expression of T-cell-mediated immunity in the lung. Multiple, often hundreds, of small calcifications in both lung fields are the telling radiographic image of healed epidemic histoplasmosis resulting from exposure to dense spore-containing dusts. Many diseases can produce a pattern of *miliary* granulomas, but multiple calcifications on healing are unique for histoplasmosis (Schwarz, 1981). Analysis of the arrested primary foci found in lungs of patients who died of causes other than histoplasmosis or tuberculosis has shown that lesions due to *H capsulatum* are larger than those containing tubercle bacilli (Schwarz, 1981).

Histoplasma capsulatum regularly elicits this pattern of calcified granulomas in the lungs. Reinfection can occur by inhalation of *H capsulatum* conidia from the environment, but also through endogenous reactivation of viable yeast forms from previously quiescent lesions. The events predisposing to reinfection are an acquired immunodeficiency, immunosuppressive drug therapy, or chronic obstructive pulmonary disease (Goodwin and Des Prez, 1978). The standoff between the yeast forms and the host unable to eradicate them implies that certain cellular components of *H capsulatum* are resistant to lysis. By persisting inside of pulmonary nodules for months or years, a slow leakage of soluble antigen outside the lesion can occur. Indeed, many persons with healed lesions continue to produce antibodies for a prolonged period (Pine, 1977). An attractive but untested hypothesis is that the cell wall of *H capsulatum* presents a barrier of resistance to macrophage microbicidal processes.

5.2 CELL WALL

The cell wall of *H capsulatum* yeast forms is a rigid, largely polysaccharide structure composed of four different glycans: soluble galactomannan (Azuma et al, 1974; Reiss et al, 1974); α-1,3-glucan (Kanetsuna et al, 1974; Reiss, 1977); β-1,3-glucan (Odds et al, 1974; Davis et al, 1977; Reiss, 1977); and a fibrillar chitin skeleton (Kanetsuna et al, 1974; Davis et al, 1977; Reiss, 1977). Several structural details are missing and there is no overall concept of the architectural arrangement of these polymers. Unlike *Candida albicans,* where typical electron-dense and lucent layers are differentiated in cross-sections with conventional stains, conventional stains fail to reveal stratifications in the *H capsulatum* yeast form cell walls (Edwards et al, 1959; Garrison and Tally, 1981). Based on immunoelectronmicroscopy, the galactomannan is believed to form an outer amorphous wall layer (Anderson, 1978) and is the only known antigenic polysaccharide in *H capsulatum* (Reiss et al, 1974; Reiss et al, 1977b). The α-1,3-glucan appears to be present as short, thick fibrils forming bundles both in Pt-shadowed walls and after precipitation of the α-1,3-glucan from alkaline solution (Kanetsuna et al, 1974). Chitin occurs as a continuous network of fine fibrils judging from their reduction in number after chitinase digestion (Kanetsuna et al, 1974; San Blas and Carbonell, 1974). The β-1,3-glucan is possibly enmeshed with chitin (San Blas and Carbonell, 1974). These findings are preliminary, and more specific probes are needed before a model for the mural architecture can be proposed.

Histoplasma capsulatum presents a most formidable challenge to cell wall investigation because of the multiple variables that may account for small but significant changes in the mural composition. Most important is the influence of dimorphism, but other factors are also important including (i) the occurrence of four surface factors and five yeast form serotypes that are recognized by immunofluorescence cross-staining and adsorptions (Kaufman and Kaplan, 1961); (ii) the two yeast form chemotypes with markedly different cell walls (Table 5.1); and (iii) the mating types that influence the conversion of isolates to the yeast form (Kwon-Chung et al, 1974).

5.2.1 Serotypes

The mural antigenic complex of the *H capsulatum* yeast forms was first recognized by Kaufman and Kaplan (1961, 1963; Kaufman and Blumer, 1966) who used immunofluorescence cross-staining and adsorptions to demonstrate four factors and five serotypes (1,2; 1,4; 1,2,3; 1,2,4; 1,2,3,4). Factors 1 and 4 are shared with *Blastomyces dermatitidis*, and the others are specific for *H capsulatum*. The chemical nature of serotypic factors is unknown, but the common occurrence of serologically crossreactive galactomannan as a surface coat polysaccharide in both genera indicates that at least one of the common factors is present in galactomannan. Although serotypes 1,2,3 and 1,2,3,4 are the most frequently encountered in the U.S., the assortment of serotypes between soil and human isolates is not known so that deducing if one of the serotypic factors correlates with virulence has not been possible.

The concentration of yeast form cell wall glycans varies according to the serotype. A 1,2,3 serotype isolate had an elevated chitin content, 30% of the dry weight, compared to 17% for a 1,4 serotype isolate (Reiss, 1977; Reiss et al, 1977b). Yeast forms of a 1,4 serotype contained 46.5% of the cell wall as α-1,3-glucan, but this component is absent in the 1,2,3 serotype.

5.2.2 Chemotypes

Chemotypes of *H capsulatum* are a major discovery stemming from studies by Domer et al (1967, 1971) and Pine and Boone (1968). Confirmation of these distinctions was provided by Kanetsuna et al (1974); Reiss (1977); and San Blas et al (1978).

Certain isolates (chemotype 2) of *H capsulatum* contain high amounts of α-1,3-glucan completely absent in other isolates. The estimates of α-1,3-glucan content of chemotype 2 walls are: *G184B*, 46.5% (Kanetsuna et al, 1974); *105*, 27.3% (Reiss, 1977). A second neutral polysaccharide, β-1,3-glucan, occurs in both chemotypes. Chemotype 1 walls of the *A811* isolate contain 18.1% glucose liberated by β-1,3-glucanase; 21.3% of the mural glucan of the *SW B* isolate is alkali- and water-soluble and is probably β-1,3-glucan (Domer et al, 1967). The value for the β-1,3-glucan in chemotype 2 varies from 20% in the *105* isolate (Reiss, 1977) to 31% for isolate *G184B* (Kanetsuna et al, 1974).

A corresponding increase in the chitin content of chemotype 1 walls can be seen in Table 5.1, varying from 29.8% to 53.7% of the wall depending on the isolate. The chitin content of chemotype 2 walls is reduced. The overall impression is that yeast form chemotype 1 walls have no α-1,3-glucan and an elevated chitin content, but that chemotype 2 walls have a large α-1,3-glucan component and a reduced chitin content. So far as is known, serotype 1,4 correlates with chemotype 2, and serotype 1,2,3 has a chemotype 1 profile (Reiss, 1977; Reiss et al, 1977b). The major shortcoming of these studies is that thus far few isolates have been examined so that only tentative conclusions can be drawn.

The cell wall components should be regarded in their macromolecular form to address properly the question of fluctuations during morphogenesis, ie, extracting and quantitating the galactomannan and soluble glucans of both α and β configurations. These processes were performed on the *G184B* isolate of chemotype 2 (Kanetsuna et al, 1974) and partial data for chemotype 1 are available (Table 5.2). The results suggest that galactomannan and β-1,3-glucan are elevated in yeast forms of both chemotypes and that the α-1,3-glucan content is elevated in yeast forms of chemotype 2, possibly at the expense of chitin. The mycelial form of chemotype 2

Table 5.1. Comparison of Selected Components of the *Histoplasma capsulatum* Yeast Form Cell Wall Chemotypes 1 and 2

	Chemotype 1 (percent dry wt)			Chemotype 2 (percent dry wt)		
	SwA	SwB	A811[c]	G184A	G184B	105[c]
Glc	19.8[a]–32.6[b]	17.2[b]–19.8[a]	39.7	17.8[a]–45.3[b]	49.1[b]–74.0[d]	72.5
Man	nd	1.0[b]–2.2[a]	3.3	5.8[b]	5.4[d]	2.9
Gal	nd	0.6[b]	1.1	2.3[b]	2.3[d]	1.3
GlcNAc	37.6[a]–46.1[b]	37.6[a]–53.7[b]	29.8	12.3[a]–18.5[b]	11.5[d]–12.3[a]	16.8
Protein	11.0	25.0[b]	14.5	22.8[b]	7.1[d]	9.4

Source: Reprinted from Reiss (1985) with permission of the publisher, Marcel Dekker, Inc.

Abbreviations: Glc, glucose; Man, mannose; Gal, galactose; GlcNAc, N-acetylglucosamine; nd, not determined.
[a] Domer 1967, 1971.
[b] San Blas et al, 1978.
[c] Reiss et al, 1977.
[d] Kanetsuna et al, 1974.

contained no α-1,3-glucan. The chitin content of yeast forms of chemotype 1 is higher than that found in the corresponding mycelial forms.

The cell walls of *H capsulatum* yeast forms have a high chitin content comparable to that found in other dimorphic fungal pathogens, and the chitin is arranged as a cage of fine microcrystalline fibrils encircling the wall, conferring rigidity and contributing to the maintenance of cell shape. In contrast, *C albicans* yeast forms have only 1.7% chitin (Chattaway et al, 1968) or 10% of that in chemotype 2 of *H capsulatum*, and only 5.7% of that reported in chemotype 1. The high chitin content of the *H capsulatum* yeast form wall is the most likely explanation for its resistance to lysis in the host, but no direct proof of this idea exists. Chitin may be considered a recalcitrant molecule and a presumptive aggressin.

Histoplasma capsulatum yeast forms can survive in macrophages after ingestion and spread hematogenously as an intracellular parasite of the reticuloendothelial system. Although histoplasmosis is primarily a pulmonary disease, others (Domer and Moser, 1980) describe it as a disease of the reticuloendothelial system in order to stress the limitations of macrophages for effectively killing this fungus. The yeast forms present in old fibrocaseous nodules from human lungs stain poorly with fluorescent antibody (Hotchi et al, 1972). Only 13 of 32 (40.6%) calcified lesions were stained with unadsorbed labeled globulins. Adsorption of the conjugate with *B dermatitidis* yeast forms removed the staining entirely. Thus yeast forms in old nodules have a surface component that is poorly antigenic and is altered from that found in acute disease. Despite this change in antigenicity, the yeast forms are not eradicated by the host.

5.2.3 Mannan

The approach used effectively by Ballou to delineate the structure of saprophytic yeast mannan (Ballou, 1980) is being applied to *H capsulatum*. Mycelial form cultures were autoclaved in neutral buffer, the supernates were acidified to remove protein, and then polysaccharide was precipitated with ethanol (Azuma et al, 1974; Anderson, 1978). The resulting mannose-based heteroglycan was subjected to acetolysis and the resistant fragments were separated by gel permeation chromatography. A peak corresponding to a tetrasaccharide was obtained, containing mannose, galactose, and glucosamine. The tetrasaccharide was an efficient inhibitor of the

Table 5.2. Fluctuations in Major Glycans of the Cell Wall in the Yeast and Mycelial Forms of *Histoplasma capsulatum*

Chemotype	Form	Percent of Mural Dry Weight			
		GM	α-1,3-Glucan	β-1,3-Glucan	Chitin
1[a]	Y[c]	2.2[e]	0	19.8[f]	37.6
	M[d]	6.6[e]	0	10.6	24.6
2[b]	Y	7.7	41.5	36.0	11.5
	M	24.7	0	18.8	25.8

Source: Reprinted from Reiss (1985) with permission of the publisher, Marcel Dekker, Inc.
[a] *H capsulatum* SwB (collated from Domer, 1971).
[b] *H capsulatum* Gl84B (collated from Kanetsuna et al, 1974).
[c] Yeast form.
[d] Mycelial form.
[e] Mannose content of intact walls, (galactose not quantitated).
[f] Glucose content of intact walls assuming all glucose in the M form is β-1,3-glucan.

immune precipitation of the parent polysaccharide in the Landsteiner hapten-inhibition technique. The linkage and sequence of the sugars in the acetolysis fragment or the original galactomannan have not been characterized, so that a role for glucosamine has not been ruled out. This possibility has appeal because it suggests a basis for antigenic specificity. Galactomannan was also extracted from defatted mycelial forms of *H capsulatum* with dilute (0.25M) alkali at ambient temperature (Reiss et al, 1974). The alkali-extracted glycan was fractionated into a galactomannan and a galactomannan protein. In both fractions the ratio of galactose:mannose is 2:5, but a portion of the total carbohydrate could not be accounted for. The galactomannan and galactomannan-protein both had similar serologic activities. The galactomannan-protein was a potent elicitor of migration inhibitory factor in peritoneal exudates of guinea pigs sensitized with an alkali extract of defatted *H capsulatum* mycelia in complete Freund adjuvant.

Azuma et al (1974) also characterized galactomannan extracted with 1M NaOH from *H capsulatum* mycelial forms and precipitated with Fehling solution $[\alpha]_D = +84°$ (water). Methylation-fragmentation analysis showed that this galactomannan has a linear 1,6-linked backbone substituted with 1,2-mannosyl oligosaccharides and galactofuranosyl nonreducing termini. The site of insertion of galactofuranosyl residues is probably as single side substituents. No provision was made in this analysis for amino sugar determination in the galactomannan. The galactomannans of three genera of systemic mycotic pathogens, *Histoplasma, Blastomyces,* and *Paracoccidioides* crossreacted extensively in quantitative precipitin and Ouchterlony immunodiffusion analyses. Terminal mannosyl and galactofuranosyl epitopes were implicated by using, as a probe, antiserum produced against *Alternaria kikuchiana*. That mold produces a mannan devoid of galactose.

The preparation of isolated cell walls of *H capsulatum* yeast forms with multiple washings in sodium dodecyl sulfate solution excludes the presence of H and M factors (Reiss, 1977). Attempts have been made to probe the antigenic structure of yeast form walls with a series of enzymes consisting of α-1,3-glucanase, β-1,3-glucanase, pronase, and chitinase. After each stage, soluble, nondialyzable fragments are released as the surrounding wall matrix is dissolved. These soluble fragments are serologically active both with human antisera and antisera obtained from experimentally infected goats. The antigen extracted from chemotype 1 walls after β-1,3-glucanase digestion had less tendency to crossreact with human anti-*B dermatitidis* and anti-*C immitis* globulins so it was further characterized. The antigen released from chemotype 1 (serotype 1,2,3) may carry serotype specificity because it produced an additional immunoprecipitin arc that was absent in a similar preparation from serotype 1,4 walls. The mural antigen obtained from chemotype 1 contained mannose, 17.8%; glucose, 35.8%; galactose, 2.7%; glucosamine, 5.4%; and 11% protein by dry weight. Further characterization of the *H capsulatum* mural heteroglycan is needed to delineate the nature of the epitopes that direct serotype specificity.

5.3 HISTOPLASMIN

5.3.1 H and M Factors

The filtrate of aged (3- to 6-month-old) mycelial form cultures of *H capsulatum* on synthetic asparagine medium (Smith et al, 1948) is histoplasmin. The factors in histoplasmin that are the specific antigens were identified by the pediatrician Heiner in 1958. He described H-factor as associated with active histoplasmosis. The M, for mycelial, as the antigen that stimulates antibodies in previously sensitized individuals after skin testing with histoplasmin. Substantial progress has been made in the purification and characterization of these factors, in determining their specificity for this fungus, and in determining their relationship to the stage of infection in humans. Nonetheless, the biologic functions of these proteins in the life cycle of *Histoplasma* are as yet unknown.

When young mycelial forms cultured on agar slants are immersed in Merthiolate-saline overnight, enough H and M antigens are solubilized so that the culture can be positively identified by immunodiffusion. This "exoantigen" test illustrates the regularity with which these two factors are associated with *H capsulatum*, and their absence from other pathogenic fungi, or from morphologically similar soil saprophytes of the genera *Arthroderma, Chrysosporium, Corynascus, Rennispora,* or *Sepedonium* (Standard and Kaufman, 1976; Kaufman and Standard, 1978). Filtrates of *H capsulatum* yeast forms also produce H and M antigens but in lower concentrations than in histoplasmin. Isolates of *H capsulatum* vary in the proportion of H and M factors recoverable from histoplasmins and yields also depend on the medium for growth. Although Smith's synthetic asparagine medium is most frequently used to produce histoplasmin, Pine's synthetic medium is a better stimulant of M antigen (Ehrhard and Pine, 1972b; Pine et al, 1977).

H and M factors begin to appear in the culture supernate when growth has reached a plateau and the carbon source is depleted. A precipitous drop in the cell yield occurs in some isolates, correlated with acute disintegration of the hyphae and release of the M factor. These findings were typified by results obtained with isolate 6623, a particularly good source of both H and M factors. Such findings predict the existence of an autolytic process in *H capsulatum*, and the suggestion was made that M-antigen is released only as a result of enzymatic degradation of the cell wall (Ehrhard and Pine, 1972a, 1972b). Two other observations support this idea: (i) immunofluorescent probes failed to localize M on the surface of yeast forms, but the fixation of complement damages them, releasing M into the fluid phase (Green et al, 1976); (ii) yeast form cultures of *H capsulatum* do not secrete appreciable amounts of M and H factors (Ehrhard and Pine, 1972). The mycelial mode of growth favors autolysis and liberation of M and H, whereas yeast forms may be protected from autolysis. It remains to be seen if the secretion of M or H is unrelated to the mechanism of dimorphism or whether these factors are involved as enzymes in mural synthesis or autolysis.

Culture filtrates, cell walls, and cytoplasm of *H capsulatum* yeast forms contain a β-1,3-glucanase (Davis et al, 1977). Activity could not be demonstrated on homologous cell walls but the enzyme could depolymerize a standard laminarin (β-1,3-glucan) preparation. This enzyme, when properly activated, may function to plasticize the cell walls at the site of budding in yeast forms, or at points of apical growth of filaments. Such an activity, if it were mandatory for yeast form cell division, would be expressed in the host. Testing H and M preparations for β-1,3-glucanase therefore seems reasonable.

The time necessary to induce *H capsulatum* to secrete H and M antigens can be reduced from months to days by using yeast forms as the starter culture and incubating cultures at 25 °C on a gyratory shaker. Both of these antigens are proteins (Pine et al, 1977), and they give a characteristic pattern in immunoelectrophoresis; both are anionic, but H is more acidic (Pine et al, 1977) (Figure 5.1).

The H and M proteins are potent stimulants of humoral and T-cell-mediated immunity. Complement-fixing antibodies to yeast forms are generally the first to become evident in acute pulmonary histoplasmosis. The H and M factors are not readily accessible on the yeast form surface but the fixation of complement alters the permeability of the yeast forms leading to escape of H and M into the fluid phase (Green et al, 1976). The first precipitins detectable in acute histoplasmosis are anti-M. In many instances anti-M is the only precipitin that is observed, and it may endure for years after the infection is in remission (Tompkins, 1965). Antibodies to H detected in immunodiffusion are less frequent and more often are associated with extrapulmonary dissemination. Anti-H levels are rarely as high as those of anti-M and they decline more rapidly after recovery than levels of anti-M. These observations, and the selective ability of M to boost circulating antibodies after skin testing in previously sensitized persons (Food and Drug Administration, 1977), suggest that H is not as antigenic as M.

Efforts to designate histoplasmin as a World Health Organization standard for immunodiffusion tests have compelled consideration of the effects of storage on its antigenic activity (Pine et al, 1981a). The H-antigen is the more stable of the two important factors in histoplasmin. When kept in lyophilized form, or as a liquid at 5 °C, or frozen at -20 °C, the H-antigen showed no loss of potency after 2 years. The activity of the M-antigen stored under the same conditions declined by 20%. If histoplasmin is kept at 37 °C the serologic activity in radial immunodiffusion of the M-factor declined by 50% in 10 days, but the H-factor retained full activity (Pine et al, 1981b).

5.3.1a Purification of H and M Antigens of Histoplasmin

These two proteins are the most important antigens to be delineated in *H capsulatum* because they are: (i) expressed during infection and the complement-fixing and precipitating antibodies they evoke provide specific evidence of the disease; (ii) invariably associated with isolates of this fungus and are not found in morphologically similar saprophytes; and (iii) responsible for eliciting the delayed cutaneous hypersensitivity reaction to histoplasmin. For these reasons, purification and further characterization are vital to progress in understanding the molecular immunology of this disease. Up to now, progress in purification of these factors has been largely the work of Pine and his group at the Centers for Disease Control (Ehrhard and Pine,

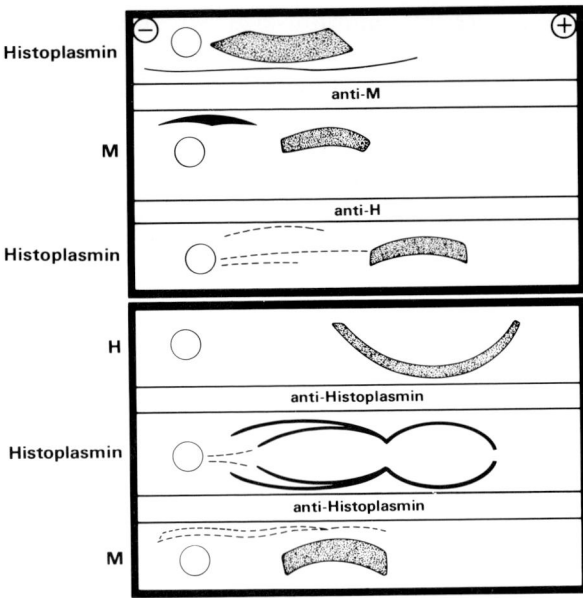

Figure 5.1. Immunoelectrophoretic patterns obtained with histoplasmin, antigenic factors H and M purified from histoplasmin, and rabbit antibodies obtained by immunization with histoplasmin, or with immunoprecipitin arcs containing the H-factor (anti-H) or the M-factor (anti-M). *Upper panel:* Top left shows the major M-antigen. The homologous rabbit anti-M contains a small amount of antibody against a factor "Bd" that is crossreactive with *Blastomyces dermatitidis*. The corresponding "Bd" immunoprecipitate appears below the antigen well and also in the "purified" M-antigen. This diffusely migrating antigen may be a heat stable polysaccharide common to both *Histoplasma capsulatum* and *B dermatitidis*. The bottom portion of the upper panel illustrates the greater anionic mobility of the H- as compared with the M-factor. Here too, a trace of the "Bd" reaction is visible. *Lower panel:* The top pattern illustrates the H-factor purified free of any antigen detectable by this means. The middle pattern depicts the classical gull wing-shaped fused precipitin arcs between the H and M reactions. Note the obvious additional immunoprecipitate closely associated with the M-factor that Pine et al (1977) referred to as resulting from the "non-M" antigen. "Non-M" is missing from both the purified H- and M-factors. *Source:* Tracing based on Figure 6 from Pine et al, 1977 and used with the author's permission.

1972a, 1972b; Bradley et al, 1974; Gross et al, 1975; Green et al, 1976; Pine et al, 1977; Pine, 1977 (review); Pine et al, 1978).

Unpurified histoplasmin is first concentrated with polyethylene glycol, and then gel permeation chromatography excludes colored impurities and nucleic acid. Two cycles of anion exchange chromatography on diethylaminoethyl gels are desirable to achieve maximal separation of H and M factors by this means. In the first run, Tris buffer with a salt gradient is used; the M-factor is weakly bound and elutes at a low salt concentration, but elution of the H-factor requires 0.15M NaCl. Rechromatography on DEAE with a phosphate buffer and a pH gradient further separates these two factors because the M activity elutes in the breakthrough volume at pH 8 in 0.1M phosphate buffer, and H is eluted by lowering the pH to 4.0. Purification to the point of mutual exclusion is accomplished by two cycles of preparative discontinuous polyacrylamide gel electrophoresis. The purified proteins give a positive PAS reaction indicating a closely associated glycosidic moiety.

Electrophoretically purified H-factor contained 32% carbohydrate and its serologic activity is abolished by pronase digestion. The M-antigen had 55% associated carbohydrate and was pronase-resistant.

The persistence of a relatively constant carbohydrate : protein ratio in the purified preparations of H and M has been linked to the crossreactivity observed with antiserum to *B dermatitidis*. This reaction is believed to be caused by the presence of the heat stable (100 °C) factor referred to as "Heiner's C" (Heiner, 1958), a presumptive galactomannan common to *H capsulatum, B dermatitidis,* and *Paracoccidioides brasiliensis* (Azuma et al, 1974). Possibly the galactomannan is carried through the entire purification as a contaminant, and another separation technique (eg, immunoaffinity chromatography) will be required to remove it. Otherwise, a covalent glycosidic moiety of both H and M factors may share determinants with the more abundant and distinct galactomannan. A third possibility is that the association of a large carbohydrate component occurs because it is involved in a precursor–product relationship with H and M factors. In this view, H and M could be enzymes responsible for synthesis of the *H capsulatum* cell wall. These questions are matters for research, and evidence bearing on them is not yet available.

Long-term storage of these antigens has shown that the purest preparations (after PAGE) undergo decomposition, but that if purification proceeds only through the DEAE stage, and when stored with preservatives (Pine et al, 1981b) stability is increased.

5.3.1b Polyacrylamide Gel Electrophoresis and Monoclonal Antibodies
Because of its importance as a protein antigen capable of evoking both cutaneous and humoral immunity, the M antigen was selected as a model for the induction of monoclonal antibodies (MAbs). Some of the details of this line of research are discussed in Chapter 16. In the first round of fusions, mice immunized with the M factor of histoplasmin gave rise to spleen cells that, after hybridization, all produced IgM against a polysaccharide impurity, probably galactomannan (Knowles et al, 1983). The second round of fusions employed, as the antigen for immunizing mice, periodate-treated M-antigen (Knowles et al, 1984). The hybridomas recovered after this fusion all recognized periodate-treated M, but not native M. The third round of fusions insured against a repetition of this problem by basing selection on the ability to bind both periodate-treated M-protein antigen (to rule out selection of polysaccharide reacting clones) and untreated M. The resulting five IgG producing clones all reacted with both native and periodate-treated M-antigen in microtiter plates and gave a characteristic doublet of activity in the enzyme-linked immunoelectrotransfer blot technique with gradient SDS-polyacrylamide gels (Figure 5.2). This 71,000 da molecular weight doublet was encountered when human and rabbit antisera were reacted in the immunoblot technique. However, the MAbs were not able to precipitate in agar gel immunodiffusion, or to inhibit the reaction between M and rabbit-anti-M. At present, whether or not the M-antigen is indeed this 71,000 da protein cannot be determined.

5.4 HUMORAL RESPONSES

5.4.1 Complement Fixation and Precipitins

In primary pulmonary infection, circulating antibodies are detectable 2 to 4 weeks after exposure to the fungus at the same time that symptoms appear. These early antibodies fix complement to yeast forms. Antibodies to the H and M factors of histoplasmin arise 1 to 2 weeks later in primary infection but the levels of these complement-fixing antibodies are lower than those to whole yeast forms. The first precipitins in acute histoplasmosis are anti-M and in many instances these are the

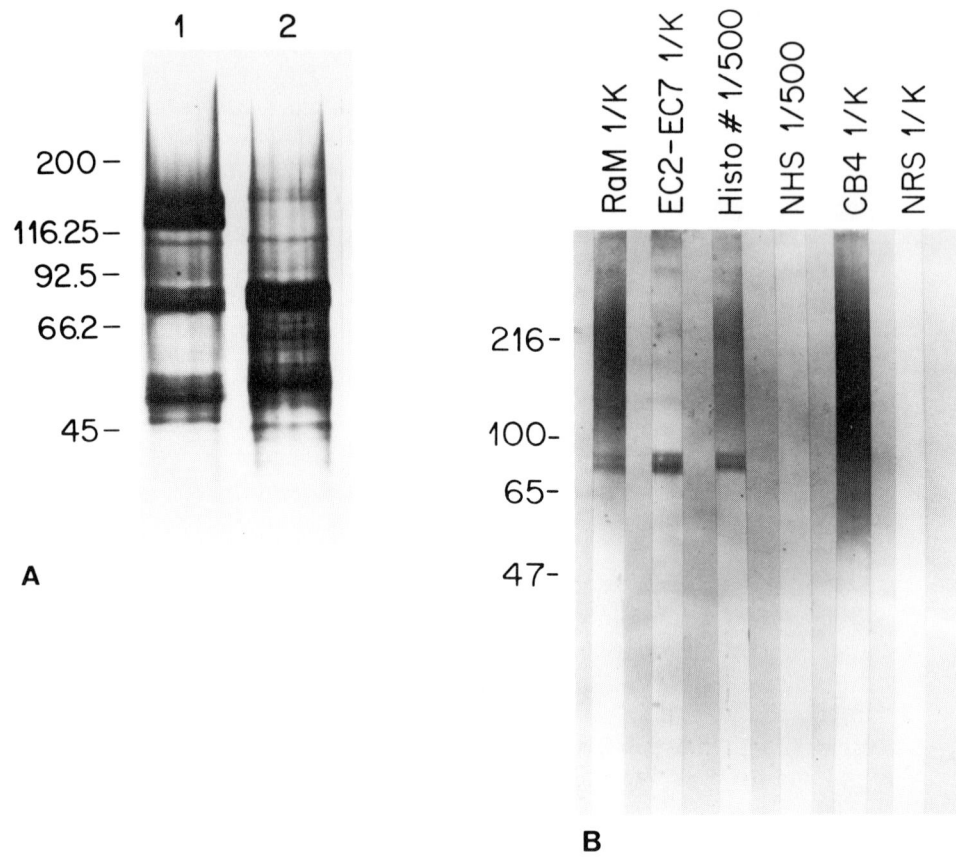

Figure 5.2. The M-factor of histoplasmin is found in a nonantigenic high molecular weight complex form of approximately 150,000 da (Panel A, lane 1) that upon treatment with dithiothreitol reduces to a pair of monomers in the 70,000 to 75,000 da range (lane 2) that are the actual antigens (SDS-PAGE in a 5% to 20% acrylamide gradient) (Knowles et al 1984). Panel B shows an enzyme-linked immunoelectrotransfer blot assay of the dithiothreitol-reduced and electrophoresed M-factor. The first lane shows the reaction of rabbit anti-M serum, the second lane the reaction of an IgG monoclonal antibody; the third lane shows the reaction of a culture-proven human histoplasmosis serum, followed in lane 4 by a normal human serum control. Lane 5 shows the reaction of the monoclonal IgM, "CB4" that recognizes a carbohydrate impurity in histoplasmin. NRS, normal rabbit serum. All sera were diluted as indicated (Knowles et al, 1984).

only precipitins observed, enduring for years after the infection is in remission (Picardi et al, 1976). Antibodies to the H-factor, detected in immunodiffusion, are less frequent and more often are associated with extrapulmonary dissemination. Anti-H levels are rarely as high as those to the M-antigen and decline more rapidly after recovery than anti-M (Picardi et al, 1976). In general, complement-fixing antibody titers of 1/16 to whole yeast forms are considered positive, and higher titers increase the likelihood of a positive case. Similarly, rising titers are most often indicative of progressive disease, and their eventual decline coincides with clinical improvement (Schwarz, 1981). However, serologic findings are not as unequivocal an indicator for diagnosis and prognosis in histoplasmosis as in coccidioidomycosis (Domer and Moser, 1980).

Precipitins reactive with histoplasmin factors were detected in 63 of 70 (90%) proven cases, and complement-fixing antibodies to *H capsulatum* yeast forms were significantly elevated in 94% of the patients in one series (Bauman and Smith, 1975). These same workers screened sera from 1843 healthy coal miners residing in Kentucky. The extent of delayed cutaneous hypersensitivity to histoplasmin in this group was 19.4%, but only a fraction of these were serologically reactive. Picardi et al (1976) have reported that of sixteen healthy residents of the U.S. endemic area, five had complement-fixing antibodies reactive at a sixteenfold serum dilution.

The complement-fixing antibodies to histoplasmin factors arise later in infection than those directed against whole yeast forms and the number of cases in which these antibodies can be detected varies from 3% (Bauman in Schwarz, 1981, p. 165) to 54% (Pine et al, 1978) of cases. Nevertheless, purification of H and M factors has afforded a means of introducing greater specificity into this diagnostic variation. Experiments to measure complement-fixing antibodies to purified H and M factors of histoplasmin were carried out with sera from 126 proven or strongly suspected cases of histoplasmosis. The diagnosis rested on cultural or histologic proof, or on grounds of high antibody levels and radiographic findings (Pine et al, 1978). Information was not available about the stage of infection, whether acute or convalescent, or the type of histoplasmosis, whether acute pulmonary, disseminated, or chronic cavitary. This lack of clinical correlations is typical of many serologic studies. Results showed that complement-fixing titers of >4 against purified H and M factors indicate the presence of specific antibodies even in the absence of positive precipitins. Overall complement-fixation titers with histoplasmin, well-balanced in its content of H and M factors, generally reflect the level of anti-M. This statement is supported by the poor correlation between titers obtained with purified H and histoplasmin, but maximal complement-fixing titers occur when antibodies to both H and M are present, probably because the multiplicity of antibodies reflect extrapulmonary disease and hence an increased humoral response.

Repeated skin tests with histoplasmin do not induce delayed cutaneous hypersensitivity, but a single positive skin test may reinforce the concentration of antibodies to the M-factor. Several reports have reviewed this aspect (Schwarz, 1981; Food and Drug Administration, 1977). A significant increase in the concentration of complement-fixing antibodies to histoplasmin occurred in 20% to 30% of patients studied. The titers ranged up to 1/256, but were usually 1/8 to 1/32. The reinforcing effect may appear 5 days after the skin test and reach a maximum after 2 to 3 weeks.

Mycelial histoplasmin deficient in the M-factor elicited positive delayed cutaneous responses in sensitized individuals without increasing the concentration of complement-fixing antibodies (Kaufman et al, 1969). The drawbacks inherent in measuring complement-fixing antibodies against whole yeast forms are also avoided in precipitin (ie, immunodiffusion) tests even when unfractionated histoplasmin is the source of antigen, because precipitins are less likely to occur in the healthy residents of the endemic area. In the case of blastomycosis or other mycoses, when crossreactions do occur, they can be ruled out with positive control antisera as crossreactions yielding lines of intersection with the immunoprecipitin arcs containing H or M factors. Investigators have sought to increase the sensitivity of precipitin tests, perhaps leading to earlier diagnosis, and at the same time introduce a more rapid method by adapting counterimmunoelectrophoresis, CIE, to the measurement of anti-H and anti-M. A series of 54 patients was studied by Kleger and Kaufman (1973) who found anti-M precipitins in 42 and anti-H in only six patients. No additional benefit was achieved using CIE to measure these precipitins, in place of immunodiffusion. But a later investigation (Picardi et al, 1976) used the same CIE conditions and reagents in a different series of patients and found that anti-H precipitins were more easily demonstrated in CIE, providing a useful prognostic sign because the H-precipitins disappeared after adequate therapy, but anti-M persisted throughout the follow-up period.

The diagnostic usefulness of complement-fixing antibodies against whole *H capsulatum* yeast forms is limited by (i) crossreactions in patients with other mycoses, (ii) increased incidence of positive reactions in patients hospitalized with unrelated diseases, (iii) positive findings in healthy persons from the endemic regions, (iv) reduced reactions in the presence of rheumatoid factor or cold agglutinins. Crossreactions arising from infection with the fungi *Blastomyces dermatitidis* and *Coccidioides immitis* have already been mentioned in connection with the occurrence of a common mural galactomannan. Of nine blastomycosis patients, six had complement-fixing anti-*H capsulatum* yeast form titers^{-1} equal to or greater than 4, but only two reacted with purified M-factor (Pine et al, 1978). Infections with *Cryptococcus neoformans* (Buechner et al, 1973) or *Aspergillus fumigatus* (Picardi et al, 1976; Terry et al, 1978) can also evoke antibodies that crossreact with *H capsulatum* yeast forms. Positive serologic findings were reviewed in patients initially suspected of having histoplasmosis but who were discovered to have other diseases (Picardi et al, 1976; Terry et al, 1978). The test most prone to produce these positive reactions was the measurement of complement-fixing antibodies to *H capsulatum* yeast forms. In one series, antibody titers^{-1} to yeast forms of equal to or greater than 32 occurred in 15% of hospitalized patients for whom a diagnosis of histoplasmosis was clinically improbable (Terry et al, 1978). This tendency towards false-positive reactions is much *reduced* when histoplasmin is used as the antigen. A few of these seropositive patients displayed anti-M precipitins, possibly owing to healed subclinical disease. Nonmycotic diseases that can contribute to positive complement-fixing antibodies to yeast forms are neoplasia, lupus erythematosus, tuberculosis, and diabetes. The bases for these false-positives are largely unexplained. The reactivation of quiescent *H capsulatum* foci has been cited to explain the increased antibodies to *Histoplasma* occasionally encountered in tuberculosis (Terry et al, 1978).

Rheumatoid factors are autoimmune antibodies, typically of the IgM class, that have the capacity to bind to immune complexes containing IgG and block the Fc receptor sites for complement. The practical result of this block is to reduce the complement-fixing antibody titers to *H capsulatum* yeast forms or to other microbial antigens (Johnson and Roberts, 1976). Cold agglutinins were also observed to exert a similar blocking effect. The blockage is removed when IgM is reduced with dithiothreitol, resulting in increased sensitivity in detecting antibodies to *H capsulatum*.

5.4.2 Radioimmunoassay and Enzyme Immunoassay

A competitive binding radioimmunoassay was compared with complement-fixation and immunodiffusion for the relative ability to detect antibodies in sera of 29 histoplasmosis patients (Reiss et al, 1977). The M-antigen, partially purified from histoplasmin by gel permeation and ion exchange chromatography, was fixed to wells of microtiter plate as the solid phase, and rabbit ^{125}I-anti-M IgG served as the indicator antibodies. Twenty-two histoplasmosis patients reacted in the RIA, 21 in complement-fixation and 16 in immunodiffusion tests. In detecting antibodies to purified M factor, the RIA was about equivalent in sensitivity to the complement-fixation test measuring antibodies to the entire histoplasmin and yeast form complexes. Radio-

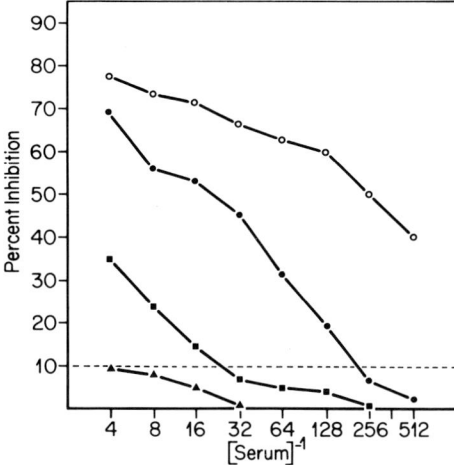

Figure 5.3. Linear regression curves show the serologic activity in histoplasmosis patients' sera reacting in the competitive binding radioimmunoassay to detect antibodies against the M antigen of histoplasmin (Reiss et al 1977). *Source:* Reprinted with permission of the authors and the American Society for Microbiology.

immunoassay was less prone to crossreactions with sera from persons with other mycoses, and the linear regression curve of RIA activity was more accurate in measuring low levels of antibodies than were determinations made by complement-fixation (Figure 5.3).

Enzyme immunoassay detection of the M-factor of histoplasmin (Figure 5.4). Since the RIA did not eliminate the problem of crossreactions, and because EIAs avoid the difficulty of the short half-life of the indicator antibody, an EIA was

Figure 5.4. Detection of antibodies against the M antigen of histoplasmin by EIA-inhibition. Wells of polystyrene microtitration plates were coated with periodate-treated M-antigen to inactivate a crossreactive polysaccharide impurity. Inhibition $\geq 20\%$ indicates a positive test (Brock et al, 1984).

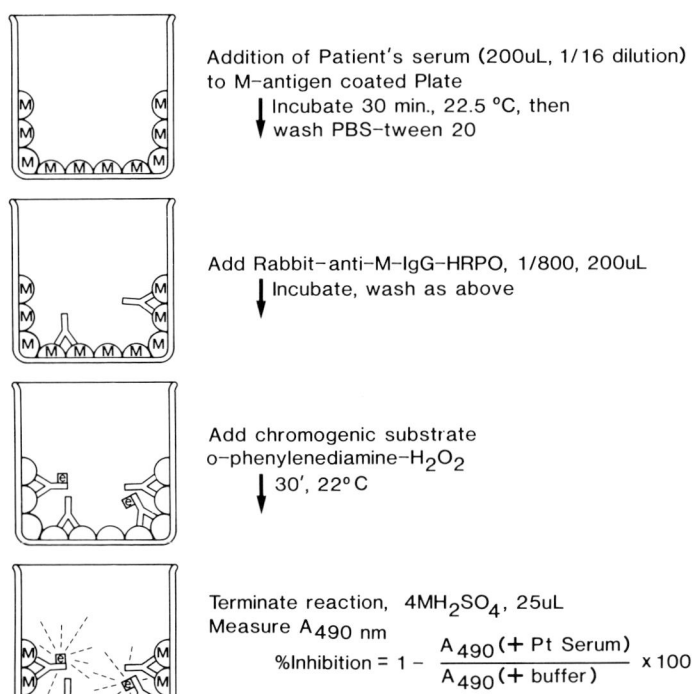

devised to measure antibodies against the M-factor (Brock et al, 1984). Two additional steps were taken to increase the specificity of the test. The M-antigen was treated with $NaIO_4$ to oxidize a polysaccharide impurity. Periodate oxidation reduced the polysaccharide content by 90%, while maintaining the reactivity of the M-factor, measured by immunodiffusion. The EIA was designed as a competitive binding or EIA-inhibition in order to take advantage of the specificity afforded by the peroxidase-labeled rabbit anti-M-IgG. This reagent was produced by immunization of rabbits with isolated M-containing precipitin arcs.

In the first step of the EIA, the M-antigen was adsorbed passively to wells of a polystyrene microtiter plate. The diluted serum of a patient and the indicator antibody (conjugate) were added sequentially, incubated for 30 minutes each, then, after washing, the binding of conjugate to the M-antigen was judged by addition of chromogenic substrate. The percentage inhibition of color development in the presence of patient's serum was calculated in comparison with a control containing conjugate and buffer diluent alone. A percentage inhibition of 20%, using a dilution of patient's serum of 1/16, separated histoplasmosis cases and sera from M-precipitin-positive individuals of the control group, and from blood donors from the endemic area.

The specificity of the EIA was improved when the M-antigen was first treated by periodate oxidation. The percentage inhibition values with proven histoplasmosis and M-precipitin positive sera were increased, and the gap between those groups and endemic area blood donors was widened. The precision of measuring negative reactions was increased, and in one endemic area blood donor a positive reaction towards M-antigen, confirmed by precipitin-in-gel analysis, was observed, whereas it was masked in the reaction with untreated M-antigen.

5.5 DELAYED CUTANEOUS HYPERSENSITIVITY

The epidemiology of histoplasmin sensitivity has been extensively studied and reviewed (Palmer and Edwards, 1960; Edwards et al, 1969; Ajello, 1971; Edwards, 1971; Schwarz, 1981). In the USA, an area of high prevalence, perhaps highest in the world, lies along the Mississippi and Ohio rivers — Missouri, Illinois, Indiana, Kentucky, Tennessee, Mississippi — where more than 80% of the population in many localities become infected by the time they reach adulthood. Irregular and scattered foci extend from this major endemic area into southeastern Canada — Ontario and Quebec — across the Appalachian mountains into Maryland, Virginia, and southwestward into the central plains (Edwards, 1971). Ajello (1971) should be consulted for a more detailed description of the distribution of H capsulatum in the United States. Other endemic regions occur in Central and South America, Africa and Asia, concentrated in the tropical and subtropical latitudes (Edwards, 1971). Persons residing in rural communities in the endemic areas are more liable to be exposed to H capsulatum than urban dwellers. In a community where the general exposure is high, Boone County, Missouri, more than 70% of children were found to be histoplasmin reactors by the age of five years (Schwarz, 1981). The hypersensitive state is durable, present 9 years after the event of histoplasmosis, according to one follow-up study (Palmer and Edwards, 1960). A general decline in hypersensitivity occurs with age among the adult population residing in the endemic area (Palmer and Edwards, 1960), probably as a result of the age-related decline in immunocompetence.

Crossreactions are common in guinea pigs infected with *B dermatitidis* and then skin-tested with histoplasmin and vice versa. The particular problem of lack of specificity of blastomycin is discussed in section 3.2. About 40% of persons infected with *C immitis* will react to histoplasmin, but only 2% of those infected with *H capsulatum* crossreact with coccidioidin. The crossreaction experienced with coccidioidin cannot be removed by replacing it with spherulin (Schwarz, 1981). Surprisingly, sensitization of guinea pigs with unrelated fungi *Aspergillus fumigatus, Aspergillus terreus, Penicillium,* or *Sepedonium* species resulted in hypersensitivity to histoplasmin (Goodman et al, 1971).

Although skin tests of apparently healthy residents of the endemic areas are often positive, in some stages of active histoplasmosis the skin test is unable to elicit a cutaneous response. The incidence of cutaneous anergy in chronic pulmonary cavitary histoplasmosis was estimated at 25%, and this figure may be even higher in persons with focal *coin* lesions, the so-called histoplasmomas (Schwarz, 1981).

The drawbacks that hinder the use of histoplasmin sensitivity as a diagnostic criterion are summarized: (i) high incidence of positive reactors among apparently healthy residents of the endemic areas; (ii) anergy that accompanies acute disseminated disease and chronic pulmonary histoplasmosis; (iii) consistent elevation of the level of complement-fixing antibodies in 10% to 50% of patients with a positive skin test (Levin, 1970) (the M-factor of histoplasmin is the antigen that evokes this humoral response, as well as contributing to the delayed cutaneous hypersensitivity reaction); (iv) crossreactions between *H capsulatum* and *B dermatitidis,* as well as with other fungi. The usefulness of histoplasmin is limited to special situations involving (i) young children, ie, under five years; (ii) known negative reactors with evidence of recent exposure, ie, laboratory workers, speleologists; (iii) residents outside of endemic areas with a recent history of travel in a highly endemic area. Histoplasmin reactivity can provide (i) a prognostic indication in proven disease because the return of cutaneous responsiveness signals a favorable response to therapy; (ii) a tool to determine the source of outbreaks; (iii) a test of general immunocompetence before beginning immunosuppressive therapy.

An alternative to histoplasmin for measuring delayed cutaneous hypersensitivity to *H capsulatum* has been proposed (Levine et al, 1979). Suspension of yeast forms in distilled water resulted in the leaching of sufficient antigen to use for skin tests in humans. This controlled yeast lysate, or CYL antigen, had sensitivity comparable to that of mycelial histoplasmin, but was less prone to reactions in residents of the *C immitis* endemic areas. Unlike histoplasmin, the CYL skin test did not appear to elevate complement-fixing antibody levels.

5.6 INTERACTION WITH MACROPHAGES

The cells of central importance in histoplasmosis are abundant tissue-fixed macrophages engorged with yeast forms that are the commonly observed histologic finding in this disease (Schwarz, 1981). *Histoplasma capsulatum* is considered to be an intracellular pathogen of the reticuloendothelial system (Domer and Moser, 1980). Alveolar macrophages seem incapable of freeing the lung of *Histoplasma* yeast forms and clearance seems to depend on the entry of a large number of specifically activated macrophages in pulmonary granulomas (Collins, 1978). Activation of blood-borne monocytes probably depends on the generation of sensitized T-lymphocytes in the hilar lymph nodes and spleen (Collins, 1978). Alveolar macrophages from mice immunized with *Mycobacterium bovis* BCG were able to phagocytose and kill *H capsulatum* mycelial forms, but yeast forms were resistant (Kimberlin et al, 1981).

Activated peritoneal macrophages from immune mice can restrict the intracellular growth of yeast forms, and this effect is augmented by supernates of specifically sensitized T-cells (Howard and Otto, 1977). Moreover, lymphocytes from sublethally infected mice can arm macrophages from naive animals against yeast forms, but in that event the requirement for specific T-cells cannot be spared by their culture supernates or by supernates of mitogen-stimulated normal T-cells. Thus the requirement for arming normal macrophages against yeast forms requires physical contact with specifically sensitized T-cells. Even though the replication of yeast forms ingested by immune macrophages is arrested, the fungus is not killed and resumes growth after 24 hours in culture (Howard, 1973).

The failure of macrophages to efficiently kill yeast forms of *H capsulatum* is not an isolated finding and is encountered in the facultative intracellular bacteria such as tubercle bacilli, *Listeria monocytogenes*, and *Brucella abortus* (Collins, 1978). In some instances ingested pathogens have been shown to inhibit the fusion of lysosomes to the phagosome membrane, thus avoiding lysis (Goren et al, 1976). It is not known whether this pathogenic mechanism is applicable to *H capsulatum* or whether the yeast forms are resistant to lysis because no efficient lytic enzymes in macrophages are capable of penetrating the yeast cell wall.

Macrophages produce an array of enzymes and factors, some of which are lysosomal, and others are also secreted into the fluid phase. Among those with direct and indirect microbicidal power are lysozyme, β-glucuronidase, other glycosidases, plasminogen activator, elastase, collagenase, complement proteins, peroxidase, and H_2O_2 (Unanue, 1976; Gordon, 1978; Johnston, 1978). The activity of lysosomal enzymes is greatly increased when macrophages are activated. Activation can occur through immunization with sublethal doses of live agents by injection of irritants, or by coculture of *normal* macrophages with specifically sensitized syngeneic T-lymphocytes. In some cases, lymphokines from sensitized T-lymphocytes can substitute for the cells, but so far this has not been possible to demonstrate with *H capsulatum* (Howard and Otto, 1977). Activated macrophages are enlarged with spreading borders, ruffled outer membranes, and greater motility. These physical and biochemical changes result in increased microbial killing or, in the case of facultative intracellular pathogens, slowing or cessation of replication.

When suppression of growth of yeast forms ingested by macrophages does occur, no convincing evidence exists that the myeloperoxidase-H_2O_2-halide pathway is the principal effector function. Guinea pig peritoneal macrophages contain myeloperoxidase but cannot suppress ingested yeast forms in vitro (Howard, 1973a). *Histoplasma* yeast forms are not completely resistant to oxidative fungicidal mechanisms because human and cavine neutrophils can efficiently phagocytose and kill these targets (Howard, 1973b; 1975). Purified human myeloperoxidase and halide ions (especially iodide) prevented germination of yeast forms and conidia of *H capsulatum*, forming a potent antifungal system.

Lysates of granules from cavine neutrophils are lethal for yeast forms, but this activity does not depend on the presence of potassium iodide and H_2O_2 (Howard, 1981). The concentration of catalase in acetone-permeabilized yeast forms of *H capsulatum* was surveyed, but no relationship was found between this enzyme and virulence or susceptibility to killing of yeast forms by H_2O_2 (Howard, 1983), although catalase concentrations varied widely among various isolates from 3 to 27 mEq H_2O_2 destroyed $\cdot min^{-1} \cdot mg^{-1}$.

The relationship between oxidative metabolism and microbicidal activity has been securely established for the neutrophil, and perhaps the blood monocyte, but not for the human tissue macrophage (Johnston, 1978). Phagocytosing macrophages of various species produce luminescence, an indication of the respiratory burst, but the signal is weaker than for neutrophils, and the resulting changes are less pronounced than those with neutrophils studied under the same conditions. The relation between oxidative metabolism and killing of microbes by macrophages is still not clearly defined. Actively phagocytosing macrophages show an increase in the production of superoxide anion and H_2O_2 (Johnston, 1978). Extracts of normal rabbit alveolar macrophages are inhibitory to *H capsulatum* yeast forms in vitro as measured by reduced uptake of amino acids, loss of structural integrity in the cytoplasm, and after 48 hours, a 50% reduction in viability (Calderone and Peterson, 1979), but the lysosomal contents that are active against *Histoplasma* have yet to be characterized.

The phagocytic role of macrophages is augmented in immune animals by cytophilic antibodies that enhance phagocytosis of the target cell, permitting these macrophages to acquire some degree of specificity. There is evidence that serum from rabbits immunized with *H capsulatum* cell walls will enhance phagocytosis of yeast forms by normal macrophages (Bahar de Sanchez and Carbonell, 1975).

The macrophage functions as an antigen processing cell, expressing part of the ingested microbe on its surface in association with an immune response gene product. In that form, the antigen generates a more powerful stimulus to the T- and B-cells. Collaboration among these cell types results in elaboration by T-lymphocytes of antigen-specific helper factors. Macrophages fractionated to 95% purity from nu/+ euthymic mice immunized with a live, but low-virulence, isolate of *H capsulatum* could adoptively transfer immunity to nu/nu, athymic recipients (Williams et al, 1981).

5.7 IMMUNOMODULATION

5.7.1 Human Infections

A general finding applicable to many infections is that the delayed cutaneous hypersensitivity responses wane as the active infection develops, despite evidence of continuing in vitro cellular responses. Collins (1978) described this phenomenon as "desensitization" or "infection anergy." The proliferation of microbes in the regional and central lymphoid tissues appears to depress peripheral hypersensitivity. This suppression of delayed cutaneous hypersensitivity may be specific for antigens of the infecting agent, or broad, extending to unrelated sensitins. The antigen-specific anergy may be a consequence of a central antigen overload preventing sensitized lymphocytes from recognizing the skin test antigen.

Cutaneous anergy and suppression of blastogenic responses to histoplasmin and to whole yeast forms are common findings in acute pulmonary histoplasmosis (Cox, 1979), as compared to healthy adults with positive delayed cutaneous hypersensitivity to histoplasmin. Responses to the T-dependent phytomitogens, PHA and concanavalin A, are likewise suppressed. However, no correlation has been possible between cutaneous anergy and lymphocyte proliferative responses. To some extent the suppression can be explained by a serum inhibitor of blastogenesis, most likely anti-*H capsulatum* immunoglobulin or soluble immune complexes (Cox, 1979), but culturing of patients' peripheral blood lymphocytes with normal serum does not completely restore histoplasmin-elicited blastogenesis. Other more direct evidence for circulating immune complexes in acute histoplasmosis was obtained in a patient presenting with hematuria and glomerulonephritis. Soluble complexes were detected, along with bumpy glomerular deposits of IgA, IgM, and C3 (Bullock et al, 1979).

Chronic pulmonary histoplasmosis, often involving cavitation, may span a course of up to 20 years. Typically, persons with this disease have underlying chronic obstructive pulmonary disease with emphysema (Goodwin and Des Prez, 1978; Alford and Goodwin, 1972). The measurement of cutaneous hypersensitivity and blastogenic responses to histoplasmin in chronic pulmonary histoplasmosis patients has yielded some contradictory results.

Reviewing the experience with histoplasmin skin tests in 183 cases of chronic pulmonary cavitary histoplasmosis, Schwarz (1981) reported that 25% were cutaneously anergic. Newberry et al (1968) found that histoplasmin elicited blastogenesis in this disease was invariably suppressed. Only a minority of patients with reduced blastogenic responses and cutaneous anergy were detected by Alford and Goodwin (1972). Suppressor T-lymphocytes were characterized in a human case of histoplasmal meningitis with no pulmonary involvement (Couch et al, 1978). Although peripheral blood lymphocytes were transformed to lymphoblasts by histoplasmin, this reaction was specifically inhibited by autologous T-lymphocytes from cerebrospinal fluid. Two disseminated histoplasmosis patients with no known immunologic abnormalities had suppressed blastogenic responses to mitogens and histoplasmin (Stobo et al, 1976). A purified T-cell population was fractionated that could suppress blastogenesis in reactive cell cultures. This provided evidence of a T-suppressor set that was activated by *H capsulatum* infection.

Three disseminated histoplasmosis patients had a reduced percentage of peripheral T-lymphocytes, cutaneous anergy to histoplasmin, and broad anergy to a battery of recall antigens and mitogens (Artz et al, 1980). An attempt was made to estimate T-suppressor cells in the peripheral blood by activating them with concanavalin A and then reconstituting them with patients' own lymphocytes and histoplasmin. All patients were found to have subnormal T-suppressor activity. One explanation for this apparently low activity was the overall deficit of T-lymphocytes in the peripheral blood and the depressed ability of the remaining T-cells to respond to either Con A or histoplasmin. The sequestration of T-lymphocytes and macrophages in the lymph nodes and spleen at sites of granulomatous pathology may provide an explanation for the paucity of T-suppressor cells in the peripheral blood capable of responding to Con A. Hyperactive T-suppressor cells and their concentration in lesions are known to occur in another granulomatous disease, the lepromatous form of leprosy (Van Voorhis et al, 1982).

The potential for endogenous reactivation of treated histoplasmosis was reported in a series of 58 immunosuppressed patients who developed histoplasmosis (Kauffman et al, 1978). The underlying conditions included cancer, autoimmune disease, and maintenance of kidney allografts. In this group the rate of dissemination was high, and most patients were cutaneously anergic. The apparent cause of disease in these patients was dormant but viable yeast forms in quiescent lesions that could no longer be contained because of a nonfunctioning immune system.

5.7.2 Presumed Ocular Histoplasmosis Syndrome

This section is concerned with the relationship of presumed ocular histoplasmosis (POHS) to the immune response linked to human leukocyte antigen (HLA) haplotypes. The evidence linking *H capsulatum* with granulomatous uveitis involving macular disciform scars and peripheral retinal lesions is indirect because yeast forms are infrequently or rarely identified in the choroid or retina, and because ocular pathology is many years removed from the event of usually benign histoplasmosis.

Domer and Moser (1980) cited some of the reports in which yeast forms were found in eye lesions. Positive histoplasmin skintests occur in greater than 88% of POHS patients and complement-fixing antibodies against *H capsulatum* are present in a smaller and more variable segment of those afflicted with POHS (Kaplan and Waldrep, 1983). Lymphocyte blastogenic responses to histoplasmin have shown increased blastogenic indices among POHS patients with respect to matched controls. Moreover, this association extends to patients with inactive macular scars who have, as a group, higher blastogenic indices compared with lower reactions observed in those with peripheral scars. Ocular pathology is estimated to occur in 1.6% to 2.5% of people in the USA endemic regions (Kaplan and Waldrep, 1983).

Kaplan and Waldrep (1983) proposed that persistent lymphocytic infiltration in the choroid or retina reflects periodic seeding of the eye by recurrent fungemia, with other foci elsewhere in the body serving as a source of antigen. Supporting this argument is the observation in nonhuman primates that intravenous injection of *H capsulatum* yeast forms produced chorioretinitis, which can evolve into an atrophic scar.

To account for the rare incidence of POHS after disseminated or chronic pulmonary histoplasmosis in humans, Kaplan and Waldrep (1983) proposed that in this disease genetically determined immune responses are faulty. They summarized the status of HLA association with POHS showing a 77% frequency of HLA B-7 versus 26% in controls. The HLA-DR w2 allele is expressed in 81% of patients with disciform scars compared with 62% with peripheral atrophic scars and 28% of controls. This HLA association does not implicate *H capsulatum* in the etiology of POHS and the disease may have multiple etiologies.

The available evidence linking *H capsulatum* with POHS implies that the immune response to antigens of *H capsulatum* causes the disease. The heightened blastogenic responses to *H capsulatum* in the POHS patients might indicate faulty down-regulation of the immune response, possibly representing a failure of T-cell-mediated suppression.

5.7.3 Murine Infections

Splenic enlargement and thymic involution occur in mice during the acute phase of self-limited *H capsulatum* infection (Artz and Bullock, 1979a, 1979b). After recovery, these lymphoid organs return to their normal size. Mitogen- and histoplasmin-induced blastogenesis of splenocytes are initially depressed, but gradually increase during recovery. These findings are probably a reflection of the *anergy of infection* that occurs in human histoplasmosis.

Mice infected with *H capsulatum* develop suppressor cell activity in their spleens coinciding with maximal yeast form proliferation (Deepe et al, 1982). The suppression measured as the inhibition of the primary in vitro plaque forming cell (PFC) response to sheep erythrocytes was associated with T-lymphocytes and a macrophage-like cell that is poorly adherent to glass or plastic (Nickerson et al, 1981). Suppressor cell activity is long-lived, lasting up to 60 days, depending on the inbred strain of mice. Supernates of suppressed spleen cell cultures could also modulate the primary in vitro PFC response to sheep erythrocytes. This activity could not be mimicked by cultivating *H capsulatum* yeast forms with spleen cells from normal mice and is not likely to result from some byproduct of *H capsulatum*. Supernates from infected C3H/HeJ mice suppressed the PFC responses of normal mouse spleen cells, but a supernate from C57B1/6 mice considerably enhanced these responses. Moreover, the helper factor from C57B1/6 could spare the requirement for T-cells in driving normal splenic B cells to become PFC. These soluble modulators produced from spleens of *H capsulatum*-infected mice acted in the induction phase of the primary in vitro immune response to sheep erythrocytes. The soluble suppressor factor is a 25,000 da protein and the helper factor's molecular weight was estimated at 29,000 da. The paradoxical finding that a helper factor is liberated by immunosuppressive lymphocytes suggested to Deepe et al (1982) that soluble mediators can serve as messengers to shift the balance in the immune response from suppression towards amplification.

5.7.4 Ribosomal Vaccines

Ribosomes from *H capsulatum* emulsified in incomplete Freund adjuvant protected 90% of mice against a lethal intravenous challenge with yeast forms (Feit and Tewari, 1974). If the adjuvant was omitted, protection declined to 60%. Nevertheless, mice protected with ribosomes had pulmonary lesions. Optimal protection was achieved with a dose of 100μg of ribosomal protein, and digestion of ribosomes with proteinase reduced the level of protection from 70% to 25–30%. RNase digestion was even more inhibitory, lowering protection to 10%. Ribosomes represent a potent vaccine, but the preparations require further characterization because they were isolated as high speed pellets and not from a density gradient. The use of SDS in the extraction buffer decreased the level of membrane contamination, but this detergent may also alter the ribosomal integrity. The lineage of lymphocytes required to transfer protective immunity was investigated in female C3H mice primed with *H capsulatum* ribosomes in adjuvant, or with sublethal doses of live yeast forms. Splenocytes, or peritoneal exudate cells, were adoptively transferred to syngeneic recipients (Tewari et al, 1978). Mice receiving splenocytes or peritoneal exudate cells from donors primed with live yeast forms were 90% to 100% protected against lethal challenge, whereas those receiving cells from donors primed with ribosomes, were 80% to 90% protected. Immunity could not be transferred with immune serum. Recipients irradiated with 500 rad before adoptive transfer of primed peritoneal exudate cells were also protected against lethal challenge. This control was used to exclude active immunization due to inadvertent transfer of live yeast forms from the donors. Peritoneal cells gave greater protection than splenocytes even though the splenocytes contained three times more T-cells. Negative selection experiments were done to characterize the effector cells. Removal of adherent cells or treatment of donor cells with anti-immunoglobulin and complement did not reduce their ability to transfer immunity. If donor peritoneal exudate cells or splenocytes were treated with anti-thy-1 serum and complement they could no longer protect recipients from lethal challenge, thus showing that ribosomal vaccines confer protection against H capsulatum via T-lymphocytes. Positive selection techniques are indicated to extend this line of research.

5.8 SUMMARY

Overwhelming parasitism of alveolar macrophages is characteristic of acute dissemination, because H capsulatum yeast forms survive in macrophages after ingestion and spread hematogenously via the reticuloendothelial system. Calcified pulmonary lesions are characteristic for healed histoplasmosis, and are a mature expression of type IV delayed hypersensitivity. Larger lesions with central necrosis contain yeast forms in a dormant but probably viable state. Certain, as yet undefined, components of the yeast forms confer resistance to lysis in the host. Many persons with healed lesions continue to produce antibodies for prolonged periods suggesting the continued leakage of antigens from these foci.

The cell wall of H capsulatum consists of four glycans, a soluble galactomannan and insoluble α-1,3-glucan, β-1,3-glucan, and chitin. Galactomannan is a minor component by weight of the yeast forms, but is a major antigen. It occupies a peripheral site in the wall. The α-1,3-glucan is an intermediate layer of short, thick fibrils, whereas the inner layer of the wall is an interwoven mesh of fine fibrils of β-1,3-glucan and chitin. Chemotype 1 yeast forms lack α-1,3-glucan and have a higher chitin content than chemotype 2, which contains α-1,3-glucan as a major cell wall constituent. Four surface factors and five serotypes have been described by immunofluorescent cross-staining and adsorptions. Factors 1 and 4 are shared with B dermatitidis and presumably reside in galactomannan. The galactomannan of mycelial forms is extractable in hot neutral buffer or with dilute alkali at ambient temperature and forms a Cu^{++} complex with Fehling solution. Mannose, galactose, and glucosamine were characterized in an acetolysis fragment of the galactomannan. The galactomannan (Gal:Man ratio 2:5) consists of a linear, 1,6-linked mannan backbone with 1,2-mannosyl and galactosyl side substituents. Galactomannan has also been extracted from the cell wall of yeast forms with β-1,3-glucanase. The resulting polysaccharide from chemotype 1 contained: mannose, 17.8%; galactose, 2.7%; glucosamine, 5.4%; and protein, 11.0%.

Histoplasmin is the filtrate of mycelial form cultures on chemically defined medium. The principal histoplasmin antigens, H and M factors, are potent and specific elicitors of humoral and cell-mediated immunity to H capsulatum and do not crossreact with the antigens of other pathogens or morphologically similar saprophytes. Young mycelial form cultures of H capsulatum can be rapidly identified without the necessity of converting them to the yeast form by using the "exoantigen" test for the elaboration of H and M factors. The M factor is a heat-labile protein resistant to trypsin, proteinase K, and pronase. It elutes before the H antigen from anion exchange columns. Precipitins and complement-fixing antibodies to M arise early in the acute phase of infection, but are also stimulated by skin tests in previously exposed persons. H factor is a more strongly acidic, heat-stable, but proteinase susceptible antigen. Antibodies to H are less frequent in acute histoplasmosis, and are more often associated with extrapulmonary dissemination. The histoplasmin skin test is validly applied as an indication of recovery from anergy in proven disease, as a tool to determine the source of outbreaks, in very young children, and as an immunocompetence test before beginning immunosuppression. Conventional methods to purify H and M antigens do not succeed in removing an antigenic but crossreactive polysaccharide, presumably the galactomannan. This contaminant does not affect precipitin tests because few humans produce galactomannan-specific precipitins, but it interferes with primary binding assays such as RIA and EIA. This interference can be minimized by treating the M-factor preparation with sodium meta-periodate. Monoclonal antibodies against the polysaccharide impurity in histoplasmin have been developed that should aid in the further purification of histoplasmin factors.

Measurement of complement-fixing antibodies to yeast forms is complicated by crossreactions from other mycoses, from nonmycotic diseases, and from healthy persons in the endemic areas. Lower than expected results arise in a small number of cases from interference by cold agglutinins and rheumatoid factor.

Activation of blood borne monocytes by specifically sensitized T-cells is the probable mechanism for clearing yeast forms out of the lung or preventing their replication. Extracts of normal rabbit alveolar macrophages are inhibitory to yeast forms. Macrophages from euthymic mice immunized with live *H capsulatum* were found to adoptively transfer immunity to athymic recipients.

Cutaneous anergy and suppression of blastogenic responses to histoplasmin and to yeast forms are common findings in acute pulmonary histoplasmosis. Responses to phytomitogens are likewise often suppressed. Blocking factors, including soluble immune complexes, have been described in histoplasmosis, but substitution of sera from normal humans does not completely restore blastogenic indices. Suppressor T-lymphocytes have been characterized in a few histoplasmosis patients and subnormal numbers of T-suppressor cells in an even smaller group. The latter observation may result from sequestration of T-cells at the sites of infection in the reticuloendothelial system.

The mouse is a model for immunosuppression in histoplasmosis. T-cells active in suppressing histoplasmin- and mitogen-induced blastogenesis peak at a time when there is maximal yeast form proliferation. Generalized suppression is also measured as a reduction in splenic PFC responses against sheep erythrocytes. The suppression is a durable state associated with T-cells and with a soluble substance that can modulate murine plaque forming cell responses positively or negatively depending on the inbred strain of mice. Further evidence for the importance of T-cell-mediated immunity in histoplasmosis is that the protective effect of ribosomal vaccines in mice can be adoptively transferred with T-cells, but not with serum. Finally, endogenous reactivation of quiescent lesions poses a threat to persons undergoing immunosuppression.

REFERENCES

Ajello L, 1971. Distribution of *Histoplasma capsulatum* in the United States. pp 103–122. In Histoplasmosis—Proceedings of the Second National Conference. Ajello L, et al eds. Springfield: C.C. Thomas.

Alford RH, Goodwin RA, 1972. Patterns of immune response in chronic pulmonary histoplasmosis. J Infect Dis 125:269–275.

REFERENCES

Anderson KL, 1978. Immunochemical and electron microscopy studies of mannoprotein from the yeast-form of *Histoplasma capsulatum*. First International Histoplasmosis Conference, Atlanta, GA, American College of Chest Physicians. unpublished.

Artz RP, Bullock WE, 1979a. Immunoregulatory responses in experimental disseminated histoplasmosis: lymphoid organ histopathology and serological studies. Infect Immun 23:884–892.

Artz RP, Bullock WE, 1979b. Immunoregulatory responses in experimental disseminated histoplasmosis: depression of T-cell dependent and T-effector responses by activation of splenic suppressor cells. Infect Immun 23:893–902.

Artz RP, Jacobson RR, Bullock WE, 1980. Decreased suppressor cell activity in disseminated granulomatous infections. Clin Exp Immunol 41:343–352.

Azuma I, Kanetsuna F, Tanaka Y, Yamamura Y, Carbonell LM, 1974. Chemical and immunological properties of galactomannans obtained from *Histoplasma duboisii, Histoplasma capsulatum, Paracoccidioides brasiliensis,* and *Blastomyces dermatitidis.* Mycopathol et Mycol Appl 54:111–125.

Bahar de Sanchez S, Carbonell LM, 1975. Immunological studies on *Histoplasma capsulatum.* Infect Immun 11:387–394.

Ballou CE, 1980. Genetics of yeast mannoprotein biosynthesis. pp 1–14. In Fungal Polysaccharides. Sandford PA, Matsuda K, eds. Washington, DC: Amer Chem Soc.

Bauman DS, Smith CD, 1975. Comparison of immunodiffusion and complement fixation tests in the diagnosis of histoplasmosis. J Clin Microbiol 2:77–80.

Bradley G, Pine L, Reeves MW, Moss CW, 1974. Purification, composition and serological characterization of histoplasmin-H and M antigens. Infect Immun 9:870–880.

Brock EG, Reiss E, Pine L, Kaufman L, 1984. Effect of periodate oxidation on the detection of antibodies against the M-antigen of histoplasmin by enzyme immunoassay (EIA)-inhibition. Current Microbiol 10:177–180.

Buechner HA, Seabury JH, Campbell CC, Georg LK, Kaufman L, Kaplan W, 1973. The current status of serologic, immunologic and skin tests in the diagnosis of pulmonary mycoses. Chest 63:259–270.

Bullock WE, Artz RP, Bhathena D, Tung KS, 1979. Histoplasmosis. Association with circulating immune complexes, eosinophilia, and mesangiopathic glomerulonephritis. Arch Intern Med 139:700–702.

Calderone RA, Peterson E, 1979. Inhibition of amino acid uptake and incorporation into *Histoplasma capsulatum* by a lysosomal extract from rabbit alveolar macrophages. J Reticuloendothel Soc 26:11–19.

Chattaway FW, Holmes MR, Barlow AJE, 1968. Cell wall composition of the mycelial and blastospore forms of *Candida albicans.* J Gen Microbiol 51:367–376.

Collins FM, 1978. Cellular antimicrobial immunity. CRC Critical Rev in Microbiol 7:27–91.

Couch JR, Abdou NI, Sagawa A, 1978. Histoplasma meningitis with hyperactive suppressor T-cells in cerebrospinal fluid. Neurology 28:119–123.

Cox RA, 1979. Immunologic studies of patients with histoplasmosis. Amer Rev Respir Dis 120:143–149.

Davis TE Jr, Domer JE, Li Y-T, 1977. Cell wall studies of *Histoplasma capsulatum* and *Blastomyces dermatitidis* using autologous and heterologous enzymes. Infect Immun 15:978–987.

Deepe GS Jr, Watson SR, Bullock WE, 1982. Generation of disparate immunoregulatory factors in two inbred strains of mice with disseminated histoplasmosis. J Immunol 129:2186–2191.

Domer JE, 1971. Monosaccharide and chitin content of cell walls of *Histoplasma capsulatum* and *Blastomyces dermatitidis.* J Bacteriol 124:1489–1501.

Domer JE, Hamilton JG, Harkin JC, 1967. Comparative study of the cell walls of the yeast-like and mycelial phases of *Histoplasma capsulatum.* J Bacteriol 94:466–474.

Domer JE, Moser SA, 1980. Histoplasmosis—a review. Rev Med Vet Mycol 15:159–182.

Edwards LB, Acquaviva FA, Livesay VT, Cross FW, Palmer CE, 1969. An atlas of sensitivity to tuberculin, PPD-B and histoplasmin in the United States. Amer Rev Respir Dis 99:1–132.

Edwards MR, Hazen EL, Edwards GA, 1959. The fine structure of the yeast-like cells of *Histoplasma* in culture. J Gen Microbiol 20:496–503.

Edwards PQ, 1971. Histoplasmin sensitivity patterns around the world. pp 97–102. In Histoplasmosis: Proceedings of the Second National Conference, Springfield, IL: C.C. Thomas.

Ehrhard H-B, Pine L, 1972a. Factors influencing the production of H and M antigens by *Histoplasma capsulatum*: development and evaluation of a shake culture procedure. Appl Microbiol 23:236–249.

Ehrhard H-B, Pine L, 1972b. Factors influencing the production of H and M antigens by *Histoplasma capsulatum*: effect of physical factors and composition of medium. Appl Microbiol 23:250–261.

Feit C, Tewari RP, 1974. Immunogenicity of ribosomal preparations from yeast cells of *Histoplasma capsulatum*. Infect Immun 10:1091–1097.

Food and Drug Administration. 1977. Skin test antigens, proposed implementation of efficacy review. Federal Register 42:52674–52687.

Garrison RG, Tally JF, 1981. Electron cytochemical evidence for lysosomal-like equivalents in *Histoplasma capsulatum*. Mycopathol 73:183–190.

Gell PGH, Coombs RRA, Lachmann PJ, 1975. Clinical Aspects of Immunology. 761 pp. Oxford: Blackwell Scientific Publications.

Goodman NL, Larsh HW, Palmer CE, 1971. Cross reactivity in skin testing with histoplasmin. Amer Rev Respir Dis 104:258–260.

Goodwin RA Jr, DesPrez RM, 1978. State of the art—histoplasmosis. Amer Rev Respir Dis 117:929–955.

Gordon S, 1978. Regulation of enzyme secretion by mononuclear phagocytes: studies with macrophage plasminogen activator and lysozyme. Fed Proc 37:2754–2757.

Goren MB, D'Arcy Hart P, Young MR, Armstrong JA, 1976. Prevention of phagosome-lysosome fusion in cultured macrophages by sulfatides of *Mycobacterium tuberculosis*. Proc Nat Acad Sci (U.S.) 73:2510–2514.

Green JH, Harrell WK, Gray SB, Johnson JE, Bolin RC, Gross H, Bradley-Malcolm G, 1976. H and M antigens of *Histoplasma capsulatum*: preparation of antisera and location of these antigens in yeast-phase cells. Infect Immun 14:826–831.

Gross H, Bradley G, Pine L, Gray S, Green JH, Harrell WK, 1975. Evaluation of histoplasmin for the presence of H and M antigens: some difficulties encountered in the production and evaluation of a product suitable for the immunodiffusion test. J Clin Microbiol 1:330–334.

Heiner DC, 1958. Diagnosis of histoplasmosis using precipitin reactions in agar gel. Pediatrics 22:616–629.

Hotchi J, Schwarz J, Kaplan W, 1972. Limitations of fluorescent antibody staining of *Histoplasma capsulatum* in tissue sections. Sabouraudia 10:157–163.

Howard DH, 1973a. Further studies on the inhibition of *Histoplasma capsulatum* within macrophages from immunized animals. Infect Immun 8:577–581.

Howard DH, 1973b. Fate of *Histoplasma capsulatum* in guinea pig polymorphonuclear leukocytes. Infect Immun 8:412–419.

Howard DH, 1975. The role of phagocytic mechanisms in defense against *Histoplasma capsulatum*. pp 50–59. In Mycoses, Proc Third Inter Conf on the Mycoses. Sci Pub No 304. Washington, DC: Pan American Health Organization.

Howard DH, 1981. Comparative sensitivity of *Histoplasma capsulatum* conidiospores and blastospores to oxidative antifungal systems. Infect Immun 22:381–387.

Howard DH, 1983. Studies on the catalase of *Histoplasma capsulatum*. Infect Immun 39:1161–1166.

Howard DH, Otto V, 1977. Experiments on lymphocyte-mediated cellular immunity in murine histoplasmosis. Infect Immun 16:226–231.

Johnson JE, Roberts GD, 1976. Blocking effect of rheumatoid factor and cold agglutinins on complement fixation tests for histoplasmosis. J Clin Microbiol 3:157–160.

Johnston RB Jr, 1978. Oxygen metabolism and the microbicidal activity of macrophages. Fed Proc 37:2759–2764.

Kanetsuna F, Carbonell LM, Gil F, Azuma I, 1974. Chemical and ultrastructural studies on the cell walls of the yeastlike and mycelial forms of *Histoplasma capsulatum*. Mycopath Mycol Appl 54:1–13.

Kaplan HJ, Waldrep JC, 1983. Immunological basis of presumed ocular histoplasmosis. Int Ophthalmol Clin 23:19–31.

Kauffman CA, Israel KS, Smith JW, White AC, Schwarz J, Brooks GF, 1978. Histoplasmosis in immunosuppressed patients. Amer J Med 64:923–932.

Kaufman L, Blumer S, 1966. Occurrence of serotypes among *Histoplasma capsulatum* strains. J Bacteriol 91:1434–1439.

Kaufman L, Kaplan W, 1961. Preparation of a fluorescent antibody specific for the yeast phase of *Histoplasma capsulatum*. J Bacteriol 82:729–735.

Kaufman L, Kaplan W, 1963. Serological characterization of pathogenic fungi by means of fluorescent antibodies. 1. Antigenic relationships between yeast and mycelial forms of *Histoplasma capsulatum* and *Blastomyces dermatitidis*. J Bacteriol 85:986–991.

Kaufman L, McLaughlin D, Terry RT, 1969. Immunological studies with an M-deficient histoplasmin skin-test antigen. Appl Microbiol 18:307–309.

Kaufman L, Standard P, 1978. Immunoidentification of cultures of fungi pathogenic to man. Current Microbiol 1:135–140.

Kimberlin CL, Hariri AR, Hempel HO, Goodman NL, 1981. Interactions between *Histoplasma capsulatum* and macrophages from normal and treated mice: comparison of the mycelial and yeast phases in alveolar and peritoneal macrophages. Infect Immun 34:6–10.

Kleger B, Kaufman L, 1973. Detection and identification of diagnostic *Histoplasma capsulatum* precipitates by counterelectrophoresis. Appl Microbiol 26:231–238.

Knowles JB, Reiss E, Brock EG, Bragg SL, Pine L, Aloisio CH, 1983. Characterization of monoclonal antibodies to histoplasmin factors. Abstr Ann Mtg Amer Soc Microbiol p 389.

Knowles JB, Reiss E, Aloisio CH, 1984. Characterization of monoclonal antibodies (MAbs) against histoplasmin antigens: effect of periodate oxidation. Abstr Ann Mtg Amer Soc Microbiol p 299. Abst No F40.

Kwon-Chung KJ, Weeks RJ, Larsh HW, 1974. Studies on *Emmonsiella capsulata* (*Histoplasma capsulatum*). II. Distribution of the two mating types in 13 endemic states of the United States. J Epidemiol 99:44–49.

Levin S, 1970. The fungal skin test as a diagnostic hindrance. J Infect Dis 122:343–345.

Levine HB, Scalarone GM, Campbell GD, Graybill JR, Kelly PC, Chaparas SD, 1979. Histoplasmin-CYL, a yeast phase reagent in skin test studies with humans. Amer Rev Respir Dis 119:629–636.

Mandell GL, Douglas RG Jr, Bennett JE, 1979. Principles and Practice of Infections Diseases. Vol 2. pp 1979–2084. New York: John Wiley and Sons.

McGinnis MR, 1980. Laboratory Handbook of Medical Mycology. 661 pp. New York: Academic Press.

Newberry WM Jr, Chandler JW Jr, Chin TDY, Kirkpatrick CH, 1968. Immunology of the mycoses. I. Depressed lymphocyte transformation in chronic histoplasmosis. J Immunol 100:436–443.

Nickerson DA, Havens RA, Bullock WE, 1981. Immunoregulation in disseminated histoplasmosis: characterization of splenic suppressor cell populations. Cell Immunol 60:287–297.

Odds FC, Kaufman L, McLaughlin D, Callaway C, Blumer SO, 1974. Effect of chitinase complex on the antigenicity and chemistry of yeast form cell walls and other fractions of *Histoplasma capsulatum* and *Blastomyces dermatitidis*. Sabouraudia 12:138–149.

Palmer CE, Edwards PQ, 1960. The histoplasmin skin test. pp. 189–210. In Histoplasmosis. Sweany HC, ed. Springfield, IL: CC Thomas.

Picardi JL, Kauffman CA, Schwarz J, Phair JP, 1976. Detection of precipitating antibodies to *Histoplasma capsulatum* by counterimmunoelectrophoresis. Amer Rev Respir Dis 114:171–176.

Pine L, 1977. Histoplasma antigens: their production, purification and uses. pp 138–168. In Proc 21st OHOLO Biological Conf, Ma'alot, Israel. Beemer AM, Ben-David A, Klingberg

MA, Kuttin ES, eds. Contribution Microbiol and Immunol. Vol. 3. Basel: S Karger.

Pine L, Boone CJ, 1968. Cell wall composition and serological reactivity of *Histoplasma capsulatum* serotypes and related species. J Bacteriol 96:789–798.

Pine L, Gross H, Bradley-Malcolm G, George JR, Gray SB, Moss CW, 1977. Procedures for the production and separation of H and M antigens in histoplasmin: chemical and serological properties of the isolated products. Mycopathol 61:131–141.

Pine L, Gross H, Malcolm GB, Green JH, et al, 1981a. Evaluation of candidate international reference reagents and a microimmunodiffusion test for the identification of precipitins to the H and M antigens of histoplasmin. J Biol Standard 9:513–530.

Pine L, Malcolm GB, Gross H, Gray SB, 1978. Evaluation of purified H and M antigens of histoplasmin as reagents in the complement fixation test. Sabouraudia 16:257–269.

Pine L, Smith SJ, Gross H, Barbaree JM, Malcolm GB, 1981b. Studies on the thermal degradation of the H and M antigens of lyophilized histoplasmin. Sabouraudia 19:55–70.

Reiss E, 1977. Serial enzymatic hydrolysis of cell walls of two serotypes of yeast-form *Histoplasma capsulatum* with $\alpha(1 \to 3)$glucanase, $\beta(1 \to 3)$glucanase, pronase and chitinase. Infect Immun 16:181–188.

Reiss E, Hutchinson H, Pine L, Ziegler DW, Kaufman L, 1977a. Solid-phase competitive-binding radioimmunoassay for detecting antibody to the M antigen of histoplasmin. J Clin Microbiol 6:598–604.

Reiss E, Miller SE, Kaplan W, Kaufman L, 1977b. Antigenic, chemical and structural properties of cell walls of *Histoplasma capsulatum* yeast-form chemotypes 1 and 2 after serial enzymatic hydrolysis. Infect Immun 16:690–700.

Reiss E, Mitchell WO, Stone SH, Hasenclever HF, 1974. Cellular immune activity of a galactomannan-protein complex from mycelia of *Histoplasma capsulatum*. Infect Immun 10:802–809.

Rippon JW, 1982. Medical Mycology—The Pathogenic Fungi and the Pathogenic Actinomycetes 2nd ed. 842 pp. Philadelphia: WB Saunders, Co

San Blas G, Carbonell LM, 1974. Chemical and ultrastructural studies on the cell walls of the yeastlike and mycelial forms of *Histoplasma farciminosum*. J Bacteriol 119:602–611.

San Blas G, Ordaz D, Yegres FJ, 1978. *Histoplasma capsulatum*: chemical variability of the yeast cell walls. Sabouraudia 16:279–284.

Schwarz J, 1981. Histoplasmosis. 472 pp. New York: Praeger.

Smith CE, Whiting EG, Baker EE, Rosenberger HG, Beard RR, Saito MT, 1948. The use of coccidioidin. Amer Rev Tuberc 57:330–360.

Standard PG, Kaufman L, 1976. Specific immunological test for the rapid identification of members of the genus *Histoplasma*. J Clin Microbiol 3:191–199.

Stobo JD, Paul S, van Scoy RE, Hermans PE, 1976. Suppressor thymus-derived lymphocytes in fungal infection. J Clin Invest 57:319–328.

Terry PB, Rosenow EC III, Roberts GD, 1978. False-positive complement-fixation serology in histoplasmosis. A retrospective study. JAMA 239:2453–2456.

Tewari RP, Sharma DK, Mathur A, 1978. Significance of thymus-derived lymphocytes in immunity elicited by immunization with ribosomes or live yeast cells of *Histoplasma capsulatum*. J Infect Dis 138:605–613.

Tompkins VN, 1965. Soluble antigenic constituents of yeast-phase *Histoplasma capsulatum*. Amer Rev Respir Dis Suppl 92:126–133.

Unanue ER, 1976. Secretory function of mononuclear phagocytes. Amer J Pathol 83:396–417.

Van Voorhis WC, Kaplan G, Sarno EN, Horwitz MA, et al, 1982. The cutaneous infiltrates of leprosy. Cellular characteristics and the predominant T-cell phenotypes. New England J Med 307:1593–1597.

Williams DM, Graybill JR, Drutz DJ, 1981. Adoptive transfer of immunity to *Histoplasma capsulatum* in athymic nude mice. Sabouraudia 19:39–48.

Chapter 6

Paracoccidioides brasiliensis

6.1 INTRODUCTION

Paracoccidioides brasiliensis is a dimorphic primary systemic pathogen, having a filamentous saprophytic form and large, round to oval yeast cells, 5 to 60 μm, that develop in tissue. Multiple buds attached to the mother cell by slender necks typically form a pilot wheel shape. The daughter cells occur in two types: (i) uniformly smaller than the mother cell or (ii) variable in size, but larger than the mother cell (Chandler et al, 1980). The disease is endemic from Mexico to Argentina, in tropical and subtropical forested regions high in humidity. Brazil is at the center of the endemic zone. Males are most often infected (male : female cases = 15 : 1), although skin tests show equal exposure of the sexes.

Paracoccidioidomycosis is a complex clinical entity with the common theme of pulmonary involvement (Giraldo et al, 1976). There may be a long latent period between exposure and clinical illness. Mucocutaneous-lymphangitic involvement is characteristic of *P brasiliensis* infection, and is the result of dissemination from a primary pulmonary focus. Ulcers often occur in the mouth, including the gums. These granulomatous lesions drain into the cervical lymph nodes, which may become massive. The replicating yeast forms in lymph nodes preferentially destroy the paracortical regions, as occurs in leprosy (Mendes, 1975).

Patients younger than 30 years have more involvement of the reticuloendothelial system (lymph nodes, spleen, liver, gut associated lymphoid tissue, bone marrow) (Figure 6.1) and a more acute disease. In older patients, the trend is toward a decline in obvious infection of the reticuloendothelial system and a more chronic, progressive course generally. Dual infection with *M tuberculosis* is common (26% of 46 patients in one series) and may complicate assessment of immunocompetence.

6.2 CELL WALL

The main glucosyl polymer of yeast-form cell walls is an α-1,3-glucan that can account for up to 45% of the mural dry weight (San Blas and San Blas, 1977). Short

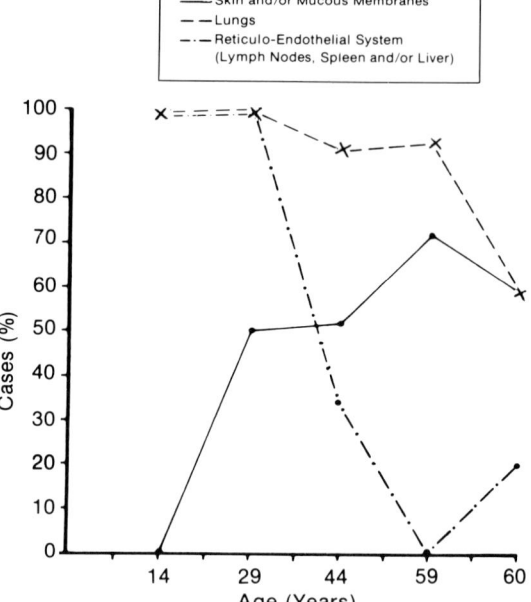

Figure 6.1. Organ involvement in paracoccidioidomycosis as a function of the age of the patient. In younger persons under 30 years the reticuloendothelial system is severely affected and the disease takes an acute course. In older persons, a chronic, protracted course is more typical and skin or mucosal lesions are often present (Giraldo et al, 1976). *Source:* Reprinted with permission from Dr. W. Junk, Publishers.

fibrils of the α-1,3-glucan help to maintain the ellipsoid shape of yeast forms. When the temperature is shifted from 37 °C to 20 °C, conversion from yeast to mycelial forms occurs, α-1,3-glucan synthesis is repressed, and the content of long, thin, interwoven β-1,3-glucan fibrils increases. The different optimal temperatures for α- and β-1,3-glucan synthetases affect their relative rates of synthesis and may contribute to the determination of cell shape. Synthesis of α-1,3 glucan occurs in old mycelial form cultures, and it may represent a storage carbohydrate. The α-1,3-glucan can be separated by its alkali solubility from an insoluble matrix of β-1,3-glucan and chitin. Upon acidification, the α-1,3-glucan is precipitated, leaving a soluble galactomannan.

The culture referred to as *P brasiliensis 7193* (Kanetsuna et al, 1969; 1972) is more recently known as *IVIC Pb 9* or *ATCC 36324* (San Blas and San Blas, 1977, 1982). More recent cell wall analysis of a second wild type, *IVIC 73* (San Blas and San Blas, 1982), indicates that revisions will be necessary in our concept of the cell wall composition of this and, by extension, of other systemic dimorphic fungal pathogens. Milder conditions of acid hydrolysis of the wall fractions appear to preserve a substantial galactose component not detected in the earlier analyses (Table 6.1). Galactose is the most prevalent hexose in the mycelial form wall of *P brasiliensis IVIC 73*, and it occurs in both alkali-soluble and insoluble fractions. Under conditions that preserve galactofuranoside linkages, the amino sugar content of the mycelial form walls is revised to only 5.8% of the dry weight. Morphogenesis to the yeast form causes an increase in the chitin component which becomes *five times* higher, or 29.8% of the cell wall.

The protein content reported for mycelial walls of *P brasiliensis* is high (32.9%, San Blas and San Blas, 1982), and attention should be paid to its source, whether as structural protein, periplasmic enzymes, or extraneous membrane-bound proteins.

Mycelial form walls are 80 to 150 nm thick and only one layer is seen in electron micrographs, whereas the wall of yeast forms is much thicker, 200 to 600 nm, with two or three layers seen in cross-sections. The broad, inner, electronlucent layer of yeast form walls consists of long bundles of chitin and glucan, and the outer layers are more electrondense. A thin, superficial coat of galactomannan is exterior to a broad layer of short, thick α-1,3-glucan fibrils (San Blas and San Blas, 1977).

6.2.1 α-1,3-Glucan

The role of α-1,3-D-glucan in conferring resistance to lysis is an intriguing question. Its occurrence is not mandatory for mycelial to yeast conversion. α-1,3-D-Glucan deficient mutants are dimorphic, but have diminished mouse virulence. Repeated subculture of *P brasiliensis* IVIC Pb 9 resulted in a variant that was attenuated in virulence and reduced in α-1,3-D-glucan content from 40% (wild type) to 3% of the yeast form mural weight. When passaged in hamsters or cultured in vitro with fetal calf serum, the α-1,3-D-glucan content increased, but not to the level present in the wild type. Another induced mutant incapable of α-1,3-D-glucan production in any of the attempted conditions of culture was avirulent. Evidence exists (Goihman-Yahr et al, 1980) that human polymorphonuclear neutrophils can kill α-1,3-D-glucan-deficient mutants more efficiently than wild-type yeast forms. Electron micrographs of phagocytosed yeast show a gradual lysis of protoplasm while the cell wall remains largely intact. This ultrastructural analysis of the wild-type bears repeating in glucan-deficient mutants. The available evidence implicates α-1,3-D-glucan as a factor that protects yeast forms from destruction. Otherwise, glucan synthetases could possibly be genetically linked to a separate unknown virulence factor. Occam's razor favors α-1,3-D-glucan as the cause of lysis resistance, but more direct evidence for this conclusion is needed.

6.2.2 Galactomannan

Galactomannan (GM), the only mural glycan known to be antigenic, comprises 5% of the wild-type, yeast form walls, but 16.6% to 31.4% of mycelial form mural weight. This polysaccharide was extracted from mycelial forms of four dimorphic

Table 6.1. Composition of *Paracoccidioides brasiliensis* IVIC Pb 73 Cell Walls

Component[a]	Total Wall		Alkali-Insoluble Fraction		Alkali-Soluble, Water-Soluble		Alkali-Soluble Acid-Precipitable	
	M	Y	M	Y	M	Y	M	Y
Neutral sugar	50.1	33.5	33.1	6.8	13.0	0.4	3.9	26.3
Glucose	18.6	33.1	11.6	6.8	3.0	0	3.9	26.3
Galactose	23.8	0	16.1	0	7.7	0	0	0
Mannose	7.6	0.4	5.4	0	2.2	0.4	0	0
Amino sugar	5.9	29.8	5.8	29.8	0	0	0	0
Amino acid	20.7	23.3	3.4	22.6	17.3	0.3	<1	<1
Recovery (%)	76.7	86.6						

Source: Reprinted from Sabouraudia with permission of the copyright holder, International Society for Human and Animal Mycology.

Abbreviations: M, mycelial; Y, yeast.
[a] Calculated as percent dry weight of cell walls (San Blas and San Blas, 1982).

fungal pathogens (*Histoplasma capsulatum, Histoplasma duboisii, Blastomyces dermatitidis* and *P brasiliensis*) (Azuma et al, 1974). The GMs were compared in immunodiffusion versus anti-*H capsulatum* globulins and all formed lines of identity suggesting that these polysaccharides were not species-specific. Quantitative precipitin reactions between anti-*H capsulatum* immunoglobulin and four GMs also showed a high degree of relatedness, although *P brasiliensis* GM was less crossreactive than that from *H duboisii*. A 1,6-linked mannan from *Alternaria kikuchiana* having 1,2- and 1,3-mannosyl side chains failed to react implying that galactose was an antigenic determinant. Galactomannan from *Alternaria zinniae* crossreacted with antiserum produced in animals immunized with whole, boiled *P brasiliensis*. The *A zinniae* 1,6-mannan had galactofuranose residues linked to C-3 in the main chain. These data suggest extensive crossreactions among GMs isolated from four fungal pathogens. This does not mean that all GMs are crossreactive; for example, GM from *Aspergillus fumigatus* did not crossreact with antiserum produced against *P brasiliensis* or *H capsulatum* stressing the need for more sequence and linkage analysis to define haptenic groups.

The method of preparing GM (Azuma et al, 1974; Gorin and Spencer, 1968) by boiling in 1M NaOH followed by Fehling precipitation was not designed to conserve antigenic groups and alternative methods to minimize alkaline degradation should be considered. Passage of GM through diethylaminoethyl-cellulose (Azuma et al, 1974) may appear advisable to remove contaminating protein, but mannan fractions of potent antigenicity are now known to be bound by anion exchange chromatography. The combined effects of alkali degradation and work-up of the neutral GM fraction may inadvertently select the most crossreactive fraction for comparison. Assessment of crossreactions between GMs should also make provision for testing homologous and heterologous antisera.

The results of methylation analysis of the neutral portion (DEAE-effluent) of GM from *H capsulatum* and *P brasiliensis* (Azuma et al, 1974) are summarized in Table 6.2. These results suggest a 1,6-mannan backbone with a small amount of 2-man 1- ($<1.0\%$), possibly as a contaminant. These mannans are highly branched, *P brasiliensis* GM having twice as many branch points as that from *H capsulatum*. Galactofuranose exists as terminal nonreducing single side substituents. Other side substituents containing mannosyl residues are also possible. Elsewhere (Chapter 5) $GlcNH_2$ was mentioned as having been identified in acetolysis fragments of *H capsulatum* GM.

6.3 SECRETED ANTIGENS

6.3.1 Factor "E"

Immunoelectrophoresis of culture supernatants from *P brasiliensis* mycelial forms revealed a distinctive cathodic precipitin arc, "E," formed in the reaction with sera from paracoccidioidomycosis patients (Yarzabal et al, 1976). Alkaline phosphatase activity was localized in this arc by an immunoenzyme test. Precipitin arcs were emulsified in complete Freund's adjuvant, and immunization of rabbits resulted in specific antisera. The IgG fraction was bound to CNBr-activated agarose and *P brasiliensis* mycelial form culture supernatants were percolated through this immunoaffinity column. Purified "E" antigen was eluted by pH reduction. Affinity-

Table 6.2. Methylation Analysis of Galactomannans from Mycelial Forms

Component, Linkage	P brasiliensis		H capsulatum	
	Molar Ratio	Mole Percent	Molar Ratio	Mole Percent
Galactofuranose 1 →	3.16	27.5	5.17	39.4
Mannose 1 →	1.00	8.7	1.00	7.6
2-Mannose 1 →	0.13	1.1	0.12	0.9
3-Mannose 1 →	0	0	0	0
6-Mannose 1 →	2.80	24.3	4.30	32.8
di-O-methyl derivatives of 6-Mannose 1→ $\uparrow^3$ or 6-Mannose 1→ $\uparrow^2$	4.40	38.3	2.50	19.0

Source: Redrawn from Azuma et al, (1974).

column purified "E" did not crossreact with anti-*H capsulatum* or anti-*B dermatitidis* sera. The antiserum specific for "E" was used to detect human antibodies as follows: The immunoprecipitate in immunodiffusion between monospecific anti-E immunoglobulin and *P brasiliensis* culture supernatant formed a line of identity with the arc in adjacent wells containing patients' sera. This reaction was observed in sera from 43 paracoccidioidomycosis patients. Reacting immunoaffinity column-purified "E" directly with patients' sera would be easier, but this has not been reported. This antigen has been found to be identical to factor 1 described by Restrepo and Moncada (1974) on grounds of identical immunodiffusion results and electrophoretic mobility.

The significance of the "E" or "1" factor in the serologic profile in this disease is discussed below.

Fifty-four freshly diagnosed paracoccidioidomycosis patients were monitored for precipitins and complement-fixing antibodies for periods of up to 2 or 3 years (Restrepo and Moncada, 1974). The antigen for immunodiffusion and complement-fixation was paracoccidioidin from yeast forms (Restrepo-Moreno and Schneidau, 1967). Three characteristic precipitin arcs were recognized in immunodiffusion and their persistence was studied. The reaction designated as "1" occurred in most patients and persisted for the longest time. The "2" and "3" immunodiffusion reactions observed in comparison with pooled human reference antiserum were seen less frequently and were more transitory. A positive correlation was found between the number of precipitins observed and the concentration of complement-fixing antibodies. No attempt was made to classify patients according to disease progression, ie, pulmonary, mucocutaneous, lymphangitic, or disseminated. Such correlation would aid in interpreting antibody levels as prognostic aids. In this disease, precipitins are persistent. After 6 months' treatment, of 18 patients, 14 showed no change in the number of precipitins. In the remaining four, one less precipitin was observed. After two years' treatment, 13 of 18 patients' sera still showed precipitins.

Confusion exists in the interpretation of positive precipitins in paracoccidioidomycosis patients' sera that react with histoplasmin. Whether these are crossreactions or evidence of dual infection is not clear. In the Colombian area endemic for *P brasiliensis*, *H capsulatum* and histoplasmosis are also encountered so that the potential for dual infections is real.

6.3.2 Paracoccidioidin

Paracoccidioidins obtained from mycelial or yeast form culture supernatants were capable of evoking cutaneous hypersensitivity in guinea pigs and human volunteers (Restrepo-Moreno and Schneidau, 1967). Four-week-old cultures grown on trypticase soy dialyzate medium at either 22° or 35 °C were killed with formalin and the supernates were dialyzed and precipitated with 5 vols of ethanol. All of the cutaneous reactivity was present in the ethanol precipitate. The nitrogen content of the yeast or mycelial ethanol-precipitated antigens was 2% to 4%. When these preparations were deproteinized by the Sevag method, the protein, some carbohydrate, and all the cutaneous reactivity were extracted into the chloroform phase. Guinea pigs were skin-tested 4 weeks after receiving a total of 1.5×10^6 yeast forms intratesticularly. They responded equally well to mycelial or yeast form ethanol-precipitated antigens at doses of 100 μg or higher. Peak reactions occurred at 24 hours but the type of cellular infiltrate was not ascertained. Yeast form antigens were less prone to crossreact in animals sensitized with *S schenckii* or *H capsulatum*. This may be a moot point since this antigen and that from mycelial culture supernates elicited positive delayed cutaneous hypersensitivity in 12 of 18 humans who had positive histoplasmin reactions. These were US residents, where *P brasiliensis* is not found.

The optimal dose for eliciting delayed cutaneous hypersensitivity in humans (10 μg) was much lower than that required in guinea pigs, but the crossreactivity between the *P brasiliensis* antigen and *H capsulatum* presents a problem in estimating the significance of reactors to paracoccidioidin. The active portion of paracoccidioidin behaves like a glycoprotein on the basis of the Sevag experiment, but further characterization and relation to the antigens of Yarzabal (1976) or Restrepo and Moncada (1974) have not appeared.

6.4 CELL-MEDIATED RESPONSES

Analysis of T-cell effector mechanisms continues to reveal a variety of immunologic deficits, both antigen-specific and in general. Many of the cell-mediated immune reactions that are compromised resemble the situation in other systemic mycoses. In the instance of paracoccidioidomycosis, a parameter of cell-mediated immunity that remains in the normal range in serious infection is difficult to find. It is not clear if these acquired deficiencies are the result of an immune system that has lost its ability to function in the face of a fulminant infection, or whether the T-cells are marshalled in the infected tissues and are unavailable in the peripheral blood, whence their functions are most readily analyzed. Another derangement of cell-mediated immunity may operate at the level of humoral inhibitors of blastogenesis, possibly related to immune complexes in antigen excess (see below). Studies of cell-mediated immunity in this disease have not yet penetrated to the level of T-cell subsets. Knowledge of the presence and role of T-helper and T-suppressor cells may help to understand the nature of the suppression of T-cell functions in paracoccidioidomycosis.

Three series of paracoccidioidomycosis patients have been monitored for humoral and cell-mediated immunity. Musatti et al (1977) studied 19 Brazilian patients.

Thirty-six Colombian patients were selected on the basis of those who were symptomatic, compared with those whose disease was in remission (Mok and Greer, 1977). Sixteen other Colombian patients were evaluated at the time of diagnosis and after six months' treatment (Restrepo et al, 1978).

Ethanol-precipitated paracoccidioidin from mycelial forms was used in two studies (Mok and Greer, 1977; Restrepo et al, 1978). Musatti et al (1976) filtered the sonicate of *P brasiliensis* yeast forms for in vivo and in vitro tests. Evidently more effort is needed to characterize these antigens, particularly because the amount of blastogenesis evoked by them in patients' lymphocytes is of a low order (see section 6.4.2).

6.4.1 Delayed Cutaneous Hypersensitivity

6.4.1a Specific

Cutaneous anergy to *P brasiliensis* antigens occurred in slightly more than half of the patients studied (Musatti et al, 1976; Mok and Greer, 1977). This finding was most striking in freshly diagnosed cases (Restrepo et al, 1978); only 2 of 16 patients were initially responsive, rising to 6 of 16 after treatment. Positive delayed cutaneous hypersensitivity to paracoccidioidin was lowest in incidence among patients whose disease was longest in duration. A few patients were followed throughout treatment. Two developed delayed cutaneous hypersensitivity after nine months; however, in two others, decreasing delayed cutaneous hypersensitivity and deteriorating clinical conditions ensued (Mok and Greer, 1977). Over half of the patients in one series (Mok and Greer, 1977) had positive cutaneous reactions to histoplasmin indicative of either dual infection or crossreactivity between antigens. Another sign of defective immunoregulation was an increased incidence of immediate hypersensitivity to *P brasiliensis* antigens, positive in 13 of 18 patients tested (Musatti et al, 1976).

6.4.1b General

Hypersensitivity responses to common microbial antigens, mitogens, and contact sensitin are impaired in this disease. Cutaneous responses to purified protein derivative of tuberculin (PPD), and to *Candida albicans* were often weaker than normal (Musatti et al, 1976). A broad state of immunosuppression existed with respect to PPD (Restrepo et al, 1978). Initially, reactors accounted for three of 16 patients, rising to nine of 16 after treatment. Mok and Greer (1977) noted a dissociation between in vivo and in vitro blastogenic responses to PPD. Cutaneous and blastogenic responses to PHA were diminished (Musatti et al, 1976; Mok and Greer, 1977). The proportion of patients capable of sensitization with dinitrochlorobenzene was depressed even lower than that observed in leprosy (Musatti et al, 1976). This type of suppression was corroborated by Mok and Greer (1977) who found only half of the patients they studied were capable of being sensitized.

6.4.2 In Vitro Cellular Responses

6.4.2a Blastogenesis

Blastogenic indices to antigens of *P brasiliensis* were flatly unresponsive in 11 of 17 patients (Musatti et al, 1976). In another study by Mok and Greer (1977) blastogenic indices to paracoccidioidin were, in all instances, less than 4. Restrepo et al (1978) found that the number of cells transformed to lymphoblasts was low initially and rose during treatment but to levels below that of healthy controls. All researchers thus far have reported dissociation between delayed cutaneous hypersensitivity to paracoccidioidin and the analogous blastogenesis assay. The reason for the meager in vitro lymphocyte responsiveness may be purely technical, for example, less than ideal lymphocyte isolation and culture conditions. Certainly there has not yet been a concerted effort to compare *P brasiliensis* antigenic fractions for their ability to evoke blastogenesis. Moreover, the combination of ficoll-Hypaque separation of peripheral blood lymphocytes, microtitration plate cultures, radiometric measurement of blastogenesis, and multiple automated sample harvesting (MASH) conditions are only now being applied to this problem. Technical considerations aside, there may be a genuine lack of antigen-responsive lymphocytes in the circulation, or as Musatti et al (1976) have described, a humoral blastogenic inhibitory factor may be operating. In seven patients they demonstrated such an inhibitor by the enhancement of blastogenic indices to paracoccidioidin when lymphocytes were cultured in homologous instead of autologous plasma and, on the other hand, phytohemagglutinin (PHA) responses of lymphocytes from normal persons were inhibited when cultured in plasma containing inhibitory factor. When clinical improvement occurred, blastogenic inhibitory factor decreased or disappeared, suggesting that it was related to antigen overload (Musatti, 1975). In addition to the reduction of cutaneous responses to PHA, blastogenic indices evoked by this mitogen as a functional test for polyclonal activation of T-cells were diminished in some symptomatic patients (Mok and Greer, 1977, Musatti et al, 1976).

6.4.2b Leukocyte Inhibitory Factor

Cutaneous reactions to *P brasiliensis* correlated better with leukocyte inhibitory factor (LIF) than with blastogenesis assays (Musatti et al, 1976). Mok and Greer (1977) found that eight patients displayed > 20% inhibition in the LIF assay and nine did not; but there was no correlation between LIF and delayed cutaneous hypersensitivity in their study.

6.4.2c E-Rosettes

The percentage of E-rosetting lymphocytes in nine patients was lower than in the lowest value found in normal persons (Musatti et al, 1976). This was corroborated by Restrepo et al (1978) who found the percent of E-rosettes was depressed to 44% in these patients as compared with the 53% mean value in normal controls.

6.4.2d Gammopathy

An inverse relation was observed between the presence of precipitins and positive blastogenic indices to *P brasiliensis* antigens, analogous to that investigated in more detail in coccidioidomycosis (Restrepo et al, 1978, also see Chapter 3). Polyclonal gammopathy occurred in 16 patients and the total IgG concentration rose slightly during six months of treatment to a mean value of 1.69%, wt/volume, compared to the mean in normal control subjects of 1.05 percent.

6.4.2e Killing by Granulocytes

Freshly diagnosed paracoccidioidomycosis patients not as yet receiving antifungal therapy, were tested for the ability of their polymorphonuclear neutrophils (PMN) to ingest and kill *P brasiliensis* yeast forms (Goihman-Yahr et al, 1980). The phagocytic indices of normal and patients' PMN were not significantly different. The mean percent decrease in killing of phagocytosed yeast forms by PMN from patients was 32% lower than killing by PMN from healthy control subjects. When the effector:target ratio was decreased from 1:1 to 1:5, there was an increase in the number of "ghost" (ie, killed) yeast forms, but the disparity between normal PMN and those from paracoccidioidomycosis patients remained constant.

Further delineation of the defective killing by patients PMN can be anticipated. For instance, any effect exerted by antibody-complement would be interesting to measure. Phagocytosis in the experiments cited above occurred in autologous or pooled normal human serum and consequently the concentrations of specific antibodies or of blastogenic inhibitory factor were uncontrolled variables. The potential of lymphocytes to act as either natural killer cells, as K-cells involved in antibody-dependent cell-mediated cytotoxicity, or as producers of lymphokines that influence PMN, requires a more rigorous sorting out of cell populations in further studies. The fungicidal ability of PMN from paracoccidioidomycosis patients should be monitored during therapy to see if any observed defect in this parameter is reversible. When specific antifungal therapy is absent because patients delay seeking treatment or because they cannot afford it, their immune systems may not be up to the task of a successful defense. In that event, the disease may progress in the presence of increasing complement-fixing antibodies, cutaneous anergy, and defective killing of yeast forms by PMN.

6.4.2f Immunotherapy

A single injection of transfer factor obtained from human spleen was attempted as immunotherapy in five anergic patients with paracoccidioidomycosis (Musatti et al, 1976). Only one patient showed any clinical improvement, accompanied by conversion to positive delayed cutaneous hypersensitivity. Clearly, studies in coccidioidomycosis have shown that transfer factor therapy requires multiple dosages (see Chapter 4). Levamisole treatment 2 anergic paracoccidioidomycosis patients resulted in recovery of cutaneous hypersensitivity in both and clinical improvement in one case.

The impression gained from these studies is that among patients in better clinical condition there is a positive correlation between delayed cutaneous hypersensitivity to *P brasiliensis* and to other antigens, normal PHA responses, and absence of humoral blastogenic inhibitor. A second group, with anergy, with other depressed cell-mediated immune functions, and with circulating blastogenic inhibitor, had progressive disease and died.

6.5 SUMMARY

The cell wall of *P brasiliensis* yeast forms is thickened by the presence of a broad median layer of short fibrils of α-1,3-glucan that is reduced or absent in mycelial forms. A thin layer of antigenic galactomannan is peripheral, and the innermost mural layer consists of abundant chitin fibrils interwoven with a small amount of β-1,3-glucan. Mutants deficient in α-1,3-glucan are attenuated in virulence and are more readily killed by granulocytes. The *P brasiliensis* galactomannan has a 1,6-linear mannan backbone extensively substituted with single galactofuranosyl and mannosyl antennae. Crossreactions occur between this galactomannan and those of other primary systemic fungi but not with *A fumigatus* galactomannan. A secreted alkaline phosphatase, factor E or 1 evokes precipitins in paracoccidioidomycosis that persist even after remission is achieved. Precipitins and total serum IgG are inversely proportional to positive blastogenic responses to *P brasiliensis*. Paracoccidioidin crossreacts in skin tests of persons exposed to *H capsulatum*, a consideration in places where the endemic zones of the two fungi overlap. Cutaneous anergy is common in acute disease, and skin tests become positive as patients respond favorably to therapy. A broad state of immunosuppresion exists in paracoccidioidomycosis patients with respect to PPD responses. Blastogenic responses to *P brasiliensis* antigens, to PHA, and to contact sensitins are often weakened or flat in active disease. A factor present in the plasma of some patients inhibits blastogenic responses to *P brasiliensis* antigens. The total E-rosetting lymphocytes in the peripheral blood of paracoccidioidomycosis patients is reduced to the bottom of the normal range.

REFERENCES

Azuma I, Kanetsuna F, Tanaka Y, Yamamura Y, Carbonell LM, 1974. Chemical and immunological properties of galactomannans obtained from *Histoplasma duboisii, Histoplasma capsulatum, Paracoccidioides brasiliensis,* and *Blastomyces dermatitidis.* Mycopathol Mycol Appl 54:111–125.

Brummer E, Morizumi PA, Vo PT, Stevens DA, 1982. Protection against pulmonary blastomycosis: adoptive transfer with T lymphocytes but not serum from resistant mice. Cell Immunol 73:349–359.

Catanzaro A, Spitler L, 1976. Clinical and immunologic results of transfer factor therapy in coccidioidomycosis. pp 447–494. In Transfer Factor, Basic Properties and Clinical Applications. Ascher MS, Gottlieb AA, Kirkpatrick CH, eds. New York: Academic Press.

Giraldo R, Restrepo A, Gutiérrez F, Robledo M, Londoño F, Hernández H, Sierra F, Calle G, 1976. Pathogenesis of paracoccidioidomycosis: a model based on the study of 46 patients. Mycopathol 58:63–70.

Goihman-Yahr M, Essenfeld-Yahr E, de Albornoz MC, Yarzabal L, de Gomez MH, SanMartin B, Ocanto A, Gil F, Convit J, 1980. Defect of in vitro digestive ability of polymorphonuclear leukocytes in paracoccidioidomycosis. Infect Immun 38:557–566.

Gorin PAJ, Spencer JFT, 1968. Galactomannans of *Trichosporon fermentans* and other yeasts: proton magnetic resonance and chemical studies. Can J Chem 46:2299–2304.

Kanetsuna F, Carbonell LM, Azuma I, Yamamura Y, 1972. Biochemical studies on the thermal dimorphism of *Paracoccidioides brasiliensis.* J Bacteriol 110:208–218.

Kanetsuna F, Carbonell LM, Moreno RE, Rodriguez J, 1969. Cell wall composition of the yeast and mycelial forms of *Paracoccidioides brasiliensis.* J Bacteriol 97:1036–1041.

Mendes NF, 1975. Lymphocytes and lymph nodes in patients with paracoccidioidomycosis. pp 30–35. In Proc Intern Conf on the Mycoses, III, Sao Paulo. Scient Publ No 304. Washington, DC: Pan Amer Health Org.

Mendonça L, Gorin PAJ, Lloyd KO, Travassos LR, 1976. Polymorphism of *Sporothrix schenckii* surface polysaccharides as a function of morphological differentiation. Biochem USA 15:2423–2431.

Mok PN, Greer DL, 1977. Cell-mediated immune responses in patients with paracoccidioidomycosis. Clin Exp Immunol 28:89–98.

Musatti CC, 1975. Cell-mediated immunity in patients with paracoccidioidomycosis. pp 23–29. In Proc Intern Conf on the Mycoses, III, Sao Paulo. Scient Publ No 304. Washington, DC: Pan Amer Health Org.

Musatti CG, Rezkallah MT, Mendes E, Mendes NF, 1976. In vivo and in vitro cell mediated immunity in patients with paracoccidioidomycosis. Cell Immunol 24:356–378.

Restrepo M, Moncada LH, 1974. Characterization of the precipitin bands detected in the immunodiffusion test for paracoccidioidomycosis. Appl Microbiol 28:138–144.

Restrepo A, Restrepo M, de Restrepo F, Aristizábal LH, Moncada LH, Vélez H, 1978. Immune responses in paracoccidioidomycosis. A controlled study of 16 patients before and after treatment. Sabouraudia 16:151–163.

Restrepo-Moreno A, Schneidau JD Jr, 1967. Nature of skin-reactive principle in culture filtrates prepared from *Paracoccidioides brasiliensis*. J Bacteriol 93:1741–1748.

San Blas G, San Blas F, 1977. *Paracoccidioides brasiliensis* cell wall structure and virulence. Mycopathol 62:77–86.

San Blas G, San Blas F, 1982. Variability of cell wall composition in *Paracoccidioides brasiliensis:* a study of two strains. Sabouraudia 20:31–40.

Yarzabal LA, Andrieu S, Bout D, Naquira F, 1976. Isolation of specific antigen with alkaline phosphatase activity from soluble extracts of *Paracoccidioides brasiliensis.* Sabouraudia 14:275–280.

Chapter 7

Sporothrix schenckii

7.1. INTRODUCTION

Sporothrix schenckii is a dimorphic, primarily subcutaneous pathogen that can at times disseminate to the deep tissues. *S schenckii* is worldwide in distribution occurring in nature as a soil saprophyte that is associated with woody plants. Most infections result from traumatic implantation, but pulmonary disease can arise from inhalation of conidia. At room temperature, off-white mold form colonies develop that become pigmented with age, turning yellow, brown, or black depending on the isolate. Simple, oval conidia ($2-3 \times 3-6\ \mu m$) are borne on elongate, tapering conidiophores. The cluster of conidia resembles a palm tree or flower. Sporulation intensifies in older cultures and conidia are produced directly from small sterigmata arrayed along the septate branching hyphae. Some isolates of *S schenckii* are morphologically similar to *Ceratocystis* species, especially *C stenoceras*. *Sporothrix schenckii* undergoes temperature-sensitive morphogenesis producing a yeast form at 37 °C in culture and in the infected host. Round, oval, or cigar-shaped cells are formed at the restrictive temperature that vary in size from $1-3 \times 3-10\ \mu m$, sometimes with several buds appearing simultaneously on one mother cell (McGinnis, 1980; Chandler et al, 1980; Rippon, 1982).

The presence of ulcerative lesions on an extremity with or without a linear chain of nodules along the lymphatic vessels requires consideration of sporotrichosis in the differential diagnosis. The fungus can be identified in biopsied tissue by culture and direct immunofluorescent or immunoperoxidase staining techniques (Russell et al, 1979). Because of the ready availability of tissue for examination, antibody levels are less important in the diagnosis, but serology can be useful because typically few fungal cells are observed in dermal tissue sections (Chandler et al, 1980). The humoral response in cutaneous-lymphatic sporotrichosis is relatively weak in the sense that precipitins and complement-fixing antibodies may not be detected in almost half (42% to 47%) of the cases (Blumer et al, 1973). Persons with marked lymphadenitis and suppurating lesions are more serologically active than those with epidermal plaque-like lesions (Karlin and Nielsen Jr, 1970).

Tests for antibodies are diagnostically important in the less frequent and more problematic deep tissue *S schenckii* infections, because the patterns of organ involvement mimic other diseases and tissue or exudates for direct examination are more difficult to obtain. Disseminated infection may take the form of widespread hard, moveable, subcutaneous nodules. Other forms of the disease where skin lesions or nodules may not be found are pulmonary, osseous, and articular. Ocular or meningeal infections rarely occur (Wilson et al, 1967). Extracutaneous sporotrichosis must not only be differentiated from other systemic mycoses, but also from syphilis, tularemia, glanders, tuberculosis, and amoebic abscesses (Blumer et al, 1973).

7.2 CELL WALL

Cell walls of yeast and mycelial forms show only minor differences in their glycosidic components (Previato et al, 1979a) (Table 7.1). Levels of protein and lipid in the mycelial form walls are higher than those reported for other mycotic pathogens. Amino acids known to participate in glycopeptide linkages (serine, threonine, alanine, aspartic, and glutamic acids) are enriched in the mycelial walls. It would be interesting to determine what proportion of these proteins are covalent structural wall components. The conditions necessary to extract wall-associated proteins (ie: detergents, proteinases or extremes of pH) have not been determined for *S schenckii*. The lipid content, often used as a criterion for wall purity, is also high in the *S schenckii* walls of both dimorphic forms. Cytoplasmic membranes may still be adherent to the walls since no detergents were used in their preparation.

Alkali-soluble and insoluble glucans were characterized from *S schenckii* cell walls (Previato et al, 1979b). Only quantitative differences were found between the two solubility types. These glucans are β-linked containing 1,3-, 1,4-, and 1,6-linked glucose residues in molar ratios of 1 : 1.5 : 1, respectively, for the soluble β-glucans of yeast form walls, and 6 : 13 : 1 for the insoluble glucans. The configuration of linkage was established by polarimetry and susceptibility of the glucans to endo β-1,3- and β-1,6-glucanases of *Bacillus circulans* WL-12.

The insoluble fraction of mycelial and yeast form walls are similar, containing a glucan : chitin ratio of approximately 3 : 1. There are no obvious differences in the amount of soluble glucan between the dimorphic forms so there is no evidence that glucan influences cell shape in *S schenckii*. The mixture of the three glycosidic linkages in cell wall glucans is rare and it would be useful to know if these linkages reside on the same or different molecules. The insoluble *S schenckii* glucan injected into mice at a dose of $10 \text{ mg} \cdot \text{kg}^{-1} \cdot \text{day}^{-1}$ caused regression of sarcoma 180 (Previato et al, 1979b). The fibrillar wall matrix thus has some immunomodulating activity.

7.2.1 Peptido-L-rhamno-D-mannan

Peptidorhamnomannan complexes occur as the readily soluble surface coat of *S schenckii*. Different L-rhamnose : mannose ratios are found depending on the dimorphic state of the fungus. Rhamnomannans are isolated from the medium after growth or are extracted from whole cells or cell walls. A rhamnose-containing mannan from *S schenckii* was selectively precipitated from a hot buffer extract of yeast forms with borate-cetyltrimethylammonium bromide (Lloyd and Bitoon, 1971). It could also be prepared from an alkali extract and precipitated as the Cu^{++}

Table 7.1. Composition of *Sporothrix schenckii* Cell Walls[a]

	Percent of Wall Dry Weight	
Component	Yeast Form	Mycelial Form
Glucose	29.3	24.2
Mannose	21.9	12.7
Galactose	trace	trace
L-Rhamnose	9.6	7.1
Glucosamine	7.0	7.0
Protein	14.4	21.7
Lipid	18.0	26.0
PO$_4$	0.7	1.2
Recovery	100.9	99.9

Source: Reprinted from Reiss (1985) with permission of the publisher, Marcel Dekker, Inc.
[a]Isolate 1099.18, data calculated from Previato et al, 1979.

complex with Fehling solution, in which case a coproduced galactomannan remained soluble. The peptidorhamnomannan was bound to DEAE anion exchanger because of a large proportion of glutamic and aspartic acids in the peptido-moiety. Quantitative analysis of the column-purified heteroglycan showed these constituents as a percent of the dry weight: L-rhamnose, 33.5%; mannose, 50%; protein, 18.2%. The L-rhamnose:mannose ratio is thus 1:1.6. Gel permeation on Sepharose 4B indicated a polydisperse molecular weight range of between 50,000 and 150,000 da.

Methylation-fragmentation, gas-liquid chromatography–mass spectrometry (GC-MS) showed that all the rhamnose existed in terminal positions, whereas only 20% of the mannose units were end groups. These data point to a linear α-1,6-mannan backbone with single rhamnose antennae. The type of rhamnomannan produced, whether mono- or dirhamnosyl, depends on the morphology of the culture (Figure 7.1). Yeast forms produce only monorhamnosyl mannan. This is not solely a temperature restriction because on a defined medium Mendonça et al (1976) induced a yeast form growth at either 25 or 37 °C. In cultures that contain both filaments and conidia, dirhamnosyl mannan is found. The polysaccharide isolated from a strain of *S schenckii* that grew as unsporulated mycelia contains no rhamnose, instead only a galactomannan. The major structural features of the galactomannan are 6-*O*- and 2,6-di-*O*-substituted α-mannosyl units with terminal β-galactofuranose residues accounting for 7% to 10% of the carbohydrate (Mendonca et al, 1976).

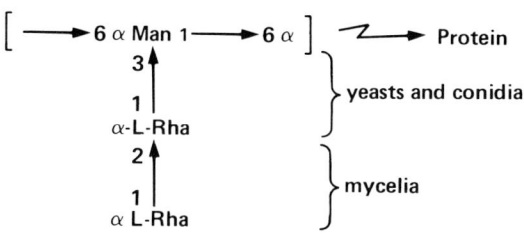

Figure 7.1. Proposed repeat unit structure of peptido-L-rhamno-D-mannan of *Sporothrix schenckii*. Man, mannose; Rha, rhamnose. All sugars in the pyranose form. Monorhamnosyl mannans were synthesized by yeast forms irrespective of the incubation temperature (Travassos and Lloyd, 1980) *Source:* reprinted with permission of the authors and the publisher, American Chemical Society.

The dirhamnosyl mannan of the mycelial forms is thus synthesized mainly by conidia or by conidia-producing mycelia. Mono- and dirhamnosyl mannans can also be differentiated on the basis of their carbon-13 nuclear magnetic resonance, ^{13}C-NMR spectra. Complete assignment of the signals has been accomplished for the monorhamnosylmannan (Table 7.2, Figure 7.2).

Concanavalin A binds to rhamnomannan indicating that terminal mannose is α-linked. Both rhamnose and mannose are present in the Con A precipitate, giving evidence of covalent linkages between the two sugars. Mannan from unsporulated mycelia is separated to some extent from rhamnomannan due to their different affinity for Con A. The serologic specificity of the two variant rhamnomannans of *Sporothrix* was tested using rhamnose oligosaccharides to inhibit immune precipitation (Lloyd and Travassos, 1975). These fragments were prepared by partial acetolysis of the rhamnomannan. The precipitation of dirhamnosylmannan by its homologous rabbit antiserum was best inhibited by the trisaccharide α-L-rhamnose $1 \rightarrow 2$-α-L-rhamnose $1 \rightarrow 3$-mannose, and L-rhamnose itself inhibited only about 1/1000 as well as the trisaccharide. Otherwise, L-rhamnose is a very powerful inhibitor of the reaction between monorhamnosylmannan and its homologous antiserum. Of the oligosaccharides tested, α-L-rhamnose $1 \rightarrow 3$-mannose was a slightly better inhibitor than the trisaccharide. The immunodiffusion reaction between dirhamnosylmannan and its antiserum was strong and showed a line of partial identity that spurred over the line of the reaction with monorhamnosylmannan.

Two kinds of microheterogeneity are known in the peptidorhamnomannan from yeast forms fractionated with cetyltrimethylammonium bromide (Lloyd and Bitoon, 1971). One component precipitated directly and a second type, with a higher mannose content, required borate complexation to achieve the negative charge necessary to precipitate. Another variant in the peptidorhamnomannan of the mycelial form was a 2,4-di-O-substituted mannose detected by methylation-fragmentation GC-MS. This variation is attributed to a minor structural feature: α-1,4-linear man-

Table 7.2. Complete Assignments of Signals in the ^{13}C NMR Spectrum of *S schenckii* Strain 1099.18 Rhamnomannan Obtained at 30 °C[a]

Signal, δ_c (70 °C) (ppm)[b]	Assignment
101.1	C-1 of 3,6-di-O-substituted α-D-mannopyranose units
98.3	C-1 of α-L-rhamnopyranose nonreducing end units
76.6	C-3 of 3,6-di-O-substituted α-D-mannopyranose units
73.6	C-4 of α-L-rhamnopyranose nonreducing end units
72.4	C-5 of 3,6-di-O-substituted α-D-mannopyranose units
72.0–71.9	C-2 and C3 of α-L-rhamnopyranose nonreducing end units
70.4	C-5 of α-L-rhamnopyranose nonreducing end units
67.6	C-2 of 3,6-di-O-substituted α-D-mannopyranose units
67.3	C-3 of 3,6-di-O-substituted α-D-mannopyranose units
66.3	C-4 of 3,6-di-O-substituted α-D-mannopyranose units
62.8 (trace)[c]	C-6 of O-6 unsubstituted α-D-mannopyranose units
18.4	CH$_3$ of α-L-rhamnopyranose units

Source: Reprinted from Travassos and Lloyd, 1980, with permission of the authors and of the publisher, American Society for Microbiology.

[a] See Figure 7.2.
[b] Signals in the δ_c 66.3 to 76.6 region have their values corrected for 70 °C (+0.6 ppm), as they were originally obtained at 33 °C.
[c] This rhamnomannan contains only trace amounts of 4-O-substituted α-D-mannopyranose units.

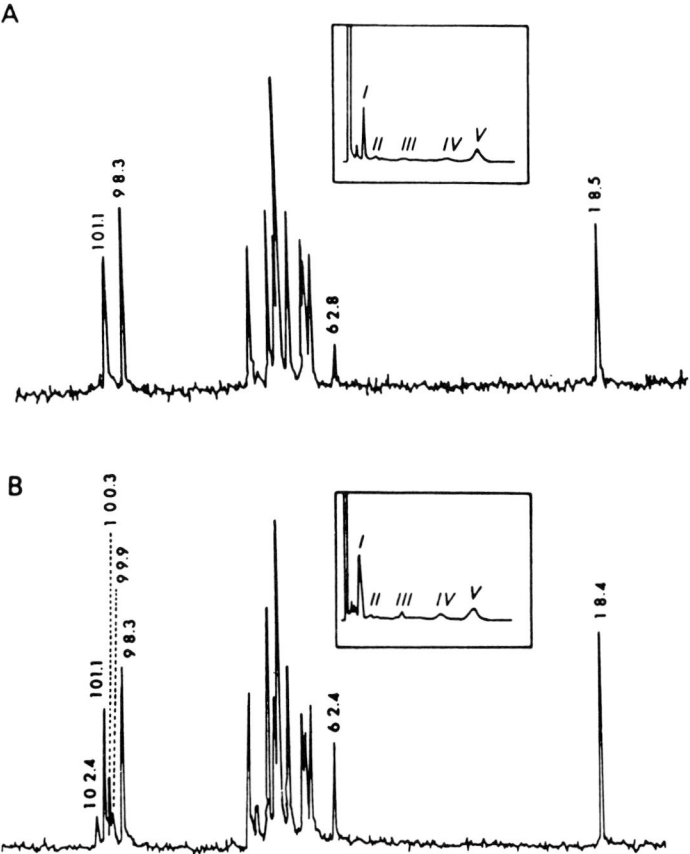

Figure 7.2. Partial ^{13}C-nuclear magnetic resonance spectra and gas-liquid chromatography (*insets*) of partially methylated alditol acetates from methylation analysis of cell-bound polysaccharides from *Sporothrix schenckii* (strain 1099.18) growing in the yeast form at 37 °C (A) or at 25 °C (B), yeast forms also occur at this temperature with this isolate. A complete assignment of signals in *S schenckii* rhamnomannans is shown in Table 7.2. Alditol acetates: I, 2,3,4-tri-*O*-methylrhamnitol; II, 3,4-di-*O*-methylrhamnitol; III, 2,3,6-tri-*O*-methylmannitol; IV, 3,6-di-*O*-methylmannitol; V, 2,4-di-*O*-methylmannitol. *Source:* Reprinted from Biochemistry 15:2423–2431, with permission of the publisher. Copyright 1976, American Chemical Society.

nosyl sequences substituted at C-2 with α-L-rhamnose-1 → 4-α-man. This component is absent from yeast forms (Gorin et al, 1977; Travassos and Lloyd, 1980).

The peptido portion of the rhamnomannan is important structurally, and as the presumed source of determinants capable of evoking cell-mediated immunity. Elimination of the peptide is accomplished by alkaline cleavage (2% KOH, 100 °C, 2 hours). The L-rhamnose:mannose ratios in the rhamnomannan after β-elimination are 1:1 to 1:1.2 and the nitrogen content is less than 1% (Travassos et al, 1973).

An alternative approach to producing polysaccharides from mycelial mats of *S schenckii* is extraction with phenol/water in the Westphal procedure (Shimonaka et al, 1975). A polysaccharide-protein complex is extracted into the aqueous phase consisting of 87.1% carbohydrate and 12.5% protein. The carbohydrate component consists of L-rhamnose:mannose:galactose in ratios of 2.1:5.6:1.0. Based on the work of others, reviewed by Travassos and Lloyd (1980) this composition most likely reflects a mixture of peptidorhamnomannan and galactomannan. The latter polysaccharide is crossreactive with the galactomannan from *Aspergillus fumigatus* and other fungi (Travassos and Lloyd, 1980). The complex was chromatographed on Sephadex G100, and material eluting in the void volume was pooled. It is not known if all of the protein component present in this fraction is covalently linked to carbohydrate or is a mixture of other cellular proteins.

Digestion of the peptidorhamnomannan with papain caused a 50% reduction in the nitrogen content (Shimonaka et al, 1975). The entire polysaccharide component was labile to periodate oxidation, consistent with major 1,6- and 1,2-glycosidic bonds. Precipitin tests showed that the serologic activity of the complex was unaffected by papain but abolished by periodate oxidation. A similar result was obtained using passive cutaneous anaphylaxis. The carbohydrate component was essential to the ability to trigger degranulation of mast cells. Carbohydrate is either the epitope recognized by IgE or serves as a carrier for protein determinants unaffected by papain digestion.

Three sera of human sporotrichosis cases were compared in the quantitative precipitin test: one was precipitated to a much greater extend by dirhamnosylmannan, one was slightly more reactive and one was about equally reactive with either variant (Lloyd and Travassos, 1975). Of the small sample of human sporotrichosis cases tested for precipitins, all reacted with rhamnomannan. Concerning crossreactivity, human anti-*Cladosporium (Exophiala) werneckii*, anti-*Trichophyton mentagrophytes*, and anti-*Blastomyces dermatitidis* sera did not react with rhamnomannan.

Results of direct migration inhibitory factor and delayed cutaneous hypersensitivity tests in guinea pigs sensitized by injection of the peptidorhamnomannan in complete Freund adjuvant agree that papain digestion drastically reduces the ability of the antigen to evoke cell-mediated immunity, whereas periodate oxidation has only a slight adverse effect on this activity.

The surface location of peptidorhamnomannan and its ability to evoke specific precipitin and IgE responses are good evidence that it is the major antigen responsible for specific humoral immunity to *S schenckii*. On the basis of a much smaller amount of data, it is possible to propose that the peptido-portion of the rhamnomannan confers a dual ability to evoke T-cell-mediated reactions. Further such studies with the purified peptidorhamnomannan are indicated.

Sporothrix schenckii coexists in its natural habitat of plants with *C stenoceras*, and the latter fungus was proposed as the perfect ascigerous state of *S schenckii* (Mariat, 1971). *Ceratocystis stenoceras* also produces a rhamnomannan in both neutral and acidic form. The presence of glucuronic acid was first detected by a ^{13}C-NMR signal corresponding to COOH. Later, a disaccharide, GlcUA – L-rhamnose was isolated. The glucuronic acid was not an end group, but occurred as an internal component of the side chain. It was 4-*O*-substituted, as shown by the derivative 2,3-di-*O*-methyl GlcUA (Figure 7.3). The neutral rhamnomannan of *S schenckii* and *C stenoceras* could not be differentiated on the basis of methylation analysis, but the ^{13}C-NMR patterns of the two mannans are different (Travassos et al, 1974). This evidence does not support the idea of their being identical fungi.

7.3 DELAYED CUTANEOUS HYPERSENSITIVITY

7.3.1 Sporotrichin and Alternatives

Most surveys of delayed cutaneous hypersensitivity to *S schenckii* have relied on sporotrichin, a culture supernatant of yeast forms grown for several weeks on a complex medium (Schneidau et al, 1964; Steele et al, 1976). More recently, ethanol-

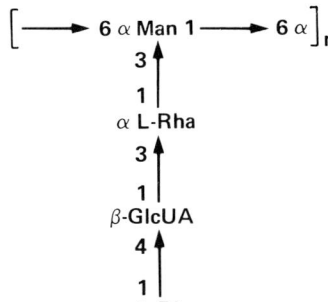

Figure 7.3. Proposed repeat unit structure of the glucuronorhamnomannan of *Ceratocystis stenoceras*. The exopolysaccharides were separated by anion exchange chromatography yielding a rhamnomannan and, at a lower pH, a polysaccharide having the structure shown above was eluted. No proof exists of glucuronosyl residues in *Sporothrix schenckii* rhamnomannans (Travassos and Lloyd, 1980).

precipitated culture supernates were used (Murillo de Linares et al, 1980). These tests have established the incidence of positive delayed cutaneous hypersensitivity in the normal population and in confirmed cases from various geographic areas. Efforts have been made to survey subcellular fractions of *S schenckii* yeast forms to refine skin test reagents (Nielsen, 1968). No subcellular fraction was found that had sensitivity equivalent to intact yeasts or yeast cell walls. The most potent subcellular antigen, an ethanol precipitate of cytoplasmic solubles, detected delayed cutaneous hypersensitivity in 12 of 20 (60%) of sensitized guinea pigs, but in only one of five sporotrichosis patients. Although cell walls evoked potent and specific delayed cutaneous hypersensitivity in sensitized guinea pigs and in all five human cases, particulates are unsatisfactory for skin test antigens because they cannot readily be sterilized and, as foreign bodies, are capable of evoking toxic reactions. Further efforts are expected to result in concentrated subcellular fractions with dosages expressed in terms of dry weight or protein content.

Estimates of the in vitro correlates of delayed cutaneous hypersensitivity should take into account the progress in peptidorhamnomannan purification and in immunoenzyme analysis (Walbaum et al, 1978), so that homogenates of yeast forms (Steele et al, 1976) will probably be replaced by more well-characterized antigens. The likelihood that peptidorhamnomannan is a potent and specific stimulant of both humoral and cell-mediated immunity in human sporotrichosis does not exclude the possibility that other factors, proteins, in *S schenckii* yeast forms may also stimulate immune responses in vivo. Walbaum et al (1978) have shown the antigenic potential of supernates of homogenized *S schenckii* yeast forms in rabbits immunized with the antigen in incomplete Freund adjuvant. After 8 weeks of immunization, 22 precipitins were demonstrated in immunoelectrophoresis. Enzyme activities were localized in some of the precipitin arcs: β-glucosidase, acid phosphatase, and esterase. The spectrum of antigenicity of these proteins in humans is not known. Antibodies to protein antigens may be more indicative of deep-seated infection. Removal of the peptidorhamnomannan from protein antigens by Con A lectin affinity chromatography or by immunoelectrophoresis will be necessary to assess the role of other antigens.

Sporothrix schenckii can grow in the yeast form on chemically defined or dialyzate media (Travassos and Lloyd, 1980; Nielsen, 1968) so that future studies should avoid using potentially crossreactive substances like baker's yeast extract or brain–heart infusion in media designed for use in skin tests. Newer presumptive skin test fractions should be compared with sporotrichin. Methods of fractionation of material intended for skin tests should minimize denaturation by avoiding extremes of pH or heat. As an example, in one study (Nielsen, 1968) *S schenckii* culture supernates were heated at 90 °C. Such treatment is likely to cause irreversible denaturation of proteins, and in any case should not be an uncontrolled variable.

7.3.2 Skin Test Surveys

In contrast to other mycoses where cutaneous anergy is encountered in a significant number of cases, the cumulative experience in sporotrichosis is that delayed cutaneous hypersensitivity usually is present. Wilson et al (1967) collated data from 139 cases of sporotrichosis, mostly of the cutaneous-lymphatic type. Of these, 138 reacted to whole killed S schenckii yeast forms in skin tests. The durability of delayed cutaneous hypersensitivity to sporotrichin is exemplified in the strong positive responses evoked in the 65-year follow up of the second reported case (McFarland, 1966).

The incidence of positive delayed cutaneous hypersensitivity in a series of hospitalized persons and prison inmates in Louisiana was 11.2% (Schneidau et al, 1964). Another survey conducted in Arizona of 203 hospital patients with no history of sporotrichosis showed an incidence of 10.3% reactors to sporotrichin. These reactions were considered specific and not due to prior sensitization with H capsulatum (Ingrish and Schneidau, 1967). Plant nursery workers in Louisiana were a more highly reactive population to the extent of 32.3% of those tested. This finding is compatible with the concept of sporotrichosis as an occupational hazard of those who work with thorny plants. In a subgroup of 22 persons who were highly reactive to sporotrichin, 45% also reacted to histoplasmin. This was interpreted as crossreactions of persons sensitized to S schenckii because reactions to sporotrichin among histoplasmosis patients do not exceed that of those with no history of fungal disease.

A survey in El Salvador (Murillo de Linares et al, 1980) compared skin test responses between a group of 22 patients with cutaneous-lymphatic sporotrichosis and 1022 persons, the majority of whom had chronic lung disease, including tuberculosis. All sporotrichosis patients reacted to ethanol-precipitated sporotrichin and 75% had reactions at 48 hours that exceeded 10 mm diameter. Only four of the control persons reacted positively. El Salvador is not a region of high endemicity. Persons with chronic lung disease may not provide the optimal control group because of the possibility of immunosuppression.

The cutaneous and blastogenic responses to S schenckii were compared in 143 young healthy persons in Texas and one sporotrichosis patient (Steele et al, 1976). This study is noteworthy for attempting to delineate the responses on the cellular level, and for comparing antigens of S schenckii side-by-side with oidiomycin, the Hollister-Stier Company Candida albicans extract. Only 3% of the healthy adults had positive delayed cutaneous hypersensitivity to sporotrichin, but 10% had blastogenic indices greater than three evoked by a whole homogenate of S schenckii yeast forms. Of those 14 whose lymphocytes were transformed to blasts by S schenckii antigens, eleven were women. There was a high correlation coefficient (0.89) between the blastogenic index to oidiomycin and to S schenckii. This was interpreted in a provocative way meriting further investigation. The authors proposed that women, who may have a higher antigenic stimulus to C albicans because of its frequent carriage on the vaginal mucosae, are crossprotected against S schenckii, and that evidence for this state of immunity is observed in the higher incidence of blastogenic responses to S schenckii in normal women. In this view, the sex-related difference in sporotrichosis would be a product of heightened natural immunity in women and increased opportunity for infection through occupational exposure to thorny plants in men.

7.4 IMMUNODIAGNOSIS

7.4.1 Agglutinins and Precipitins

Considering the difficulty in diagnosing extracutaneous sporotrichosis, it is fortunate that simple serologic tests with unpurified *S schenckii* antigens are highly sensitive and specific. The tube agglutinin test with heat killed (65 °C, 2 hours) yeast forms was developed first by de Beurmann et al (1908) and is still highly effective (Roberts and Larsh, 1971a; Karlin and Nielsen Jr, 1970; Blumer et al, 1973). Agglutinin titers for *S schenckii* yeast forms were positive in all eight articular, pulmonary and disseminated cases in one series (Karlin and Nielsen Jr, 1970). Twenty of 25 cases in another series with pulmonary sporotrichosis had agglutinins; 17 of these reactive patients had titers higher than 1/40.

Slight agglutination of *S schenckii* yeast forms was observed at a 1/20 dilution of serum from 22 to 30 normal human donors (Welsh and Dolan, 1973). This result suggested a high incidence of nonspecific reactions, but may be artifactual because these workers centrifuged the yeast-serum suspension after overnight incubation. Experience in the Immunology Branch of the Division of Mycotic Diseases, CDC, is that such centrifugation obscures interpretation of borderline reactions. Most patients with the cutaneous-lymphatic form of sporotrichosis have 1/160 to 1/320 agglutinin titers (Karlin and Nielsen Jr, 1970). Sera from 23 of 24 cases of extracutaneous sporotrichosis were positive for agglutinins but the titers were not enumerated (Roberts and Larsh, 1971a). Sera from 1000 persons with no known fungus disease failed to react in the tube agglutinin test with *S schenckii* yeast forms (Roberts and Larsh, 1971a). The only crossreactions detected in this test, which used a panel of sera representing a variety of infections, were two of ten patients with leishmaniasis (Blumer et al, 1973). Thus a simple agglutinin test with heat-killed *S schenckii* yeast forms appears able to detect 80% to 96% of extracutaneous sporotrichosis cases while reactions in normal persons or those with other systemic infections are minimal.

Positive complement-fixation tests with culture filtrates of *S schenckii* yeast forms as the antigen were once held to differentiate cutaneous from extracutaneous forms of the disease (Jones et al, 1969) and complement-fixing antibody levels were believed to increase in proportion to disease severity. Cutaneous sporotrichosis complement-fixation reactions were positive in less than half of the cases (Blumer et al, 1973), but in the same study only 76% of extracutaneous cases had positive complement-fixing antibodies. No correlation was observed between the type of tissues involved and the height of the titer. In another series of eight extracutaneous cases (Karlin and Nielsen Jr, 1970), only four had evidence of complement-fixing antibodies. There does not seem to be a justification to use complement-fixation tests in this disease because the test lacks sensitivity and titers do not correlate with severity of the disease.

Latex agglutination tests. Further simplification of sporotrichosis serology without reducing sensitivity was achieved with the development of a microscope slide latex agglutination test (Blumer et al, 1973). Polystyrene particles were coated with a heat-treated (65 °C, 2.5 hours) filtrate of *S schenckii* yeast forms grown in brain–heart infusion broth. The latex agglutination test detected antibodies in 51 of 55 cases (92.7%) of cutaneous and in 24 of 25 cases (96%) of extracutaneous sporotrichosis. This test was not subject to false-positive reactions in sera from persons with various mycotic, bacterial, or parasitic diseases and in a small sample of normal persons.

The antigens used for immunodiffusion tests are the culture supernate of yeast forms (Blumer et al, 1973; Jones et al, 1969; Karlin and Nielsen Jr, 1970) and the supernate of sonified yeast forms (Roberts and Larsh, 1971a). The incidence of precipitins in cutaneous-lymphatic sporotrichosis ranged from a low 11% of cases (Roberts and Larsh, 1971a) to 80% (Karlin and Nielsen Jr, 1970). In the largest series (Blumer et al, 1973) of 55 patients, 42% were precipitin-positive. More patients (80% to 84%) have precipitins in extracutaneous disease (Karlin and Nielsen Jr, 1970; Roberts and Larsh, 1971a; Blumer et al, 1973). One report is at variance (McMillen, 1974); in all 15 extracutaneous cases that were studied, precipitins and complement-fixing antibodies were detected. No procedural details were given to account for this improved detection, but closeness of monitoring of patients may increase the chances for antibody detection. Precipitins do not necessarily accompany cutaneous sporotrichosis, so their absence does not conflict with the diagnosis. In extracutaneous disease, precipitins are more frequently encountered but are not as universal as agglutinins.

Successful treatment of cutaneous-lymphatic sporotrichosis results in a decline of antibodies to subclinical levels (McMillen, 1974). In chronic extracutaneous disease of the articular and osseous type, the fungus may still be cultured after 15 years duration (Karlin and Nielsen Jr, 1970) and the agglutinin titer fluctuates up to 1/640 depending on the treatment. Even after up to 7 years of clinical and cultural cure positive complement-fixation and precipitin reactions have been observed in extracutaneous sporotrichosis (McMillen, 1974).

7.5 INTERACTION WITH PHAGOCYTES

Human peripheral polymorphonuclear neutrophils are capable of efficient in vitro phagocytosis and killing of *S schenckii* yeast forms (Cunningham et al, 1979). Phagocytosis occurred to the extent of 50% at an effector:target ratio of 2:1 in 10% normal human serum. The effect of *S schenckii*-immune serum on phagocytosis was not studied. Killing of yeast forms in the presence of leukocytes was rapid, 80% in 1 hour. The possibility of lysis of *S schenckii* by T-lymphocytes or null cells was not excluded because the peripheral blood leukocyte population was not fractionated. Components of the myeloperoxidase–H_2O_2–halide microbicidal mechanism were tested singly and in combination against *S schenckii* yeast forms. Effective killing occurred only in the presence of enzyme, peroxide, and KI; chloride ions could not substitute for iodide, in contrast to experience with *C albicans* (Lehrer, 1969). This provides a potential explanation for the specific efficacy of KI therapy in cutaneous sporotrichosis. The role of human serum in phagocytosis is presumably opsonization of yeast forms through direct activation of the alternative complement pathway.

7.5.1 Asteroid Bodies

Asteroid bodies surrounding *S schenckii* are observed in tissues from humans and animals illustrating the Splendore-Hoeppli phenomenon (Chandler et al, 1980) a characteristic of the host's humoral response in sporotrichosis and schistosomiasis (Chandler et al, 1980; Lurie and Still, 1969). Stellate zones, up to 100 μm diameter of eosinophilic material are deposited around yeast or round forms in tissue. These deposits may be composed of immune complexes (Lurie and Still, 1969). When asteroid bodies are encountered in cutaneous and pulmonary sporotrichosis, they usually occur in the centers of granulomas or abscesses. The ultrastructural development of asteroid bodies containing *S schenckii* yeast forms was studied in tissue sections from experimentally infected hamsters (Lurie and Still, 1969). At 10 days postinfection, yeast form cell walls had an inner lucent layer distinct from a thick granular outer layer. Sections from older lesions showed broadening of the outer layer and host-induced changes in the form of irregular deposits of amorphous material on the mural outer edge. Mature asteroid bodies had a swollen mural outer layer fused to the abundant external raylike projections. Yeast forms grown in vitro and resuspended in rabbit anti-*S schenckii* serum, but not normal serum, also became coated with a thick extramural layer. Further definition of the deposits formed in vivo and in vitro are needed. Fluorescent anti-Ig and anti-C3 reagents would help to corroborate their nature as immune complexes. The viability and rate of phagocytosis of such coated yeast forms and asteroid bodies is not known. The significance of asteroid body formation and its relation to the immune status of the host or to immunopathology remains to be investigated.

7.6 EXPERIMENTAL INFECTIONS AND IMMUNITY

Estimating the relative contribution of the humoral and cell-mediated effector limbs towards protection against *S schenckii* would be premature. Two studies, one in athymic nu/nu mice (Shiriashi et al, 1979) and one in normal hamsters (Charoenvit and Taylor, 1979) have implicated both functions. Thymus-bearing heterozygote nu/+ mice cleared an intravenous challenge dose of 10^6 yeast forms rapidly but spleen, liver, and kidney counts rose in the 3 week period after infection in nu/nu mice. No paradoxical resistance was observed in athymic mice towards *S schenckii* as has been reported for *C albicans* (see section 11.6). Resistance to *S schenckii* seems to require intact thymic function, as is the case with other primary systemic fungi. The reaction of hamsters to footpad inoculation of *S schenckii* yeast forms was dose-dependent. After 5000 yeast forms were injected, the result was inflammation confined to one limb. Draining abscesses formed and resolved in 3 weeks, providing a parallel to the localized cutaneous-lymphatic disease of humans. When the challenge dose contained 5×10^6 yeast forms, dissemination occurred to the liver but not the spleen. This too was a self-limited process resolving in 4 weeks. Hamsters immunized with cell walls in complete Freund adjuvant were the most highly protected against dissemination, and ribosomes in adjuvant also had a significant protective effect. The ribosomal vaccine was obtained by differential centrifugation of a whole cell homogenate and criteria for purity were not provided. The antigen would more accurately be described as a microsomal fraction. Agglutinin titers were boosted in challenged animals that had been immunized. This result was observed in the group receiving ribosomes, and especially in response to a cell wall vaccine. In this hamster model, agglutinin titers were in direct proportion to protection. This seems important because a protective role for antibodies in mycotic infections has not been definitely established.

The antibody response to *S schenckii* on the cellular level was estimated by immunocytoadherence of yeast forms to lymphocytes from infected guinea pigs (Roberts and Larsh, 1971b). The guinea pig is generally quite resistant to *S schenckii*, nonetheless, infection was induced with a large dose of (7×10^8) yeast forms injected intratesticularly. Stress was placed on the animals by keeping them at 10 °C. The resulting orchitis suppurated, drained, and spontaneously healed by 4 weeks after infection. Peripheral blood lymphocytes were purified on nylon wool columns and the nonadherent population was used in the experiments. This fraction should be enriched in T-cells, but the ratio of T : B lymphocytes was not reported. Immunocytoadherence, judged by the formation of *S schenckii* yeast form rosettes, correlated positively with agglutinins, peaking at 2 weeks after infection, and declined sharply or disappeared after four weeks. No precipitins or complement-fixing antibodies could be detected. The rosetted lymphocytes were specific for *S schenckii* and no rosettes were formed with yeast forms of *C albicans, H capsulatum,* or *B dermatitidis* with lymphocytes from *S schenckii* infected animals. It would be of interest to follow this study with nylon wool-adherent B-cells and nylon wool-nonadherent T-lymphocytes. The hamster would be a more suitable model because it is more susceptible to low doses of *S schenckii*, and agglutinins remain high after cure (Charoenvit and Taylor, 1979). Immunocytoadherence measurements are well-suited to the yeast forms of dimorphic fungi, and further studies could be practical and informative.

7.7 SUMMARY

The tube agglutinin test is capable of detecting infection in many cutaneous-lymphatic and most extracutaneous cases of sporotrichosis, with few crossreactions in sera of patients with other mycoses. Precipitins are frequent in the extracutaneous form of the disease and persist for years after clinical and cultural cure.

Cell walls of *S schenckii* have a higher protein and lipid content than those of other mycotic pathogens. The protein component is high in hydroxy- and acidic amino acids involved in glycopeptide bonds. All the glucans of the cell wall are β-linked containing 1,3, 1,4, and 1,6 bonds. The alkali-insoluble wall fraction can cause regression of tumors in mice and thus has immunostimulatory properties. The major surface antigen of *S schenckii* is peptidorhamnomannan (percent composition: L-rhamnose : mannose : protein = 33.5 : 50 : 18.2). It is specifically precipitated with quaternary ammonium detergent as the borate complex and as the Cu^{++} complex with Fehling solution. The rhamnomannan consists of a linear α-1,3-mannan backbone with single 1,3-rhamnose antennae in yeast forms and dirhamnosyl side chains in conidia-producing mycelial forms. The serologic and passive cutaneous anaphylactic activity is removed by periodate oxidation but not by proteinase treatment. Exposure to papain inhibits the antigen's ability to evoke delayed cutaneous hypersensitivity and MIF production. Peptidorhamnomannan has a dual ability to act as a stimulant of humoral and T-cell-mediated functions.

Taxonomic considerations have led to comparisons between *S schenckii* and the saprophyte, *Ceratocystis stenoceras*; however, the ^{13}C-NMR patterns of rhamnomannans of these fungi are distinctly different.

Sporotrichin is the supernate of old yeast form cultures. Ethanol-precipitated sporotrichin is used in epidemiological skin test surveys. Another source of *S schenckii* antigens is the supernate of homogenized yeast forms. Immunoenzyme analysis of precipitin arcs has identified β-glucosidase, acid phosphatase, and esterase activity in reactions between this antigenic complex and rabbit antiserum.

Positive delayed cutaneous hypersensitivity is usually present in cutaneous-lymphatic sporotrichosis. Surveys in normal humans and those hospitalized with no evidence of fungal disease estimate the extent of subclinical exposure at about 10%. Occupational exposure of plant nursery workers is higher, 32%. Positive correlations have been reported between blastogenic indices to *C albicans* and *S schenckii* in normal persons, indicating that exposure to the former confers crossprotection to *S schenckii*.

Human PMN effectively phagocytose and kill *S schenckii* yeast forms when myeloperoxidase, peroxide, and KI are present; however, chloride ions do not substitute for iodide.

Asteroid bodies are stellate zones of eosinophilic material deposited around some *S schenckii* yeast forms in granulomas or abscesses. These deposits are probably immune complexes because the reaction can be reproduced in vitro with specific antiserum.

Resistance to *S schenckii* requires intact thymic function, and athymic mice cannot contain the infection. Hamsters immunized with subcellular fractions and then challenged with live yeast forms respond with a spike in agglutinin production that correlates with protection. Humoral immunity may therefore also contribute to host resistance.

Lymphocytes from infected guinea pigs are capable of binding several *S schenckii* yeast, forming rosettes in vitro and demonstrating specific immunocytoadherence. The lineage of these lymphocytes is not established.

REFERENCES

Blumer SO, Kaufman L, Kaplan W, McLaughlin DW, Kraft DE, 1973. Comparative evaluation of five serological methods for the diagnosis of sporotrichosis. Appl Environ Microbiol 26:4–8.

Chandler FW, Kaplan W, Ajello L, 1980. A Colour Atlas and Textbook of Histopathology of Mycotic Diseases. 333p. London: Wolfe Medical Publications.

Charoenvit Y, Taylor RL, 1979. Experimental sporotrichosis in Syrian hamsters. Infect Immun 23:366–372.

Cunningham KM, Bulmer GS, Rhoades ER, 1979. Phagocytosis and intracellular fate of *Sporothrix schenckii*. J Infect Dis 140:815–817.

de Beurmann L, Gougerot H, Vaucher L, 1908. Diagnostic retrospectif de la sporotrichose par la sporoagglutination. Bull Mem Soc Med Hop Paris 25:75.

Gorin PAJ, Haskins RH, Travassos LR, Mendonca-Previato L, 1977. Further studies on the rhamnomannans and acidic rhamnomannans of *Sporothrix schenckii* and *Ceratocystis stenoceras*. Carbohydr Res 55:21–33.

Ingrish FM, Schneidau JD Jr, 1967. Cutaneous hypersensitivity to sporotrichin in Maricopa County, Arizona. J Invest Dermatol 49:146–149.

Jones RD, Sarosi GA, Parker JD, Weeks RJ, Tosh FE, 1969. The complement fixation test in extracutaneous sporotrichosis. Ann Intern Med 71:913–918.

Karlin JV, Nielsen HS Jr, 1970. Serologic aspects of sporotrichosis. J Infect Dis 121:316–327.

Lehrer RI, 1969. Antifungal effects of peroxidase systems. J Bacteriol 99:361–365.

Lloyd KO, Bitoon MA, 1971. Isolation and purification of a peptido-rhamnomannan from the yeast form of *Sporothrix schenckii*. Structural and immunochemical studies. J Immunol 107:663–671.

Lloyd KO, Travassos LR, 1975. Immunochemical studies on L-rhamno-D-mannans of *Sporothrix schenckii* and related fungi by use of rabbit and human antisera. Carbohydr Res 40:89–97.

Lurie HI, Still WJS, 1969. The capsule of *Sporotrichum schenckii* and the evolution of the asteroid body. A light and electron microscopic study. Sabouraudia 7:64–70.

Mariat F, 1971. Adaptation de *Ceratocystis* a la vie parasitaire chez l'animal. Etude de l'aquisition d'un pouvoir pathogène comparable a celui de *Sporothrix schenckii*. Sabouraudia 9:191–205.

McFarland RB, 1966. Sporotrichosis revisited. 65-Year follow-up of the second reported case. Ann Intern Med 65:363–366.

McGinnis MR, 1980. Laboratory Handbook of Medical Mycology. p. 278. New York: Academic Press.

McMillen S, 1974. The serology of sporotrichosis. pp. 401–409. In The Diagnosis and Treatment of Fungal Infections. Robinson HM Jr, ed. Springfield, Ill: C. C. Thomas.

Mendonça L, Gorin PAJ, Lloyd KO, Travassos LR, 1976. Polymorphism of *Sporothrix schenckii* surface polysaccharides as a function of morphological differentiation. Biochemistry 15:2423–2431.

Murillo de Linares LI, Mena E, Ulloa AJ, Schneidau JD, 1980. Intradermal reactions with polysporo Y (*Sporothrix schenckii* antigen) in persons with and without sporotrichosis. Proc Fifth Inter Conf Mycoses. Caracas, Venezuela, 1980. Scient Pub No 396. pp. 300–307. Washington, DC: Pan American Health Organization.

Nielsen HS, 1968. Biological properties of skin test antigens of yeast form *Sporotrichum schenckii*. J Infect Dis 118:173–180.

Previato JO, Gorin PAJ, Travassos LR, 1979a. Cell wall composition in different cell types of the dimorphic species *Sporothrix schenckii*. Exp Mycol 3:83–91.

Previato JO, Gorin PAJ, Haskins RH, Travassos LR, 1979b. Soluble and insoluble glucans from different cell types of the human pathogen *Sporothrix schenckii*. Exp Mycol 3:92–105.

Rippon JW, 1982. Medical Mycology — The Pathogenic Fungi and the Pathogenic Actinomycetes. p. 264. Philadelphia: W. B. Saunders, Co.

Roberts GD, Larsh HW, 1971a. The serologic diagnosis of extracutaneous sporotrichosis. Amer J Clin Pathol 56:597–600.

Roberts GD, Larsh HW, 1971b. A study of the immunocytoadherence test in the serology of experimental sporotrichosis. *J Infect Dis* 124:264–269.

Russell B, Becket JH, Jacobs PH, 1979. Immunoperoxidase localization of *Sporothrix schenckii* and *Cryptococcus neoformans*. Arch Dermatol 115:433–435.

Schneidau JD, Lamar LM, Hairston MA, 1964. Cutaneous hypersensitivity to sporotrichin in Louisiana. J Amer Med Assn 188:371–373.

Shimonaka H, Noguchi T, Kawai K, Hasegawa J, Nozawa Y, Ito Y, 1975. Immunochemical studies on the human pathogen *Sporothrix schenckii*: Effects of chemical and enzymatic modification of the antigenic compounds upon immediate and delayed reactions. Infect Immun 11:1187–1194.

Shiriashi A, Nakagaki K, Arai T, 1979. Experimental sporotrichosis in congenitally athymic (nude) mice. J Reticuloendothel Soc 26:333–336.

Steele RW, Cannady PB Jr, Moore WJ, Gentry LO, 1976. Skin test and blastogenic responses to *Sporotrichum schenckii*. J Clin Invest 57:156–160.

Travassos LR, Gorin PAJ, Lloyd KO, 1973. Comparison of the rhamnomannans from the human pathogen *Sporothrix schenckii* with those from *Ceratocystis* species. Infect Immun 8:685–693.

Travassos LR, Gorin PAJ, Lloyd KO, 1974. Discrimination between *Sporothrix schenckii* and *Ceratocystis stenoceras* rhamnomannans by proton and carbon-13 magnetic resonance spectroscopy. Infect Immun 9:674–680.

Travassos LR, Lloyd KO, 1980. *Sporothrix schenckii* and related species of *Ceratocystis*. Microbiol Rev 44:683–721.

Walbaum S, Duriez T, Dujardin L, Biguet J, 1978. Etude d'un extrait de *Sporothrix schenckii* (forme levure): analyse electrophoretique et immunoelectrophoretique; characterisation des activites enzymatiques. Mycopathol 63:105–111.

Welsh DD, Dolan T, 1973. *Sporothrix* whole yeast agglutination test. Low-titer reactions of sera of subjects not known to have sporotrichosis. Amer J Clin Pathol 59:82–85.

Wilson DE, Mann JJ, Bennett JE, Utz JP, 1967. Clinical features of extracutaneous sporotrichosis. Medicine 46:265–279.

Chapter 8

Aspergillus fumigatus

8.1 INTRODUCTION

Aspergillosis has been cited as the third most common systemic mycosis in the United States and as the one with the most rapidly increasing incidence (Fraser et al, 1979). This finding is attributed to the double thrust of an aging population afflicted more often with immunosuppressive conditions. Allergic aspergillosis is also important, especially as it illustrates how the immune response to a fungus can cause disease. *Aspergillus fumigatus* is the major causative agent and to a lesser extent *Aspergillus nidulans, Aspergillus niger* and *Aspergillus flavus.*

The antigenic stimulus provided by *A fumigatus* conidia arises from their common occurrence in house dust in many regions of the world (Bardana et al, 1972a; 1972b). Once inhaled, the fate of conidia is determined by the immunologic status of the host and the integrity of the lung parenchyma. Normal persons are known to produce natural antibodies to *A fumigatus* demonstrable by radioimmunoassay (RIA) (Bardana et al, 1972b; Bardana et al, 1975; Sandhu et al, 1978). Occupational exposure results in precipitins to *A fumigatus* in sera from 20% of farmers, and 45% of pigeon breeders (Malo et al, 1977c).

In persons with atopy, inhalation of *A fumigatus* can contribute to bronchial asthma. About 15% of atopics react positively to scratch tests with *A fumigatus* extract (Longbottom, 1978). When these spores germinate and grow in the airways of atopics, a more specific disease, allergic bronchopulmonary aspergillosis (ABPA), can result. Even in the absence of atopy, *A fumigatus* can colonize preexisting active or healed tuberculous pulmonary cavities, forming a "fungus ball" or aspergilloma. There is some evidence that the pulmonary damage due to histoplasmosis can lead to increased susceptibility to *A fumigatus* colonization of the respiratory tree (Walter and Jones, 1968), but this sequence of mycoses has not received much attention, probably because of its rarity.

The conidiophore of *A fumigatus* is a single stalk with a club-shaped end called a vesicle. Many phialides arise directly from the vesicle to produce straight chains of

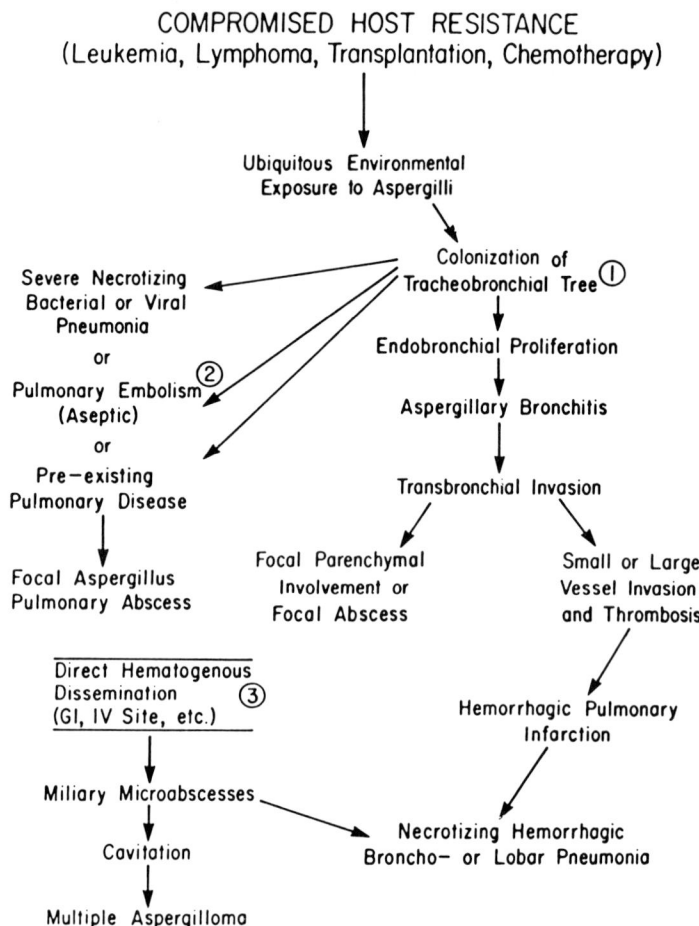

Figure 8.1. Diagram of three major hypothetical pathogenic mechanisms in the development of invasive pulmonary aspergillosis (Bardana 1980b). *Source:* Reprinted from CRC Critical Reviews of Clinical Laboratory Science 13:85–159 with permission from the publisher CRC Press, Inc., copyright 1980.

greenish conidia, 2.5 to 3 μm in diameter (McGinnis, 1980). The conidial heads are compact and conidial production is profuse. The mold is thermotolerant and survives at 48 °C.

Aspergillus fumigatus invades the lung parenchyma in the immunocompromised, often granulocytopenic, host. The typical tissue form consists of short septate hyphae 3 to 6 μm in diameter that branch regularly at a 45° angle. The resulting tissue reaction is frequently granulocytic, with necrosis thought to result from elaboration of toxins by the mold (Bardana, 1980a). Further dissemination usually depends on hematogenous spread when the hyphae grow through the walls of blood vessels, causing occlusion and thrombosis (Figure 8.1). A granulomatous response to *A fumigatus*, with lesions circumscribed by Langhans giant cells, is more often observed as a hypersensitivity reaction following exposures to dusts heavily laden with conidia (Bardana, 1980b).

8.2 POLYSACCHARIDE ANTIGENS

The complexity of the antigenic structure of *A fumigatus* was illustrated in the number of secreted proteins that can induce antibodies in humans (Longbottom, 1978). Yet, in allergic bronchopulmonary aspergillosis the fungus does not invade the lung parenchyma so that antigens must either diffuse into the tissue or, in the case of the fungal cell wall, must adhere to alveolar macrophages or to brush border cells.

Several polysaccharides are present in the cell walls of aspergilli. An antigenic role has been described for some of these, but for others there is only speculation at this time. The surface coat is an alkali-soluble galactomannan (Sakaguchi et al, 1968; Azuma et al, 1971; Reiss and Lehmann, 1979). A crystalline glucan, nigeran, with alternating α-1,3 and α-1,4 linkages can be extracted with hot water from *A niger* (Bobbitt and Nordin, 1978). Nigeran has been looked for but not found in *A fumigatus*. A highly dextrorotatory α-1,3-glucan that is alkali-soluble and acid-precipitable ($[\alpha]_D + 232°$, 1M NaOH) was isolated from *A niger* (Johnston, 1965). About 54% of the cell walls of *A niger* is insoluble in alkali and is a chitin-associated β-1,3-glucan (Stagg and Feather, 1973). The glucan–chitin component is a fibrillar component of the wall. Some of the glucan may be soluble.

A great deal is known about the *A niger* cell wall, but few definitive reports of the *A fumigatus* wall composition have appeared. Compositional analysis of the cell walls of the aspergilli are shown in Table 8.1. The analyses of the three species shown are different enough so that cell wall analysis could be a taxonomic aid. The conditions of growth and the preparation of cell walls are likely to influence the analysis. Until several isolates of *A fumigatus* and other species are analyzed under common conditions, these analyses are considered preliminary. Of particular interest is the large galactosyl component in the walls of *A fumigatus*. The absence of nigeran, the glucan with alternating α-1,3 and α-1,4 linkages, in *A fumigatus* is reflected in the low glucose content of the latter fungus' cell wall.

Table 8.1. Composition of Cell Walls of *Aspergillus* Species, as Percent Dry Weight

Component	*Aspergillus niger* 17454[a]	*Aspergillus fumigatus* NCPF2109[b]	*Aspergillus nidulans* 13.1.OL[c]
Glucose	65.4	17.0	28.9
Mannose	3.1	11.0	2.8
Galactose	12.1	30.0	3.8
Hexosamine	12.5	22.0	25.1[d]
Protein	1.1	10.0	9.2
Lipid	2.0	7.2	9.0
Phosphate	0.03	0.4	0.9
Acetyl	3.2	nd	nd
Glucuronic acid	nd	nd	3.5
Melanin	nd	nd	0[e]
Recovery	99.4	97.6	83.2

Source: Reprinted from Reiss (1985) with permission of the publisher, Marcel Dekker Inc.

Abbreviations: nd, not determined.
[a] Johnston, 1965.
[b] Hearn and Mackenzie, 1979.
[c] Bull, 1970, pigmentless mutant.
[d] Glucosamine + N-acetylglucosamine = 14.3%, galactosamine = 10.8%.
[e] Wild type = 17.3%.

8.2.1 Galactomannan

Different approaches have been applied in attempts to characterize galactomannan from *A fumigatus* (Table 8.2). Typically these studies are carried out without regard for previous findings, thus complicating efforts to trace the development of this research. The ability of galactomannan to evoke cutaneous Arthus reactions in immunized animals (Suzuki and Hayashi, 1975) suggests that it may provide an antigenic stimulus in allergic aspergillosis. Invasive aspergillosis in the immunocompromised host is an acute disease in which the level of circulating antibodies is too low to allow discrimination between natural antibody levels and infection. The galactomannan is postulated to circulate in the ng/ml range and can be detected by counterimmunoelectrophoresis (Lehmann and Reiss, 1978; Reiss and Lehmann, 1979). An identical antigen or one that shares antigenic determinants with galactomannan is detected by a Farr-type RIA in infected rabbits and humans (Weiner and Coats-Stephen, 1979).

Galactomannans have been recovered from *A fumigatus* culture filtrates (Suzuki et al, 1967; Sakaguchi et al, 1968; Sakaguchi et al, 1969), or from supernates of homogenized mycelia (Weiner and Coats-Stephen, 1979; Kim and Chaparas, 1978; Kim et al, 1968). Galactomannan was extracted from the Sabouraud broth filtrates of week-old cultures of *A fumigatus*, and from 45% phenol extracts of macerated mycelia (Suzuki et al, 1967). The galactomannan was purified by gel filtration and removal of impurities by passage through anion and cation exchange columns. The product had a galactose:mannose ratio of 1:1, and no nitrogen or phosphate was detected. The $[\alpha]_D^{20}$, $H_2O = -2$ ° for the galactomannan from the filtrate and $[\alpha]_D^{20}$,

Table 8.2. Antigenic Polysaccharides of *Aspergillus fumigatus*

Antigenic Polysaccharide	Source	Extraction	Reference
Galactomannan	Culture filtrate and mycelia	45% Aqueous phenol	Sakaguchi et al, 1968
Glucan Galactomannan	Defatted mycelia	Warm aqueous pyridine	Azuma et al, 1971
Glycoprotein	Mycelial homogenate	Trichloroacetic acid soluble, 50%-saturated $(NH_4)_2SO_4$-soluble	Bardana et al, 1972b
Two different polysaccharides (unknown composition)[a]	Mycelial homogenate	50% or 75% saturated $(NH_4)_2SO_4$ soluble	Kim et al, 1978
Periodate-sensitive, binds to concanavalin A	Mycelial homogenate	Mechanical disruption	Weiner and Coats-Stephen, 1979
Galactomannan Galactan Glucan	Mycelia	Alkaline borohydride	Reiss and Lehmann, 1979
Glycoprotein	Cell walls	Triton X-100 soluble	Hearn and Mackenzie, 1981

[a] Fraction ASII did not bind to DEAE, contained one coomassie blue-positive protein (disc electrophoresis), and had a protein:polysaccharide ratio of 1:12.5 weight/weight.

$H_2O = -17.5°$ for the galactomannan obtained from mycelia. Mild acid hydrolysis (0.01 N H_2SO_4, 100°C) removed galactofuranosides and the acid stable mannan core was devoid of serological activity. An oligosaccharide fragment, with a molar ratio of galactose:mannose = 4:5 was recovered from the dialyzate after mild acid treatment (Sakaguchi et al, 1968).

Azuma et al (1971) used a different method to separate galactomannan and glucan of *A fumigatus*. Partially defatted mycelium was extracted with warm aqueous pyridine. The water-soluble portion of the extract was precipitated with acetone: one half volume of acetone precipitated glucan and 1.5 vol precipitated the galactomannan. The galactose:mannose molar ratio was 1:1.5; with a total N, 1%; and $[\alpha]_D^{25} = +7.6°$ (Azuma et al, 1971). A structure was proposed for this galactomannan consisting of an α-1,2-linked mannan backbone substituted with side chains of galactofuranose units linked by α-1,6-bonds and other side chains of α-1,6-linked oligomannosides (Azuma et al, 1971). The glucan that also was extracted was nitrogen-free and had an $[\alpha]_D^{25}, = +170.4°$.

Galactomannan of *A fumigatus* and *A niger* resist extraction with hot neutral buffer (Reiss and Lehmann, 1979; Bardalaye and Nordin, 1977). The galactomannan and glucan were extracted from the hot buffer resistant-residue with alkaline borohydride (0.4M NaOH and 0.1M $NaBH_4$) in the cold overnight (Reiss and Lehmann, 1979). Both polysaccharides were further purified by adsorption to concanavalin A-agarose. The eluate was sized on a Sephacryl S200 column and the glucan was excluded, whereas the galactomannan resolved in the column volume. Reiss and Lehmann (1979) also proposed structural details of the galactomannan of *A fumigatus* extracted with alkaline borohydride. The molar ratio of galactose:mannose was 1:1.2 and the molecular weight was estimated at between 25,000 to 75,000 da. Methylation-fragmentation–gas-liquid chromatography suggested that the galactomannan consisted of a 1,6-linked backbone with oligogalactoside chains three residues long terminating in galactofuranose (Figure 8.2). The internal linkages in the side chains were either 1,4 or 1,2. More definitive studies are needed to resolve the various interpretations of galactomannan structure, especially identification of methylated monosaccharides by mass spectrometry.

Antigens from *A fumigatus* homogenate-supernates were radiolabeled by Bardana et al (1972a), and by Weiner and Coats-Stephen (1979). They are presumed to have immunodominant glycosidic determinants. Possibly both groups were working with the same galactomannan-protein antigen, but definitive chemical characterization is lacking. Bardana et al (1972a) have described the active portion of the radiolabeled antigen as glycosidic on grounds of its lability to periodate oxidation, resistance to proteinases, and staining with alcian blue. Weiner and Coats-Stephen (1979) found their antigen was susceptible to periodate, formed lines of identity with a polysaccharide extracted from *A fumigatus* cell walls, and was bound by Con A. Antibodies to this carbohydrate antigen were found in normal persons, and at elevated levels in allergic aspergillosis (Table 8.3).

8.2.1a Antigenemia
Antigenemia is known to occur in invasive aspergillosis, and results from three laboratories (Reiss and Lehmann, 1979; Shaffer et al, 1979; Weiner and Coats-Stephen, 1979) are in agreement that the antigen is a polysaccharide, probably galactomannan.

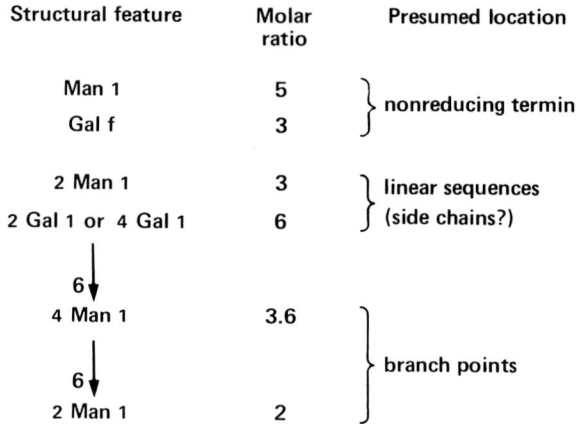

Figure 8.2. Structural features of *Aspergillus fumigatus* galactomannan inferred by methylation-fragmentation-gas-liquid chromatography. Man, mannopyranose; Gal, galactopyranose; Gal-f, galactofuranose (Reiss and Lehmann, 1979). *Source:* Reprinted with permission of the authors and the publisher, American Society for Microbiology.

The evidence that galactomannan circulates in invasive aspergillosis in heavily immunosuppressed animals and in human patients is based on the rabbit model (Lehmann and Reiss, 1978). The assumption is that in an immunosuppressed host some fungal antigen might escape phagocytosis and digestion by the host long enough to be detected in the body fluids. This proved to be true in rabbits that were immunosuppressed with cortisone and cyclophosphamide and then injected with *A fumigatus* conidia. They developed disseminated aspergillosis that was uniformly fatal within 8 days. Sera were drawn from rabbits 50 to 60 hours after infection, the reasoning being that this serum would be a source of circulating antigen. The serum was emulsified in Freund adjuvant and used to immunize a second group of normal rabbits. The resulting antiserum reacted in CIE with antigen in the sera of rabbits

Table 8.3. Measurement of IgG Against a Radioiodinated Glycoprotein of *Aspergillus fumigatus* in Normal and Diseased Populations[a]

		^{125}I Antigen Precipitated (percent)	
Clinical Category	Number/Group	Percent, Mean[b]	Standard Deviation
Normal adult	538	13.3	5.7
Bronchial asthma	98	14.4	5.2
Cystic fibrosis	53	16.0	5.3
Tuberculosis	55	18.1	6.2
Farmer's lung	29	26.4	16.5
Allergic bronchopulmonary aspergillosis	87	46.2	8.6
Aspergilloma	76	41.8	9.6

[a] Abstracted from Bardana, 1980.
[b] Radioactivity precipitated by a 1/5 dilution of serum.

with invasive aspergillosis. The antiserum recognized a galactomannan extract from *A fumigatus* mycelia and absorption with galactomannan removed the activity specific for circulating antigen. Furthermore, a galactan present in the alkaline borohydride-extracted polysaccharide interfered with the immune precipitation of galactomannan but could not, by itself, form an immune precipitate with the reference antiserum.

An immunoprecipitin arc formed between antigen in the serum of a leukemic child with invasive aspergillosis and the rabbit reference antiserum. This arc was not present in serum drawn from the patient 4 months before. When reference serum was absorbed with a purified polysaccharide of the fungus, the line disappeared. This antigen was later characterized as galactomannan after column chromatography and methylation-fragmentation–gas-liquid chromatography analysis (Reiss and Lehmann, 1979) (Table 8.4). In antigenemia, the galactomannan is complexed in serum, probably to immunoglobulin, and dissociation of soluble immune complexes increases the chances of detecting antigen.

A modified Farr-type RIA for antigenemia was developed by Weiner and Coats-Stephen (1979). A mycelial extract of *A fumigatus* was chromatographed on Sephacryl S-200 and fractions reactive with antiserum produced against whole cell homogenate, or to a cell wall extract, were pooled, activated with CNBr, conjugated to tyramine, and radioiodinated. The radiolabeled *A fumigatus* preparation was further purified by immunoadsorption on an agarose column conjugated with rabbit anti-*A fumigatus*. In performing the RIA, immune complexes were dissociated by heating at acidic pH. The limits of sensitivity of the assay are 3 ng/ml for isolated *A fumigatus* antigen and 110 ng/ml for the antigen in serum. Antigenemia in experimentally infected rabbits was present at 12 days after infection, and tests remained positive in a majority of animals until 19 days. By that time only 20% of the infected group remained alive. Subsequently, *A fumigatus* polysaccharide was detected in four of seven patients with invasive aspergillosis (Weiner, 1980). Studies in two additional series of patients with acute leukemia have increased confidence in the ability of the RIA to detect a carbohydrate-containing antigen in sera during invasive aspergillosis (Weiner et al, 1983). The test is highly specific in that no antigen has been found in acute leukemics without aspergillosis. The most recent impression of the test's sensi-

Table 8.4. Summary of Properties of the *Aspergillus fumigatus* Extract That Behaves Serologically Like the Antigen That Circulates in Invasive Aspergillosis

1. Extracted from mycelia with cold alkaline borohydride
2. Serologic activity abolished by periodate oxidation or by 0.01M HCl at 100 °C
3. Serologic activity heat-stable at neutrality and insensitive to pronase
4. Comigrates in immunoelectrophoresis with circulating antigen
5. Absorbed by antibody produced by immunization of rabbit with serum from an immunosuppressed, infected rabbit
6. Bound by concanavalin A
7. Included by Sephacryl S-200 gel permeation
8. Estimated molecular weight > 25000, < 75000 da
9. Mannose : Galactose : Glucose : $= 1 : 1.16 : 0.14$

Source: Reiss and Lehmann, 1979.

tivity is that antigenemia was detected in 71% of seven patients shown histologically to have invasive aspergillosis (Weiner et al, 1983).

Andrews and Weiner (1981) showed the feasibility of detecting *A fumigatus* polysaccharide antigen by their Farr-type of RIA in bronchoalveolar lavage fluid obtained from rabbits infected with *A fumigatus*. The heat-stable carbohydrate was detected in the lavage fluid from 10 of 11 rabbits which were scored, on histologic grounds, as having major pulmonary involvement after intravenous challenge with *A fumigatus* conidia. When pulmonary involvement was scored as minor, antigen was detected in only 40% of the animals, and only after fivefold concentration of the sample. Antigenemia was detected in three fourths of rabbits with disseminated aspergillosis.

In a small subgroup of infected rabbits, antigen could be detected in bronchoalveolar lavage fluid but not in the serum. Conversely, antigenemia was demonstrated in a larger subgroup but no antigen was found in the lung wash fluid. This latter subgroup typically showed minor pulmonary involvement. Examination of such lavage fluids from humans for antigen may provide the most appropriate indicator of an infection that is confined to the lungs. In that event, antigenemia may not be present in detectable concentration and cultural evidence would be equivocal.

Shaffer et al (1979) also devised an RIA to detect *A fumigatus* galactomannan. The antigen recovered from the culture filtrate of *A fumigatus* and purified by gel permeation and anion exchange chromatography contained 78% carbohydrate and 10% protein. The antigen was radioiodinated by activating the polysaccharide with cyanogen bromide before conjugation with tyramine. The solid phase for the RIA was protein A containing staphylococci to which anti-whole *A fumigatus* IgG was adsorbed. Preparation of patient's sera included a deproteinization step (i) so that interference of human IgG with protein A could be excluded and (ii) to dissociate immune complexes containing *Aspergillus* species antigen. Protein A is now known to provide a sensitive means of sequestering soluble immune complexes. Perhaps this could be used to advantage by redesigning this RIA omitting the deproteinization step. Inhibition of the binding of radiolabeled *A fumigatus* galactomannan by antigen in the patient's sera enables detection of antigen in the 10 to 100 ng/ml range. At lower concentrations of antigen, nonspecific reactions occurred in sera from patients with cryptococcosis, histoplasmosis, and from normal humans. One patient with disseminated aspergillosis had levels of galactomannan antigenemia that completely inhibited binding of the radiolabeled reference antigen.

A high potential exists for false-positive reactions in the double antibody sandwich enzyme immunoassays for antigenemia resulting from *Candida* or *Aspergillus* species if sera are not dissociated (Warren et al, 1979). Interference is reduced by dithiothreitol treatment, supplying indirect evidence that small amounts of rheumatoid factor react with the rabbit IgG adsorbed as the solid phase. A specificity control using normal rabbit IgG as the first layer would help to uncover nonspecific reactions. A panel of 33 sera from cancer patients with invasive or local aspergillosis was tested and six to twelve were considered to have antigenemia, the number of positives depending on the fraction of IgG used as the conjugate. Three to six positive reactions occurred in 13 controls composed of normal humans or hospitalized patients without aspergillosis. This finding underscores the necessity to dissociate and denature antibodies in the patients' sera. This may be especially important when convalescent sera from rabbits infected with *A fumigatus* are used to produce enzyme-conjugates because C-substance has been demonstrated in *A fumigatus* and C-reactive protein may thus be present in the conjugate.

8.2.2 C-Substance

Longbottom and Pepys (1964) first described the serologic activity of *A fumigatus* polysaccharides extracted from mycelia with hot aqueous phenol. C-substance was detected in the culture filtrate during autolysis and was absent during primary growth. The culture filtrate also contained polysaccharides with specific antigenic activity. Polysaccharides were partially purified by adsorption of contaminating proteins on DEAE-cellulose. Proof that the polysaccharide was present in immunoprecipitin arcs was obtained with periodic acid-Schiff stain. Adsorption of the preparations with C-reactive protein and washing precipitin arcs with citrate buffer were used to separate C-substance activity from more specific reactions involving *A fumigatus* polysaccharide.

The classic C-substance is the species-specific antigen of *Streptococcus pneumoniae*, which is a teichoic acid with a repeating unit of glucose, diaminodideoxyhexose, N-acetylgalactosamine, and ribitol phosphate with the immunodominant group, phosphorylcholine, attached to galactosamine (Watson and Baddiley, 1974). Antibodies produced by injecting mice with pneumococci have restricted clonality and are of an idiotype that reacts with phosphorylcholine, even when presented in a different linkage and on a structurally unrelated carrier (Glaudemans et al, 1977). At present, no *A fumigatus* antigen has been identified that has either the same composition as C-teichoic acid, or has phosphorylcholine as a determinant.

C-substance activity is bothersome from a serodiagnostic standpoint, but the earlier report (Longbottom and Pepys, 1964) of its presence in *A fumigatus* should still be corroborated because C-reactive protein can activate complement directly and enhance phagocytosis (Lazda, 1977). Far from being inert, fungal C-substance could contribute to an inflammatory response.

8.2.3 β-1,3-Glucan

This type of fungal glucan from *Lentinus edodes* and *Poria cocos* was shown to activate complement by the alternative pathway, and activation occurred with both soluble or insoluble components of the same polysaccharide. The β-glucan with a predominance of 1,3 linkages described in *A niger* (Stagg and Feather, 1973) may also be present in *A fumigatus*.

Delineation of the contribution of polysaccharides to the antigenic structure of *A fumigatus* is still at an early stage. The ability of fungal polysaccharides to activate complement by the alternative pathway and to stimulate C-reactive protein may contribute to the host–fungus interaction in aspergillosis.

8.3 PROTEIN ANTIGENS

What are the properties desired for *A fumigatus* antigens? For scratch or prick tests, aspergillins, the filtrate of autolyzed mycelia, or homogenate-supernates are adequate. More highly purified proteins would need to be used with caution because their high potency could trigger strong local or even systemic flares (Turner-Warwick, 1975). The adequacy of antigens for precipitin testing is called into question because of the weak or solitary precipitins that are typical of allergic aspergillosis. Assurance of maximal coverage would depend on pooling culture filtrates or mycelial extracts from different *A fumigatus* isolates and related species. Progress in quantitation of specific IgG levels which are believed to correlate with disease activity (Malo et al, 1977c) requires selection and purification of marker antigens, such as the chymotrypsin-like proteinase found in culture filtrates (Tran Van Ky et al, 1966) and there are successful efforts in that direction (Senet et al, 1978a,b). Assessment of the contribution of *A fumigatus*-specific IgE to the large elevation of serum IgE arising during pulmonary episodes requires in vitro duplication of the kind and amount of antigens expressed in the host.

8.3.1 Secretion During Growth and Autolysis

Aspergillus fumigatus follows a pattern similar to many fungi by showing active secretion of a relatively low number of proteins, presumably enzymes, during physiological youth. Thereafter, the pH of the medium falls and nutrients become exhausted, primary growth ceases over a period of 9 (Kim and Chaparas, 1978a) or 10 days (Kauffman and deVries, 1980a), depending on the conditions of culture. Over the next 20 to 30 days the pH rises, and autolysis occurs with a loss of mycelia dry weight of up to 40% (Kauffmann and deVries, 1980a), and some early secreted proteins are replaced by autolysis products, many of which are serologically active. Most antigenic analyses pool the proteins actively secreted during primary growth along with the autolysis products. This procedure tends to obscure the contribution of each phase to the total antigenic mosaic. It would be desirable to remove the culture supernate after primary growth, and resuspend the culture in buffer to further induce and recover the autolysis products. This method would provide a preliminary separation of antigens that could then be purified by size, charge, and affinity.

When grown on synthetic asparagine–glycerol–salts medium the 6 to 8-week-old culture filtrate is termed *aspergillin* (Kim and Chaparas, 1978; Walter and Jones, 1968). The culture filtrates are presumed to contain antigens similar to those secreted by *A fumigatus* in vivo. Verification that some of these factors are a reliable index of sensitization is provided by the studies of Longbottom and Pepys (1964); Bardana et al, 1975, and especially by the French groups who have examined these antigens in detail (Biguet et al, 1964; Tran Van Ky et al, 1966; Drouhet et al, 1972).

8.3.2 Immunoenzyme Analysis

Enzymatic activity was localized in precipitin arcs after immunoelectrophoresis (Tran Van Ky et al, 1966) of the culture filtrate "metabolic antigen" versus antiserum of rabbits immunized with the antigen in incomplete Freund adjuvant. After 11 weeks of immunization up to 22 precipitins were detected. Five of the antigens with defined enzymatic functions were reactive with sera from allergic bronchopulmonary aspergillosis patients: catalase, "chymotrypsin I," chymotrypsin II, and glucuronidases I and II (Tran Van Ky et al, 1969). These factors were highly antigenic for rabbits, and precipitins to them were detected early in the immunization period, anti-catalase as early as 3 days postimmunization denoting that in rabbits the response was anamnestic. The substrates used to localize the enzymatic activity of the principal antigens were N-acetyl-D,L-phenylalanine-β-naphthyl ester for chymotrypsin and 6-bromo-2-naphthyl-β-D-glucopyranoside for glucuronidase. The two chymotrypsin-like proteinase antigens had cathodic mobility indicating either basicity or highly glycosylated proteins migrating by electroendosmosis (see Figure 8.3).

Since the first demonstration of multiple precipitins in aspergilloma and allergic aspergillosis patients, further resolution of these antibodies was accomplished by two-dimensional crossed rocket immunoelectrophoresis (2D-IEP). Longbottom (1978) showed that pooled human aspergillosis sera could be used as a standard to evaluate different culture filtrates and mycelial extracts. Kauffmann and deVries (1980b) used sera from APBA patients in the intermediate gel variation thus conserving reagents, improving the sensitivity, and providing a prototype clinical immunoassay. Five antigens in the *A fumigatus* culture filtrate reacted with 80% of the patients' sera. Trypsin-like proteinase activity was localized in factors 13 and 18 out of a maximum of 22 recognized by rabbit antiserum (Kauffman and deVries, 1980b). Antibodies to the proteinase occurred in about 90% of the patients' sera. The results of different laboratories agree that proteinases of *A fumigatus* stimulate antibody production in allergic aspergillosis. These proteinases probably are determinants of pathogenicity.

8.3.3 Mycelial Extracts

The lack of standardization of allergenic extracts prompted a study to purify and characterize soluble cytoplasmic proteins of *A fumigatus* (Kim and Chaparas, 1978; Kim et al, 1978). Four-day-old mycelia grown on asparagine–glycerol–salts medium were disrupted in a Braun homogenizer and centrifuged. A portion of this mycelial extract was precipitated with 75% saturated ammonium sulfate and was fractionated on Sephadex G75. The nonprecipitable portion was chromatographed on DEAE cellulose with the following elution profile: A large amount of polysaccharide was present in the effluent comprising 40% of the original mycelial dry weight. Another complex of proteins was bound to the DEAE column and eluted with a high salt concentration. Rabbits were immunized with *A fumigatus* mycelia until fifty-two precipitins were detectable by 2D-IEP. Fraction "A" of the ammonium sulfate-precipitated material contained 21 factors reacting with rabbit antiserum. This antigenic complex was a potent elicitor of blastogenesis of lymph node cells obtained from guinea pigs immunized with the antigen in adjuvant. Fraction "A" was coupled to CNBr-activated paper discs to form an *A fumigatus*-specific radioallergosorbent test that gave a positive reaction with pooled atopic sera. A protein fraction was not precipitable with ammonium sulfate, but did bind to DEAE and was also analyzed. It produced 16 Laurell rockets in 2D-IEP although only ten proteins were detected in PAGE.

Bardana and coworkers (reviewed in Bardana, 1980a; 1980b) developed an RIA based on the Farr technique with an antigen derived from a supernate of sonicated 3-week-old heat-killed mycelia. The portion of the supernate not precipitable with 10% trichloroacetic acid was recovered and radiolabeled with ^{125}I. Polyacrylamide gel electrophoresis analysis showed it contained three major proteins. The antigens in this fraction were not precipitable with half-saturated ammonium sulfate and did not form immunoprecipitin arcs with rabbit anti-*A fumigatus*. Immunologic activity could be demonstrated with this rabbit antiserum, and with human sera by the Farr

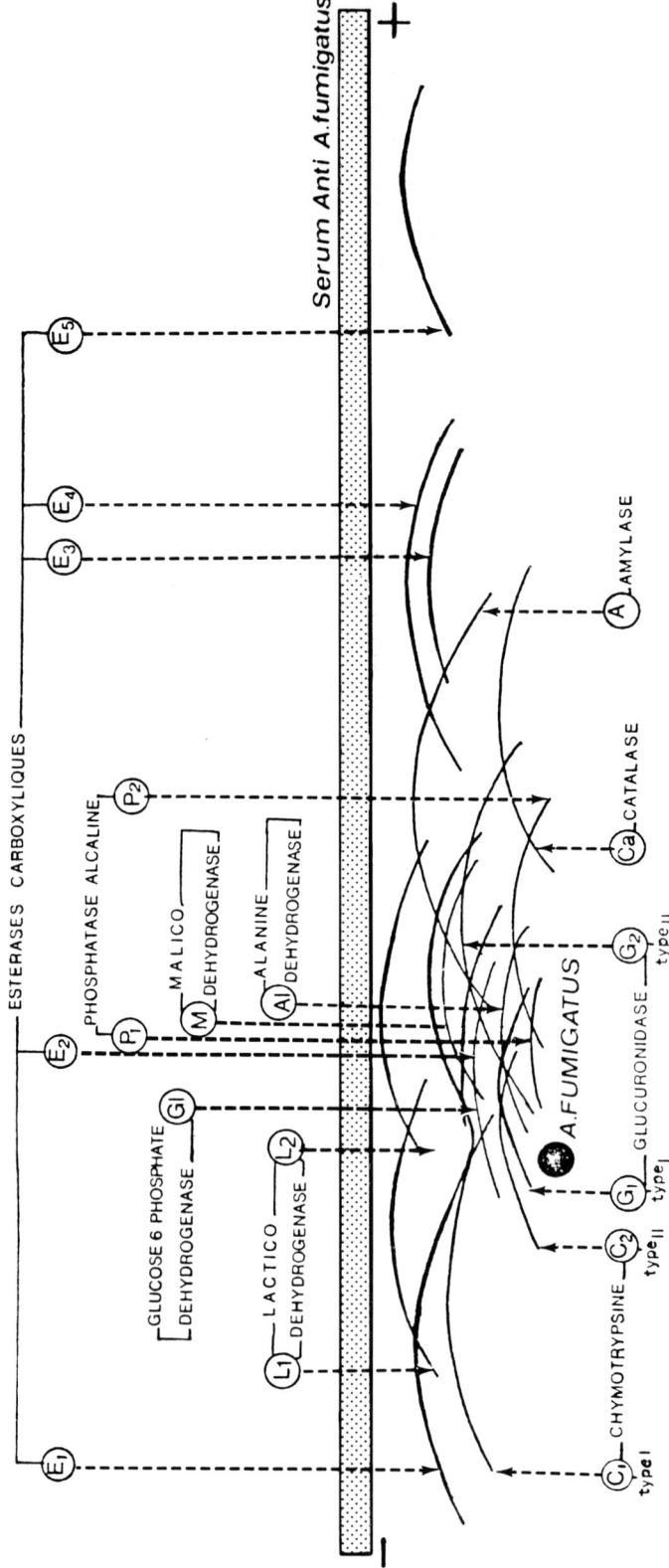

Figure 8.3. Localization of enzymatic activity in immunoprecipitin arcs. The antigenic complex used was a mixture of equal parts of a culture supernatant and mycelial extract of *Aspergillus fumigatus*. The antiserum was from a hyperimmunized rabbit (Tran Van Ky et al, 1969). *Source:* Reprinted from *Sabouraudia* with permission of the copyright holder, the International Society for Human and Animal Mycology.

assay. The antigenic activity was not sensitive to pepsin but was inactivated by sodium metaperiodate and $NaBH_4$, implicating a glycosidic component as the important epitope. Approximately 40% of the radioactivity did not bind even when concentrated antisera were used, so that a large immunologically inert labeled component was present. The failure of the antigen to produce immunoprecipitin arcs may be explained if the level of the circulating antibodies is low, or if the antigen is depolymerized during its extraction. The class of antibody involved was shown to be IgG because precipitin arcs were made visible in immunodiffusion when developed with anti-human IgG (Bardana et al, 1972b). The test is structured so that the percent of binding is calculated by mixing 0.01 μg N/ml of labeled antigen with 0.5 ml of a 1:5 dilution of serum; then precipitating the resulting immune complexes with an equal volume of saturated ammonium sulfate.

Under these conditions, the percent of radioactivity bound by sera from normal persons is 14.1% ± 5.5% (Bardana et al, 1975). Elevated binding significant for ABPA occurs at two standard deviations higher than that found with normal sera. These results show that natural antibodies to aspergilli are universal. In a series of 63 APBA patients at the Delhi Chest Institute a correlation was made between the number of precipitins, percent of sputa that contained *A fumigatus*, and the extent of binding in the Farr assay (Sandhu et al, 1978). In another series (Bardana et al, 1975), a higher number of ABPA patients had significantly elevated percent binding in the RIA than had demonstrable precipitins. Several serial specimens obtained from an ABPA patient followed through flare and remission periods fluctuated in the RIA (Bardana, 1978). The available evidence suggests that the Farr RIA is a reasonable attempt to quantitate IgG to an antigen that may be overlooked by others who screen for activity solely by means of precipitin tests. The labeled products could benefit from additional chemical characterization to remove inert material.

Protein profiles in homogenate-supernates of *A fumigatus* were compared by gradient polyacrylamide gels and two-dimensional crossed immunoelectrophoresis (Odds et al, 1983). The variables of homogenization conditions and age of mycelia were studied. Many antigenic proteins were recovered in all preparations tested. Line and crossed immunoelectrophoresis with a reference antigen in the intermediate gel revealed only quantitative differences between antigens prepared from 1-day-old and 7-day-old mycelia. Clearly, more qualitative comparisons of proteins based on the physicochemical properties (ie, mass, isoelectric point and enzymatic function) are needed as markers to standardize antigenic extracts.

A major cytoplasmic protein was isolated from a fermenter-grown 4-day-old culture of *A fumigatus 507* (Calvanico et al, 1981). This isolate was selected because, of those tested, it had the highest reactivity with aspergillosis patients' sera. The supernate of homogenized mycelia was isolated in the presence of the proteinase inhibitor, phenylmethylsulfonyl fluoride, an important consideration in preserving protein antigens in view of the proteinase that is found in *A fumigatus*. Purification was effected by gel permeation chromatography, and a similar procedure was used to prepare an antigen from the culture filtrate. The SDS-PAGE profiles of the semipurified cell extract indicated a major protein of 182,000 da that produced subunits of 45,000 da after reduction of disulfide bonds. The analogous culture filtrate preparation had a slightly lower molecular weight both in the aggregate and subunits. The workers inferred that the major antigen in the mycelial extract is a higher molecular weight precursor of one that occurs in the culture filtrate in a somewhat degraded form. The antigens in these cell extract and culture filtrate preparations produced identical immunoprecipitin arcs, but this still is not direct evidence that the proteins identified in sodium dodecyl sulfate gels are in fact the antigens that participate in the serologic reaction. The cell extract elicited immunoprecipitins in about 73% of the aspergilloma and allergic aspergillosis patients that were tested. The culture filtrate-derived antigen had equivalent reactivity.

The approaches taken to characterize the *A fumigatus* antigens are difficult to reconcile because few investigators have aligned their research to build on that of others. On one hand, there is the approach of analyzing autolyzed cultures—aspergillins; however, more recently the emphasis has shifted to analysis of extracts of young mycelia. The work of Bardana et al (1972a) and Kim and Chaparas (1978) suggests that ammonium sulfate fractionation of mycelial extracts provides a means of separating neutral polysaccharides, presumably glucans and galactomannans, that remain in the supernate, from antigenic proteins. Hearn et al (1980) have studied the behavior of *A fumigatus* antigenic proteins that partition at various concentrations of ammonium sulfate. The extract of *A fumigatus* NCPF2109 was obtained as a supernate of homogenized 3-day-old mycelia. Most of the proteins (90% of the total protein) that reacted in immunoelectrophoresis occurred in the 50% cut. One fraction of interest precipitated at 15% saturation. This material was predominantly carbohydrate and reacts strongly with rabbit antisera and human aspergillosis sera. Most of the antigenic activity in this fraction was bound by Con A, suggesting a highly antigenic glycoprotein. Apart from this interesting glycoprotein, the bulk neutral polysaccharide occurred in the supernate after 50% saturation, consistent with the work of others (Bardana, 1980a; Kim et al, 1978). Two ammonium sulfate cuts, 15% and 50% saturation provide a ready means of separating proteins, and a glycoprotein, from the bulk neutral polysaccharide.

In analyzing the soluble supernates for antigens, after homogenizing mycelia, scant attention has been paid to any proteins that may be adherent to the cell walls. A notable exception is the work of Hearn and Mackenzie (1979, 1981), who extracted cell walls with triton X 100 to release five or more proteins that were found to be serologically reactive with sera from aspergillosis patients. The antigens were resolved by 2D-IEP and their immunologic activity was ablated by pronase digestion, or by placing Con A or wheat germ agglutinin in the intermediate gel.

8.3.4 Purified Antigens

The chymotrypsin-like proteinase was isolated from an *A fumigatus* thirty-day culture filtrate (Girault et al, 1977) by ammonium sulfate precipitation and chromatography on a column containing the ligand ϵ-aminocaproyl-D-tryptophan methyl ester bound to agarose. The enzyme retained activity after elution with 0.1M acetic acid. Polyacrylamide gel electrophoresis analysis resolved four components in the column fractions. Catalase activity was isolated from the same starting material by a combination of gel permeation and ion exchange chromatography. The purified *A fumigatus* proteinase was coupled to sheep erythrocytes with glutaraldehyde, and these sensitized cells were agglutinated by rabbit antiserum to a 1/20,000 endpoint (Girault et al, 1977). Sera from human aspergillosis patients also reacted with this antigen whereas sera from normal persons did not (Senet et al, 1978a). A direct correlation existed between the number of precipitins in the patients' sera and the hemagglutination endpoint. A similar direct relationship was observed between agglutination of erythrocytes coated with purified *A fumigatus* catalase and the number of precipitins (Senet et al, 1978b).

Efforts continue to be made to purify *A fumigatus* homogenate-antigens to find a suitable fraction for indirect EIA that will discriminate patients with allergic bronchopulmonary aspergillosis from asymptomatic exposed controls (Schønheyder and Andersen, 1983). In this instance the controls were patients with chronic obstructive pulmonary disease with lung infiltrates due to nonmycotic infections. Gel permeation chromatography on Sephadex G200 resulted in most of the carbohydrate component eluting in the void volume. The fraction with the most promise of test specificity had an approximate molecular weight of 70,000 and had no detectable carbohydrate. This fraction may contain a specific antigen that should be compared to the 182,000 da protein characterized by Calvanico et al (1981).

8.3.5 Summary of Secreted and Cell Extract Proteins

The state of knowledge of these antigens is shown in Table 8.5. This is a rapidly developing area of research. Further studies should be oriented to the isolation and purification of the individual factors listed in the table to produce international standards for use in monitoring class-specific antibody and antigenemia responses in the three types of aspergillosis: allergic, aspergilloma, and invasive.

8.4 ALLERGIC BRONCHOPULMONARY ASPERGILLOSIS

ABPA is found in some atopic individuals and is characterized by asthma, episodes of recurrent, fleeting, pulmonary shadows, fever, sputum plugs containing *A fumi-*

Table 8.5. Antigens Present in Aspergillin and in Mycelial Extracts of *Aspergillus fumigatus*

Antigens	References
Aspergillin	
Glucuronidase	Biguet et al, 1964
Chymotrypsin–proteinase,	Tran Van Ky et al, 1966
Catalase	Drouhet et al, 1972
	Girault et al, 1970
pI 3 to 4 glycoprotein	Kurup et al, 1983
Mycelial extracts	
Glycoprotein extracted from cell walls with Triton × 100 detergent	Hearn and Mackenzie 1981
Glycoprotein from cytoplasm precipitated with 15% saturated ammonium sulfate	Hearn et al, 1980
Protein complex precipitated with half-saturated ammonium sulfate	Hearn et al, 1980 Kim and Chaparas, 1978
Major cytoplasmic antigen, 182,000 molecular weight	Calvanico et al, 1981
Ammonium sulfate-soluble antigens	
precipitinogens	Kim et al, 1978
nonprecipitating	Bardana, 1980a

gatus hyphae, and hemoptysis. The reader is referred to the monograph by Bardana, (1980a, 1980b) for a complete discussion of pathogenesis, immunopathology, and therapy. Over a long period (median duration in one series was 10.9 years, Malo et al, 1977a), airway obstruction occurs with decreased respiratory capacity. Cumulative damage to the lungs occurs as central or saccular bronchiectasis, the latter meaning dilated bronchi ending in blind sacs (Turner-Warwick, 1975).

Immunologic and other laboratory findings include a dual type I and type III immediate cutaneous hypersensitivity to *A fumigatus* antigens, and culture of the fungus from sputum plugs. Weak to moderate precipitins to *A fumigatus* can be detected in 80% of patients' sera. Very high levels of serum IgE occur during episodes of pulmonary eosinophilia. More sensitive methods for demonstrating antibodies to *A fumigatus* will be discussed, including a Farr-type RIA, 2D-IEP, and a specific RAST.

The rate of deterioration of pulmonary function is generally slow in ABPA. One series followed 33 patients for more than 5 years, and nine patients showed progressive permanent lung shadows (Malo et al, 1977c). After 4 more years of follow-up, 24 showed no further radiographic changes. In a minority of patients, a rapidly deteriorating severe lung disease occurs (Imbeau et al, 1978). Current therapy consists of oral steroids and sodium cromoglycate. Corticosteroids block the type III but not the type I cutaneous and bronchial reactions (Pepys, 1977; Imbeau et al, 1978). Sodium cromoglycate blocks the immediate asthmatic reaction. Although the rate of deterioration in ABPA, measured radiographically, is slow, the cumulative pathology is considerable (Turner-Warwick, 1975; Malo et al, 1977b). Fixed shadows, consisting of bronchial dilations, can progress to apical fibrosis, bronchiectasis, progressive airway obstruction, and respiratory failure. Other patients, over a period of 20 years, stopped having flares due to *A fumigatus* and were treated as moderate to severe asthmatics.

Pitfalls in the laboratory diagnosis of ABPA abound. Sputum cultures are not always positive for *A fumigatus*. In one longitudinal series of 50 ABPA patients, a history of mucoid sputum plugs was recorded in 37 and cultures positive for *A fumigatus* in 23 (Malo et al, 1977a). Patients in remission have IgE levels that are often nearly normal (Imbeau et al, 1978). The occurrence of precipitins in sera of ABPA patients is not universal, and oral steroid therapy can suppress precipitins. In one study (McCarthy and Pepys, 1971) 60% to 64% of ABPA patients had weak precipitins in immunodiffusion. The use of counterimmunoelectrophoresis can increase the detection of precipitins as can concentration of the sera fourfold before immunodiffusion (Malo et al, 1977c).

8.5 HUMORAL IMMUNITY AND IMMUNOPATHOLOGY

8.5.1 Immunoglobulin E

Type I immediate cutaneous hypersensitivity to *A fumigatus* antigens is a necessary part of the diagnosis of ABPA. Other elements of type I immediate hypersensitivity are involved in flares of clinical disease. Radiographic findings of fleeting pulmonary shadows are considered to be due to eosinophilic infiltrates (Turner-Warwick, 1975).

Large increases in serum IgE coincide with flare-ups (Patterson and Roberts, 1974; Imbeau et al, 1978; Rosenberg et al, 1978). Allergic bronchopulmonary aspergillosis patients as a group also have markedly higher IgD concentrations than those of normal controls or asthmatic patients (Luster et al, 1976). The proportion of total serum IgE that is specific for antigens of *A fumigatus* is disputed. Total serum IgE in allergic aspergillosis ranges from near normal values to 82,800 ng/ml (Rosenberg et al, 1978). One attempt to quantitate the *A fumigatus*-specific IgE component estimated that over 93% of the IgE was not specific for the fungus (Patterson and Roberts, 1974), leaving the possibility that *A fumigatus* growing in the airways of atopics stimulates polyclonal IgE expression. *A fumigatus*-specific IgE was proportional but lower than the total IgE in another series of patients (Dessaint et al, 1976). An attempt to determine the nature of the antigen binding to IgE was made using radioimmunoelectrophoresis with ^{125}I-anti-human IgE as the indicator. The arc that occurred in patient's serum had a different electrophoretic mobility than the chymotrypsin-like proteinase, "c_2", that was used as the reference antigen. Few attempts have been made to identify the nature of the antigens of *A fumigatus* that bind IgE in allergic aspergillosis. A successful approach has been to localize IgE in up to 15 different Laurell rockets produced after two-dimensional immunoelectrophoresis (Longbottom, 1983a; 1983b; see also section 8.5.3). In this method paired gels are stained with coomassie blue or reacted with ^{125}I-anti-IgE and visualized by autoradiography. Each allergic aspergillosis patient's serum showed a unique pattern of stained and radioactive rockets. Many of the radiolabeled precipitates were diffuse suggestive of glycoproteins or polysaccharide antigens. As controls, sera from aspergilloma patients who had no allergic component in their disease showed multiple immunoprecipitins, but gave negative autoradiograms.

Attempts to measure the contribution of *A fumigatus*-specific IgE to the total IgE level are subject to certain pitfalls: (i) Antigen adequacy. The relevant antigenic factors may not be present in the culture filtrate or mycelial extract or they may not be coupled via CNBr to the cellulose solid phase. This problem underscores the need for standardization of allergenic extracts. (ii) Interference. Specific interference with the binding of IgE to the solid phase by other immunoglobulin classes has been detected (Vervloet et al, 1974), as has nonspecific interference, which can be overcome by using more dilute serum samples.

A rapid fall in serum IgE levels occurs after an acute episode (Malo et al, 1977c; Imbeau et al, 1978). The same cannot be said for IgG against *A fumigatus*. Pathogenetic antibodies of the IgG class are observed as Arthus reactions of edematous swellings peaking at 4 to 6 hours after scratch tests, and as serum precipitins. Direct correlation was found between positive precipitins and disease activity measured radiographically as the interval since the last pulmonary shadow and the total number of shadows (Malo et al, 1977c). The formats for detecting precipitins vary and have provided insight into the fungus' antigenic structure (see below), but the general impression in allergic aspergillosis is that weak or solitary precipitins are present in up to 84% of patients (Malo et al, 1977c). Damage to lung tissue arises from the deposition of immune complexes as microprecipitates on the bronchial walls and pulmonary vascular beds with activation of complement, releasing neutrophil, and possibly eosinophil, chemotactic factors. The effector mechanisms are presumed to be the release of lysosomal enzymes from granulocytes, and direct complement-mediated damage.

8.5.2 Immune Complexes

Thorough immunopathological studies of allergic aspergillosis are lacking, but deposits of immunoglobulins (especially IgA), C3, and *Aspergillus*-species antigen were observed in and around the walls of small vessels of resected, diseased lung segments from a human patient (Katz and Kniker, 1973).

Evidence for circulating immune complexes was obtained in an allergic aspergillosis patient during a flare that required treatment for respiratory distress (Geha, 1977). Precipitins reacting with C1q were detected in immunodiffusion, but the nature of the antigen in the complexes is unknown. The soluble complexes resolved after the patient responded to oral prednisone. Another brief report of circulating immune complexes in two allergic aspergillosis patients has appeared (Dorval et al, 1979). The complexes were first trapped on a Raji cell layer and then localized with radioiodinated staphylococcal protein A. No clinical significance could be attributed to the soluble complexes, but if confirmed by others, these studies illustrate how antigens can diffuse in a noninvasive disease. It would be worthwhile to find out the nature of the responsible antigen.

8.5.3 Precipitins and Primary Binding Assays

Kurup and Fink (1978) compared four serologic tests: immunodiffusion, CIE, hemagglutination, and indirect fluorescent antibody (IFA) for the ability to detect antibodies in aspergillosis. The patients included those with aspergilloma, ABPA, asthmatics, and those with nonmycotic pulmonary conditions. Aspergilloma patients, the highest antibody responders in this group, were uniformly positive in all tests. Five ABPA patients were tested; all were reactive in immunodiffusion and three of five were positive in CIE. Six of seven asthmatics without clinical aspergillosis reacted in immunodiffusion, an incidence higher than the 22% normally expected in asthmatics with a positive prick test to *A fumigatus*, and obviously higher than the lack of response expected in asthmatics with no immediate hypersensitivity to aspergillin (Malo et al, 1977c). Hemagglutination detected reactions in all ABPA patients' sera. Four of five ABPA patients reacted in indirect immunofluorescence (IFA), a test that uses *A fumigatus* growing on cover slips as the antigen. Thus all four tests have good sensitivity, although normal controls and patients with nonmycotic pulmonary conditions are prone to crossreact in IFA. This is caused by the natural levels of antibodies present in the normal population to unpurified antigens of this fungus. Hemagglutination using a protein antigen purified from aspergillin may provide a more specific diagnosis (Senet et al, 1978b). As the tests for antibodies become more sensitive, a cut-off point to distinguish background levels of natural antibodies will be necessary, but too few data are available to firmly establish this point for the IFA and EIA tests. The drawbacks of using hemagglutination and CIE tests have been pointed out by Bardana (1980a). Precipitins in CIE have been detected in up to 20% of normal individuals and *A fumigatus* extracts contain hemolysins and hemagglutinins.

The lack of specificity in the IFA test was considered by Gordon et al (1977) to result from the method of antigen preparation. When slide cultures of *A fumigatus* were used as the substrate, specific staining was localized at the hyphal apices. A germling preparation was evaluated that used spores germinated at 42 °C for 6 to 8.5

hours and outgrowth stopped with formalin-saline. Antibody titers of 1/32 or greater occurred in aspergilloma patients and titers in excess of 1/256 in sera from patients with ABPA. The germlings may resemble the type of limited growth and lack of invasiveness characteristic of *A fumigatus* in allergic aspergillosis, thus explaining the high reactivity of those sera. Sera of normal persons had IFA titers lower than 1/16. Extended evaluation of the reactions in asthmatics, nonmycotic pulmonary disease, normal persons, and patients with allergic aspergillosis is necessary to establish the level of clinical significance.

Enzyme immunoassays enable the detection and quantitation of class-specific antibodies in aspergillosis. Sepulveda et al (1979) used a filtrate of static *A fumigatus* cultures grown on synthetic medium for 4 weeks. Proteins and polysaccharides in the antigenic complex were separated by ammonium sulfate precipitation, but the number of components in each fraction was not characterized. The IgG levels to *A fumigatus* proteins in sera of allergic aspergillosis patients correlate well with the number of precipitins in immunodiffusion and with RIA. Most of the antibodies in allergic aspergillosis were directed against the polysaccharide antigen, whereas in aspergilloma, as expected, *A fumigatus* proteins evoked the highest EIA titers because of the multiple precipitins found in that disease. Normal donors gave flat responses at the hundredfold serum dilution that is routinely used. IgE responses to both protein and polysaccharide antigens were detected in ABPA patients.

Utilizing a supernate of 4 to 5-week-old surface cultures of *A fumigatus* as the source of antigens Longbottom (1983b) compared EIA and crossed rocket immunoelectrophoresis with or without an autoradiographic overlay technique to enumerate the protein factors reactive with IgG or IgE. The sera tested derived from ABPA patients. The precipitin score correlated positively with the A_{405nm} values recorded in the EIA for a group of 29 such patients. It should be pointed out that the range of EIA values for the most prevalent (precipitin score "1") group was extremely broad, scattered over a logarithmic scale. The autoradiographic overlay technique with ^{125}I-anti IgE helped to distinguish those protein antigenic factors capable of binding *A fumigatus*-specific IgE. Two classes of antigens were recognized: (i) poor precipitinogens that nevertheless were the major allergenic components to bind IgE and (ii) strongly precipitating antigens that bound IgE only weakly. Fifteen distinct *A fumigatus* antigenic factors were enumerated that were capable of binding specific IgE in allergic aspergillosis patients' sera.

Antigens present in culture filtrates of 4 to 5-day-old *A fumigatus* were purified by preparative isoelectric focusing and Con A affinity chromatography (Kurup et al, 1983). One fraction with isoelectric points between pH 3 to 4 was eluted from a Con A column with methyl α-mannopyranoside. This glycoprotein fraction contained four precipitinogens and was adapted for indirect EIA. All allergic bronchopulmonary aspergillosis and aspergilloma patients' sera tested gave stronger positive reactions with the glycoprotein fraction than were obtained with protein fractions that did not bind to Con A.

Independent confirmation of the high reactivity of *A fumigatus* glycoproteins was reported by Wilson and Hearn (1983). The homogenate-supernate of 3-day-old *A fumigatus* mycelia grown on glucose-peptone medium was applied to a Con A–agarose column. The effluent fraction contained fewer precipitinogens and gave a lower signal-to-noise ratio in EIA than the fraction eluted from the column with methyl-α-mannopyranoside. However, tandem crossed rocket immunoelectrophoresis of both antigenic preparations revealed that both contained the same major protein or glycoprotein. The further characterization of this factor should actively be pursued.

Some difficulties emerged in using unfractionated "metabolic" antigens (ie, culture supernate) in an EIA to measure IgG in allergic aspergillosis patients and controls (Kauffmann et al, 1983). Sera from a large number of control individuals with no evidence of aspergillosis (177 of 758 samples tested) had IgG concentrations considered elevated or test-positive. These controls had few or no precipitins against *A fumigatus*. Moreover, five proven allergic aspergillosis cases with multiple (4 to 6) precipitins had EIA results that were considered test-negative. To further complicate matters, sera from some individuals demonstrated increased EIA immunoglobulin G antibodies in the face of decreasing numbers of precipitin arcs. Obviously difficulties arise when comparing a primary binding assay like EIA to precipitin tests that depend on secondary phenomena. The factors affecting passive adsorption of *Aspergillus* antigens to polystyrene are not understood. Antigens that are good precipitinogens may not adsorb readily to plastic and vice versa.

In contrast to the ease of demonstrating precipitins in aspergilloma patients and the weak or solitary precipitins in allergic bronchopulmonary aspergillosis, invasive aspergillosis patients frequently lack precipitins (Young and Bennett, 1971) because of their immunosuppressed status (Table 8.6). In the minority of invasive aspergillosis patients whose antibody-forming capacity is relatively intact, immunodiffusion tests have been helpful (Kaufman, 1982). This variation in immunocompetence probably accounts for the widely disparate results reported for immunodiffusion in invasive aspergillosis (Table 8.6).

Another variable is the closeness of monitoring; multiple specimens increase the chances for success where the index of suspicion is high (Schaefer et al, 1976). It has been shown (Table 8.6) that more sensitive tests for antibodies, in this case RIA, increase the number of positive reactors even among the immunosuppressed group (Marier et al, 1979).

The antigen used for invasive aspergillosis serology is a factor that influences the ability to detect antibodies in immunocompromised patients. Choice of antigen and

Table 8.6. Detection of Antibodies in Invasive Aspergillosis Using Unfractionated Antigenic Complexes of *Aspergillus fumigatus*

Test	Number of Patients	Status of Immuno-Suppression	Sensitivity (percent)	Specificity (percent)	References
Immunodiffusion	15	heavy	0	nd	Young and Bennett 1971
	55	heavy	29	nd	Gold et al, 1980
	10	heavy	70	94	Schaefer et al, 1976
	44	none to moderate	68	>90	Kaufman 1982
	19	moderate to heavy	26	95	Marier et al, 1979
Counterimmuno-electrophoresis	19	moderate to heavy	21	>95	Marier et al, 1979
	10	heavy	70	>95	Holmberg et al, 1980
Hemagglutination	55	heavy	27	84	Gold et al, 1980
Radioimmunoassay	19	moderate to heavy	79	88	Marier et al, 1979
Enzyme immunoassay	10	heavy	80	92	Holmberg et al, 1980

Abbreviation: nd, not determined.

careful monitoring may explain the high rate of successful detection of invasive aspergillosis in heavily immunocompromised patients by both CIE and indirect EIA (Holmberg et al, 1980). In one study, culture filtrates were pooled from two isolates of *A fumigatus* and concentrated by ammonium sulfate precipitation (Holmberg et al, 1980). Forty protein antigens were separated by two-dimensional crossed "rocket" immunoelectrophoresis in this preparation. The antigenic complex was used to measure antibodies in patients with acute leukemia or immunodeficiency disease and in control groups. In seven of ten invasive aspergillosis patients, precipitins were demonstrated with CIE, and eight of ten had antibody titers in indirect EIA that were significant. These antibodies rose in six invasive aspergillosis patients who recovered from the fungal infection but died from the primary disease. In the remaining four cases, the antibody levels did not seem a reliable indication of disease severity, and the only convenient variable was that the antibody level in aspergillosis patients was higher than that found in control groups. The difference between the mean and standard deviation of anti-*Aspergillus*-Ig present in normal healthy controls and that found in aspergillosis patients should be subjected to an analysis of variance.

8.5.4 Crossreactivity Among *Aspergillus* Species Antigens

The most potent protein antigens for precipitin reactions in the mycelial extract of *A fumigatus* are insoluble in 75%-saturated ammonium sulfate and, after solubilization in buffer, are eluted in an intermediate position from a Sephadex G75 column (Kim and Chaparas, 1979). Extracts of young *A fumigatus* mycelia produced more precipitin arcs in the Laurell rocket reference system than did the filtrate of old cultures.

The "percent of qualitative sharing" (PQS) was used by Kim and Chaparas (1979) to express the crossreactivity using analogous antigenic preparations from *A niger, A fumigatus,* and *A flavus*. A "PQS" of between 62% to 100% indicated the close antigenic relationship among strains of *A fumigatus*. *Aspergillus* species antigens reflect only a limited extent of crossreactivity. *A niger* stood apart from the other two species in being more distantly related (only <32% shared antigens). Antisera against *A flavus* showed greater overlap with *A fumigatus* antigens (<52% shared antigens) than in the reciprocal case (<26% sharing). If a human infection is due to a species other than *A fumigatus*, reliance cannot be placed on crossreactivity for detection of precipitins because over one half of *A flavus* or *A niger* infections would be undetected by this means. This prediction was borne out in a follow-up study. Chaparas et al (1980) showed that *A flavus* antigens only detected precipitins in 1 of 13 human aspergillosis cases due to *A niger*. Conversely, *A niger* antigens detected precipitins in 2 of 7 *A flavus* infections.

8.6 CELL-MEDIATED IMMUNITY

A coherent picture of the cell-mediated immune responses in allergic bronchopulmonary aspergillosis has not yet been developed. Few series have been reported and these have given somewhat contradictory findings. Positive blastogenic responses to *A fumigatus* in whole peripheral blood cultures were found in only two of six ABPA patients (Haslam et al, 1976). This reactivity was uncoupled from both positive delayed cutaneous hypersensitivity (one reactive patient out of six) and *A fumigatus*-specific inhibition of leukocyte migration. In another series (Forman et al, 1978), three of five ABPA patients' peripheral blood lymphocytes proliferated in response to *A fumigatus* antigens. In one patient the highest blastogenic index (23.7) and total serum IgE coincided. As a group, aspergilloma patients displayed a low order of blastogenesis. The antigens in these studies were most likely adequate because they contained numerous precipitating factors when tested against sera from aspergilloma patients. The phenomenon of positive immediate hypersensitivity and anergy at the level of delayed cutaneous hypersensitivity and its in vitro correlates is reminiscent of the situation that prevails in chronic dermatophytosis due to *Trichophyton rubrum*. But a contrary view that hyperproduction of IgE is the result of an overall high Th/Ts ratio that ought to find expression in a brisk blastogenic response has not been ruled out. It is certainly worthwhile to study a larger series of APBA patients to see if T-cell anergy is present or is merely the result of technical problems in the lymphocyte transformation assay alluded to by Haslam et al (1976).

Normal monocytes adhere to *A fumigatus* hyphae and destroy them, as measured ultrastructurally and by the inhibition of uptake of ^{14}C-glutamine and ^{14}C-uracil (Diamond et al, 1983). In effecting damage, the cell wall remains intact but the plasma membrane becomes detached from the wall and invaginates. It is as though the wall offers no impediment to damaging the protoplast. Damage was not potentiated by opsonins because neither autologous serum nor precipitin-positive serum increased damage. Although anaerobiosis inhibited hyphal damage resulting from normal monocytes, monocytes from chronic granulomatous disease patients worked as effectively as normal monocytes even though oxidative metabolism could not be invoked. Instead, cationic proteins were involved in mediating damage. Neutrophils from these patients adhered to but could not damage *A fumigatus*. By denying surface attachment sites to monocytes, neutrophils from chronic granulomatous disease patients may reduce the efficiency of the fungicidal process.

8.7 SUMMARY

Aspergillus fumigatus and, to a lesser extent, *A nidulans, A niger,* and *A flavus* cause a broad spectrum of diseases in animals and humans with the common theme of the respiratory route of colonization or infection and pulmonary involvement. Allergic bronchopulmonary aspergillosis in atopic patients illustrates how the immune response to a fungus can cause disease. Dual type I immediate hypersensitivity and type III immune complex deposition in the lungs are both highly probable factors contributing to immunopathology. Aspergilli secrete proteinases and lower molecular weight toxins that also probably act as aggressins to potentiate pathogenicity. Invasive aspergillosis is an opportunistic infection in the immunocompromised, granulocytopenic host, wherein the diminished immune response of the host is overrun and the hyphae spread rapidly in the lung, penetrating blood vessels and disseminating hematogenously.

Radiographic findings of fleeting pulmonary shadows from eosinophilic infiltrates coincide with high serum IgE and flares of ABPA. The proportion of total serum IgE that can be demonstrated as specific for *A fumigatus* antigens is in dispute. Much existing data cannot account for the specificity of the IgE, and it is considered to represent polyclonal activation resulting from faulty immunoregulation. Until the variables of antigen adequacy, efficiency of solid phase adsorption, and interference with measurement from other immunoglobulin classes are systematically investigated, determination of the actual proportion of *A fumigatus*-specific IgE will fluctuate considerably among laboratories.

A direct correlation has been found between precipitins and disease activity in bronchopulmonary aspergillosis. The in vivo significance of the precipitins for immunopathology are the presence of immune complexes and their deposition in the lung, but the number of patients in whom these complexes have been demonstrated is still small.

Aspergilloma patients are hyperimmune in that their serum is a rich source of precipitins against many *Aspergillus* antigens. Natural levels of antibodies to *Aspergillus* in normal individuals are detected in indirect immunofluorescence, limiting the diagnostic value of that test, but increased specificity can be achieved by the use of "germlings" as the antigenic complex. Although CIE improves the ability to detect precipitins in aspergillosis, the number of reactions among normal individuals and in asthmatics with no clinical aspergillosis is also increased. Hemagglutination tests are likewise unsuitable because *A fumigatus* extracts contain hemolysins and hemagglutinins. Most of the antibodies in bronchopulmonary aspergillosis are directed against polysaccharides as detected by indirect EIA.

Precipitin tests are frequently negative in immunosuppressed patients with invasive aspergillosis. More sensitive primary binding assays like RIA or EIA can increase the chance of detecting antibodies in the invasive form of the disease.

Glucans of the aspergilli occur in various forms: In *A niger* a hot water-soluble, crystalline glucan, nigeran, is composed of a repeated unit of α-1,3 flanking α-1,4-glucose residues. An alkali-soluble α-1,3-glucan is also present. β-1,3-Glucan occurs interwoven with chitin as the fibrillar cell wall matrix. Some of the β-1,3-glucan may be water soluble.

Cell walls of *A fumigatus* do not contain nigeran and have a high galactose content. Alkaline borohydride extraction of *A fumigatus* mycelium releases a galactomannan, galactan, and glucan. Galactomannan is the only known antigenic polysaccharide. It circulates in invasive aspergillosis of the immunocompromised host as soluble complexes that can be dissociated with heat or acidic pH. The galactomannan of *A fumigatus* binds to Con A and is separated from a large amount of soluble glucan by gel permeation chromatography. The molar ratio of galactose : mannose is 1 : 1.2. Provisional structural details include linear 1,6-mannan sequences and oligogalactoside antennae three residues long terminating in galactofuranose.

Galactomannan is detected in a Farr-type RIA in which a standard radiolabeled galactomannan is precipitated as its immune complex with reference antibody in half-saturated ammonium sulfate. This reaction is inhibited by galactomannan in dissociated patient's serum. Galactomannan antigenemia can also be detected by EIA, but the series of patients in whom antigen is successfully detected is small compared with RIA. *A fumigatus* reportedly produces a C-substance distinct from the galactomannan. At present no *A fumigatus* antigen has been identified that has structural homology with C-teichoic acid or has phosphorylcholine as a determinant.

Enzymatic activity is located in precipitin arcs containing *A fumigatus*-secreted proteins in a forerunner of the enzyme immunoassays of today. Two "chymotrypsin"-proteinases have been identified as having potential clinical value because they evoke precipitins in humans with ABPA. One of these proteinases has been purified by affinity chromatography. Mycelial extracts of *A fumigatus* contain a multiplicity of protein antigens that form precipitin arcs. One of these has been purified has a molecular weight of 182,000 da and reacts with sera from aspergillosis patients. A trichloroacetic acid-soluble, nonimmunoprecipitating glycoprotein isolated from mycelial homogenates is a potent elicitor of IgG in an RIA, and predicts the level of exposure to aspergilli in humans.

Selective ammonium sulfate precipitation of mycelial extracts of *A fumigatus* is capable of separating an antigenic glycoprotein from a major complex of proteins, and from bulk neutral polysaccharide.

Research on the cell-mediated immunity in aspergillosis is not well-advanced at this time. In ABPA, a tendency has been identified towards elevated IgE, immediate skin reactions to aspergillin, and anergy at the level of delayed cutaneous hypersensitivity and in vitro blastogenic responses to *A fumigatus*. It is too early to tell if such a relationship can be corroborated in a larger series of patients.

REFERENCES

Andrews CP, Weiner MH, 1981. Immunodiagnosis of invasive pulmonary aspergillosis in rabbits. Fungal antigen detected by radioimmunoassay in bronchoalveolar lavage fluid. Amer Rev Respir Dis 124:60–64.

Azuma I, Kimura H, Hirao F, Tsubura E, Yamamura Y, Misaki A, 1971. Biochemical and immunological studies on *Aspergillus* III: chemical and immunological properties of glycopeptide obtained from *Aspergillus fumigatus*. Japan J Microbiol 15:237–246.

Bardalaye PC, Nordin JH, 1977. Chemical structure of galactomannan from the cell wall of *Aspergillus niger*. J Biol Chem 252:2584–2591.

Bardana EJ Jr, 1978. Culture and antigen variants of *Aspergillus*. J Allergy Clin Immunol 61:225–227.

Bardana EJ Jr, 1980a. The clinical spectrum of aspergillosis — part 1: epidemiology, pathogenicity, infection in animals and immunology of *Aspergillus*. CRC Crit Rev Clin Lab Sci 13:21–83.

Bardana EJ Jr, 1980b. The clinical spectrum of aspergillosis — part 2: classification and description of saprophytic, allergic, and invasive variants of human disease. CRC Crit Rev Clin Lab Sci 13:85–159.

Bardana EJ Jr, McClatchy JK, Farr RS, Minden P, 1972a. The primary interaction of antibody to components of aspergilli I. Immunologic and chemical characteristics of a nonprecipitating antigen. J Allergy Clin Immunol 50:208–221.

Bardana EJ Jr, McClatchy JK, Farr RS, Minden P, 1972b. The primary interaction of antibody to components of aspergilli II. Antibodies in sera from normal persons and from patients with aspergillosis. J Allergy Clin Immunol 50:222–234.

Bardana EJ Jr, Gerber JD, Craig S, Cianciulli FD, 1975. The general and specific humoral immune response to pulmonary aspergillosis. Amer Rev Respir Dis 112:799–805.

Biguet J, Tran Van Ky P, Andrieu S, Fruit J, 1964. Analyse immunoelectrophoretique d'extraits cellulaires et de milieux de culture d'*Aspergillus fumigatus* par des immunserums experimentaux et des serums de malades atteints d'aspergillome bronchopulmonaire. Ann Inst Pasteur (Paris) 107:72–97.

Bobbitt TF, Nordin JH, 1978. Hyphal nigeran as a potential phylogenetic marker for *Aspergillus* and *Penicillium* species. Mycologia 70:1201–1211.

Bull AT, 1970. Chemical composition of wild-type and mutant *Aspergillus nidulans* cell walls. The nature of polysaccharide and melanin constituents. J Gen Microbiol 63:75–94.

Calvanico NJ, DuPont BL, Huang CJ, Patterson R, Fink JN, Kurup VP, 1981. Antigens of *Aspergillus fumigatus*. 1. Purification of a cytoplasmic antigen reactive with sera of patients with *Aspergillus*-related disease. Clin Exp Immunol 45:662–671.

Chaparas SD, Kaufman L, Kim SJ, McLaughlin DW, 1980. Characterization of antigens from *Aspergillus fumigatus*. V. Reactivity in immunodiffusion tests with serums from patients with aspergillosis caused by *Aspergillus flavus, A niger,* and *A fumigatus.* Amer Rev Respir Dis 122:647–650.

Dessaint JP, Bout D, Fruit J, Capron A, 1976. Serum concentration of specific IgE antibody against *Aspergillus fumigatus* and identification of the fungal allergen. Clin Immunol Immunopathol 5:314–319.

Diamond RD, Huber E, Haudenschild CC, 1983. Mechanisms of destruction of *Aspergillus fumigatus* hyphae mediated by human monocytes. J Infect Dis 147:474–483.

Dorval G, Yang WH, Osterland CK, Jennings BA, Toogood JH, 1979. Circulating immune complexes and activation of the complement sequence in acute allergic bronchopulmonary aspergillosis. J Allergy Clin Immunol 63:204 (abstract).

Drouhet E, Camey L, Segretain G, 1972. Valeur de l'immunoprecipitation et de l'immunofluorescence indirecte dans les aspergilloses broncho-pulmonaires. Ann Inst Pasteur (Paris) 123:379–395.

Forman SR, Fink JN, Moore VL, Wang J, Patterson R, 1978. Humoral and cellular immune responses in *Aspergillus fumigatus* pulmonary disease. J Allergy Clin Immunol 62:131–136.

Fraser DW, Ward JI, Ajello L, Plikaytis BD, 1979. Aspergillosis and other systemic mycoses. The growing problem. JAMA 242:1631–1635.

Geha RS, 1977. Circulating immune complexes and activation of the complement sequence in acute bronchopulmonary aspergillosis. J Allergy Clin Immunol 60:357–359.

Girault A, Boyer JP, Senet JM, Girault M, 1977. Isolement et purification de proteines d'*Aspergillus fumigatus* presentant respectivement une activite chymotrypsique et catalasique. Bull Soc Fr Mycol Med 6:75–79.

Glaudemans CPJ, Manjula BN, Bennett LG, Bishop CT, 1977. The binding of phosphorylcholine-containing antigens from *Streptococcus pneumoniae* to myeloma immunoglobulins M603 and H-8. Immunochem 14:675–679.

Gold JWM, Fisher B, Yu B, Chein N, Armstrong D, 1980. Diagnosis of invasive aspergillosis by passive hemagglutination assay of antibody. J Infect Dis 142:87–94.

Gordon MA, Lapa EW, Kane J, 1977. Modified indirect fluorescent antibody test for aspergillosis. J Clin Microbiol 6:161–165.

Haslam P, Lukoszek A, Longbottom JL, Turner-Warwick M, 1976. Lymphocyte sensitization to *Aspergillus fumigatus* antigens in pulmonary diseases in man. Clin Allergy 6:277–291.

Hearn VM, Mackenzie DWR, 1979. The preparation and chemical composition of fractions from *Aspergillus fumigatus* wall and protoplasts possessing antigenic activity. J Gen Microbiol 112:35–44.

Hearn VM, Mackenzie DWR, 1981. Analysis of wall antigens of *Aspergillus fumigatus* by two dimensional immunoelectrophoresis. J Med Microbiol 14:119–129.

Hearn VM, Wilson EV, Proctor AG, Mackenzie DWR, 1980. Preparation of *Aspergillus fumigatus* antigens and their analysis by two-dimensional immunoelectrophoresis. J Med Microbiol 13:451–458.

Holmberg K, Berdischewsky M, Young LS, 1980. Serological diagnosis of invasive aspergillosis. J Infect Dis 141:656–664.

Imbeau SA, Nichols D, Flaherty D, Dickie H, Reed C, 1978. Relationships between prednisone therapy, disease activity and the total serum IgE level in allergic bronchopulmonary aspergillosis. J Allergy Clin Immunol 62:91–95.

Johnston IR, 1965. The composition of the cell wall of *Aspergillus niger.* Biochem J 96:651–658.

Katz RM, Kniker WT, 1973. Infantile hypersensitivity pneumonitis as a reaction to organic antigens. N Engl J Med 288:233–237.

Kauffman HF, deVries K, 1980a. Antibodies against *Aspergillus fumigatus* I. Standardization of the antigenic composition. Int Arch Allergy Appl Immunol 62:252–264.

Kauffman HF, deVries K, 1980b. Antibodies against *Aspergillus fumigatus* II. Identification and quantification by means of cross immunoelectrophoresis. Int Arch Allergy Appl Immunol 62:265–275.

Kauffman HF, Beaumont F, Meurs H, van der Heide S, deVries K, 1983. Comparison of antibody measurements against *Aspergillus fumigatus* by means of double-diffusion and enzyme-linked immunosorbent assay (ELISA). J Allergy Clin Immunol 72:255–261.

Kaufman L, 1982. Immunoserology of fungal infections. pp. 130–133. In Rapid Methods and Automation in Microbiology. RC Tilton ed. Washington: American Society Microbiology.

Kim SJ, Chaparas SD, 1978. Characterization of antigens from *Aspergillus fumigatus* I. Preparation of antigens from organisms grown in completely synthetic medium. Amer Rev Respir Dis 118:547–560.

Kim SJ, Chaparas SD, Brown TM, Anderson MC, 1978. Characterization of antigens from *Aspergillus fumigatus* II. Fractionation and electrophoretic immunologic, and biologic activity. Amer Rev Respir Dis 118:553–560.

Kim SJ, Chaparas SD, 1979. Characterization of antigens from *Aspergillus fumigatus* III. Comparison of antigenic relationships of clinically important aspergilli. Amer Rev Respir Dis 120:1297–1311.

Kurup VP, Fink JN, 1978. Evaluation of methods to detect antibodies against *Aspergillus fumigatus*. Am J Clin Pathol 69:414–417.

Kurup VP, Ting EY, Fink JN, 1983. Immunochemical characterization of *Aspergillus fumigatus* antigens. Infect Immun 41:698–701.

Lazda VA, 1977. The fifth annual meeting of the mid-west immunology conference. Fed Proc 36:2465–2469.

Lehmann PF, Reiss E, 1978. Invasive aspergillosis: antiserum for circulating antigen produced after immunization with serum from infected rabbits. Infect Immun 20:570–572.

Longbottom JL, 1978. Immunological aspects of infection and allergy due to *Aspergillus* species. Mykosen Suppl 1:201–217.

Longbottom JL, 1983a. Antigens/allergens of *Aspergillus fumigatus*. Identification of antigenic components reacting with both IgG and IgE antibodies of patients with allergic bronchopulmonary aspergillosis. Clin Exp Immunol 53:354–362.

Longbottom JL, 1983b. Allergic bronchopulmonary aspergillosis: reactivity of IgE and IgG antibodies with antigenic components of *Aspergillus fumigatus* (IgE/IgG antigen complexes). J Allergy Clin Immunol 72:668–675.

Longbottom JL, Pepys J, 1964. Pulmonary aspergillosis: diagnostic and immunological significance of antigens and C-substance in *Aspergillus fumigatus*. J Pathol Bacteriol 88:141–151.

Luster MI, Leslie GA, Bardana EJ Jr 1976. Structure and biological functions of human IgD. VI. Serum IgD in patients with allergic bronchopulmonary aspergillosis. Int Arch Allergy Appl Immunol 50:212–219.

Malo JL, Hawkins R, Pepys J, 1977a. Studies in chronic allergic bronchopulmonary aspergillosis. I. Clinical and physiological findings. Thorax 32:254–261.

Malo JL, Pepys J, Simon G, 1977b. Studies in chronic allergic bronchopulmonary aspergillosis. II. Radiological findings. Thorax 32:262–268.

Malo JL, Longbottom J, Mitchell J, Hawkins R, Pepys J, 1977c. Studies in chronic allergic bronchopulmonary aspergillosis. III. Immunological findings. Thorax 32:269–274.

Marier R, Smith W, Jansen M, Andriole VT, 1979. A solid-phase radioimmunoassay for measurement of antibody to *Aspergillus* in invasive aspergillosis. J Infect Dis 140:771–779.

McCarthy DS, Pepys J, 1971. Allergic bronchopulmonary aspergillosis. Clinical immunology: (2) skin, nasal and bronchial tests. Clin Allergy 1:415–432.

McGinnis MR, 1980. Laboratory Handbook of Medical Mycology. 661 pp. New York: Academic Press.

Odds FC, Ryan MD, Sneath PH, 1983. Standardization of antigens from *Aspergillus fumigatus*. J Biol Stand 11:157–162.

Patterson R, Roberts M, 1974. IgE and IgG antibodies against *Aspergillus fumigatus* in sera of patients with bronchopulmonary allergic aspergillosis. Inter Arch Allergy 46:150–160.

Pepys J, 1977. Clinical and therapeutic significance of patterns of allergic reactions of the lungs to extrinsic agents: the 1977 J. Burns Amberson Lecture. Amer Rev Respir Dis 116:573–588.

Reiss E, Lehmann PF, 1979. Galactomannan antigenemia in invasive aspergillosis. Infect Immun 25:357–365.

Rosenberg M, Patterson R, Roberts M, Wang J, 1978. The assessment of immunologic and clinical changes occurring during corticosteroid therapy for allergic bronchopulmonary aspergillosis. Am J Med 64:599–606.

Sakaguchi O, Suzuki M, Yokota K, 1968. Effect of partial acid hydrolysis on precipitin activity of *Aspergillus fumigatus* galactomannan. Japan J Microbiol 12:123–124.

Sakaguchi O, Yokota K, Suzuki M, 1969. Immunochemical and biochemical studies of fungi XIII. On the galactomannans isolated from mycelia and culture filtrates of several filamentous fungi. Japan J Microbiol 13:1–7.

Sandhu RS, Bardana EJ, Kahn ZU, Dordevich DM, 1978. Allergic bronchopulmonary aspergillosis: studies on the general and specific humoral response. Mycopathologia 63:21–27.

Schaefer JC, Yu B, Armstrong D, 1976. An aspergillus immunodiffusion test in the early diagnosis of aspergillosis in adult leukemia patients. Amer Rev Respir Dis 113:325–329.

Schønheyder H, Andersen P, 1983. Determination of antibodies to partially purified *Aspergillus* antigens by an enzyme-linked immunosorbent assay. Int Arch Allergy Appl Immunol 70:108–111.

Senet JM, Girault A, Robert R, Girault M, 1978a. Diagnostic de l'aspergillose par hemagglutination indirecte 1. Utilisation d'une fraction chymotrypsique purifiee d'*Aspergillus fumigatus*. Bull Soc Fr Mycol Med 7:225–228.

Senet JM, Girault A, Robert R, Girault M, 1978b. Diagnostic de l'aspergillose par hemagglutination indirecte 2. Utilisation d'une fraction catalasique purifee d'*Aspergillus fumigatus*. Bull Soc Fr Mycol Med 7:229–232.

Sepulveda R, Longbottom JL, Pepys J, 1979. Enzyme linked immunosorbent assay (ELISA) for IgG and IgE antibodies to protein and polysaccharide antigens of *Aspergillus fumigatus*. Clin Allergy 9:359–371.

Shaffer PJ, Kobayashi GS, Medoff G, 1979. Demonstration of antigenemia in patients with invasive aspergillosis by solid phase (protein A-rich *Staphylococcus aureus*) radioimmunoassay. Am J Med 67:627–630.

Stagg CM, Feather MS, 1973. The characterization of a chitin-associated D-glucan from the cell walls of *Aspergillus niger*. Biochem Biophys Acta 320:64–72.

Suzuki M, Hayashi Y, 1975. Skin reaction and macrophage migration inhibition tests for polysaccharides from *Aspergillus fumigatus* and *Candida albicans*. Japan J Microbiol 19:335–362.

Suzuki S, Suzuki M, Yokota K, Sunayama H, Sakaguchi O, 1967. On the immunochemical and biochemical studies of fungi. XI. Crossreaction with the polysaccharides of *Aspergillus fumigatus*, *Candida albicans*, *Saccharomyces cerevisiae* and *Trichophyton rubrum* against *Candida albicans* and *Saccharomyces cerevisiae* antisera. Japan J Microbiol 11:269–273.

Tran Van Ky P, Torck C, Vaucelle T, Floc'h F, 1969. Etude comparee sur immunoelectrophoregramme des enzymes de l'extrait antigenique d'*Aspergillus fumigatus*, reveles par des serums experimentaux et des serums de malades atteints d'aspergillose. Sabouraudia 7:73–84.

Tran Van Ky P, Uriel J, Rose F, 1966. Caracterisation de type d'activites enzymatiques dans des extraits antigeniques d'*Aspergillus fumigatus* apres electrophorese et immunoelectrophorese en agarose. Ann Inst Pasteur (Paris) 111:161–170.

Turner-Warwick M, 1975. Immunologic lung disease due to *Aspergillus*. Chest 68:346–355.

Vervloet D, Fujita Y, Wypych JI, Reisman RE, Arbesman CE, 1974. The inhibitory effect of serum factors on measurement of IgE *Aspergillus* antibodies by RAST. Clin Allergy 4:359–369.

Walter JE, Jones RD, 1968. Serologic tests in diagnosis of aspergillosis. Dis Chest 53:729–735.

Warren RC, White LO, Mohan S, Richardson MD, 1979. The occurrence and treatment of false-positive reactions in enzyme-linked immunosorbent assays (ELISA) for the presence of fungal antigens in clinical samples. J Immunol Meth 28:177–186.

Watson MJ, Baddiley J, 1974. The action of nitrous acid on C-teichoic acid (C-substance) from walls of *Diplococcus pneumoniae*. Biochem J 137:399–404.

Weiner MH, 1980. Antigenemia detected by radioimmunoassay in systemic aspergillosis. Ann Intern Med 92:793–796.

Weiner MH, Coats-Stephen M, 1979. Immunodiagnosis of systemic aspergillosis I. Antigenemia detected by radioimmunoassay in experimental infection. J Lab Clin Med 93:111–119.

Weiner MH, Talbot GH, Gerson SL, Filice G, Cassileth PA, 1983. Antigen detection in the diagnosis of invasive aspergillosis. Utility in controlled, blinded trials. Ann Intern Med 99:777–782.

Wilson EV, Hearn VM, 1983. Use of *Aspergillus fumigatus* mycelial antigens in enzyme-linked immunosorbent assay and counter-immunoelectrophoresis. J Med Microbiol 16:97–105.

Young RC, Bennett JE, 1971. Invasive aspergillosis: absence of detectable antibody response. Amer Rev Respir Dis 104:710–716.

Chapter 9

Dermatophytes

9.1 INTRODUCTION

Dermatophyte fungi of the genera *Microsporum, Trichophyton,* and *Epidermophyton* cause infections restricted to the nonviable keratinized layers of the skin, hair, and nails. They are taxonomically related to soil-dwelling keratinophilic ascomycetes, the gymnoascaceae. The sexual ascogenous forms of dermatophytes are classified in the genera *Arthroderma* and *Nannizzia*. Dermatophytes are grouped according to their origin, whether geophilic, zoophilic, or anthropophilic. The soil-dwelling species usually cause only transient infections in animals or humans. Zoophilic fungi, for example *Microsporum canis,* transmit disease to the human host that is typically inflammatory and heals spontaneously. The large, raised, suppurative lesions associated with these infections are known as kerions. Kerions contain crusts, matted hair, and exudate. Anthropophilic species causing human infections are the most highly adapted dermatophytes and rarely cause disease in lower animals. The lesions resulting from infection have reduced inflammation, tending to become chronic and widespread, for example, infections with *Trichophyton rubrum, Trichophyton mentagrophytes* var *interdigitale,* and *Epidermophyton floccosum.*

In the USA, the major agents of prepubertal ringworm of the scalp (tinea capitis) are *Microsporum audouinii* or *Trichophyton tonsurans,* but abroad *Trichophyton violaceum* is a common cause of tinea capitis. *Trichophyton schoenleinii* causes chronic favus of the scalp and smooth skin. Favus is an exaggerated cutaneous reaction in dermatophyte lesions. These persistent lesions form crusts with weeping exudates and underlying inflammation. Favic lesions can become secondarily infected with bacteria.

Mycelia grow in the hair shaft in tinea capitis, but different patterns of hair invasion occur according to the species of dermatophyte involved. Arthroconidia, small spores released by fission, form along the outside of the hair shaft in the ectothrix type of invasion (Chandler et al, 1980). In the endothrix type, arthroconidia form inside the hair shaft. Favus infections (Rippon, 1982) of the hair produce mycelia, air spaces, and few arthroconidia. The degree of pathogenicity and hair invasion by the dermatophytes has been related by Rippon (1982) and by Grappel (1981) to the elaboration of different kinds and amounts of proteinases.

Dermatophytids are allergic reactions encountered in some dermatophytosis patients who develop sterile vesicular eruptions remote from the site of infection. In the extreme, these eruptions may cover large areas of the body and cause much discomfort.

Dermatophytes lack the capacity to invade living tissue, but the cutaneous reactions they evoke are obviously accessible and are excellent probes for studying the immunology of the skin. A predilection towards chronic dermatophytosis has been noted in congenital thymic aplasia. (Fudenberg et al, 1980), but other researchers hold that the majority of chronic dermatophytosis patients do not have a primary immunodeficit (Grappel, 1981). The commonality of exposure to dermatophytes in the general population makes these fungi suitable for inclusion in a battery of microbial antigens for immunocompetence testing (Fudenberg et al, 1980).

9.2 CELL WALL

The cell wall analyses of the dermatophytes are among the more rigorously accomplished cell wall studies for mycotic agents (Table 9.1). Objective criteria for purity were observed and levels of lipid and proteins are reasonably low. *T mentagrophytes* mycelia were macerated in a Waring blender, then in a tissue grinder. The slurry was then disrupted in a Ribi fractionator at 30,000 to 35,000 psi (Noguchi et al, 1971; 1975) and layered on a two-step discontinuous gradient of 0.4M and 1.6M sucrose. The cell wall fraction was recovered from the interface and digested with a sequence of RNAse and proteinases including papain, nagarse, and trypsin. Finally the walls were defatted with organic solvents. The use of anionic or nonionic detergents possibly could accomplish both the removal of extraneous protein and residual plasmalemma. There is controversy about the presence of galactosamine in dermatophyte antigen preparations (Young and Roth, 1979).

Kitajima and Nozawa (1975) prepared isolated cell walls of *E floccosum* by mechanical disruption. The walls then were digested with snail gut enzymes (mixed glycosidases or "glusulase" from *Helix pomatia*, the vineyard snail) and the resistant residue was examined in the electron microscope. Intact walls had a narrow elec-

Table 9.1. Cell Wall Composition of *Trichophyton mentagrophytes* and *Epidermophyton floccosum*

Component[a]	T mentagrophytes[b]	E floccosum[c]
Glucose	36.2	45.8
Mannose	11.7	6.7
Galactose	tr	nd
Glucosamine and N-acetylglucosamine	30.4	30.0
Protein	7.8	7.4
Lipid	6.6	3.3
Phosphate	0.1	nd
Galactosamine	nd	0.4
Recovery (%)	92.8	93.6

Source: Reprinted from Reiss (1985) with permission of the publisher, Marcel Dekker, Inc.

Abbreviations: tr, trace; nd, not determined.
[a] Results recorded as percent dry weight of cell walls.
[b] Noguchi et al, 1975.
[c] Shah and Knight 1968.

tron-dense outer layer and a wide electron-lucent inner layer. After enzymolysis, the inner layer was no longer evident and the preparation consisted of thin, tubular, electron-dense outer layers. Compositional analysis of glusulase-resistant outer layers revealed: protein, 65%; glucosamine, 17%; and mannose, 10%. Outer layers were subject to sodium dodecyl sulfate electrophoresis and multiple bands were seen, the most prominent was a PAS-positive, presumptive glycoprotein. It would be of interest to know if this procedure conserved antigenic determinants that would otherwise be lost or simply not extracted by hot alkali or ethylene glycol treatment.

9.3 GALACTOMANNAN POLYSACCHARIDES AND GLYCOPEPTIDES

Galactomannan antigens of the dermatophytes occur in two forms based on the method of extraction from whole mycelia. Barker et al (1962, 1963, 1967) studied glycopeptides soluble in warm ethylene glycol that have trichophytin activity. Bishop et al (1965, 1966) used hot alkali to solubilize polysaccharides that are unable to evoke delayed cutaneous hypersensitivity but retain some serologic activity.

9.3.1 Isolation

Glycopeptide antigens of *Trichophyton mentagrophytes* solubilized from surface mats with ethylene glycol are precipitated with cetyltrimethylammonium bromide in borate buffer. The galactomannan peptide (80% to 90% carbohydrate) contains galactose:mannose = 1:8.1 and 8.8% protein, $[\alpha]_D^{17} = 32.6°$, 1% water. Structural analysis was carried out on this material (Barker et al, 1963) before its heterogeneity was realized. Mild acid hydrolysis released the galactose suggesting terminal galactofuranosides. Methylation-fragmentation gas-liquid chromatography detected mannosyl linkages in the main chain that are 1,2 and 1,4. Approximately 39% of the mannose residues occurred as branch points. Later, the galactomannan peptide was separated into three fractions numbered according to their order of elution from DEAE Sephadex: I, galactose:mannose = 1:9.7, 32% protein; II, galactose:mannose = 1:9.8, 11.3% protein; and a minor component, III, galactose:mannose = 1:14.4, 12.1% protein. These galactomannan peptides are capable of eliciting trichophytin cutaneous reactions in sensitized animals and humans. Degradation of the carbohydrate portion of the trichophytin glycopeptides by periodate oxidation reduced their ability to evoke immediate hypersensitivity, whereas proteolysis with ficin reduced the delayed cutaneous hypersensitivity reaction to insignificant levels (Barker et al, 1962; 1967).

Young and Roth (1979) extracted trichophytin glycopeptides from cultures of *T mentagrophytes* using the same ethylene glycol procedure (Barker et al, 1963; Ottaviano et al, 1974). In this instance, the resulting material contained 80% carbohydrate, 65% as neutral sugar and 35% as amino sugar. The neutral sugar fraction was composed of equimolar amounts of galactose and mannose. The amino sugar component had N-acetylglucosamine and N-acetylgalactosamine in a ratio of 1.5:1. Amino sugars were not detected previously in trichophytin glycopeptides by Barker et al (1963, 1967) and the galactose:mannose ratios are quite different from those previously reported. Galactosamine was not found in the wall of *T mentagrophytes* by Noguchi et al (1971). The presence of N-acetylgalactosamine in trichophytin glycopeptides seems important to verify in view of the report that antibodies to the glycopeptides crossreact with the blood group A substance in which N-acetylgalactosamine is an immunodominant epitope. Moreover, the glycopeptides could partially inhibit the agglutination of blood group A erythrocytes by isoagglutinins.

Although a chemically defined medium for the production of trichophytin glycopeptides has been recommended by Ottaviano et al (1974) the more recent study (Young and Roth, 1979) added peptones and yeast extract to the medium. Blood group substance A can be isolated directly from peptone. Nevertheless, a sample of trichophytin glycopeptide grown on chemically defined medium also precipitated with blood group substance A in agar gel. The use of chemically defined medium for production of trichophytin glycopeptides is recommended because the standard protocol (Barker et al, 1963, 1967) uses the entire culture including the mycelia and culture supernate as the material for extraction.

The polysaccharide antigens are solubilized from defatted and trypsin-treated dermatophyte mycelia with hot 0.75M NaOH (Bishop et al, 1965; 1966). Under such conditions, alkaline degradation and fragmentation occurs. Reducing terminals can fragment releasing glyceraldehyde (Sharon, 1975). Noncarbohydrate determinants such as O-acetyl groups would be lost; O-glycosyl serine and threonine linkages would be β-eliminated (Sharon, 1975); and lipid constituents would be saponified. Nevertheless, these polysaccharides retain some biologic activity. Two galactomannans are separated on the basis of differential solubility in Fehling solution. Galactomannan I is precipitable and consists of a lightly galactose-substituted mannan (Grappel et al, 1969) with a galactose:mannose ratio of 2.5:97.5 for *Microsporum praecox*. Stripping of galactofuranose with dilute acid did not appear to alter the serologic activity. Galactomannan II is soluble in Fehling solution, binds to DEAE-cellulose and is eluted with 0.1M NaOH (Grappel et al, 1970). Galactose is more abundant in the galactomannan II than in galactomannan I: for *Trichophyton mentagrophytes* var *interdigitale*, galactose:mannose is in a ratio of 33% to 67%; for *Microsporum praecox*, galactose:mannose is in a ratio of 22% to 78%.

9.3.2 Purification

Grappel et al devised a scheme to separate glucan and the two galactomannans extracted with alkali (Grappel et al, 1970). First, addition of Fehling solution precipitates the lightly galactosylated galactomannan I. Galactomannan II and glucan in the supernate are resolved on a DEAE-cellulose column with borate buffers. Glucan appears in the effluent and galactomannan II is eluted with 0.1M NaOH. Other reports of dermatophyte polysaccharides containing glucose as well as galactose and mannose may be suspected of being mixtures of glucan and galactomannan (Moser and Pollack, 1978) (Table 9.2).

Further delineation of antigens in the immune response to the dermatophytes continues, but it should be emphasized that the validity of trichophytin glycopeptides prepared by the ethylene glycol-borate-cetyltrimethylammonium bromide procedure of Barker et al (1963, 1965) has been maintained in more recent investigations (Ottaviano et al, 1974; Kaamen et al, 1976).

The objective of purifying an antigen with trichophytin activity was approached in a different way by suspending *T mentagrophytes* mycelia in 0.5M trichloroacetic acid; the extract was then precipitated with ethanol, dialyzed, and lyophilized (Arnold et al, 1976). This material was sized on a Sephadex G200 column and the excluded portion was then fractionated on DEAE-cellulose (borate form) into five components. Two of these fractions were serologically active and resulted in a line of identity in immunodiffusion against sera from rabbits immunized with acetone

Table 9.2. Comparison of Properties of Glycopeptides and Polysaccharides of Dermatophytes

Dermatophyte	Designation of Polymer		Gal:Man	Protein as Amino Acid (percent)	Fehling Solubility	Sephadex G Series	Diethyl-amino-ethyl
Trichophyton granulosum[a]	polysaccharide	GMI	1:12.3	0	Precipitated	nd	nd
		GMII	1:2.4	0	soluble	nd	nd
Trichophyton mentagrophytes var granulosum[b]	peptido-polysaccharide	fraction 3	1:10.4	8.6	nd	G200 excluded	eluate 0.3M salt
		fraction 5	1:29.7	6.2	nd		eluate pH 5
Trichophyton mentagrophytes[c]	glycopeptide	DSI	1:10.9	5.8	nd	G200 excluded	effluent
		DSII	1:3.08	7.3	nd	G200 included	eluate
Microsporum canis[d]	glycopeptide	B1-1	1:14.4	3.4	nd	G100 excluded	effluent
		B2-2	1:2.7	2.4	nd	G100 included	eluate

Source: [a]Bishop et al, 1965, 1966; [b]Arnold et al, 1976; [c]Moser and Pollack, 1978; and [d]Basarab et al, 1968.
Abbreviation: nd, not determined.

powdered mycelia. Mannose was the predominant carbohydrate constituent (eg, the mannose:galactose:glucose ratio was 7.5:0.7:1 for fraction 3). Immunoelectrophoresis of each fraction resulted in a line of low cathodic mobility. Fraction 3 also contained 8.6% amino acids, 1.8% amino sugar, and phosphate was 1.2% of the dry weight. The antigenicity of this preparation with respect to less pure trichophytin or conventionally prepared trichophytin glycopeptides remains to be demonstrated (Table 9.3).

9.3.3 Structure

The structural analyses of dermatophyte mannans by Bishop et al (1965, 1966) remain the definitive references, but reinvestigation by combined GLC–mass spectrometry and by carbon 13 nuclear magnetic resonance is required. Methylation-fragmentation analysis revealed differences between the two galactomannans. The methylated monosaccharides derived from galactomannan I gave evidence of linear sequences of 1,6-linked mannose residues with single mannose or galactofuranose side substituents (Bishop et al, 1965) (Figure 9.1). A more complex pattern emerged for the heavily galactosylated galactomannan II (Bishop et al, 1966) (Figure 9.2): Galactofuranose units exist as nonreducing termini, relatively more 1,2-linked mannose is present than in galactomannan I, and almost all of the branch points are linked through C-1, C-2, and C-6. No oligosaccharide fragments were isolated to assist in the assignment of side chains, but Bishop et al proposed a backbone of alternating 1,2 and 1,6 bonds. If galactomannan II is amenable to partial acetolysis, oligosaccharide fragments might be recovered that would aid further structural analysis.

Table 9.3. Variations in Conditions of Preparing Trichophytin for Intradermal and In Vitro Blastogenesis Tests

Organism	Medium	Growth Period (days)	Antigen Extraction	Product
T mentagrophytes (Tagami et al, 1973)	Sab[a]	14	Dehydrated mycelia ground, extracted with water, acid insoluble precipitate removed, add 3 vol ethanol, recover precipitate, dialyze, lyophilize	Homogenate-supernate
T mentagrophytes var asteroides (Ottaviano et al, 1974)	Chemically defined	7–8	Acetone powder mycelia extracted with ethylene glycol, precipitated with ethanol, purified with borate-cetyltrimethyl-ammonium bromide	Mycelial extract
Microsporum audouinii (Allen et al, 1977)	Sab	21	Mycelia, heat-killed (60 °C), homogenized	Whole cell homogenate
T mentagrophytes (Green and Balish, 1979)	Sab	90	Acetone powder mycelia Autoclaved mycelia	Particulate Heat-stable aqueous extract
T rubrum, M. vanbreuseghemii, Epidermophyton floccosum (Hunjan and Cronholm, 1979)	Sab	21–28	Mycelia ground in saline, sonified, supernate filtered and lyophilized	Homogenate-supernate
T quinckeanum (Hunziker and Brun, 1980)	Chemically defined	16 weeks	Culture filtrate precipitated with 10 vols methanol	Culture filtrate
T rubrum MRL 1570 T mentagrophytes var interdigitale MRL 1564 (Hay and Shennan, 1982)	Sab	7	Mycelia disrupted in "Dynomill"; supernate dialyzed, lyophilized Culture filtrate, lyophilized	Homogenate-supernate Culture filtrate

[a] Sab, Sabouraud's broth.

The dermatophyte glycopeptides and polysaccharides, although extracted by different means, are probably the same mixture of two galactomannans. One is a lightly galactosylated polymer that is higher in molecular weight and does not bind anion exchangers, and the second is a more heavily galactose-substituted, lower molecular weight species that acquires a net negative charge in borate buffer sufficient to bind it to an anion exchange column. The designation glycopeptide does not mean that the molecular weight is lower than the polysaccharide, but rather that no proteinase treatment was used. Methylation analysis of the galactomannan peptides by Barker et al (1963) does not support the idea that the glycopeptides share a common carbohydrate structure with the polysaccharides, because no derivative corresponding to a 1,6 linkage was identified. This discrepancy should be reinvestigated because at the time (Barker et al, 1963) gas-liquid chromatography was not in general use and the available methods gave less than complete methylation.

9.3 GALACTOMANNAN POLYSACCHARIDES AND GLYCOPEPTIDES

$$\rightarrow 6\,\text{Man}(1\rightarrow 6)\begin{bmatrix}\alpha\,\text{Man} \\ 2 \\ \uparrow\alpha \\ 1 \\ \text{Man}\end{bmatrix}_7 (1\rightarrow 6)\begin{bmatrix}\alpha\,\text{Man}(1\rightarrow 6)\alpha\,\text{Man}\end{bmatrix}_{10}(1\rightarrow 6)\begin{bmatrix}\alpha\,\text{Man} \\ 2 \\ \uparrow\alpha \\ 1 \\ \text{Gal}\,f\end{bmatrix}(1\rightarrow 6)\alpha\,\text{Man}$$

Figure 9.1. Proposed repeat unit structure of the lightly galactosylated galactomannan I of dermatophytes. All sugars are in the pyranose form except galactofuranose. *Source:* Redrawn from Bishop et al, (1965) to exclude mannofuranose.

- *Exophiala (Cladosporium) werneckii.* A peptidophosphogalactomannan was extracted from whole cells with hot neutral buffer and selectively precipitated as the borate complex with cetyltrimethylammonium bromide (Lloyd, 1970). Quantitative analysis showed galactose, 14%; mannose, 78%; phosphate, 3.2%; and 11.3% protein. A major component eluted from an anion exchange column (fraction IV) has a galactose:mannose ratio of 1:6.5 and 5.1% phosphate. The major structural features are 1,2-linked mannose residues with lesser amounts of 1,3 and 1,6 linkages. All the galactose exists as nonreducing termini with a galactofuranose:galactopyranose ratio of 3:2. Galactomannan sequences are joined in the main chain by phosphodiesters, so that mild acid hydrolysis causes a reduction in molecular weight. (Figure 9.3) On the other hand, mild alkaline borohydride treatment cleaves glycopeptide bonds by β-elimination of hydroxyamino acid linkages, liberating mannose oligosaccharides. The resulting phosphogalactomannan (50,000 to 60,000 da, with galactose:mannose:PO_4 molar ratios of 1:3.7:0.55) can no longer bind to concanavalin A (Lloyd, 1972). Further studies have shown that one in five mannosyl residues is O-acetylated, and that these are important for serologic activity.

9.3.4 C-Substance

Cell-free filtrates of aged (6-month-old) cultures of *E floccosum* were the source of a peptidomannan with C-substance activity (Baldo et al, 1977). This preparation was chromatographed on a con A agarose column and the mannan that was eluted with methyl-α-mannoside contained 56% mannose, 31% protein and a small amount (less than 1%) of phosphorylcholine. The latter compound was detected in neutral-

Figure 9.2. Proposed structural detail of galactomannan II from *Microsporum quinckeanum*. Man, mannose; gal *f*, galactofuranose. *Source:* Redrawn from Bishop et al, (1966) to delete mannofuranose.

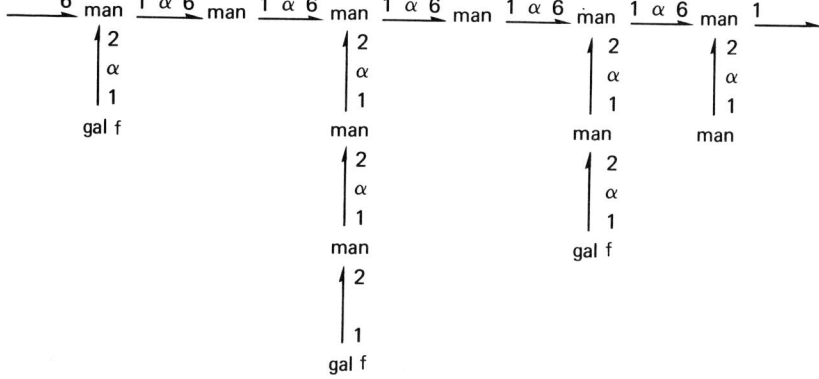

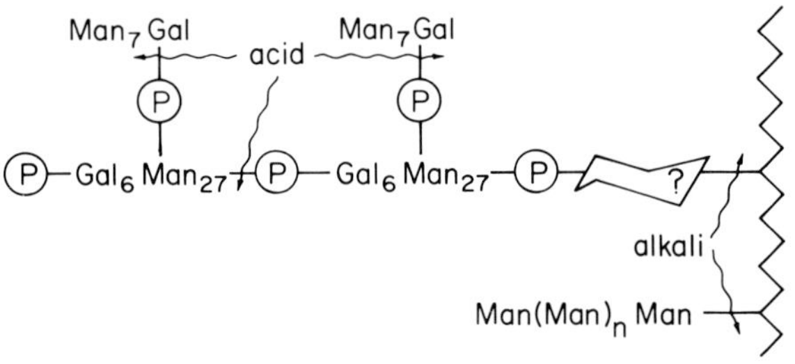

Figure 9.3. Proposed molecular organization of the peptidophosphogalactomannan of *Exophiala (Cladosporium) werneckii*. Man, mannose; Gal, galactose; P, phosphodiester linkage. The polysaccharide is linked to protein, shown as the zig-zag line at the right of the figure. *Source:* Lloyd (1972).

ized acid hydrolyzates by thin-layer chromatography. The mannan was precipitated by human serum containing C-reactive protein, washed, dissolved in sodium citrate solution, and then reprecipitated with con A. This behavior provided evidence that the C-substance activity and con A reactive sites, presumably mannan, were covalently bound. The quantitative precipitin reactions of this mannan and pneumococcal C-substance are on a different scale of activity: about 10 times more *E floccosum* mannan than pneumococcal C-substance is required to achieve maximum immune precipitation of C-reactive protein. The *E floccosum* mannan reacted with phosphorylcholine-specific mouse myeloma protein MOPC-8, and the complex was dissolved with phosphorylcholine.

The presence of small amounts of C-substance determinants in mannans derived from aged cultures of dermatophytes has no known clinical significance. Baldo et al (1977) pointed out that patients with allergic bronchopulmonary aspergillosis displayed positive immediate hypersensitivity when materials with C-substance activity were introduced into the skin in prick tests. The absence of C-substance activity in traditionally prepared dermatophytic polysaccharides is probably the result of saponification reactions during the hot alkali extraction method.

9.3.5 Crossreactivity and Taxonomic Value

The evidence summarized by Grappel et al (1974) is that cutaneous reactions evoked by dermatophytic glycopeptides in guinea pigs show no species specificity. Of the polysaccharides that were studied, serologically active galactomannan II purified from nine different dermatophytes have crossreactive and species-specific factors. Measurements were made by quantitative complement-fixation utilizing sera from rabbits immunized with killed mycelia of each species of dermatophyte (Grappel et al, 1970). The differences observed were incremental and are not a sufficient criterion to aid in the classification of these molds.

9.3.6 Humoral Responses

Measurement of the antibody responses in dermatophytosis is not an active area of research. Humoral immunity has no known role either in resistance to infection or as an aid in diagnosis. The clinical relevance of dermatophyte polysaccharides remains unknown. Precipitins in patients with tinea capitis react with saline extracts of autoclaved mycelia but not with galactomannan polysaccharides. Evidence for some recognition of these polysaccharides was obtained because patients' sera caused immune agglutination of polysaccharide-coated charcoal particles (Grappel et al, 1972). Saline extracts of autoclaved mycelia from *M audouinii*, *T mentagrophytes*, *T rubrum*, and *T tonsurans* were compared with dermatophyte galactomannan polysaccharides in serologic tests of dermatophytosis patients and apparently dermatophyte-free normal humans (Grappel et al, 1971, 1972). Charcoal particles were coated with either the saline-extracted antigen or with the homologous polysaccharides. These reagents then were used in agglutinin, complement-fixation, and immunodiffusion tests. Some patients reacted in the complement-fixation or immunodiffusion tests with the saline-extracted antigen, but the polysaccharides were not active. Normal individuals reacted with the saline extract, but had titers lower than those formed with patients' sera.

Immunoglobulin G and IgM levels were determined by indirect enzyme immunoassay in sera of dermatophytosis patients (Kaaman et al, 1981). The antigen bound as the solid phase was trichophytin glycopeptides prepared by the ethylene glycol method (Barker et al, 1963, 1967; Ottaviano et al, 1974). No significant differences in IgM levels were observed between patients and controls consisting of blood donors or dermatology clinic outpatients. Anti-trichophytin-IgG titers were elevated in dermatophytosis patients, but overlap with normal values was large, and a subgroup of patients were low responders. Of the three antigens tested from *T mentagrophytes, T rubrum* and *E floccosum,* the *T mentagrophytes* antigen was more potent, even in *T rubrum*-infected persons. No clinical significance could be ascribed to the measurement of these immunoglobulin classes to dermatophyte glycopeptides by EIA. In view of the high incidence of immediate hypersensitivity recorded in chronic dermatophytosis (Sorensen and Jones, 1976; Brahmi et al, 1980), the determination of trichophytin-specific IgE by EIA would seem to be a more rewarding maneuver. Another more potentially rewarding aspect of humoral immunity in dermatophytosis is the localization of immunoglobulin in biopsy sections of infected skin. Antibodies to keratinase II have been detected in the external sheath of previously infected hair follicles (Collins et al, 1973). This approach aids in defining what antigens are expressed in the host and may help to understand the potential contribution of immune complexes in the skin to inflammation.

9.4 GLUCANS

Glucans were extracted from defatted and trypsin-treated mycelia of five dermatophytes with hot 0.75M NaOH (Bishop et al, 1966). Addition of Fehling solution to aqueous solutions of crude polysaccharides precipitated galactomannans I. The glucans and galactomannans II were separated by diethylaminoethyl-cellulose chromatography in borate buffers. The glucans were less tightly bound and appeared in the 0.05M effluent. Methylation-fragmentation analysis showed the major linkages in the linear sequences were 1,6 and 1,3. Branching occurred through C-1, C-3, and C-6. Glucans from these dermatophytes varied in the ratio of 1,6 and 1,3 bonds: *T mentagrophytes* (*granulosum*), 1:0.72; *T mentagrophytes* var *interdigitale,* 1:0.91; *Microsporum quinckeanum,* 1:0.38; *T rubrum,* 1:4.1; and *T schoenleinii,* 1:1.32. With the exception of *T mentagrophytes* var *granulosum* the glucans were levorotatory.

A glycogen-like glucan with a majority of α-1,4 linkages was isolated from surface mats of *T rubrum* (How et al, 1972, 1973). The glucan was extracted from acetone powder mycelia with ethylene glycol and purified by gel permeation and ion exchange chromatography. Enzymic digestions with pullulanase (specific for α-1,6-glucosyl flanking α-1,4-glucosyl residues), α-amylase, and β-amylase helped to establish that at least 80% of the glucan exists as linear sequences of α-1,4-glucosyl units joined by α-1,6 branch points. The glucan evoked immediate cutaneous hypersensitivity and passive cutaneous anaphylaxis when donor guinea pigs were sensitized with killed *T rubrum* mycelia.

9.5 PROTEINASES

Proteinase production by dermatophytes has been demonstrated by several groups (Cruickshank and Trotter, 1956; Chattaway et al, 1963; Weary and Canby, 1969; Minocha et al, 1972). Zoophilic dermatophytes like *T mentagrophytes* var *mentagrophytes* incite more severe inflammation and produce higher levels of proteinases than does anthropophilic *T rubrum*. *T mentagrophytes* proteinases swabbed on stripped human skin caused vesicular eruptions (Minocha et al, 1972). Intradermal injection of this enzyme preparation in dogs caused epidermal-dermal splits. Both of these findings probably are a composite of hypersensitive and proteinolytic components.

Antigenic analysis of the pathogenic fungi reveals that many genera secrete proteinases that have a role as antigens and possibly as aggressins, factors potentiating pathogenicity. The occurrence of proteinases has been noted in *Candida albicans*, *Aspergillus fumigatus, Rhizopus oryzae*, as well as in the dermatophytes. The evidence implicating keratinase and other proteinases as determinants of pathogenicity in the dermatophytes has been reviewed by Grappel (1981), and is briefly summarized here: Keratinase can be localized by immunofluorescence in skin lesions of guinea pigs infected with *T mentagrophytes* var *granulosum* (Collins et al, 1973). The drug griseofulvin, used for systemic treatment of acute dermatophytosis, inhibits keratinase activity in vitro (Yu and Blank, 1973). Proteinase activity is lowest in chronic inflammatory lesions of the anthropophilic species, *T. rubrum*, compared with higher proteolytic activity, in isolates of the zoophilic *T mentagrophytes* var *mentagrophytes*, associated with increased inflammation (Minocha et al, 1972). Experimental infections in guinea pigs with the perfect stage of *T mentagrophytes* (the a (−) and A (+) mating types of *Arthroderma benhamiae*) showed that severe inflammation was linked to the elastase positive, a (−) mating type (Rippon and Garber, 1969). *Trichophyton schoenleinii*, causing the most severe and inflammatory favus type of infection of the scalp and smooth skin, produces many proteinases including collagenase and elastase (Rippon, 1982).

A proteinase complex was characterized from *M canis* (O'Sullivan and Mathison, 1971) that is inducible by casein in the growth medium and repressible by amino acids. Secretion of proteinases occurred during the exponential phase of growth as a true secretion, not an autolytic process. Cultures grown in casein-containing medium, washed, and replaced in fresh medium begin to secrete proteinases after a short lag period. Intracellular proteinase is membrane-bound and its activity is potentiated by the nonionic detergent, triton X-100. Proteinase activity displays pH optima at pH 6.6, 8.0, and 9.5. No free amino acids are released during proteolysis, indicating an endopeptidase action pattern. *M canis* grew poorly on soap-extracted, ethylene oxide-sterilized human hair, but a proteinase was elaborated into the medium (Takiuchi et al, 1982). The digestion of defatted guinea pig hair was monitored by the release of A_{280nm} absorbing material. An eight-hundredfold purification was achieved by ion exchange and gel permeation chromatography. The enzymatic activity was inhibited by IgG produced against the proteinase. The secretion of proteinases is consistent with the ecological role of hair digestion.

Three keratinases have been described for *T mentagrophytes* (Yu et al, 1971; Collins et al, 1973). Keratinase I has a molecular weight of 48,000 da and is secreted into the medium after growth on horse hair as the sole nitrogen source. Two other proteinases with keratinolytic activity are loosely associated with the cells and are extracted after soaking the mycelia in buffer. Keratinase II was adsorbed by DEAE-cellulose, eluted with 1M NaCl and chromatographed on Sephadex G200. It is a high molecular weight complex. The third proteinase with keratinolytic activity on guinea pig hair was adsorbed on CM-cellulose. This fraction was heterogeneous on Sephadex G100, with the peak of keratinase activity corresponding to a mass of 20,300 da. These three keratinases display partial identity in immunodiffusion.

Antibodies to keratinases have not been detected in sera, but cytophilic antibodies are present in the external sheath of infected hair follicles (Collins et al, 1973; Holden et al, 1981).

Antisera to soluble extracts of *T rubrum* localized antigens in the cell wall, mitochondria, nuclear membrane and septal pore of both in vivo and in vitro grown fungus (Holden et al, 1981). Heavy staining of the cell wall by immunoperoxidase labeling suggested that polysaccharide or keratinase are potent antigens, even when cell extracts and not walls are used as the immunogen.

9.6 CHRONIC DERMATOPHYTOSIS

Chronic dermatophytosis, especially *T rubrum* infection, is associated with immunologic hyporesponsiveness. The most apparent defect is anergy in tests for delayed cutaneous hypersensitivity with trichophytin (Hanifin et al, 1974; Jones et al, 1974; Sorensen and Jones, 1976; Kaaman, 1978; Hay, 1979; Hunziker and Brun, 1980; Brahmi et al, 1980; Hay and Shennan, 1982). This anergy is more prevalent in palmoplantar infections (Kaaman, 1978) and in tinea corporis (Tagami et al, 1977; Brahmi et al, 1980). Depressed response ratios are also observed in the in vitro correlates of delayed cutaneous hypersensitivity, lymphocyte blastogenesis, and leukocyte migration inhibitory factor production (Hay, 1979; Brahmi et al, 1980; Hay and Shennan, 1982) in comparison with more vigorous responses to *T mentagrophytes* infection in humans, and in *Microsporum*-infected guinea pigs (Hunjan and Cronholm, 1979). Several explanations for these anergic phenomena have been advanced:

1. Presence of a preexisting immunodeficit, atopy, or immunosuppressive drug therapy (Jones et al, 1974; Hanifin et al, 1974; Sorensen and Jones, 1976; Hay, 1979; Brahmi et al, 1980).
2. Induction by dermatophytes of a serum suppressor substance (Green and Balish, 1979) or a functional defect in a normal plasma component (Sherwin et al, 1979).
3. *Trichophyton rubrum* as a specific inducer of IgE and consequent suppression of delayed cutaneous hypersensitivity by the immediate reaction (Hanifin et al, 1974; Jones et al, 1974; Hunziker and Brun, 1980; Brahmi et al, 1980).
4. Evolution of *T rubrum* as a pathogen well-adapted to evade immune surveillance. The superficial layers of the skin of the feet as a site not favorable for a systemic antigenic stimulus (Kaaman et al, 1976; Kaaman, 1979).
5. Variation in antigenic potency (Ottaviano et al, 1974; Kaaman et al, 1976).
6. Low T-helper to T-suppressor ratio (Petrini and Kaaman, 1981).

Evidence both pro and con has accumulated for most of these hypotheses. Some authors claim that the observed cutaneous anergy is partly a function of an inadequate antigen (Kaaman, 1978) or inappropriate means of applying the skin test (Tagami et al, 1977). Efforts to suppress IgE responses with antihistamine have succeeded in unmasking delayed cutaneous hypersensitivity in one report (Sorensen and Jones, 1976), but this finding was not corroborated (Hanifin et al, 1974; Hay, 1979). Several cycles of experimental *T rubrum* infection in guinea pigs led not to anergy but to increased inflammatory responses (Hunjan and Cronholm, 1979). Although healing of dermatophyte lesions correlates temporally with the acquisition of delayed cutaneous hypersensitivity, this does not prove that such hypersensitivity confers protection. This aspect has been studied in guinea pigs infected with *Trichophyton mentagrophytes* (Tagami et al, 1973; Kerbs et al, 1977; Green and Balish, 1979) and with other dermatophytes (Hunjan and Cronholm, 1979), and in human infections (Sorensen and Jones, 1976). The inflammation associated with weeping dermatophyte lesions has a strong delayed hypersensitivity component that may halt the spread of infection. More severe ulcerations recurred in guinea pigs after reinfection probably because of a brisk delayed cutaneous hypersensitivity reaction, but after repeated challenges with live fungi the lesions healed more quickly. These observations agree that inflammation in dermatophyte lesions is at least in part an expression of delayed hypersensitivity and aids resolution of the lesion. Immunologically oriented reviews of progress have appeared (Grappel et al, 1974; Grappel, 1981). Since then advances have been made in antigenic analysis and in better definition of immunologic hyporesponsiveness in chronic dermatophytosis. This section is concerned with reviewing the more current trends in research.

9.6.1 Immunodeficits

About half of chronic dermatophytosis patients have underlying conditions that compromise the integrity of their immune responses (Hay, 1979), including atopy, systemic or topical corticosteroid therapy (Tagami et al, 1977), autoimmune disease (lupus erythematosus), or skin cancer. In one series (Sorensen and Jones, 1976) there was 2.9 times higher frequency of atopy among chronically infected patients compared with the noninfected controls. A group of five patients with severe, generalized dermatophytosis had profound immunologic dysfunction, including a low concentration of peripheral T-cells, and virtually complete anergy in MIF and blastogenic responses to trichophytin, to other common recall antigens, and to phytomitogens (Brahmi et al, 1980). Elevated totals of serum IgE, a familial pattern of inheritance, and T-cell dysfunction all coincided in these patients.

The thymic dependency of acquired resistance to *T mentagrophytes* was shown in athymic nude rats (RNU) (Green et al, 1983). The RNU rats were not able to eradicate the fungus, and scaling spread from the original cutaneous site. Thymus-bearing heterozygotes became culture-negative after 5 weeks, and demonstrated an accelerated rejection of the fungus upon reinfection. Acquired immunity correlates with the capacity to initiate an intense lymphocytic and macrophage infiltration of the infected skin. Despite epidermal proliferation and neutrophilic inflammation, the nude rat remains chronically infected.

9.6.2 Soluble Suppressor Substances

Serum factors are cited as contributing to the hyporesponsiveness of lymphocytes to trichophytin both in experimental dermatophytosis in guinea pigs (Green and Balish, 1979) and in human disease (Allen et al, 1977; Sherwin et al, 1979). In two human case reports, common features were severe, generalized, therapy-resistant dermatophytosis. On the immunologic level, no preexisting immunodeficit could be found, and the anergy was broad, extending to inhibition of blastogenesis to phytomitogens, to common recall antigens, and to the dermatophyte antigens. The suppression of blastogenesis by autologous serum may be a transient effect of the infection (Sherwin et al, 1979), disappearing with successful therapy. In that event, antigenemia or soluble immune complexes may be involved. Distinct from this source of suppression is a defect in a normal plasma component (Allen et al, 1977), in which case replacement of autologous plasma with that from normal donors enhanced in vitro responses, and plasma infusions were therapeutic. Complement levels and serum iron-binding capacity were normal, but a functional defect in transferrin was not ruled out. Transferrin has been implicated as a factor that stimulates lymphocytes, apart from its iron-chelating properties.

9.6.3 IgE-Mediated Suppression

When chronic dermatophytosis patients were screened to exclude those with known immunologic abnormalities, there was still a large immediate hypersensitivity component in skin tests of persons infected with *T rubrum* and a suppression of delayed cutaneous hypersensitivity (Kaaman et al, 1976; Brahmi et al, 1980). Also consistent was the more frequent isolation of *T rubrum* than *E floccosum* which in turn was found more frequently than *T mentagrophytes*. The particular association of *T rubrum* and immediate hypersensitivity has been verified. A spectrum of cutaneous reactions to trichophytin has been seen in chronically infected persons: in a study of 35 infected persons, 35% had immediate hypersensitivity, 21% displayed delayed hypersensitivity, 41% were anergic, and in 6% both immediate and delayed hypersensitivity were elicited by 10 μg of trichophytin (Sorensen and Jones, 1976). These data are in contrast to findings with noninfected controls, in 145 of whom 5% had immediate reactions, 64% delayed reactions and 31% were nonreactors. A second intradermal dose of trichophytin administered 20 minutes after the appearance of the immediate wheal and flare induced a long-lasting delayed cutaneous hypersensitivity reaction in five chronic dermatophytosis patients (Hunziker and Brun, 1980). This was interpreted as the masking of delayed reaction by complexation of the antigen with IgE. Simultaneous injection of trichophytin and the antihistamine, chlorpheniramine, increased the delayed cutaneous hypersensitivity reactions and uncovered an occult delayed hypersensitivity in some patients who previously showed only an immediate reaction (Jones et al, 1974). In a more recent attempt to corroborate this blocking effect (Hay, 1979), intradermal injection of trichophytin with chlorpheniramine (H_1-blocker) substantially reduced wheal and flare reactions to trichophytin in chronic dermatophytosis, but no delayed cutaneous hypersensitivity was uncovered.

A factor contributing to cutaneous anergy in dermatophytosis may be the requirement for a high antigenic dose to evoke delayed cutaneous hypersensitivity in the setting of immediate hypersensitivity (Jones et al, 1974). Small doses of antigen are postulated to be complexed by IgE (and possibly by other Ig classes) making them unavailable for T-cell recognition. Tagami et al (1977) have applied concentrated, purified trichophytin as a patch test on partially stripped skin. After 48 hours of contact a higher proportion of *T rubrum*-infected patients (63%) had histologically proven delayed cutaneous hypersensitivity than would be expected from intradermal tests. Positive delayed cutaneous hypersensitivity patch tests were also observed in 19% of normal controls, not an unexpected finding for a commonly occurring mold.

9.6.4 Tolerogenic Potential of *Trichophyton rubrum*

Trichophyton rubrum is by far the most prevalent agent of chronic dermatophytosis (Grappel et al, 1974; Kaaman et al, 1976; Tagami et al, 1977; Hay, 1979). Typically the lesions in these patients have low grade inflammation. Among patients with *T rubrum* infections, those whose feet are involved are less likely to display delayed cutaneous hypersensitivity than those with groin infections (Kaaman et al, 1976; Hay and Shennan, 1982). The subset with tinea corporis also have significantly reduced delayed-type patch test responses (Tagami et al, 1977). The adaptation of *T rubrum* for growth without provoking much inflammation has not been satisfactorily explained. The net effect of multiple cycles of infection in guinea pigs was an increase in inflammation combined with a decreased healing time (Hunjan and Cronholm, 1979). This increased inflammation occurred to a greater extent with *T rubrum* than with two other dermatophytes. The guinea pig model does not, therefore, reproduce the chronicity and low inflammation seen in humans. The low rate of positive delayed cutaneous hypersensitivity to trichophytin may be a local aberration of the keratinized layers of the foot as a site for antigenic stimuli (Kaaman et al, 1976). This seems unlikely because (i) 38% of chronic *T rubrum*-infected persons had precipitins demonstrable in counterimmunoelectrophoresis (Hay and Shennan, 1982) suggesting that the antigenic stimulus was present; and (ii) more widespread infections involving smooth skin, fingernails and scalp also can coexist with anergy (Brahmi et al, 1980). Further dissection of the immune responses to trichophytin in chronic dermatophytosis showed that lymphocytes from cutaneously anergic patients were capable of producing MIF, but several patients were low responders in blastogenic tests (Brahmi et al, 1980). This uncoupling of delayed cutaneous hypersensitivity and MIF may be significant because MIF reactions are generally regarded as the in vitro test most closely correlated with delayed hypersensitivity.

9.6.5 Variation in Antigenic Potency

Trichophytin historically has consisted of a ground and filtered mycelial extract of *Trichophyton* sp. A range of dermatophytic macromolecules have "trichophytin" activity as discussed by Grappel et al, 1974. Positive skin test reactions can be evoked in humans 7 to 10 days after infection, and the sensitized state may persist for up to 3 years after the infection has cleared (Grappel et al, 1974). The reaction may be immediate hypersensitivity mediated by IgE or delayed cutaneous hypersensitivity. As indicated in Table 9.3, the culture medium, conditions of growth and extraction vary considerably among recent reports. Commercial preparations are not well characterized either (Grappel et al, 1974).

Remarkably, the use of unpurified homogenates and culture filtrates as trichophytins persists even though a method of preparing glycopeptides with trichophytin activity has been known since the reports of Barker et al (1962, 1963, 1967). Indeed further studies comparing commercial trichophytin, original trichophytin glycopeptides (authentic standard from Barker et al), and freshly isolated glycopeptides substantiate the potency of the ethylene glycol-extracted trichophytin (Ottaviano et al, 1974) (Table 9.3). Yet other factors produced by dermatophytes might possibly be good elicitors of delayed cutaneous hypersensitivity, especially the extracellular proteinases (O'Sullivan and Mathison, 1971; Grappel et al, 1974). The identification and purification of the antigenic structures of *Trichophyton* species is a worthwhile objective. Well-characterized antigens would permit more comparable cutaneous and in vitro cell-mediated immunity test results among laboratories. The objectives of using a purified, characterized trichophytin that meets standards set by a working party would be to minimize antigenic variation in evaluating immune responses in chronic dermatophytosis and to provide antigens for general immunocompetency testing to which many normal individuals are sensitized and thus likely to respond. In this respect the status of trichophytin research is not unlike the situation that prevails with other mycotic antigens: even where more refined antigens have been developed in pilot studies, they are not available for general use.

The source of trichophytin influences the ability to evoke delayed cutaneous hypersensitivity reactions in normal individuals and patients with chronic dermatophytosis. Commercial trichophytin was less able to detect delayed hypersensitivity (23% positives) in chronic dermatophytosis than trichophytin purified by the ethylene glycol method (36% positives) (Kaaman et al, 1976). Ottaviano et al (1974) attributed false-positive and false-negative reactions to commercial trichophytin compared with reactions to purified trichophytin glycopeptides.

9.6.6 Ratio of T-Helper to T-Suppressor Cells

An attempt has been made to relate cutaneous anergy with the ratio of T-helper:T-suppressor lymphocytes (Petrini and Kaaman, 1981). T-cells with IgG receptors, T_G, acted as T-suppressor cells in studies with the pokeweed mitogen. In three anergic patients the T_G were greater than 30% of peripheral T-lymphocytes. Monoclonal antibodies specific for T-suppressor cells, Leu 2A, or T-helpers, leu 3A, showed a reduced T-helper:T-suppressor ratio (T_H/T_S) in one such patient. This was the only chronic dermatophytosis patient tested in which assay of the T_H/T_S by two independent methods suggested enrichment in T_S and reduced T_H. These measurements do not refer to the specific responses of T-cells to dermatophytic antigens, but provide a general assessment of an individual's immunologic status. Further studies will be necessary before these estimates can be accepted as contributing to the observed anergy.

9.7 SUMMARY

Cutaneous anergy in tests for delayed cutaneous hypersensitivity is common in chronic dermatophytosis, especially in palmoplantar and smooth skin infections. About half of chronic dermatophytosis patients have primary diseases that compromise their immunity or are receiving immunosuppressive drug therapy. Severe, generalized dermatophytosis is often accompanied by profound T-cell dysfunction, atopy, and a familial inheritance pattern. Suppressed blastogenic responses to dermatophyte antigens have been related to the presence of a soluble suppressor molecule and in another instance, to the absence of a plasma factor that permits normal blastogenesis. Typically, chronic *T rubrum* infections have a low degree of inflammation. On the immunologic level these chronic infections are characterized by suppression of delayed hypersensitivity and stimulation of immediate type I cutaneous reactions. These two phenomena may be linked, but the mechanism is not as yet known. Cutaneous anergy occurs not only when the infected site is small, but to an even greater extent in widespread tinea corporis.

Considerable variation is evident in the methods of preparing trichophytin, representing an uncontrolled variable that complicates comparison of skin test results among laboratories. Criteria of purity for trichophytin are lacking, but a standard method exists for its production on a chemically defined medium.

Trichophytin glycopeptides are solubilized from surface mats of *T mentagrophytes* with ethylene glycol and are precipitated with cetyltrimethylammonium bromide as their borate complexes. The glycopeptides are separated into three components by their order of elution from anion exchange columns. The major fractions have a galactose : mannose ratio of 1 : 9.8 and 3% to 11% protein. These molecules have a dual ability to evoke humoral and cell-mediated reactions; the peptido-component is required for the delayed cutaneous trichophytin reaction. Controversy exists about whether there is a significant N-acetylgalactosamine component in trichophytin glycopeptides. The term glycopeptide is used here to differentiate these antigens from polysaccharides extracted by harsher methods that remove or denature protein determinants.

Galactomannan polysaccharides isolated from defatted and trypsin-treated dermatophyte mycelia with hot alkali have a low nitrogen content and attenuated serologic activity. The polysaccharides include a lightly galactosylated galactomannan I that precipitates with Fehling solution. A linear 1,6-mannan backbone with single mannose or galactofuranose antennae are the known structural features for galactomannan I. Galactomannan II is more highly galactosylated and does not precipitate as the Cu^{++} complex. This polysaccharide has a more complex structure containing more 1,2-linked mannose residues. The structural analyses of trichophytin glycopeptides and polysaccharides are preliminary and confirmation by modern methods is needed.

C-substance has been detected in filtrates of old *E floccosum* cultures. The activity resides in a galactomannan that binds to phosphorylcholine-specific mouse myeloma protein.

Galactomannan II glycopeptides of the dermatophytes have crossreactive and species-specific determinants. These differences are not sufficient to aid in classifying this group of molds. Measurements of IgG and IgM responses in sera from dermatophytosis patients are not clinically significant, but cytophilic antibodies to keratinase of dermatophytes have been detected in the external sheath of infected hair follicles.

Analyses of the cell walls of dermatophytes are useful because they illustrate adherence to objective criteria for cell wall purity. Walls of dermatophytes have a high chitin content of about 30%. Anionic detergents can remove multiple proteins from the isolated cell walls of *E floccosum*.

Dermatophytes contain alkali-soluble glucans with 1,3 and 1,6 bonds. Another type of glucan with a majority of linear α-1,4 sequences and α-1,6 branches occurs in *T rubrum*. The degree of pathogenicity and extent of hair invasion by dermatophytes is directly related to the production of proteinases, especially keratinase. Three keratinases are produced by *T mentagrophytes* that differ in molecular weight. One is secreted into the medium during growth and the others are loosely bound to the cell wall.

REFERENCES

Allen DE, Snyderman R, Meadows L, Pinnell SR, 1977. Generalized *Microsporum audouinii* infection and depressed cellular immunity associated with a missing plasma factor required for lymphocyte blastogenesis. Am J Med 63:991–1000.

Arnold MT, Grappel SF, Lerro AV, Blank F, 1976. Peptido polysaccharide antigens from *Trichophyton mentagrophytes* var. *granulosum*. Infect Immun 14:376–382.

Baldo BA, Fletcher TC, Pepys J, 1977. Isolation of a peptido-polysaccharide from the dermatophyte *Epidermophyton floccosum* and a study of its reaction with human C-reactive protein and a mouse anti-phosphorylcholine myeloma serum. Immunology 32:831–842.

Barker SA, Cruickshank CND, Morris JH, Wood SR, 1962. The isolation of trichophytin glycopeptide and its structure in relation to the immediate and delayed reactions. Immunology 5:627–632.

Barker SA, Cruickshank CND, Morris JH, 1963. Structure of a galactomannan-peptide allergen from *Trichophyton mentagrophytes*. Biochim Biophys Acta 74:239–246.

Barker SA, Basarab O, Cruickshank CND, 1967. Galactomannan peptides of *Trichophyton mentagrophytes*. Carbohydr Res 3:325–332.

Basarab O, How MJ, Cruickshank CND, 1968. Immunological relationships between glycopeptides of *Microsporum canis, Trichophyton rubrum, Trichophyton mentagrophytes* and other fungi. Sabouraudia 6:119–126.

Bishop CT, Perry MB, Blank F, Cooper FP, 1965. The water-soluble polysaccharides of dermatophytes IV. Galactomannans I from *Trichophyton granulosum, Trichophyton interdigitale, Microsporum quinckeanum, Trichophyton rubrum,* and *Trichophyton schoenleinii*. Can J Chem 43:30–39.

Bishop CT, Perry MB, Blank F, 1966. The water-soluble polysaccharides of dermatophytes V. Galactomannans II from *Trichophyton granulosum, Trichophyton interdigitale, Microsporum quinckeanum, Trichophyton rubrum,* and *Trichophyton schoenleinii*. Can J Chem 44:2291–2298.

Bishop CT, Perry MB, Hulyalkar RK, Blank F, 1966. The water-soluble polysaccharides of dermatophytes VI. Glucans from *Trichophyton granulosum, Trichophyton interdigitale, Microsporum quinckeanum, Trichophyton rubrum,* and *Trichophyton schoenleinii*. Can J Chem 44:2299–2303.

Brahmi Z, Liautaud B, Marill F, 1980. Depressed cell mediated immunity in chronic dermatophytic infections. Annales Immunol (Paris) 131:143–153.

Chandler FW, Kaplan W, Ajello L, 1980. A Colour Atlas and Textbook of Histopathology of Mycotic Diseases 333 pp. London: Wolfe Medical Publications.

Chattaway FW, Ellis DA, Barlow AJE, 1963. Peptidases of dermatophytes. J Invest Dermatol 41:31–37.

Collins JP, Grappel SF, Blank F, 1973. Role of keratinases in dermatophytosis II. Fluorescent antibody studies with keratinase II of *Trichophyton mentagrophytes*. Dermatologica 146:95–100.

Cruickshank CND, Trotter MD, 1956. Separation of epidermis from dermis by filtrates of *Trichophyton mentagrophytes*. Nature (London) 177:1085–1086.

Fudenberg HH, Sites DP, Caldwell JL, Wells JV, 1980. Basic and Clinical Immunology 3rd Edition. 782 pp. Los Altos, CA: Lange Medical Publications.

Grappel SF, 1981. Immunology of surface fungi. Comprehensive Immunol 8:495–524.

Grappel SF, Blank F, Bishop CT, 1969. Immunological studies on dermatophytes IV. Chemical structures and serological reactivities of polysaccharides from *Microsporum praecox, Trichophyton ferrugineum, Trichophyton sabouraudii* and *Trichophyton tonsurans*. J Bacteriol 97:23–26.

Grappel SF, Blank F, Bishop CT, 1971. Circulating antibodies in human favus. Dermatologica 143:271–276.

Grappel SF, Bishop CT, Blank F, 1974. Immunology of dermatophytes and dermatophytoses. Bacteriol Rev 38:222–250.

Grappel SF, Blank F, Bishop CT, 1972. Circulating antibodies in dermatophytosis. Dermatologica 144:1–11.

Grappel SF, Buscavage CA, Blank F, Bishop CT, 1970. Comparative serological reactivities of twenty-seven polysaccharides from nine species of dermatophytes. Sabouraudia 8:116–125.

Green F III, Balish E, 1979. Suppression of in vitro lymphocyte transformation during an experimental dermatophyte infection. Infect Immun 26:554–562.

Green F III, Weber JK, Balish E, 1983. The thymus dependency of acquired resistance to *Trichophyton mentagrophytes* dermatophytosis in rats. J Invest Derm 81:31–38.

Hanifin JM, Ray LF, Lobitz WC Jr, 1974. Immunological reactivity in dermatophytosis. Br J Dermatol 90:1–8.

Hay RJ, 1979. Failure of treatment in chronic dermatophyte infections. Postgrad Med J 55:608–610.

Hay RJ, Shennan G, 1982. Chronic dermatophyte infections II. Antibody and cell-mediated immune responses. Br J Dermatol 106:191–198.

Holden CA, Hay RJ, MacDonald DM, 1981. The antigenicity of *Trichophyton rubrum*: in situ studies by an immunoperoxidase technique in light and electron microscopy. Acta Derm Venereol (Stockholm) 61:207–211.

How MJ, Withnall MT, Cruickshank CND, 1972. Allergenic glucans from dermatophytes. Part I. Isolation, purification and biological properties. Carbohydr Res 25:341–353.

How MJ, Withnall MT, Somers PJ, 1973. Allergenic glucans from dermatophytes. Part II. Enzymatic degradation. Carbohydr Res 26:21–31.

Hunjan BS, Cronholm LS, 1979. An animal model for cell-mediated immune responses to dermatophytes. J Allergy Clin Immunol 63:361–369.

Hunziker N, Brun R, 1980. Lack of delayed reaction in presence of cell-mediated immunity in trichophytin hypersensitivity. Arch Dermatol 116:1266–1268.

Jones HE, Reinhardt JH, Rinaldi MG, 1974. Immunologic susceptibility to chronic dermatophytosis. Arch Dermatol 110:213–220.

Kaaman T, 1978. The clinical significance of cutaneous reactions to trichophytin in dermatophytosis. Acta Derm Venereol (Stockh) 58:139–143.

Kaaman T, von Stedingk LV, von Stedingk M, Wasserman J, 1981. ELISA-determined serological reactivity against purified trichophytin in dermatophytosis. Acta Derm Venereol (Stockh) 61:313–317.

Kaaman T, von Stedingk LV, Wasserman J, 1976. An evaluation of delayed hypersensitivity in guinea pigs to various trichophytin preparations. Acta Derm Venereol (Stockh) 56:283.

Kerbs S, Greenberg J, Jesrani K, 1977. Temporal correlation of lymphocyte blastogenesis, skin test responses and erythema during dermatophyte infections. Clin Exp Immunol 27:526–530.

Kitajima Y, Nozawa Y, 1975. Isolation, ultrastructure and chemical composition of the outermost layer ("exo-layer") of the *Epidermophyton floccosum* cell wall. Biochim Biophys Acta 394:558–568.

Lloyd KO, 1970. Isolation, characterization and partial structure of peptido galactomannans from the yeast form of *Cladosporium werneckii*. Biochem 9:3446–3453.

Lloyd KO, 1972. Molecular organization of a covalent peptido-phospho-polysaccharide complex from the yeast form of *Cladosporium werneckii*. Biochem 11:3884–3890.

Minocha Y, Pastricha JS, Mohapatra LN, Kandhari KC, 1972. Proteolytic activity of dermatophytes and its role in the pathogenesis of skin lesions. Sabouraudia 10:79–85.

Miyazake T, Yadomae T, Yamada H, Hayashi O, Suzuki I, Ohshima Y, 1980. Immunochemical examination of the polysaccharides of mucorales. pp. 81–94. In Fungal Polysaccharides, Sandford PA, Matsuda K, eds. Washington, DC: American Chemical Society.

Moser SA, Pollack JD, 1978. Isolation of glycopeptides with skin test activity from dermatophytes. Infect Immun 19:1031–1046.

Noguchi T, Banno Y, Watanabe T, Nozawa Y, Ito Y, 1975. Carbohydrate composition of the isolated cell walls of dermatophytes. Mycopathol 55:71–76.

Noguchi T, Kitazima Y, Nozawa Y, Ito Y, 1971. Isolation, composition, and structure of cell walls of *Trichophyton mentagrophytes*. Arch Biochem Biophys 146:506–512.

O'Sullivan J, Mathison GE, 1971. The localization and secretion of a proteolytic enzyme complex by the dermatophytic fungus *Microsporum canis*. J Gen Microbiol 68:319–326.

Ottaviano PJ, Jones HE, Jaeger J, King RD, Bibel D, 1974. Trichophytin extraction: biological comparison of trichophytin extracted from *Trichophyton mentagrophytes* grown in a complex medium and a defined medium. Appl Environ Microbiol 28:271–275.

Petrini B, Kaaman T, 1981. T-lymphocyte subpopulations in patients with chronic dermatophytosis. Int Arch Allergy Appl Immunol 66:105–109.

Rippon JW, 1982. Medical Mycology. The Pathogenic Fungi and the Pathogenic Actinomycetes 2nd Edition. 842 pp. Philadelphia: W. B. Saunders, Co.

Rippon JW, Garber ED, 1969. Dermatophyte pathogenicity as a function of mating type and associated enzymes. J Invest Dermatol 53:445–448.

Shah VK, Knight SG, 1968. Chemical composition of hyphal walls of dermatophytes. Arch Biochem Biophys 127:229–234.

Sharon N, 1975. Complex Carbohydrates—Their Chemistry, Biosynthesis, and Functions. A set of lecture notes. 466 pp. Reading, Mass: Addison-Wesley Publishing Co.

Sherwin WK, Ross TH, Rosenthal CM, Petrozzi JW, 1979. An immunosuppressive serum factor in widespread cutaneous dermatophytosis. Arch Dermatol 115:600–604.

Sorensen GW, Jones HE, 1976. Immediate and delayed hypersensitivity in chronic dermatophytosis. Arch Dermatol 112:40–42.

Tagami H, Watanabe S, Ofugi S, 1973. Trichophytin contact sensitivity in guinea pigs with experimental dermatophytosis induced by a new inoculation method. J Invest Dermatol 61:237–241.

Tagami H, Watanabe S, Ofugi S, Minami K, 1977. Trichophytin contact sensitivity in patients with dermatophytosis. Arch Dermatol 113:1409–1414.

Takiuchi I, Higuchi D, Sei Y, Koga M, 1982. Isolation of an extracellular proteinase (keratinase) from *Microsporum canis*. Sabouraudia 20:281–288.

Weary PE, Canby CM, 1969. Further observations on the keratinolytic activity of *Trichophyton schoenleinii* and *Trichophyton rubrum*. J Invest Derm 53:58–63.

Young E, Roth FJ, 1979. Immunological cross-reactivity between a glycoprotein isolated from *Trichophyton mentagrophytes* and human isoantigen A. J Invest Dermatol 72:46–51.

Yu RJ, Blank F, 1973. On the mechanism of action of griseofulvin in dermatophytosis. Sabouraudia 11:274–278.

Yu RJ, Harmon SR, Grappel SF, Blank F, 1971. Two cell-bound keratinases of *Trichophyton mentagrophytes*. J Invest Dermatol 56:27–32.

Chapter 10

Zygomycetes

10.1 INTRODUCTION

Thermotolerant species of the *Mucorales* are ubiquitous saprophytic fungi with a low but definite pathogenic potential. Zygomycetes do not generally pose a threat to individuals with intact immune responses and are thus appropriate fungi for investigating the concept of opportunism. Although these fungi are weak pathogens, they are capable of "frightful disease with prevention, cure, and understanding of the fundamental mechanisms of invasion still out of our grasp" (Baker, 1975). Since the time of that remark by Roger D. Baker, a renowned scholar of disease caused by the pathogenic *Mucorales*, a modest amount of progress has been made.

Members of the genus *Rhizopus, (Rhizopus oryzae, Rhizopus rhizopodiformis)* are the only zygomycetes known to cause the rhinocerebral form of infection. The authenticated agents of pulmonary-systemic zygomycosis are a larger group including the *Rhizopus* species, *Mucor pusillus, Absidia corymbifera,* and other rarely encountered species discussed elsewhere (Chandler et al, 1980). Cutaneous and subcutaneous infections are less frequent, but represent an emerging type related to burns or contaminated bandages (Chandler et al, 1980; Marchevsky et al, 1980). A series of 32 zygomycosis patients representing 20 years of experience in one medical center suggests the incidence of zygomycosis is increasing (Marchevsky et al, 1980) (Figure 10.1, Table 10.1). More cases are being diagnosed during life, with pulmonary infections considered the most difficult to recognize.

Rhizopus and *Mucor* species appear in tissue as aseptate or sparsely septate, broad hyphae, 6 to 24 μm in diameter (average 12 μm), that are irregularly branched. The hyphae have a strong tendency to invade large and small arteries, resulting in mycotic emboli, infarction, and necrosis (Figure 10.2). This characteristic pattern is widely illustrated in general textbooks, which depict masses of hyphae in the lumens of arteries (Chandler et al, 1980; Rippon, 1982; McGinnis, 1980).

10.2 CELL WALL

The cell wall of *Mucor rouxii* was completely analyzed and serves as a model for the *Mucorales* (Bartnicki-Garcia and Nickerson, 1962). This cell wall type differs in several respects from other genera of frank zoopathogenic fungi (Table 10.2). There

178 ZYGOMYCETES

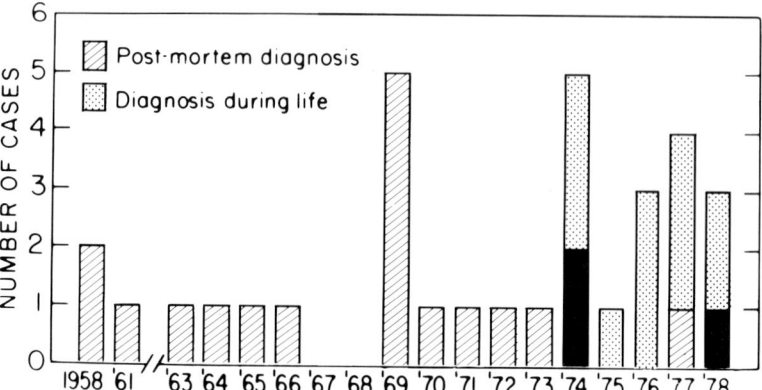

Figure 10.1. Incidence of zygomycosis at Mount Sinai Hospital, New York City, during the period 1958 to 1978. *Source:* Reprinted from Marchevsky et al (1980) with permission of the publisher W. B. Saunders Co.

is a high content of amino sugars, occurring mainly as chitosan (poly-1,4-β-glucosamine). Peptido-L-fuco-D-mannan also occurs in *Mucorales* cell walls (Miyazaki et al, 1980), and although a minor mural component by weight, it is implicated as a major antigen. The sites of insertion of the large phosphate component are believed to be inorganic polyphosphate, which titrates the basicity of chitosan.

10.2.1 Mucoran

Glucuronic acid is present as a structural component of *M rouxii* (Bartnicki-Garcia and Reyes, 1968) and *Absidia cylindrospora* (Miyazaki et al, 1980), and is probably universal in the *Mucorales*. Fucose-, mannose-, and glucuronic acid-containing polysaccharides were recovered from culture supernatants of *Mucor* and *Rhizopus* species by Martin and Adams (1956). They also detected these sugars in whole mycelial hydrolyzates. Two forms of glucuronic acid polymer are found: a heteroglycan, mucoran, extracted with cold dilute alkali and the more resistant mucoric acid. Mucoran was chemically characterized (Bartnicki-Garcia and Lindberg, 1972), but a single attempt to demonstrate serologic activity failed (Miyazaki et al, 1980), although its lack of activity has no ready explanation. Mucoran, solubilized with alkali and selectively precipitated with Fehling solution, comprised 12% of the *M rouxii* cell wall and 20% of the *A cylindrospora* cell wall (Miyazaki et al, 1980). The heteroglycan from *M rouxii* contained a glucuronic acid : mannose : L-fucose ratio of 5 : 3 : 2, $[\alpha]_D^{23} = +48°$ (c 1.0, water). Purification on anion and cation exchangers and by gel permeation chromatography allowed an estimation of molecular weight as 119,000 da. The major structural feature is a linear repeating sequence of equimolar amounts of mannose and glucuronic acid:

$$\text{D-man 1} \rightarrow (4\text{-}\alpha\text{-D-glcUA-1} \rightarrow 3\text{-D-man} \rightarrow)_n$$

Mucoric acid could be extracted from *M rouxii* in alkali after vigorous acid hydrolysis pretreatment. It contains about 85% glucuronic acid and has no known antigenicity (Bartnicki-Garcia and Reyes, 1968).

In more recent work, exposure of defatted walls of *Mucor mucedo* to nitrous acid (pH 3, 22 to 25 °C) resulted in 60% solubilization of walls, half of the solubles being dialyzable (Datema et al, 1977). The remainder was a soluble glycouronan containing L-fucose:mannose:galactose:glucuronic acid in a ratio of 5:1:1:2 and some polyphosphate. The basis of this extraction is the ionic interaction between the two polyanions glycouronan and polyphosphate, and the substantial chitosan component. Glucosaminyl residues are converted to 2,5-anhydromannose by nitrous acid oxidation that also cleaves the adjacent glycosidic bond. Thus deaminative depolymerization of chitosan, under mild conditions, solubilizes the major glycouronan which may also contain the sequences of glucuronic acid previously referred to as mucoric acid. The mild conditions of extraction should favor preservation of antigenic determinants, but no immunologic assessment of the glycouronan has been reported.

Dow et al (1983) reinvestigated the occurrence of polyuronides in cell walls of *M rouxii*. They found all of the polyuronides could be removed from yeast form cell walls with mild conditions. Polyuronide I was extracted with 5.6M LiCl and then purified by anion exchange and gel permeation chromatography. Polyuronide I was water-soluble and contained 58% glucuronic acid, 28.8% L-fucose, 9.1% mannose, and 4.1% galactose. Polyuronide II, containing virtually all glucuronic acid, was extracted with 1M KOH at 25 °C for 30 minutes. It was soluble above pH 5, but insoluble in LiCl or $CaCl_2$. The molecular weights of polyuronides I and II were determined by sedimentation-equilibrium-ultracentrifugation to be between 32,000 to 35,000 da. Polyuronide I is similar to mucoran and polyuronide II resembles mucoric acid. Polyuronide I is lower in molecular weight than the soluble glycouronan of Datema et al (1977).

10.2.2 Peptido-L-fuco-D-mannan

Peptido-L-fuco-D-mannan is a minor component of the cell walls of *Absidia cylindrospora* (less than 5%), but is a major antigen of that mold and is common to other genera of *Mucoraceae*. The L-fucose content of these fungi varies, being higher in *Mucor* species and *Absidia* species than in *Rhizopus* species (Miyazaki et al, 1980). Peptidofucomannan can be extracted from the culture filtrate, from the cytosol after mechanical disruption of mycelia, and from the cell wall. The proportion of protein, carbohydrate, and phosphate vary depending on the source and method of extraction, but the L-fucose:mannose ratio of 1:2.5 seems independent of these conditions. Hot aqueous 45% phenol treatment of *A cylindrospora* mycelia or cell walls results in partition of the peptidofucomannan into the aqueous phase. Exhaustive removal of mural peptidomannan requires a sequence of aqueous phenol, hot water, and 1% KCl. Peptidofucomannan-containing fractions from the culture filtrate and cytosol are fractionated on concanavalin A-agarose. Column effluents had a different serologic specificity from peptidofucomannan purified by desorption from the lectin, which suggests that the antigenic structure of these molds is, at present, only poorly understood. The composition of con A-purified peptidofucomannans of *A cylindrospora* are given in Table 10.3 (Miyazaki et al, 1980).

Smith degradation of *A cylindrospora* peptidofucomannan resulted in extensive oxidation of fucosyl and mannosyl residues, releasing glycerol and a tetrahydric alcohol. The peptidofucomannan isolated from the culture filtrate and sized on Sephadex G200 was partially acid hydrolyzed to produce oligomannosides. The resulting mannotriose to mannoheptaose fragments were purified on Biogel P-2 columns and by preparative paper chromatography. This dextrorotatory series contained only 1,6 linkages as determined by methylation-fragmentation analysis indicating long linear 1,6 sequences are the major structural feature (Figure 10.3).

Table 10.1. Mucormycosis: Clinical and Laboratory Parameters in 32 Patients

Patient Number	Age/Sex	Anatomic Site of Involvement	Underlying Disease	Diagnostic Procedure	Species Isolated	Survival
1	36 M	Rhinocerebral	Acute myelogenous leukemia	Autopsy	None	No
2	68 F	Rhinocerebral	Diabetes mellitus, miliary tuberculosis	Autopsy	None	No
3	67 F	Rhinocerebral	Diabetes mellitus, carcinoma of uterus	Autopsy	None	No
4	55 M	Lung	Mitral stenosis, valve replacement	Autopsy	None	No
5	65 F	Lung	Acute myelogenous leukemia	Autopsy	None	No
6	68 M	Lung, spleen	Diabetes, nephrotic syndrome	Autopsy	None	No
7	78 F	Lung	Diabetes, acute myelogenous leukemia	Autopsy	None	No
8	12 F	Lung	Diabetes	Autopsy	None	No
9	32 M	Lung	Acute myelomonocytic leukemia	Autopsy	None	No
10	67 M	Lung	Acute myelogenous leukemia	Autopsy	None	No
11	36 F	Lung, small intestine	Breast carcinoma	Autopsy	None	No
12	57 M	Skin of chest wall, thoracic outlet vessels, lung	Angina, triple coronary bypass, thrombocytopenia (Pronestyl-related)	Autopsy	None	No
13	55 F	Jejunum	Macroglobulinemia	Autopsy	None	No
14	60 F	Lung, spleen	Carcinoma of ovary	Autopsy	None	No
15	49 F	Esophagus	Cushing's disease, diabetes, basal cell carcinoma of skin	Autopsy	None	No
16	38 F	Larynx, stomach	Ulcerative colitis	Autopsy	None	No
17	58 M	Colon	Diverticulitis, colonic resection	Autopsy	None	No

#	Age/Sex	Site	Underlying condition	Specimen	Organism	Outcome
18	18 day old F	Jugular vein	Prematurity	Autopsy	None	No
19	53 M	Rhinocerebral	Renal transplant	Sinus biopsy	*R arrhizus*	No
20	71 F	Orbital	Diabetes mellitus	Nasal biopsy	*R arrhizus*	No
21	48 F	Orbital	Diabetes mellitus	Biopsy of ethmoid sinus	*R arrhizus*	Yes
22	58 F	Orbital	Diabetes mellitus	Palatal biopsy	*R arrhizus*	Yes
23	55 F	Facial cellulitis	Acute lymphocytic leukemia	Scraping from lip lesion	*Absidia corymbifera*	No
24	49 M	Orbital	Acute myelogenous leukemia	Nasal biopsy	*Rhizopus* sp	No
25	49 M	Perinasal orbital	Acute lymphocytic leukemia	Nasal biopsy	*R rhizopodiformis*	No
26	14 M	Perinasal orbital	Acute lymphocytic leukemia	Scraping from palate	*Rhizopus* sp	No
27	25 F	Orbital	Diabetes mellitus	Necrotic tissue from sinus	Failed to grow	Yes
28	54 M	Skin	Acute myelogenous leukemia	Scraping from eschar	*R rhizopodiformis*	No
29	30 M	Skin and subcutaneous tissue over renal biopsy tract	Renal transplant	Biopsy of necrotic subcutaneous tissue	*R rhizopodiformis*	Yes
30	67 M	Skin	Cardiac surgery	Skin biopsy	*R rhizopodiformis*	Yes
31	45 M	Lung	Renal transplant	Bronchoscopic aspirate and biopsy	*R rhizopodiformis*	Yes
32	49 M	Lung	Acute myelogenous leukemia	Percutaneous lung biopsy	*R rhizopodiformis*	No

Source: Reprinted from Marchevsky et al (1980) with permission of the publisher, W. B. Saunders Co.
Abbreviations: M, male; F, female.

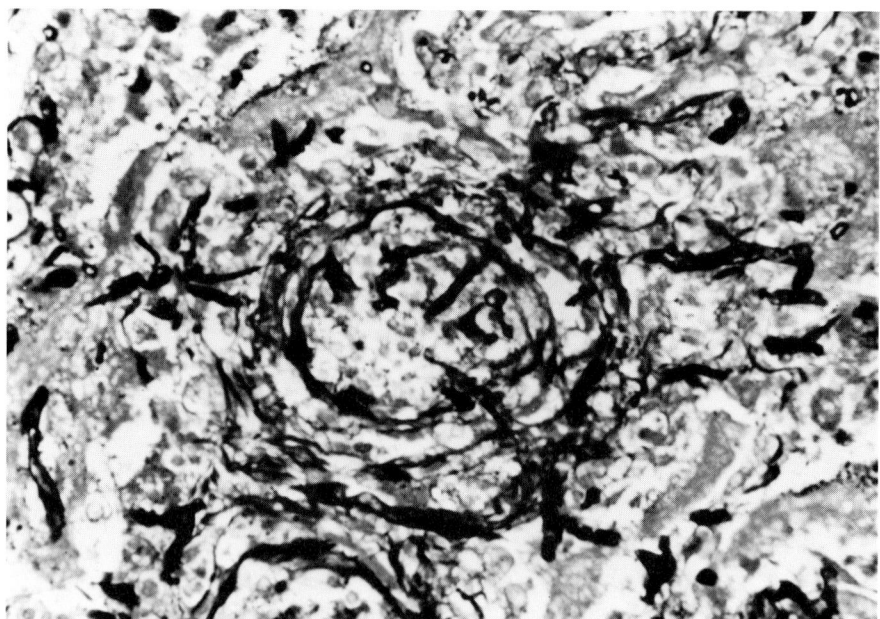

Figure 10.2. Human pulmonary zygomycosis. Hyphae of a zygomycete have invaded the lumen of a blood vessel resulting in thrombosis. The irregularly branched hyphae are infrequently septate and have nonparallel contours. Gomori's methenamine silver stain, 352 ×. *Source:* Dr. F. W. Chandler, CDC.

Rabbits immunized with *A cylindrospora* peptidofucomannan in complete Freund adjuvant produce antisera that crossreacts in immunodiffusion with cytosols and culture filtrates from four *Mucor* species and from *Rhizopus nigricans*. A single line of identity was observed, favoring the view that peptidofucomannans are a major antigen common among the zygomycetes. The lack of crossreactivity with mannan of *Candida albicans* and with baker's yeast mannan, or with the galactomannan from *Penicillium chrysogenum*, is preliminary evidence that determinants in peptidofucomannan are specific for the *Mucorales*. The immunodominance of carbohydrate determinants was shown by the stability of peptidofucomannan activity in the presence of proteinases and its lability to periodate oxidation. These results suggest that peptidofucomannan or antibodies to it may provide a specific indicator of exposure, or of active zygomycosis.

Table 10.2. Hyphal Wall Composition of *Mucor rouxii*

Component	Percent Dry Weight
Chitin	9.4
Chitosan[a]	32.7
Mannose	1.6
L-Fucose	3.8
Galactose	1.6
Glucuronic acid	11.8
Protein	6.3
Lipid	7.8
Phosphate	23.3

Source: Bartnicki-Garcia and Nickerson, 1962.
[a] poly-β-1,4-glucosamine

Table 10.3. Composition of *Absidia cylindrospora* Peptido-L-fuco-D-mannan Purified by Affinity for Con A, as Percent of Dry Weight

Source	Carbohydrate	Protein	Phosphate	L-Fucose:Mannose Molar Ratio
Cytosol	63.0	22.7	7.3	1:2.5
Culture filtrate	55.6	7.7	15.8	1:2.5
Aqueous phase of phenol-water cell extract	80.7	8.9	1.9	1:2.3

Source: Reproduced from Miyazaki et al (1980) in Fungal Polysaccharides, P. A. Sandford and K. Matsuda, (eds.), with permission from the publisher, American Chemical Society.

Peptidofucomannan oligomannoside fragments, described above, are haptens that inhibited the precipitation of con A-purified peptidofucomannan by anti-*A cylindrospora* culture filtrate. Mannoheptaose exerted the strongest (75%) inhibition, indicating a major role for mannosyl determinants. Fucosyl residues were also probably haptenic, but their contribution to the total reactivity is not yet known.

Thermotolerant *Rhizopus* species that cause rhinocerebral zygomycosis have not been studied for the occurrence of peptidofucomannans. Many structural details of peptidofucomannan are lacking, preventing construction of a minimum repeating unit. Cell walls of medically important zygomycetes have not been tested as immunogens. Thus the apparent inertness of the mucoran, glycouronan, which also contains L-fucose and galactose, is not resolved. Assessment of natural antibodies to peptidofucomannans in humans needs to be made.

10.3 RHINOCEREBRAL ZYGOMYCOSIS

This opportunistic infection occurs most frequently in poorly regulated diabetics in a state of ketoacidosis, but it also occurs in renal disease among recipients of renal allografts and in acute leukemia (Marchevsky et al, 1980; Baker, 1975). Seventy percent of cases reviewed by Baker (1975) were among diabetics.

Asexual spores of fungi of the genus *Rhizopus* (*R oryzae*, also called *R arrhizus, R rhizopodiformis*) inhaled from common environmental sources can germinate on the nasal septum and, in the absence of a normal host defense, can spread to the paranasal sinuses, through the cribriform plate to the meninges and brain. Alternatively, invasion of the ethmoid sinuses to the retroorbital region can occur. The process is rapid, with fungi growing in arteries and cartilage. Successful therapy has required prompt diagnosis, restoration of metabolic balance, craniotomy, and a full course of amphotericin B.

L — Fuc : Man = 1:2.5

Phenol-Water extract purified by Con A affinity

Protein, 8.9%; Carbohydrate, 80.7%, PO_4, 1.9%

Known structural feature:

$$\left[\longrightarrow 6\ \alpha\ \text{Man}\ 1 \longrightarrow \right]_7$$

Figure 10.3. Characteristics and structural feature of peptido-L-fuco-D-mannan of *Mucorales*, summarized from Miyazaki et al (1980). Fucose and mannose were detected in culture supernates and cell extracts of ten *Mucorales* species by Martin and Adams (1956).

The anti-*Rhizopus* factor is present in sera from normal persons and from well-controlled diabetics, and is absent in ketotic diabetes of humans and in the acute alloxan-induced disease of rabbits or rodents. It is the only known specific in vitro reaction to help distinguish the normally effective resistance against zygomycetes from increased susceptibility.

Defective immune functions of diabetics may help explain their susceptibility to zygomycetes. Humoral immunity is considered intact with mildly decreased IgG in insulin-dependent human adult diabetics (Galbraith, 1979). Mitogen-responses can be decreased in poorly controlled diabetics, but no generalized defect in cell-mediated immunity has been detected in human diabetes (Galbraith, 1979). Diabetic rodents and rabbits show marked impairment of a variety of cell-mediated responses (Handwerger et al, 1980) including T-dependent mitogenesis, prolonged skin graft survival, weaker mixed lymphocyte responses, and suppressed delayed cutaneous hypersensitivity to microbial antigens, but the responses in diabetic humans to zygomycetes or to any fungal agent are largely unknown.

Phagocytic functions that are impaired in diabetics are diapedesis (migration of polymorphonuclear neutrophils from the vascular compartment) which is delayed in ketoacidosis (Galbraith, 1979), lowered PMN chemotactic indices, reduced ingestion of bacteria by PMN in ketoacidosis (Galbraith, 1979), and decreased nitroblue tetrazolium reduction after phagocytosis of *C albicans* (Okuda et al, 1974). The lack of serum anti-*Rhizopus* factor coupled with depressed phagocytic activity probably accounts for the reduced nonspecific immunity in diabetic ketoacidosis. T-cell-mediated immune mechanisms active against zygomycetes have not yet been identified.

The only property of the fungus as yet characterized that could account for the spread from a mucosal focus through anatomical barriers to the brain is an alkaline proteinase, inducible in all pathogenic *Rhizopus* species that were tested (Reinhardt et al, 1979).

Laboratory diagnosis has relied on recovery of the fungus in nasal discharge, a problematic procedure. Immunodiffusion tests for antibodies were proposed (Jones and Kaufman, 1978) and despite the difficulty in assembling a large series of patients to test, this approach appears promising for general use.

10.4 *RHIZOPUS* INHIBITORY FACTOR

Human serum at concentrations above 20% inhibits the growth of *Rhizopus* species when added to a normally adequate medium (Gale and Welsh, 1961; Owens et al, 1965). Sera from ketoacidotic human diabetics, heavily immunosuppressed patients (Baker, 1975), and from alloxan-induced diabetic rabbits lack the anti-*Rhizopus* factor, whereas well-regulated diabetic patients or alloxan-induced, chronic, nonketotic diabetic rabbits have normal anti-*Rhizopus* activity.

The occurrence of precipitins and serum fungistatic factor were compared in six zygomycosis patients (Marchevsky et al, 1980). Four had precipitins specific for *Rhizopus* species. Sera from two patients failed to inhibit germination of *R rhizopodiformis* conidia despite the simultaneous presence of precipitins. Inhibition of conidial germination is consistently observed in normal human serum, and conversion to loss of serum fungistasis was related to the severity of infection in three ultimately fatal zygomycosis cases.

Several possible sources of *Rhizopus*-inhibitory factor have been ruled out: (i) inactivation of complement (56 °C, two hours) (Gale and Welch, 1961); (ii) hyperglycemia without ketosis does not ablate anti-*Rhizopus* activity (Owens et al, 1965). Dialysis of serum versus pH 7.2 Tris buffer removed the anti-*Rhizopus* factor. Additions of glucose, acetoactate, β-hydroxybutyrate, oleic acid, and loading with Fe^{++} or Cu^{++} were tested (Reinhardt, personal communication). Only those additives that lowered the pH of serum were able to ablate anti-*Rhizopus* activity. Normal serum, a pathological fluid, has a pH on standing of 8.0 to 8.5. The dialyzable and heat-stable anti-*Rhizopus* factor observed in vitro is at least partly due to failure to maintain high enough pH to inhibit growth. The simplest in vivo extrapolation is that depression of blood pH favors growth of *Rhizopus* species, but no direct supporting evidence is available.

10.5 NEUTROPHIL CIDAL MECHANISMS

Damage to *Rhizopus oryzae* hyphae incubated in vitro with human PMN, without serum, was detected in electron micrographs and by inhibition of uptake of the RNA precursor ^{14}C-uracil (Diamond et al, 1978a). *Rhizopus oryzae* hyphae are too large for endophagocytosis; instead PMN adhere to and spread along the hyphal walls. Degranulation occurs followed in 2 hours by a marked loss of fungal cytoplasmic organization. The mean inhibition of ^{14}C-uracil uptake by *R oryzae* hyphae exposed to PMN purified from several normal humans was 59%. This assay may prove useful as a sensitive index of the fungicidal potential of PMN from diabetic subjects. Myeloperoxidase-H_2O_2-halide was implicated as the effector mechanism, judging from the sparing of damage by catalase, cyanide, and azide, used at levels below that required to suppress overall PMN oxidative respiration. This pathway also mediated damage to *Aspergillus fumigatus* (Diamond et al, 1978a) and to pseudohyphae of *C albicans* (Diamond et al, 1978b). In the instance of *Rhizopus,* indirect evidence was obtained for damage due to cationic proteins of PMN granules because of the sparing effect of the polyanion, poly-L-glutamic acid. Lysozyme, also present in PMN granules, could damage hyphae in distilled water, but not in isotonic solution. The primary pathway of hyphal destruction appears to be myeloperoxidase, with cationic proteins and lysozyme as secondary contributors to damage. Taken together, they help explain the basis of innate resistance to zygomycetes.

10.6 EXPERIMENTAL ZYGOMYCOSIS

Sporangiospores from different genera of zygomycetes were instilled intranasally into acute alloxan-diabetic rabbits to determine their potential for rhinocerebral infection. All 13 isolates of five *Rhizopus* species were cerebral pathogens, whereas brain involvement occurred in infection by only one of five isolates of *A corymbifera* and two of five *M pusillus* isolates (Reinhardt et al, 1981). These findings accord with *Rhizopus* species as the most pathogenic of the zygomycetes, in fact the only known cause of human rhinocerebral zygomycosis (Baker, 1975; Reinhardt et al, 1981). The other genera of mucorales are encountered in human pulmonary disease. Pathogenic isolates of *Rhizopus* species all produced an alkaline proteinase that is active at physiologic pH (Reinhardt, personal communication). This enzyme is the only presumed virulence factor in zygomycosis.

Further characterization of the pathogenic and antigenic potential of the alkaline proteinase is needed. For example: (i) Can this enzyme be detected in secretions or body fluids?; (ii) Do antibodies to the proteinase occur in acute infection?; (iii) Can immunization with proteinase confer resistance to zygomycosis in laboratory animals?

Absidia ramosa = corymbifera is an example of a normally saprophytic zygomycete that can cause animal disease, including mycotic abortion (Cordes et al, 1967). Spores of *A corymbifera* injected intravenously into mice can result in central nervous system involvement in the form of continuous circling movement, cerebral dissemination, and death. Mice primed with killed bacteria (tubercle bacilli, or *Brucella abortus*), or with endotoxin in order to stimulate the reticuloendothelial system, instead displayed increased susceptibility to *A corymbifera* infection. Large doses of priming substances resulting in macrophage blockade, instead of activation, are the probable cause of this susceptibility (Eades and Corbel, 1976b).

Aged NZB mice, despite their C5 deficiency and autoimmune disease, are not more susceptible to zygomycetes than CBA mice (Corbel and Eades, 1976a). The situation with *C albicans* and *Cryptococcus neoformans* is different: aged NZBs have a higher rate and incidence of fatal outcome than the CBA strain. This is taken as evidence of the greater importance of specific immunity in resistance to the more pathogenic organisms. The inference is that nonspecific, or innate immunity (serum inhibitors, alternative complement pathway, neutrophil cidal mechanisms) is sufficient to resist zygomycetes without the need for T-cell-mediated functions.

Of the three molds tested, *R oryzae* is more pathogenic than either *M pusillus* or *A corymbifera* (Corbel and Eades, 1976a). Hyphae were observed in the brain, eliciting there a mixed inflammatory response consisting of PMN, macrophages, and plasma cells. Similar but larger lesions occurred in the kidneys. Deaths occurred 2 to 4 days after intravenous injection of 2×10^6 viable conidia. This time frame did not permit life table analysis of survival nor did it allow the development of T-cell-mediated immunity.

Normal anatomic barriers prohibit invasion by zygomycetes, and the factors that maintain proper surveillance may be innate, ie, brisk diapedesis of phagocytes, alternative pathway activation, opsonization, and phagocytosis. Conidia of *A corymbifera* do not cause lethal infection when injected subcutaneously, but when an intravenous route is used, the fungus germinates, rapidly disseminating to the brain, heart, and kidney of both athymic, nu/nu, and euthymic heterozygotes (Corbel and Eades, 1977). Deaths began to occur in less than 1 week after infection in both groups, but more extensive and severe lesions were found in the nu/nu mice. At 2 weeks after infection, when T-cell-mediated immunity should be in evidence, there were additional deaths within the nu/nu group. The more vigorous natural killer cell and macrophage responses known to exist in nu/nu mice may compensate for a lack of T-cells in the early phase of infection, but the ability to eradicate zygomycetes from internal foci may depend on a T-cell-mediated granulomatous response.

Mice primed with low doses of viable *A corymbifera* conidia acquired resistance to challenge with higher doses that were lethal for normal mice (Corbel and Eades, 1976b). Both normal and infected mice produced specific, histologically proven, delayed cutaneous hypersensitivity when mycelial extract was injected into their footpads. Priming with killed *A corymbifera* mycelia resulted in serum precipitins but did not increase the level of delayed cutaneous hypersensitivity, nor promote resistance to graded doses of live spores. This finding shows that humoral immunity to zygomycetes does not prevent cerebral infection.

Early attempts to determine the host response in zygomycosis compared the histopathology of *R oryzae* lesions in normal and alloxan-induced diabetic animals

(Sheldon and Bauer, 1959, 1960). More recent assessment of the interaction between PMN and *R oryzae* hyphae in vitro have delineated the nature and mechanism of damage to the fungus (Diamond et al, 1978a). *Rhizopus oryzae* conidia inoculated intradermally in normal rabbits produce small granulomatous lesions containing macrophages and Langhans-type multinucleate giant cells (Sheldon and Bauer, 1959) that heal spontaneously. This characteristic feature is also observed in mice injected subcutaneously with *A corymbifera* conidia (Corbel and Eades, 1976b). Induction of acute diabetic ketoacidosis with alloxan activates quiescent lesions which grow beyond the granuloma wall into local blood vessels. Morphologic abnormalities were noted in the PMN during fungal outgrowth (Sheldon and Bauer, 1958). Tardy and diminished acute inflammatory responses occurred in alloxan-diabetic rats upon intradermal injection of *R oryzae* conidia (Sheldon and Bauer, 1960), and this correlated with the failure of mast cell degranulation. This effect was simulated by treating normal rats with the histamine liberator 48/80 prior to conidia injection. Defective gatekeeper function of mast cells in ketoacidotic rats, but not in chronic alloxan nonketotic controls, was coincident with increased susceptibility to zygomycosis.

10.7 SUMMARY

Innate immunity such as normal anatomic barriers, serum *Rhizopus* inhibitory factor, and neutrophil cidal mechanisms appear sufficient for successful host defense against the zygomycetes. The *Rhizopus* inhibitory factor is present in normal serum and absent in ketoacidotic human diabetics, heavily immunosuppressed patients, and in alloxan-diabetic rabbits. The factor is removed from serum by dialysis. Alkaline proteinases are produced by zoopathogenic *Rhizopus* species at physiologic pH.

Athymic nude mice have more severe lesions in the late stages of experimental *A corymbifera* infection than euthymic heterozygotes, giving indirect evidence of a role for T-cell-mediated immunity in eradicating the hyphae from established lesions. Humoral immunity induced in mice by injections of killed mycelia do not increase resistance to dissemination to the brain after challenge with conidia.

Dormant *R oryzae* enclosed in small granulomas of rabbits are activated when diabetes is induced with alloxan. The PMN from alloxan-diabetic rats react slowly and produce less inflammation in response to *R oryzae* conidia. This sluggishness may be related to inhibition of mast cell degranulation, which slows the rate of increased vascular permeability and retards chemotaxis. It may also be influenced by the reduced diapedesis that occurs in diabetes. After PMN adhere to hyphae, they discharge granules that mediate damage to *Rhizopus oryzae*. The effector molecules are myeloperoxidase-H_2O_2-halide, with cationic proteins and lysozyme playing secondary roles.

One third of the cell wall of the *Mucorales*, for example, *M mucedo*, consists of poly-β-1,4-glucosamine (chitosan) which is complexed to inorganic polyphosphate. Mucoran is a heteroglycan composed of glucuronic acid : mannose : L-fucose in proportions of 5 : 3 : 2. The major structural feature in this polymer is the repeating disaccharide 4-α-glcUA 1 $\rightarrow$ 3-mannose. A second alkali-soluble polymer, mucoric acid, contains up to 85% of glucuronic acid. Exposure of *M mucedo* walls to nitrous acid-induced deaminative depolymerization solubilizes a glycouronan containing L-fucose : mannose : galactose : glucuronic acid in proportions of 5 : 1 : 1 : 2. Immunochemical characterization of this complex is lacking.

Peptido-L-fuco-D-mannans are a minor component of the *Mucorales* cell walls by weight, but are the major antigen in these molds. The polymer is purified by affinity for con A and has an L-fucose : mannose ratio of 1 : 2.5. Major linear sequences are 1,6-linked mannose residues. Peptidofucomannans from *Absidia, Mucor,* and *Rhizopus* species are serologically identical in their immunodiffusion reactions and do not crossreact with mannan from *C albicans*. The serologic activity is unaffected by proteinase.

REFERENCES

Baker RD, 1975. Mucormycosis (opportunistic zygomycosis) pp. 204–214. In Opportunistic Fungal Infections. 2nd Inter Conf. Springfield, Ill: C. C. Thomas.

Bartnicki-Garcia S, Nickerson WJ, 1962. Isolation, composition and structure of cell walls of filaments and yeast-like forms of *Mucor rouxii*. Biochim Biophys Acta 58:102–119.

Bartnicki-Garcia S, Lindberg B, 1972. Partial characterization of mucoran: the glucuronomannan component. Carbohydr Res 23:75–85.

Bartnicki-Garcia S, Reyes E, 1968. Polyuronides in the cell walls of *Mucor rouxii*. Biochim Biophys Acta 170:54–62.

Chandler FW, Kaplan W, Ajello L, 1980. A Colour Atlas and Textbook of Histopathology of Mycotic Diseases 333 pp. London: Wolfe Medical Publications.

Corbel MJ, Eades SM, 1976a. The relative susceptibility of New Zealand black and CBA mice to infection with opportunistic fungal pathogens. Sabouraudia 14:17–32.

Corbel MJ, Eades SM, 1976b. Experimental phycomycosis in mice: examination of the role of acquired immunity in resistance to *Absidia ramosa* syn. *corymbifera*. J Hyg 77:221–233.

Corbel MJ, Eades SM, 1977. Experimental mucormycosis in congenitally athymic (nude) mice. Mycopathol 62:117–120.

Cordes DG, Royal WA, Shortridge EH, 1967. Systemic mycosis in neonatal calves. New Zeal Vet J 15:143–149.

Datema R, van den Ende H, Wessels JGH, 1977. The hyphal wall of *Mucor mucedo* 1. Polyanionic polymers. Eur J Biochem 80:611–619.

Diamond RD, Krzesicki R, Epstein B, Yao W, 1978a. Damage to hyphal forms of fungi by human leukocytes in vitro: a possible host defense mechanism in aspergillosis and mucormycosis. Amer J Pathol 91:313–323.

Diamond RD, Krzesicki R, Jao W, 1978b. Damage to pseudohyphal forms of *Candida albicans* by neutrophils in the absence of serum in vitro. J Clin Invest 61:349–358.

Dow JM, Darnall DW, Villa VD, 1983. Two distinct classes of polyuronide from the cell walls of a dimorphic fungus, *Mucor rouxii*. J Bacteriol 155:1088–1093.

Eades SM, Corbel MJ, 1976. Enhancement of susceptibility to experimental phycomycosis by agents producing reticuloendothelial stimulation. Br Vet J 131:622–624.

Galbraith RM, 1979. Immunological Aspects of Diabetes Mellitus. 80 pp. Boca Raton, FL: CRC Press.

Gale GR, Welch AM, 1961. Studies of opportunistic fungi I. Inhibition of *Rhizopus oryzae* by human serum. Amer J Med Sci 241:604–612.

Handwerger BS, Fernandes G, Brown DM, 1980. Immune and autoimmune aspects of diabetes mellitus. Hum Pathol 11:338–352.

Jones KW, Kaufman L, 1978. Development of an immunodiffusion test for diagnosis of systemic zygomycosis (mucormycosis): preliminary report. J Clin Microbiol 7:97–101.

Marchevsky AM, Bottone EJ, Geller SA, Giger DK, 1980. The changing spectrum of disease, etiology, and diagnosis of mucormycosis. Hum Pathol 11:457–464.

Martin SM, Adams GA, 1956. A survey of fungal polysaccharides. Can J Microbiol 2:715–721.

McGinnis MR, 1980. Laboratory Handbook of Medical Mycology. 661 pp. New York: Academic Press.

Miyazaki T, Yadomae T, Yamada H, Hayashi O, Suzuki I, Ohshima Y, 1980. Immunochemical examination of the polysaccharides of mucorales, pp. 81–94. In Fungal Polysaccharides, Symposium Series no. 126. Sandford PA, Matsuda K, eds. Washington, DC: American Chemical Society.

Okuda K, Tadokoro I, Noguchi Y, Okuda K, 1974. Nitroblue tetrazolium (NBT)-dye test and myeloperoxidase reaction of human leukocytes. Japan J Microbiol 18:337–342.

Owens AW, Shacklette MH, Baker RD, 1965. An antifungal factor in human serum I. Studies of *Rhizopus rhizopodiformis*. Sabouraudia 4:179–186.

Reinhardt DJ, Licata I, Kaplan W, Ajello L, Chandler FW, Ellis JJ, 1981. Experimental cerebral zygomycosis in alloxan diabetic rabbits: variation in virulence among zygomycetes. Sabouraudia 19:245–255.

Reinhardt DJ, Hon PJ, Abdelal AT, 1979. Isolation and characterization of an alkaline protease from *Rhizopus oryzae*. Los Angeles CA: Abstr. Ann. Mtg. Amer. Soc. Microbiol. #F8, p. 364.

Reinhardt DJ, (Personal communication).

Rippon JW, 1982. Medical Mycology—The Pathogenic Fungi and the Pathogenic Actinomycetes, 2nd edition. 842 pp. Philadelphia, PA, W. B. Saunders Company.

Sheldon WH, Bauer H, 1958. Activation of quiescent mucormycotic granulomata in rabbits by induction of acute alloxan diabetes. J Exp Med 108:171–178.

Sheldon WH, Bauer H, 1959. The development of the acute inflammatory response to experimental cutaneous mucormycosis in normal and diabetic rabbits. J Exp Med 110:845–852.

Sheldon WH, Bauer H, 1960. Tissue mast cells and acute inflammation in experimental cutaneous mucormycosis of normal, 48/80 treated and diabetic rats. J Exp Med 112:1069–1083.

Chapter 11

Candida albicans

11.1 INTRODUCTION

Candida albicans is a commensal yeastlike fungus whose habitat is the mucosae of warm-blooded animals and humans. In individuals whose immune and endocrine systems are intact the organism is typically benign, but *C albicans* is a better clinician and can discover metabolic or immunologic abnormalities earlier in their development than we can with our chemical tests. This sentence speaks volumes about the tendency of *C albicans* and other pathogenic candidae to be opportunists. A successful defense against these organisms seems to require a high order of communication between regulator and effector cells. The standard texts such as Rippon (1982) have much background material on the *Candida* species. Great depth of coverage is provided by Odds (1979). Reviews of the immunology of candidiasis (Wilton and Lehner, 1980) and perspectives on histopathology (Chandler et al, 1980) should also be consulted.

Most of the literature emphasized in this chapter on molecular and cellular immunology pertains to *C albicans*, but other species are also involved in human disease: *Candida parapsilosis* is isolated from paronychia, endocarditis, and otitis externa; *Candida tropicalis* from vaginitis, onychomycosis, bronchopulmonary, and disseminated disease; *Candida guilliermondii* from endocarditis, dermatomycosis, and onychomycosis; *Candida pseudotropicalis* from vaginitis; and *Candida krusei* rarely from endocarditis and vaginitis (Rippon, 1982). *Torulopsis (Candida) glabrata* and *C tropicalis* are encountered in systemic disease in immunocompromised hosts (de Repentigny and Reiss, 1984). *C tropicalis* has emerged as the major species isolated in disseminated candidiasis in the U.S.A. (Wingard et al, 1979; Meunier-Carpentier et al, 1981). *C tropicalis* is found in the human intestinal flora and on human skin. It also survives in foodstuffs and in treated sewage, whereas *C albicans* does not (Ahearn et al, 1968).

C albicans has a diploid multipolar budding yeast form (blastoconidia, 3×4 μm diameter) and also produces pseudomycelia, which are elongated cells joined together at the ends by constrictions. A true septate mycelia is produced, especially in tissues of the host. Dimorphism of *C albicans* can be regulated in vitro by the temperature of growth on a chemically defined medium (Table 11.1) (Lee et al, 1975). A pure mycelial form grows at 37 °C provided that, under the conditions, cells are harvested between the 18th and 27th hours. Afterwards, blastoconidia begin to appear. Cultures maintained on this medium at 25 °C grow in a pure yeast form. The response to temperature elevation is a function of the isolate, ie, some tend to produce mixed cultures of blastoconidia, pseudomycelia, and mycelia (Manning and Mitchell, 1980a).

Several factors are well-recognized as creating opportunities for the conversion of *Candida* species from a commensal existence into a pathogen, invading the tissues of the host from mucosal foci (Odds, 1979). Hormonal fluctuations in the third trimester of pregnancy alter the vaginal secretions, elevating the glycogen content, which favors the growth of *C albicans*. Prolonged broad spectrum antibacterial therapy, tetracyclines in particular, results in a replacement flora of resistant organisms, in cluding the yeasts. Hyperalimentation with solutions of high osmotic pressure creates an environment for yeast vegetations at the tips of indwelling catheters. Systemic corticosteroid therapy causes a cyclic fluctuation in peripheral lymphocytes, which exerts an immunosuppressive effect that increases the risk of candidiasis. In the guinea pig model, systemic long-acting glucocorticosteroids, like cortisone, suppress established *C albicans*-specific cell-mediated immunity and increase renal colony counts and mortality (Hurley et al, 1975). The granulocytopenia and immunosuppression resulting from management of hematologic malignancy with cytotoxic drugs allows endogenous commensal organisms like *Candida* to invade the

Table 11.1. Amino Acid Synthetic Medium for *Candida albicans*

Chemicals	Grams
$(NH_4)_2SO_4$	5.0
$MgSO_4 \cdot 7 H_2O$	0.2
K_2HPO_4 (anhydrous)	2.5
NaCl	5.0
Glucose	12.5
L-Amino acids:	
Alanine	0.5
Leucine	1.3
Lysine	1.0
Methionine	0.1
Ornithine	0.0714
Phenylalanine	0.5
Proline	0.5
Threonine	0.5
Biotin[a]	0.001
Distilled water to 1000 ml	
Combine ingredients and autoclave at 110 °C, 10 lb, 20 minutes	

Source: Reproduced from Lee et al (1975) with permission of the copyright holder, the International Society for Human and Animal Mycology.

[a] Add biotin, filter sterilize; pH 6.8 ± 0.05.

deep tissues. The onset of adult diabetes mellitus is sometimes heralded by the appearance of mucocutaneous candidiasis.

Hematogenous spread of *Candida* species frequently seeds the yeasts in the kidneys and a pyelonephritis can develop (Odds, 1979; Rippon, 1982). The renal focus of *C albicans* after intravenous infection is well-characterized. In the rabbit, multiple microabscesses are found that cause renal failure. The initial lesions occur in the cortex, then extend to the medulla and into the renal pelvis. The early lesions, meanwhile, become granulomatous (de Repentigny et al, 1984).

11.2 CELL WALL ARCHITECTURE

11.2.1 Protein

C albicans blastoconidia adhere to buccal or vaginal epithelial cells to a much greater extent than other yeasts of the genus (King et al, 1980). Electron micrographs show threadlike fibrils extending from *C albicans* to the epithelial cells. Adherence is inhibited by digestion of the blastoconidia with trypsin, other broadly specific proteinases, or with sulfhydryl reagents (Lee and King, 1983), but not with α-mannosidase. When cell walls are prepared, however, α-mannosidase gains access to the yeast cell wall and disturbs adherence to vaginal epithelium. The available evidence thus implicates a dual recognition of surface coat protein and mannan.

Other researchers have also recognized a superficial outer coat on *C albicans* yeast and mycelia (Pugh and Cawson, 1978). These "secretions" were detected in electron micrographs by staining cells grown in serum or in chick embryos with ruthenium red. This outer material contained protein, polysaccharide and enzymes, but was not a true capsule and could not readily be detected in the light microscope. Two enzymes were localized in the outer coat, acid phosphatase and N-acetylglucosaminidase. Further evidence of an outer protein layer of *C albicans* is derived from studies with the sulfhydryl reagent, dithiothreitol (DTT). Exposure of yeast to DTT caused the release of 42% of total surface acid phosphatase (Chattaway et al, 1974). The enzyme was liberated as a soluble mannan complex after treatment with DTT-glucanase (Chattaway et al, 1976).

The role of protein in maintenance of the integrity of *C albicans* cell walls is illustrated by the necessity of adding sulfhydryl reagents and proteinase in order to improve the efficiency of protoplast formation with mixed glycosidases (Torres-Bauza and Riggsby, 1980). A pretreatment of *C albicans* with DTT, Na$_2$EDTA, and pronase was beneficial. Pretreated yeast resuspended in diluted snail digestive juice rapidly and completely converted to protoplasts. In one isolate tested, DTT treatment alone removed an outer wall layer. Another study (Cassone et al, 1978) reported that DTT and pronase could remove two outer mural layers and reduce wall thickness from 135 to 85 μm. Moreover, the layered appearance of the inner wall cross-section was destroyed by this treatment.

11.2.2 Mannan

The agglutination of *C albicans* by concanavalin A and the blockage of agglutination by methyl-α-mannoside is evidence for the presence of mannan on the cell surface (Cassone et al, 1978). Another yeast, *Rhodotorula glutinis*, used as a control, does not contain α-mannan on its surface and was not agglutinated by Con A. The augmentation of adherence of *C albicans* blastoconidia to buccal epithelial cells by bacteria with mannose-sensitive pili is indirect proof of the surface location of mannan (Centeno et al, 1983).

The outer mural layer appears as an electron-dense spiky coat. Alkali digestion does not change wall thickness, but the stratification is lost and electron density decreases. This treatment is in contrast to DTT-pronase, which removes two outer wall layers, and suggests that protein, or the protein component of mannan, is more important than polysaccharide in maintenance of wall integrity. Although some glucan and all chitin resist both pronase and alkali treatment, these fibrillar components do not determine the layering seen in electron microscope cross-sections stained by conventional techniques (ie, lead citrate, uranyl acetate, and OsO_4). The agglutinability of yeast by Con A, even after digestion with glusulase and production of spheroplasts, suggests that mannan occurs in deep as well as superficial layers, and that laminations are due to mannan–protein in the electron-dense areas and glucan–chitin in the electron-lucent layers.

11.2.3 Glucan

Pretreatment with DTT is necessary for the subsequent action of *endo-β-1,3*-glucanase from *Cytophaga johnsonii* (Chattaway et al, 1976). The action of DTT opens up outer mural layers, exposing the inner glucan layer. Protoplasts in high yield (up to 90%) were obtained with *Cytophaga* L_1 enzyme complex containing *endo-β-1,3*-glucanase and a proteinase. Purified *endo-β-1,3*-glucanase alone induced only partial hydrolysis, resulting in spheroplasts. Chitinase is helpful for high efficiency protoplasting, but the *Cytophaga L* preparation, which lacks chitinase, also results in protoplasts. Although many of *C albicans* linkages are β-1,6, cleavage of β-1,3 bonds is sufficient for protoplasting, in combination with DTT and protease.

11.2.4 Chitin

Chitin is localized in bud scars of baker's yeast, more particularly in a fibrillar ring surrounding the bud scar where it functions to protect the channel between mother and daughter cell during passage of the nucleus into the bud (Cabib and Bowers, 1971). In filamentous fungi such as *Neurospora*, the skeletal layer containing chitin is adjacent to the plasmalemma as a thin microcrystalline sleeve (Trinci, 1978; Farkas, 1979). Chitin was identified in *C albicans* by X-ray diffraction and chemical analysis (Chattaway et al, 1976). The chitin content of yeast forms, 1.7%, was less than the amount, 7.6%, found in mycelial form walls. The walls of myelial forms are half as thick as yeast forms (Cassone et al, 1978) so that the higher chitin content of mycelia implies that chitin has more influence in maintaining their rigidity and shape. Chitin not present in bud scars forms a thin layer uniformly distributed over the lateral walls in *Saccharomyces cerevisiae* and also in *C albicans*.

The composition of *C albicans* walls (Chapter 2, Table 2.3), which is often cited, partly because of the paucity of other reports, is subject to verification. Such verification is especially needed for mycelial forms because the recovery of components is relatively low and the medium for growth, ox serum, may not be optimal for normal expression of the cell wall mosaic. Two of the *C albicans* cell wall studies utilized acidic conditions during defatting that are now known to lead to loss of phosphodiester-linked mannose residues. Lowered recovery of wall components also results from the use of alkali to fractionate the cell walls, with consequent loss of base-labile oligomannosides. For these reasons the values calculated for the mannan content of the cell walls, ranging from 15.2% to 22.9% of the mural dry weight, may be underestimates.

The incorporation of U-^{14}C-glucose into blastoconidia and mycelia of *C albicans* grown on synthetic medium permitted an estimate of the relative amounts of mannan and chitin (Elorza et al, 1983). Radiolabeled cells were subjected to a cycle of hot dilute alkali–hot dilute acetic acid, plus a second alkaline extraction. The residue after extraction, mainly chitin, contained 9% of the radioactivity of the blastoconidial cell, and 21% of the mycelial radioactivity. Mannan was that portion of the alkali extract that precipitated as the Cu^{++} complex with Fehling solution. It constituted only 4% of the mycelial form and 19% of the radioactivity of the blastoconidia. A question remains about whether these data would be similar to those from a conventional analysis determined by first isolating cell walls. No direct comparison of the two methods has yet been made.

11.2.5 Summary of Ultrastructure

Conventional fixation and staining of thin sections for electron microscopy reveal laminations in the cell walls of yeast, but whether these are artifactual owing to the extent of penetration of reagents is questionable (Djaczenko and Cassone, 1971). More specific probes have become essential in order to localize regions of the cell wall that vary in their composition. Progress has been reported with the use of labeled lectins and with an adaptation of the periodic-acid oxidation technique to electron microscopy (Thiery and Rambourg, 1974). Labeled Con A has been used as a probe for mannan (Cassone et al, 1978), and wheat germ agglutinin as a chitin localizing reagent (Horisberger and Rosset, 1976).

The model of *C albicans* mural architecture that emerges from available chemical and electron microscope evidence is a mosaic of glucans, mannan, and chitin, but one in which protein plays an important structural role. An outermost spiky or threadlike layer of protein mediates adherence through either an immunologic or lectinlike mechanism. Beneath this protein lies a layer of readily soluble mannan that is the substrate for anti-*Candida*-agglutinins. This mannan layer is sloughed off into the tissues and body fluids during proliferation. Mannan also occurs in deeper layers of the wall, embedded in the more fibrillar glucan. This occurrence is suspected because spheroplasts are still agglutinable by Con A (Cassone et al, 1978). Glucan occurs as a broad electron-lucent and probably fibrillar layer (Poulain et al, 1978). It is inaccessible to glucanases unless outer mural layers are breached by a reducing agent and proteinase. Once that occurs the glucan is susceptible to *endo-β-1,3* attack resulting in spheroplasts. Chitin is believed to occur as a thin crystalline layer enmeshed in glucan near the plasmalemma (Chapter 2, Figure 2.6). Evidence for this layer is the action of chitinase in effectively removing the inner wall layers (Torres-Bauza and Riggsby, 1980; Chattaway et al, 1976).

Localization of wall layers by periodate oxidation provides another line of evidence compatible with the above model. In this method (Thiery and Rambourg, 1974; Poulain et al, 1978; Cassone et al, 1978) susceptible carbohydrates are oxidized and the resulting aldehydes are coupled with thiosemicarbazide (Figure 2.1). This ligand will bind silver proteinate, depositing electron-dense granules or confluent layers. The technique has been given the acronym PAT-Ag. The outermost mural layer of *C albicans* reacts strongly as predicted from the susceptible 1,2- and 1,6-mannosyl linkages in mannan (Poulain et al, 1978). This broad PAT-Ag positive layer revealed scattered large granules in yeast grown on synthetic medium. When grown in horse serum, a change to a dense band with fine outer microfibrils was observed (Poulain et al, 1978). A broad, median, lucent layer was not oxidized by periodate and may be the site of glucan–chitin. Proximal to the plasmalemma was a narrow PAT-Ag positive layer that may be the site of polysaccharide synthesis.

Exposure to a reducing agent, a chelator, pronase, and mixed glycosidases containing chitinase, results in a high yield of protoplasts. The mycelial form wall is considered to contain the same arrangement of glucans, but is half the diameter of that of yeast forms.

The wall serves a multiplicity of functions: maintenance of cell shape is an obvious one. The occurrence of a superficial protein layer mediating adherence in *C albicans* increases the pathogenic potential of this yeast. Mannan linked to structural protein in which disulfide bridges play a conformational role creates an outer barrier of lysis resistance. Mannan, glucan, and chitin seem to be arranged in concentric layers of increasing structural rigidity. Within this framework reside periplasmic enzymes, perhaps some with aggressive designs on the host.

11.3 MANNAN IMMUNOCHEMISTRY AND IMMUNOLOGY

11.3.1 General Plan

The general plan of yeast mannan is a linear α-1,6 backbone heavily substituted with α-1,2- and α-1,3-linked oligomannoside chains. The length of side chains, location of 1,3 bonds, and substitution with phosphodiesters determine antigenic specificity. Two domains recognized in yeast mannans are the inner core and the outer chain (Ballou, 1976; 1980) (Figure 2.2). The inner core structure is a common property of yeast, plant, and mammalian glycoproteins. It consists of structural protein linked to core mannan (average degree of polymerization, 12) through sequences composed of asparagine-x-serine (threonine). These sites of asparagine glycosylation occur via chitobiosylaspartamido bonds (Nakajima and Ballou, 1974b). A second type of glycopeptide is composed of oligomannosides linked to the serine or threonine of the same protein via O-glycosidic hydroxyamino acid esters. These linkages are subject to β-elimination under alkaline conditions used either to extract mannan from yeast (Gorin and Spencer, 1968, see 11.3.3) or in the Fehling precipitation step (Peat et al, 1961, see below) and consequently these oligomannosides may not be preserved in the extracted mannan.

The polymannosyl outer chain region distinguishes yeast mannans from other glycoproteins. The outer chain is a trimer to pentamer of a repeat unit consisting of approximately 15 mannose residues (Ballou, 1976). This region is of primary concern because it contains the immunodominant epitopes and the structure versus activity relationships expressed by the outer chain will be emphasized herein.

If the three glycosidic elements (outer chain, inner core, base-labile oligosaccharides) of a model mannan, ie, *S cerevisiae* X2180 (Ballou, 1980), are considered, the approximate degree of polymerization of the overall repeat unit is 88, giving a molecular weight for the carbohydrate portion estimated at 1 to 5×10^4 da. This estimate gives an impression of the size and complexity of yeast mannan.

Ballou and associates contributed much to knowledge of mannan structure through chemical, immunochemical, and genetic studies. The derivation and exploitation of mutants blocked at different stages of mannan synthesis (Ballou, 1980) should be consulted because that aspect is outside the scope of this book.

11.3.2 Serotypes

The A and B serotypes that allow discrimination between *C albicans* isolates were first described in agglutinin reactions (Hasenclever and Mitchell, 1961) and these differences were found to reside in mannan (Summers et al, 1964). Serotype A is the most abundant isolate, accounting for 74% of the *C albicans* clinical isolates in one large series from Canada (Auger et al, 1979). Serotype A has been found to correlate positively with 5-fluorocytosine sensitivity, whereas half of serotype B isolates showed primary resistance (Auger et al, 1979). In 1970, serotype C was described (Nishikawa et al, 1970). A multiplicity of surface factors of yeast were formulated into a scheme of numerical taxonomy based on agglutinin adsorption reactions (Tsuchiya et al, 1974) to differentiate *Candida* species. This has permitted further discrimination between the serotype A (factors 6, occasionally 13b) and serotype C (factors 6 and 7). Serotype B (factor 13b, occasionally 7) is distinct from serotype A and factor 13b serum was obtained by adsorption of anti-*C albicans* B Ig with *C tropicalis* yeast. Difficulties in preparing factor 7 antibodies made it desirable to combine serotype C with serotype A. In this discussion the focus is on those epitopes present in mannans, but it should be kept in mind that noncarbohydrate determinants may be responsible for some of these factors. Serologic relationships among various *Candida* species and baker's yeast are important taxonomically as well as clinically. Serology is a sensitive indicator of small changes in the structure of mannans.

Adsorption of type A antiserum with type B yeast leaves A-specific globulins that can be used as an A-typing reagent (Hasenclever and Mitchell, 1961). In the reciprocal adsorption of type B antiserum with type A yeast, all antibodies capable of agglutinating *C albicans* B were removed. On this basis, it was claimed that type B yeast contained all the determinants present in type A except for an A-specific factor (Hasenclever and Mitchell, 1961). When mannans extracted from these two types were compared in immunodiffusion and quantitative precipitin reactions, other differences were found that support the existence of type B-specific factors (Tsuchiya et al, 1974). Immunodiffusion results indicate that anti-A serum reacts weakly with type B mannan (Reiss et al, 1974; Lehmann and Reiss, 1980) and only 21% of type A antibody N was precipitated by type B mannan compared with precipitation by the homologous type A mannan (calculated from Sunayama and Suzuki, 1970). Moreover, subfractions of type B mannan enriched in a B-specific antigen were prepared by anion exchange chromatography (Okubo et al, 1980, see below). The serologic behavior of *C albicans* mannan of type A or B in the presence of *S cerevisiae* reagents gave evidence of further qualitative differences. Baker's yeast antiserum does not crossreact with type A mannan, but does with that from type B. In the reciprocal case, baker's yeast mannan was precipitated by anti-*C albicans* A globulins to less than 10% of the homologous *C albicans* A reaction (Summers et al, 1964). The result obtained with anti-*C albicans* B globulins, about 50% of the antibody was precipitated, revealing greater crossreactivity with baker's yeast (Hasenclever and Mitchell, 1964).

Quantitative precipitin reactions (Sunayama, 1970; Summers et al, 1964) found that *C tropicalis* mannan is closely related to type A and that *C albicans* A serum adsorbed with type B yeast was still reactive with *C tropicalis* mannan. In that sense, *C tropicalis* is closer to *C albicans* A mannan than is *C albicans* B. Reciprocal cross adsorptions between *Candida stellatoidea* and *C albicans* B revealed no differences in agglutinin tests. Quantitative precipitin reactions of isolated *C stellatoidea* mannan with type B antiserum precipitate less antibody N than homologous type B or type A mannan so that *C stellatoidea* and *C albicans* B may not be serologically identical (Sunayama and Suzuki, 1970). Another difference between *C albicans* B and *C stellatoidea* mannan was evident in their acetolysis fingerprints.

The oligomannoside haptens of *C parapsilosis* were studied because that yeastlike fungus contains agglutinin factor 13b, which is shared by *C albicans* serotype B and to a minor extent by *C albicans* type A (Funayama et al, 1983; Tsuchiya et al, 1974). The mannan prepared by hot dilute alkali extraction of whole blastoconidia was precipitated by Fehling solution and then subjected to acetolysis. Acetolysis fragments were sized by gel permeation chromatography with M6 being the most prevalent fragment and best inhibitor of immune precipitation of the mannan–anti-*C parapsilosis* reaction. The structure of M6 was deduced by methylation-fragmentation and ^{1}H-nuclear magnetic resonance spectrometry. It consisted of a straight chain α-1,2-linked sequence with a single α-1,3 bond at the penultimate sugar from the nonreducing terminus. Antiserum prepared against *C parapsilosis* was adsorbed with *C albicans* A cells. The adsorbed serum could still agglutinate *C parapsilosis* cells and presumably contained antibodies against factor 13b. However, this adsorbed antiserum could not precipitate the mannan of *C parapsilosis*, indicating that the 13b epitope is a nonmannan constituent, or more likely is lost during the preparation of the mannan, possibly through alkaline degradation.

Antiserum to *C albicans* A reacted with galactomannan from *Wangiella (Hormodendrum) dermatitidis* in quantitative precipitin tests (Takeda et al, 1979). This reaction was stable after removal of galactofuranosides with mild acid hydrolysis. The reciprocal reaction between anti-*W dermatitidis* and *C albicans* mannan did not occur. Two other related dematiaceous fungi showed no crossreactions. Weak crossreactions between anti-*C albicans* and galactomannan from *Aspergillus fumigatus* and *Trichophyton rubrum* have also been observed (Suzuki et al, 1967).

As a rule, immunochemical studies with yeasts have used rabbit antisera produced by intravenous injection of whole heat- or formalin-killed *Candida* species. There is no difficulty in inducing antibodies by this means, but factors regulating antibody synthesis or the relative affinity of these antibodies have not been determined. Studies in which more than one isolate of a given serotype have been compared are rare. The methods of preparing bulk mannan vary in harshness (Gorin and Spencer, 1968; Peat et al, 1961; Nakajima and Ballou, 1974a; Reiss et al, 1974) and although mannan is heat-stable and will tolerate alkaline conditions to some degree, too little effort has been made to optimize extraction conditions for maximum preservation of antigenic determinants and molecular weight. Moreover, to think of extracted mannan as a single entity is an oversimplification because several factors contribute to heterogeneity: (i) polydispersity resulting from method of biosynthesis and fractional depolymerization during extraction; (ii) occurrence of three domains in mannan: outer chain, inner core, and base-labile oligosaccharides; (iii) fractionation based on charge resulting from *O*-phosphonomannan; (iv) presence of glucan-mannan complexes. Subject to these concerns, remarkable progress has been made in elucidating the antigenic structure of mannan. The techniques most valuable for that purpose are reviewed below.

11.3.3 Preparation

Mannan may be prepared by several means, which vary in harshness. For immunologic purposes, exposure to extremes of pH and heat should be minimized or avoided for maximum preservation of antigenic determinants and the degree of polymerization. In some instances, when it is desirable to know the minimal repeating unit necessary to elicit a response, harsh methods of preparation may be justified.

- Fehling precipitation. Detailed protocols based on the original Peat et al (1961) procedure have been published (Colonna and Lampen, 1974a; Okubo et al, 1980). In general, whole packed yeast are autoclaved in dilute neutral citrate buffer. The soluble extract is acidified to precipitate protein or deproteinized with chloroform-n-butanol (Okubo et al, 1980). Protein-depleted mannan is neutralized and precipitated with ethanol. The polysaccharide is dissolved in water and separated from glycogen with Fehling solution, in the cold, to give the blue Cu^{++} mannan complex. The complex is recovered by centrifugation, resuspended in water and dissociated with cation exchange resin. Mannan is neutralized, concentrated in vacuo, precipitated with ethanol, dialyzed and lyophilized. Fehling solution is a 1:1 vol/vol solution of (i) 6.93% $CuSO_4 \cdot 5H_2O$ and (ii) 34.6% sodium potassium tartrate (Rochelle salt) in 13% NaOH. (Fieser, 1957).

- Cetyltrimethylammonium bromide complexation (Lloyd, 1970; Nakajima and Ballou, 1974a). Mannan extracted from whole yeast by autoclaving in neutral buffer is precipitated with ethanol and dialyzed. Cetyltrimethylammonium bromide is added to precipitate acidic proteins and nucleic acids. The supernate receives 1% boric acid and is titrated to pH 8.8, with resultant borate complexation of vicinal hydroxyls in the mannan giving a net negative charge to the molecule, which then forms an insoluble CTAB complex. The complex is recovered and dissociated at acidic pH (Nakajima and Ballou, 1974a) or with concentrated NaCl (Cherniak et al, 1980). In this way mannan is prepared without exposure to the alkaline conditions of Fehling solution.

- Hot alkali extraction (Gorin and Spencer, 1968). Whole yeast are suspended in 2% KOH and boiled for 2 hours. The cooled solution is then neutralized with acetic acid and centrifuged. The supernates are concentrated and precipitated with ethanol. Mannan is then selectively precipitated by Fehling complexation.

- Cold alkali. Mannan has also been extracted with cold alkali from isolated cell walls of *C albicans* (Reiss et al, 1974). Walls prepared by mechanical disruption were cleaned of membranes by washing in sodium dodecyl sulfate solution. A further defatting step consisted of chloroform–methanol. Walls were then extracted with ice cold dilute alkali for 24 hours resulting in solubilization of 25% of the wall dry weight. Mannan and glucan were present in the extract (mannose:glucose = 3:2) but it is not known if they occur as a covalent complex. Mannan prepared in this way is immunogenic, whereas mannan prepared from the same isolate by Fehling precipitation is serologically active but not immunogenic (Lehmann and Reiss, 1980a).

- Properties of the extracted mannans. Mannan isolated from baker's yeast by CTAB precipitation, avoiding exposure to alkali, had a molecular weight of 133,000 da, as determined by sedimentation-equilibrium-ultracentrifugation (Nakajima and Ballou, 1974a). Further treatment of this mannan with 0.1M NaOH at ambient temperature reduced the size to 40,000 da. The mildness of this procedure was unlikely to cleave phosphodiesters, but could saponify acyl esters linking oligomannosides to serine or threonine of the peptide moiety. More drastic basic conditions, ie, boiling in 1M NaOH leads to extensive degradation by breakage of O-phosphonomannan, O-mannosidic hydroxyamino acid and N-glucosaminylaspartamide. Even mild alkali digestion can reduce the molecular size through β-elimination of O-mannosidic hydroxyamino acids.

11.3.4 Methylation-Fragmentation

Methylation-fragmentation permits the determination of glycosidic bond arrangements and the degree of substitution in a polysaccharide, often making it possible to propose a minimal repeat unit. The technique consists of permethylation, fragmentation, identification, and quantitation of methylated monosaccharides. Some early studies of *C albicans* mannan (Yu et al, 1967a; Sunayama and Suzuki, 1970) were done before the technique was refined. Early methods required large amounts of sample, risked incompleteness of methylation, and complicated the assignment of linkages. At present, only 10 mg or less of pure polysaccharide is usually sufficient for analysis and the Hakomori reagent (dimethylsulfinyl carbanion and methyl iodide) enables complete methylation in many instances. Gas-liquid chromatography–mass spectrometry affords unequivocal identifications of methylated derivatives. Advances in the preparation of volatile derivatives of sugars eliminate the multiple peaks arising from the α and β anomers. The difficulties in interpreting data derived before the advent of these improved methods have been discussed elsewhere (Stewart et al, 1968). Early reports of mannofuranose in *C albicans* mannan (Yu et al, 1967) based on the presumed detection of 3,5-dimethoxymannose were actually due to 2,4-dimethoxymannose (Reiss et al, 1981), and correct identification awaited the general availability of GLC-MS. Various estimates have been made of the extent of branching in *C albicans* mannan as indicated by the di-O-methoxy sugars. The extent to which these discrepancies result from strain variation, method of extraction, or efficacy of methylation analysis is not known. The report of a lower abundance of 1,2 linkages in serotype A mannan (Yu et al, 1967a) has been contradicted by others (Reiss et al, 1981).

Two sources of *C albicans* mannan were subjected to methylation analysis: bulk mannan (Reiss et al, 1981), and a mannohexaose acetolysis fragment (Sunayama and Suzuki, 1970). Mannans from *C albicans* and from *Torulopsis (Candida) glabrata* prepared by Fehling precipitation (Peat et al, 1961) were methylated, fragmented, and the peracetyl aldononitrile derivatives were analyzed by GLC-MS (Reiss et al, 1981) (Table 11.2). There was a larger amount of 1,2-linked mannose in *C albicans* serotype A mannan compared with serotype B, indicating longer side chains in serotype A. This finding accords with Sunayama (1970), who isolated oligomannosides from *C albicans* A by acetolysis. The mannoheptaose was the strongest inhibitor of immune precipitation of serotype A mannan. This finding does not prove that the only differences between serotypes is the length of the side chains. A larger number of mannosyl branch points linked through C-1, C-3, and C-6 were found in serotype B mannan. This branch point was undetected in previous analysis of *C albicans* mannan and its location is unknown. Ballou (1976) shows this linkage in the penultimate side chain residue in the inner core region of baker's yeast mannan. Seymour et al (1976) detected a large amount of this branch point in mannan from *Pachysolen tannophilus NRRL-Y2461* accounting for 36.2 mol%.

The ratio of 1,2 to 1,3 linkages in serotype A mannan of 9.8:1 (Reiss et al, 1981) can be accounted for if two types of side chains are present: a type with five α-1,2-linked mannose residues and one with four α-1,2 linkages and one 1,3 linkage, an explanation consistent with other reports (Suzuki and Sunayama, 1970; Fukazawa et al,

Table 11.2. Relative Mol% of Peracetylated Nitriles of Methyl Ethers of Mannans Derived From Medically Significant Yeasts

Peak	Methyl Ether	C albicans (serotype A)	C albicans (serotype B)	T glabrata	Type Sugar
B	2,3,4,6-Tetra	23.0	28.6	36.6	Teminal nonreducing sugar
C	-Tetra	2.2	2.4	3.5	—
D	2,4,6-Tri-	4.7	3.5	2.4	(1-3) Linked internal sidechain sugar
E	3,4,6-Tri-	46.2	37.1	33.1	(1-2) Linked internal sidechain sugar
F	2,3,4-Tri-	7.6	7.9	1.3	(1-6) Unbranched back-bone
G	2,4-Di-	5.3	9.2	1.0	(1-3-6) Branched
H	3,4-Di-	11.0	11.4	22.2	(1-2-6) Branched back-bone

Source: Reproduced from Reiss et al (1981), copyright 1981 by John Wiley and Sons, Ltd.

1980, see below). The large number of nonreducing terminal mannosyl residues cannot be accommodated in the outer chain region of mannan and probably occur as short oligosaccharides linked to the peptide moiety through O-glycosidic hydroxyamino acid ester bonds. The occurrence of this structural feature in the inner core region is discussed by Ballou (1976).

Methylation analysis of the mannohexaose isolated after acetolysis from a *C albicans* isolate of unspecified serotype, probably type A (Sunayama and Suzuki, 1970) reported mannose derivatives in the following molar ratios: terminal nonreducing (1); 1,3-linked (1); and 1,2-linked (4). This distribution is consistent with a straight chain with a terminal 1,3 linkage and has received corroboration (Fukazawa et al, 1980) as the immunodominant hapten of serotype A.

The overview of the outer chain region of *C albicans* mannan suggested by methylation analysis and compatible with serologic findings is of a linear 1,6-linked backbone heavily substituted with oligosaccharides having a degree of polymerization of 5 or 6 in which α-1,2 linkages predominate, but with terminal 1,3-linked mannose residues in every other side chain. The weak crossreactivity between serotypes A and B and the resistance of serotype A mannan to exo-α-mannan hydrolase (see below) imply that chain length alone cannot account for serotypes.

The oligomannosides produced from *Saccharomyces kluyveri 1b* cell mannan by either acetolysis or by β-elimination were compared (Zhang and Ballou, 1981). Although acetolysis fragments have proven a valuable aid in fingerprinting yeast mannans, in this instance two immunochemically important structural features were labile to this treatment. Terminal β-1,2-linked mannose and single 1,6 branches in the oligomannosides were lost. These structural features remain intact in the β-eliminated oligomannosides. The 1,6 branches were detected by the presence of the methylated fragment 2,4-di-O-methylmannose. The β-linked mannose in the otherwise α-linked mannan was detected in the proton magnetic resonance spectrum as a signal at $\delta = 4.75$ ppm. The β-linkage was labile to β-mannosidase, indicating that it occupied a terminal position. The removal of 1,6 branches and nonreducing β-1,2-mannosyl termini greatly reduced the hapten inhibitory power of oligomannosides from *S kluyveri*. This study has important implications for the immunochemistry of *C albicans* mannan. Antiserum to *S kluyveri 1b* cells crossreacted strongly with *C albicans B311A* (serotype A) mannan but not at all with mannan of *C albicans B792* (serotype B). Mannan of serotype A was previously shown to be resistant to exo-α-mannan hydrolase (Reiss et al, 1981), so β-linked mannosyl termini appear to play an immunodominant role in type A mannan (Figure 11.1).

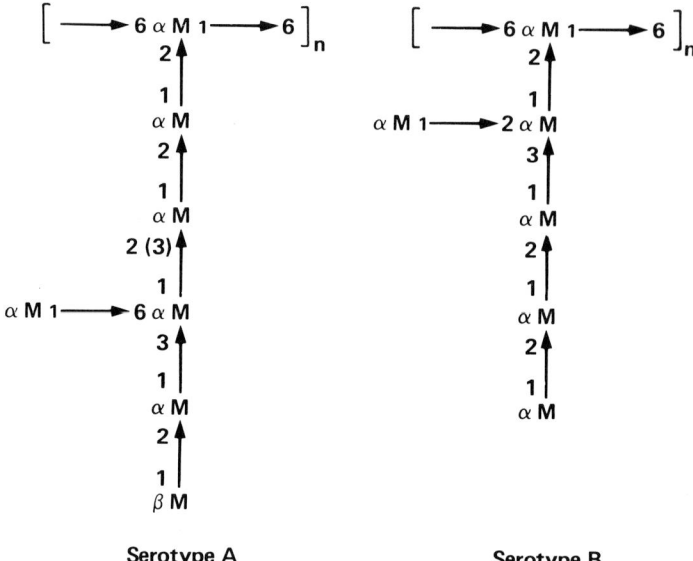

Figure 11.1. Hypothetical models of the immunodominant epitopes of *Candida albicans* serotypes A and B mannan. The model in the left panel is based on the assumption that the difference between serotypes is not chain length alone, a theme that is developed in the text. Zhang and Ballou (1981) have shown that nonreducing terminal β-1,2-mannosyl residues and single 1,6 branches in the oligomannosides are labile to acetolysis in *Saccharomyces kluyveri* mannan. These linkages may also occur in *C albicans* serotype A because antiserum to *S kluyveri* reacts with *C albicans* A but not with serotype B. The epitope shown in the right panel for serotype B is based on the methylation-fragmentation analysis of Fukazawa et al (1980).

A somewhat different interpretation of the nature of β-linkages in *C albicans* mannan is given by Suzuki and Fukazawa (1982). They proposed that the determinant specific for serotype A (factor "6") in *C albicans* is terminal 1,3-linked mannose and that the density of this determinant is low because the corresponding methylated fragment, 2,4,6-tri-*O*-methyl mannose is found in only trace amounts. Proof is based on the high inhibitory power of the acetolysis fragments (see below) mannopentaose and mannohexaose, which contain this linkage, and low inhibitory power of similar fragments of *C krusei*, which lack this linkage. Further studies showed a positive correlation between a β-linkage in mannan of serotype A and serologic activity. Evidence for this is that mannans of type B are more dextrorotatory than those of serotype A. The PMR spectra of *C albicans* type A, and *C tropicalis* display a signal at $\delta = 4.85$ ppm that is assigned to a β-1,6 linkage. This signal is conserved after mild acetolysis based on the premise that β-1,6 is more resistant to acetolysis than α-1,6. Lability of serotype A mannan to CrO_3 oxidation, which selectively cleaves β-linkages is additional proof, by negative selection.

11.3.5 Acetolysis

Controlled partial acetolysis is a fragmentation method that provides a fingerprint of the oligomannoside chains removed from mannan (Kocourek and Ballou, 1969). (Figure 11.2) Success depends on the increased acid lability of the 1,6 linkages in the mannan backbone to the reagent (acetic anhydride : glacial acetic acid : concentrated sulfuric acid 1 : 1 : 0.1 vol/vol, ice cold for three days or 40 °C for 2 to 12 hours) and the relative resistance of the 1,2 and 1,3 linkages in the oligomannoside chains, which are released intact, then de-*O*-acetylated and separated by gel permeation

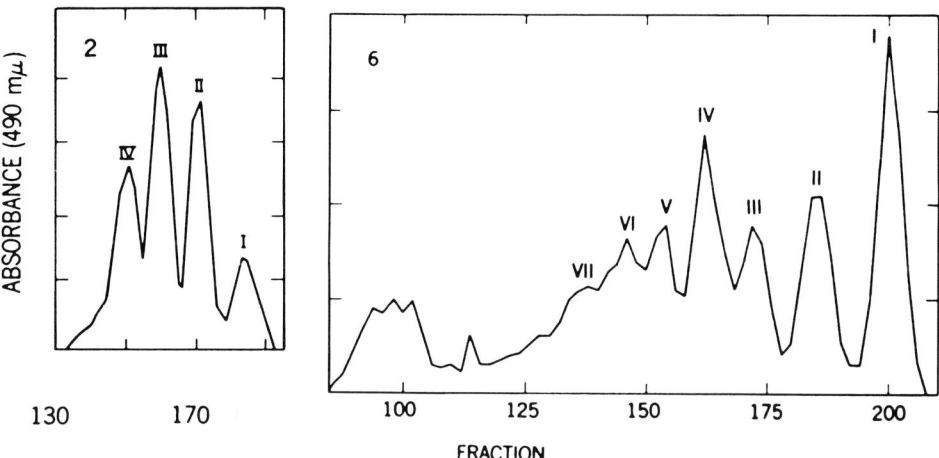

Figure 11.2. Acetolysis patterns of bulk mannans isolated from (2) *Saccharomyces cerevisiae*, Red Star strain 2; and (6) *Candida albicans* B311, serotype A. Note the increased complexity and high degree of polymerization of fragments from *C albicans*. Roman numerals refer to the degree of polymerization of mannosyl residues (Kocourek and Ballou, 1969).

chromatography. The longest oligosaccharides isolated from *C albicans* serotype A mannan were heptaoses and from serotype B, hexaoses (Sunayama and Suzuki, 1970). Among the saprophytic yeasts, oligomannosides are shorter, containing four, or rarely, five mannose residues (Kocourek and Ballou, 1969). Acetolysis fragments provide fingerprints of taxonomic value, for use in sequence analysis, and as probes of serologic relatedness. Separated oligomannosides selectively inhibit the quantitative immunoprecipitation of mannan and the strength of this inhibition aids in the identity of the configuration and length of the immunodominant haptenic groups. In the homologous mannan A–anti-A reaction, the order of inhibitory power was heptaose greater than hexaose (Sunayama, 1970). On the other hand, in the reaction between serotype B mannan and anti-A globulins, the order of inhibitory power of A-oligomannosides was hexaose greater than heptaose. It was concluded that serotype specificity is a function of the length of the oligomannosides, but as is described below, the resistance of serotype A mannan to exo-α-mannanase makes it unlikely that length is the only determining factor. Fukazawa et al (1980) compared the structures of haptenic oligomannosides produced by acetolysis of bulk mannan from *C albicans* mannan serotypes A and B. The oligomannosides were studied by PMR, methylation-fragmentation and quantitative precipitin analysis. Acetolysis fingerprints of serotype A and B oligomannosides chromatographed on Biogel P-2 were very similar. PMR spectra of mannopentaoses (M5) were similar, containing 1,2-linked residues and signals indicating a 1,3 linkage. Serotype A-M6 gave rise to four anomeric proton signals including one referrable to a terminal α-1,3 linkage, whereas serotype B-M6 had only three such signals because of the absence of a 1,3-terminal residue. Methylation-fragmentation analysis showed that serotype A-M5 is a mixture of a 1,2-linked pentaose and one with three 1,2 and one 1,3 linkages. Serotype A-M6 was also a mixture of one purely 1,2-linked hexaose and one with a 1,3 terminal. The major difference in serotype B-M6 was the occurrence of the 1,3 linkage as an internal branch point (Figure 11.1). Inhibition of the homologous reaction between serotype A mannan and anti-*C albicans* A was maximal with serotype A-M7 but M6 was nearly as good an inhibitor. Oligomannosides of type B

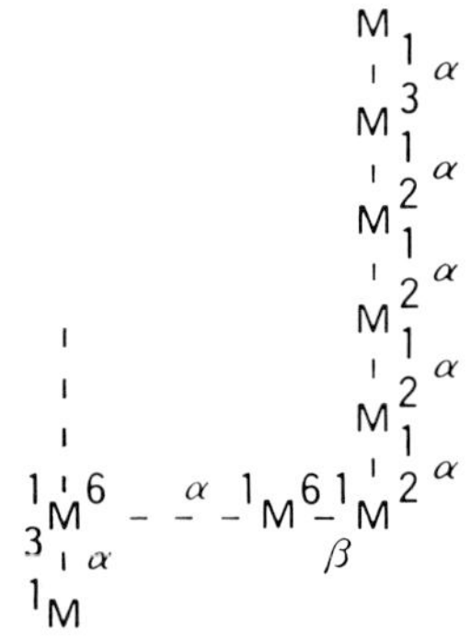

Figure 11.3. Alternative model for the type specific epitope "6" of *Candida albicans* serotype A proposed by Suzuki and Fukazawa (1982). These authors also find support for isolated β-linkages in mannan, but propose that such residues do not play a direct role in directing type specificity but rather link the sparse, linear mannohexaose hapten to an arboreal backbone.

$$\begin{array}{c} M\underset{1}{\overset{}{|}}\alpha \\ M\underset{1}{\overset{3}{|}}\alpha \\ M\underset{1}{\overset{2}{|}}\alpha \\ M\underset{1}{\overset{2}{|}}\alpha \\ M\underset{1}{\overset{2}{|}}\alpha \\ M\underset{1}{\overset{2}{|}}\alpha \\ \underset{3\,|\,\alpha}{\overset{1\,|\,6}{M}} \xrightarrow{\alpha}{^1M^6}{^1M^2}\,\alpha \\ {}^1M \end{array}$$

including M6 had comparatively low ability to inhibit the serotype A–anti-A reaction. The reaction between serotype B mannan and anti-*C albicans* B was inhibited equally well by M6 and M7 of serotype B or by M6 of serotype A.

A Farr type of radioimmunoassay was developed to assess the ability of ^{3}H-oligomannosides of *C albicans* to inhibit the reaction of *S cerevisiae* mannan with antibody. Linear 1,2-α-mannotriose from *C albicans* was ineffective as an inhibitor but the triose or tetraose with a 1,3-α-mannosyl nonreducing terminal was inhibitory (Nakajima et al, 1979).

Considering all three types of analysis (acetolysis, methylation-fragmentation, hapten inhibition) serotype specificity results from the sequence in the mannohexaose from these two serotypes. The immunodominant hapten of serotype A consisted of a straight chain 1,2-linked oligosaccharide with a 1,3 terminal. Type B-M6 is more complex, lacking a terminal 1,3 linkage but having instead a single O-1, O-2, O-3 branch point and an additional internal 1,3 linkage. In order to account for these structural details Suzuki and Fukazawa (1982) propose an arboreal rather than comblike structure for *C albicans* type A mannan composed of a protein trunk with a linear branch of alternating 1,6-mannose residues two to three units long flanking 1,2-linked mannobiose. Type specificity was considered by them to reside in sparse terminal sub-branches of 1,2-mannopentaose joined to the main branch by crucial β-1,6-linked mannose (Figures 11.3, 11.4).

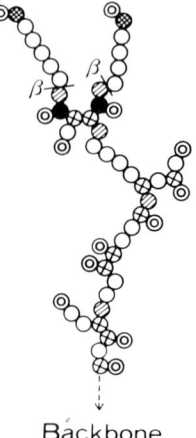

Figure 11.4. View of a larger segment of the A-specific epitope than is shown in the previous figure (Suzuki and Fukazawa, 1982). Concentric circles, nonreducing termini; shaded circles, 6-mannose-1; open circles, 2-mannose-1; crosshatched circles, 3-mannose-1; filled circles, mannosyl residues branched through O-1, O-3, and O-6; four quadrant circles, branching through O-1, O-2, O-6.

11.3.6 O-Phosphonomannan

The crossreactivity between exocellular O-phosphonomannan of *Hansenula* and antiserum to *C albicans* was noticed some years ago (Summers et al, 1964). The phosphate content expressed as molar ratio of mannose/phosphate in *C albicans* serotype B mannan (mannose/phosphate = 18) is higher than that in baker's yeast mannan (mannose/phosphate = 78-120) (Stewart and Ballou, 1968). The total phosphate content of bulk mannans has been measured for *C albicans B792*, 0.86%; *C albicans 311A*, 0.25%; *S cerevisiae*, <0.1% (Hasenclever and McAtee, 1977). Fractionation of bulk mannans on DEAE-Sephadex columns resulted in neutral mannan in the effluent and acidic mannan subfractions enriched in phosphodiesters eluted with a salt gradient according to phosphate content (Thieme and Ballou, 1972; Colonna and Lampen, 1974a). Such heterogeneity of charged species is a common property of mannans and occurs in *C albicans* (Okubo et al, 1980). Serologic activity has been concentrated in fractions enriched in phosphate, compelling consideration of the form in which phosphate occurs in *C albicans* mannan, and whether this linkage is itself an important antigenic determinant. Much information concerning the molecular organization of O-phosphonomannan of saprophytic yeasts has accumulated that provides a model for analysis of *C albicans* mannan.

Four kinds of O-phosphonomannans have been described. In baker's yeast (Ballou, 1976) phosphodiesters link mannose-1,3-α-mannose disaccharide to a mannotriose at the C-6 position of the reducing terminal mannose residue joined to the linear α-1,6-mannan backbone (Figure 11.5). The α-mannosyl phosphate groups of some *Saccharomyces* isolates are important antigenic determinants (Ballou, 1976). A second type of O-phosphonomannan was discovered in *Kloeckera brevis*; the largest subfraction after DEAE-Sephadex chromatography had a mannose/phosphate ratio of 8.6 (Thieme and Ballou, 1972). Acid cleavage of diesters did not significantly decrease molecular size. Alkali degradation to the monoester form and gel permeation chromatography of the products did not reveal low molecular weight carbohydrates. Thus the phosphodiesters in this mannan did not link short mannosyl sequences to a main chain nor did phosphodiesters join together large segments of mannan. The core mannan resistant to exo-α-mannan hydrolase was enriched in phosphate. These lines of evidence are compatible with phosphate linked to the α-1,6-mannan backbone. The nature of the second ligand is unknown, and might be an amino acid of a structural protein or of an enzyme (Stewart and Ballou, 1968).

Exocellular O-phosphonomannans occur in *Hansenula* species. In *Hansenula capsulata* repeat units of β-1,2-mannobiosyl units linked through α-1,6-phosphodies-

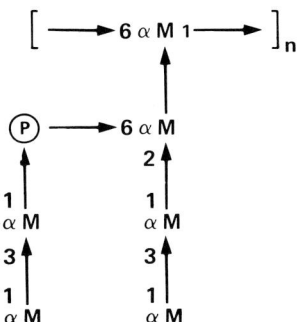

Figure 11.5. O-phosphonomannan in the outer chain region of *Saccharomyces cerevisiae* mannan (Ballou, 1976, 1980).

Figure 11.6. O-phosphonomannan of Hansenula capsulata. Repeat units of β-1,2-mannobiosyl units linked through α-1,6-phosphodiesters (Slodki, 1980). Source: reprinted with permission of the publisher from Fungal Polysaccharides, P. A. Sandford and K. Matsuda, eds. Copyright 1980, American Chemical Society.

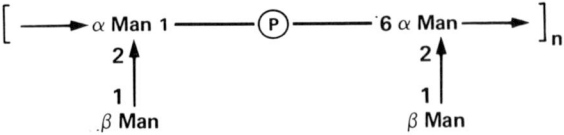

6-O-phosphono [2-O-β-mannopyranosyl]-mannose

$$\left[\text{(P)} - 6-\alpha\text{M 1} \longrightarrow 3\alpha\text{M 1} \longrightarrow 3\alpha\text{M 1} \longrightarrow 3\alpha\text{M 1} \longrightarrow 3\alpha\text{M} \right]_n$$

Figure 11.7. The pentasaccharide monophosphate of *Hansenula holstii* is the autohydrolysis product of the parent O-phosphonomannan. The fragment is linked to a phosphorylated and highly branched core (Slodki, 1980). Source: reprinted with permission of the publisher from Fungal Polysaccharides, P.A. Sandford and K. Matsuda, eds. Copyright 1980, American Chemical Society.

ters account for up to 90% of the structure (mannose/phosphate is about 2.5) (Slodki, 1980). The repeat unit shown in Figure 11.6 is consistent with interpretation of the ^{13}C-NMR spectrum (Gorin, 1973). Another O-phosphonomannan appearing in *H holstii* (mannose/phosphate approximately 5) was delineated when autohydrolysis liberated a mannopentaose phosphate monoester. Methylation-fragmentation analysis indicated it was 1,3-linked (Figure 11.7). Successive α-1,3 side chains are appended via phosphodiesters to a highly branched and phosphorylated core that was resistant to autohydrolysis.

Bulk mannan prepared from *C albicans* by autoclaving and Fehling precipitation was separated into five fractions on DEAE-Sephadex columns according to their phosphate content (Okubo et al 1979, 1980) (Table 11.3, Figure 11.8). The three known serotypes of *C albicans* were compared by this means. Serotype B mannan was the most highly phosphorylated, 33% of the bulk mannan was eluted at high ionic strength, and serotype C mannan was the least phosphorylated. The phos-

Table 11.3. Chemical Composition of the Mannan Subfractions Isolated by Diethylaminoethyl Anion Exchange Chromatography from *Candida albicans* A and B Serotypes

Mannan Subfractions[a] (percent)	Carbohydrate[b] (percent)	Protein[c] (percent)	Phosphate[d] (percent)	Mannose:Phosphate Molar Ratio	Yield[e] (percent)
Serotype A					
I	96	0.8	0.00	>1000	14.5
II	94	0.8	0.18	224	23.5
III	96	1.1	0.44	94	19.0
IV	93	1.4	0.84	48	19.5
V	92	2.3	1.20	34	3.5
Serotype B					
I	97	1.3	0.00	>1000	12.5
II	95	1.1	0.21	198	17.0
III	95	1.3	0.60	69	19.5
IV	96	1.4	1.15	36	21.0
V	93	1.4	1.83	22	12.0

Source: Okubo et al, 1979.
[a] See figure 11.8 for sample elution profile.
[b] Phenol-sulfuric acid.
[c] Lowry et al method.
[d] Ames-Dubin method.
[e] Based on the weight of the bulk mannan starting material.

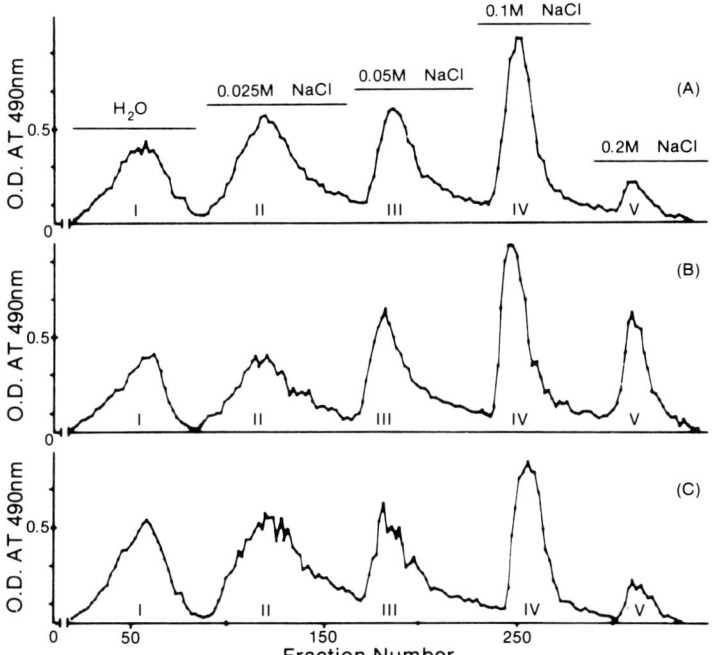

Figure 11.8. Fractionation of bulk mannans of *Candida albicans* serotypes A and B by diethylaminoethyl Sephadex using a stepwise elution consisting of water and 0.025, 0.05, 0.1, and 0.2M NaCl. Fractions were assayed for total carbohydrate with phenol-sulfuric acid reagent (Okubo et al, 1979). *Source:* reprinted with permission of the authors and the publisher, American Society for Microbiology.

phate linkages are believed to be diesters, but oligomannosides from *C albicans* containing *O*-phosphonomannan have not been isolated. Quantitative precipitin reactions showed that the amount of antibody precipitated was proportional to the phosphate content of column-derived fractions. The addition of inorganic phosphate did not inhibit immune precipitation, providing no direct evidence for the participation of phosphate in the haptenic group. Short sequences of mannose residues containing a phosphodiester linkage would be a stronger probe for this purpose but have not yet been tested.

Crossreactions among serotypes A, B, and C of *C albicans* and *S cerevisiae* were studied using fractions obtained from bulk mannans separated according to phosphate content. Anti-*C albicans* A crossreacted strongly with serotype C mannan, and weakly with that of serotype B. Anti-*C albicans* B reacted strongly with all mannans tested except weakly with baker's yeast mannan. Anti-*C albicans* C was strongly crossreactive with serotype A. Thus, serotypes A and C are closely related. Anti-*S cerevisiae* reacted strongly with mannans of serotypes B and C.

The higher reactivity of phosphate-enriched fractions of all yeasts suggests that the greatest density of haptenic groups is concentrated in acidic mannan. The anion exchange technique did not separate crossreactive from serotype-specific determinants with one exception: a fraction of intermediate phosphate content from serotype B mannan was reactive with homologous antiserum but not with anti-*C albicans* A or C. This fraction comprised 19.5% of the bulk mannan. No clues about the nature of the A-specific determinant were found in this analysis. In general the protein content (1.1% to 4.1% of the dry weight) also varied directly with the phosphate content.

Table 11.4. Analysis of Carbohydrate and Phosphate in Fractions Obtained by Treating Mannan of *Saccharomyces cerevisiae* A1640B with Alcian Blue

Fraction	Total Carbohydrate[a] in Sample (mg)	Glucose Content (percent)	Mannose[b] Content (percent)	Phosphate:Hexose Molar Ratio[c]
Total mannan	18.1	—	—	1:191
Alcian blue supernatant	18.2	0.5	99.5	1:212
Alcian blue precipitate	0.2	43	57	1:17.5

Source: Friis and Ottolenghi, 1970.
[a] Anthrone concentrations.
[b] Mannose content, total reducing sugar—reducing value of glucose.
[c] Hexose, total reducing sugar.

Further studies with baker's yeast mannan (Okubo et al, 1980) showed that the fractions highest in phosphate also contained oligomannosides with the longest chain length. Phosphate, protein, and longer oligomannosides are concentrated in the acidic fractions of mannan separated by anion exchange chromatography. The location of O-phosphonomannan of *C albicans* with respect to the inner core-outer chain model of Ballou (1976) is unknown. It is also not known what strata observed in cell wall cross sections are enriched in O-phosphonomannan.

The phthalocyanine dye, alcian blue, binds to polyanions and to mannan preparations from some *S cerevisiae* isolates (Friis and Ottolenghi, 1970). A stable mannan-alcian blue precipitate was formed at pH 1.7 involving only a small portion of bulk mannan (ie, one percent). The precipitate had a high mannose:phosphate molar ratio (17.5:1 compared with 191:1 in bulk mannan) and an increase in glucose:mannose equal to 43%:57% (Table 11.4). When the Korn and Northcote (1960) procedure for solubilizing cell walls with anhydrous ethylenediamine was used, only the soluble fraction containing a glucomannan complex was capable of binding alcian blue. No definitive proof has been provided to establish the presence of a covalent glucomannan in the yeast cell wall. The result with alcian blue is compatible with that proposition and should revive interest in further defining glycosidic linkages and potential antigenicity of the phosphonoglucomannan complex. Further exploitation of the alcian blue probe may prove useful.

Baker's yeast or *C albicans* mannan is not appropriately thought of as a single entity in view of the finding that fractions, prepared by anion exchange chromatography or by alcian blue complexation, vary qualitatively and in epitope density. The effects of strain variation and substrate concentrations during growth on microheterogeneity of mannan need to be assessed so that the impact of phosphorylation on mannan structure can be more fully appreciated.

11.3.7 Exo-α-Mannan Hydrolase

The characterization of an inducible exo-α-mannanase from a soil bacterium, *Arthrobacter* GJM provided a valuable probe for structural analyses (Jones and Ballou, 1969a; 1969b). This organism is very likely an actinomycete and a nonmotile variant of *Oerskovia xanthineolytica* because type cultures were induced to secrete exo-α-mannanase, and otherwise conformed bacteriologically to the isolate in the original report (Lechevalier, 1972). The enzyme acts on baker's yeast mannan to cleave mannose residues from the nonreducing ends of oligomannoside chains having α-1,2 and α-1,3 linkages, but the α-1,6-mannan backbone is not attacked. The mannan core resistant to the enzyme was enriched in phosphate (Colonna and Lampen, 1974a, b). Significant differences were found in the susceptibility to exo-α-mannanase of *C albicans* serotypes A and B (Reiss et al, 1981). Serotype A mannan was 5.4% digested based on reducing sugar released and 17.4% based on the dry weight of the resistant mannan core, whereas the corresponding values for serotype B were 33.5% and 43.2% digestion. Certain substituents such as phosphodiesters and N-acetylglucosamine (Raschke and Ballou, 1972) are thought to block the action of exo-α-mannanase. Phosphate content alone is not an absolute indicator of susceptibility because high phosphate *Kloeckera brevis* mannan was 54% digested (Jones and Ballou, 1969a; 1969b). In the instance of *C albicans*, serotype B mannan is higher in phosphate content than that of serotype A (Hasenclever and McAtee, 1977; Okubo et al, 1980). Other isolates of each serotype need to be examined to determine if the association of serotype with enzyme susceptibility is a consistent finding.

Inducible enzymes of *O xanthineolytica* that lyse yeast cell walls were studied by Scott and Schekman (1980). This organism was previously known as *Arthrobacter luteus* by the Kirin Breweries, Gunma, Japan and the commercial lytic enzyme from it is called Zymolyase. "Teichozyme Y" is a similar commercial enzyme from *O xanthineolytica*. The shift in nomenclature of these bacteria and the different trade names for their enzymes causes some confusion. When baker's yeast glucan is the sole carbon source for *O xanthineolytica*, an *endo*-β-1,3-glucanase, a minor amount of alkaline proteinase, and chitinase are produced. The glucanase has affinity for Sephadex dextran gels and the proteinase is also retarded by this gel matrix. This finding suggests a means for removal of chitinase and DNAse from the preparation. The glucanase and proteinase are then resolved on Biogel P150. The glucanase has a molecular weight of 55,000 da. Its pH optimum for lysis of yeast cell walls is 8.0 with a steep drop between pH 7.5 and 7.0. The *endo* action pattern releases oligomers especially glucopentaoses. No lysis of yeast cells occurs in the absence of 2-mercaptoethanol, but the proteinase can spare the requirement for 2-ME and acts synergistically to potentiate lysis. Neither pronase nor trypsin can substitute for the native proteinase. Commercial enzyme preparations were compared with the purified *O xanthineolytica* preparations (Table 11.5). Glusulase is comparatively a very poor

Table 11.5. Comparison of Activities Found in Various Enzyme Preparations Lytic for *Saccharomyces cerevisiae*

Lytic Enzyme Source	Activity (U/mg)				Ratio		
	Lytic	Protease	Chitinase	DNase	Lytic/Protease	Lytic/Chitinase	Lytic/DNase
Zymolyase 500	3,200	120	1.9	4900	28	1,700	0.66
Zymolyase 60000	4,700	50	0.4	14000	94	12,000	0.34
Glusulase	15[a]	0.2	1.2	—[b]	66	12	—[b]
Teichozyme Y	240	30	6.1	51000	8	38	4.6×10^{-3}
Fraction II[c]	32,000	29	22	<100	1100	1,400	>320
Fraction III	22,000	13	1.4	<100	1800	16,000	>220
Fraction IVa	31,000	$<1 \times 10^{-2}$	<0.5	<100	$>3 \times 10^6$	>62,000	>310

Source: Reprinted with permission from Scott and Schekman (1980) and the American Society for Microbiology.
[a] The published pH optimum of this preparation (6.0) gave an even lower value.
[b] Intrinsic absorbance of glusulase made the assay impossible.
[c] Purified *O. xanthineolytica* preparations.

lytic agent. The *O xanthineolytica* fraction devoid of both chitinase and proteinase is a most effective lytic system.

11.3.8 Mannan as an Immunomodulator

As immunologic approaches to the fungi move from the descriptive to the analytical, reports begin to appear concerning the interactions of antigens in various stages of purity with cells of the immune system. The value of this research is that it builds a framework for understanding the nature of the stimuli presented by the antigenic mosaic of *Candida* and the responses they evoke. Such insight helps reveal the regulation of immune responses that contribute to successful surveillance against *Candida* and the critical responses that fail in derangements such as chronic mucocutaneous candidiasis.

11.3.8a Regulation of Blastogenesis

- Experimental endocarditis. Mannan evoked blastogenesis of popliteal lymph node cells from rabbits with *C albicans* endocarditis, but higher blastogenic indices were stimulated by whole formalin-killed yeast. In severe infections involving multisystem dissemination, a trend to anergy was observed in the form of a suppression of in vitro blastogenesis. This anergy was dissociated from mitogenic responses to Con A, which remained normal (Skerl et al, 1980).
- Normal human peripheral lymphocytes. Human B lymphocytes require T-lymphocytes to respond to a *C albicans* polysaccharide fraction "Mangion purified polysaccharide", (MPPS) (Piccolella et al, 1980). This is the deproteinized, ethanol-precipitated supernate of autoclaved whole yeast, which contains glucan, mannan, and 3.1% polypeptide. The requirement for T-cells could be spared by adding soluble helper factor supplied as the culture supernate of MPPS-stimulated peripheral blood lymphocytes (PBL). Continued in vitro cultivation of human PBL with MPPS results in an abrupt shift from stimulation to 80% suppression of blastogenesis. This was not due to loss of viability or to a soluble inhibitor. The source of suppression was an Ia^- T-cell subset. These T_s could block MPPS stimulation of PBL even in the presence of soluble helper factor. The T_s did not inhibit blastogenesis by phytomitogens but could partially suppress mixed lymphocyte culture responses. Cultivation of human PBL with MPPS for 2 to 4 days rendered them unable to respond to a second challenge with the same antigen. This unresponsiveness was not due to T_s because these were not evident until the sixth day of culture. Responsiveness was restored by fresh mitomycin-treated PBL or by soluble helper factor. Treatment of fresh PBL with monoclonal anti-Ia immunoglobulin inhibited their ability to supply help. The target cell that could provide help was not a T-cell or macrophage, but rather a novel role for B-lymphocytes to stimulate T_h cells was postulated (Piccolella et al, 1981a, 1981b). The implications of this series of experiments are important in understanding T-cell anergy in candidiasis. However, better criteria for purity of the antigen must be provided. This material was not chromatographically pure and contained both glucan and mannan. Key experiments would benefit from corroboration with more well-defined antigens. The requirement for T-cell help in driving blastogenic responses to polysaccharides is perhaps unusual, but it should be recalled that structural protein is covalently bound to *C albicans* mannan, although as a low percentage by weight. The unresponsiveness of human PBL to *C albicans* polysaccharides seems to occur on two levels: (i) generation of T_s;

and (ii) lack of helper activity resulting from a lack of secondary stimulation. The relation of these findings to the dynamic of *Candida* infection is unknown.

• The spectrum of immunologic defects in chronic mucocutaneous candidiasis (CMC) is complex (see 11.6.2) and there is likely more than one reason for the development of this disease. Some of the common findings are cutaneous anergy to *C albicans* and depression of *Candida*-induced in vitro blastogenesis. Fischer et al (1978) found a serum blocking factor in six children with CMC, out of a series of 23 such patients. In the test for inhibition, peripheral blood lymphocytes were reconstituted in medium with 10% pooled human serum and *C albicans* metabolic antigens. Then 1% to 10% of serum containing blocking factor was added, displacing the normal human serum. The inhibitory activity of sera was in direct proportion to its concentration and exceeded 50% inhibition. Sera from normal humans lacked the blocking factor and the inhibitory sera from children with CMC were not, per se, toxic for lymphocytes. Blocking factor is specific for *Candida* and does not suppress responses to PPD or Con A. The factor is present during relapses and disappears with antifungal therapy. It is heat stable, nondialyzable, and could be adsorbed to either anti-*C albicans* IgG or to Con A. The factor is not precipitable with half-saturated ammonium sulfate, and on that ground it is not considered to be an immune complex. Mannan extracted from *C albicans* mimics the inhibitory activity of serum factor from patients in blocking blastogenesis of human lymphocytes stimulated with *C albicans* antigens. Human CMC sera that was adsorbed onto a Con A agarose column and then eluted with methyl-α-mannoside was enriched in blocking factor. Fischer et al (1978) proposed that mannan circulates in flares of chronic CMC and that the reversible specific defect in this disease, anergy, is secondary to an overload of *Candida* polysaccharide contributing to chronicity by inducing immunologic unresponsiveness.

• Purified human T-lymphocytes did not respond to an unpurified metabolic extract of *C albicans* in the absence of monocytes (Fischer et al, 1982). The pivotal role of human monocytes in presenting antigenic mannan to T-lymphocytes is illustrated in the defective handling of this antigen in some children with CMC (Fischer et al, 1982). Although mannan radiolabeled by tritium gas exchange is taken up normally by monocytes, in two of these patients the subsequent release of mannan occurred very slowly. This slow rate of metabolism was independent of the stage of the disease, whether active or infection-free. When monocytes were first primed with mannan their ability to collaborate with T-cells in the proliferative response to *C albicans* metabolic extract was seriously impaired. The universal presence of *Candida* antibodies in normal human sera (Lehmann and Reiss, 1980b) makes it possible for immune complexes to be formed in vitro when mannan or blocking factor from chronic CMC patients is added to lymphocyte cultures stimulated with *Candida* antigen. Thus inhibition of blastogenesis by immune complexes has not been ruled out; to do this would require steps to assure removal of the globulin fraction of serum that is to be used for in vitro blastogenesis. Further chemical characterization of blocking factor is necessary to confirm its identification as mannan.

11.3.8b Complement Activation
Some yeast mannans can activate complement after their injection in mice, thus leading to increased vascular permeability and even fatal anaphylaxis (Kind et al, 1972). *Candida albicans* mannan is active after injection of 10 to 50 μg whereas

mannans from baker's yeast and *Rhodotorula glutinis* were only weakly effective at 100 times higher doses. Proof of the role for complement as the source of anaphylatoxin was provided by the absence of anaphylaxis in C5-deficient SWR mice primed with mannan. Others have shown that C5-deficient mice are hyper-susceptible to *Candida* (Morelli and Rosenberg, 1971a).

Candida albicans mannan activates complement directly via the alternative pathway measured by immunoelectrophoretic detection of split produts of C3 and properdin factor B in C4-deficient human serum (Ray et al, 1979). Further evidence for complement activation was the generation of polymorphonuclear neutrophil chemotactic factors in normal serum but not in heat inactivated serum (56 °C, 30 minutes). Some refinement of the antigen will be required to gain confidence in these findings because fucose was found in the antigen, and it had not been detected before in mannan preparations. The protein content of the antigen was not measured and no purifications were done by chromatographic or selective precipitation techniques.

A glucomannan protein complex from *C albicans* was found to stimulate degranulation and histamine release from isolated rat mast cells (Nosal, 1974). Injection of the complex in rats resulted in hypotension and bronchoconstriction, which could be prevented by pretreatment of the animals with the mast cell degranulating agent 48/80 (Svec, 1974).

11.3.9 Mannanemia

The conversion of *Candida* species from a commensal existence on the mucosa to a pathogen invading the deep tissues occurs in the setting of serious immune compromise, most frequently encountered in persons receiving systemic corticosteroids and cytotoxic drugs for management of neoplastic or autoimmune processes, or for maintenance of renal allografts. Candidiasis as a major contributing cause of death has been well-documented in several series of leukemia patients (Table 11.6). Some iatrogenic factors are implicated as increasing the risk of candidiasis. The prolonged use of broad spectrum antibacterial antibiotics allows overgrowth of a resistant

Table 11.6. Incidence of Systemic Candidiasis in Leukemia Patients

Reference	Period of Study	Number of Patients	Incidence	Percent
Craig and Farber (1953)	pre-1953	175	11	6.3
Baker, (1962)	1930–1953	143	3	2.1
	1954–1961	118	14	11.9
Schumacher et al (1964)	1957–1962	205	25	12.2
Bodey (1966)	1954–1958	157	11	7.0
	1959–1964	302	60	19.9
Armstrong et al (1975)	1973	nd[a]	36	nd
Myerowitz et al (1977)	1963–1970	82	2	2.4
	1971–1975	47	16	34.0

[a] nd, not determined.

intestinal flora in which *Candida* species can flourish. Long term use of steroids, like prednisone, results in cyclic fluctuations in peripheral blood lymphocytes, especially T-cells (Bellanti, 1985). Cyclophosphamide, a DNA alkylating agent, selectively suppresses the most actively dividing lymphocytes, B-cells (Barrett, 1978). Neutropenia brought about by drug-induced bone marrow depression reduces the innate phagocytic functions. The high osmotic pressure of intravenous solutions creates a selective medium for yeasts at the tips of indwelling catheters. These circumstances permit *Candida* species to disseminate from mucosal foci to the kidneys, lungs, heart, and other organs. The successful treatment of bacterial infections and a trend towards more aggressive anti-cancer therapy, including whole body γ-irradiation and bone marrow reconstitution, has increased the incidence of candidiasis in leukemia (Table 11.6).

Physical diagnosis of systemic or disseminated candidiasis is difficult, although some researchers have emphasized the value of examination of the eyes for endophthalmitis (Edwards et al, 1974). Multiple isolations from normally sterile sites and body fluids are suggestive of tissue invasion. Increased frequency of yeast isolations from sputum, urine, or feces should increase the index of suspicion in high risk individuals. The most potentially useful yeast isolation, from the blood, has often proved unrewarding because of negative cultures in the presence of disseminated infection (Weiner and Yount, 1976; Meckstroth et al, 1981). The delay of approximately 6 days that is often necessary to allow recovery of yeasts in "biphasic" culture bottles limits the usefulness of this technique (Roberts et al, 1975). Improvements in blood culture methods, especially radiorespirometry of $^{14}CO_2$ liberated during growth of yeast on radiolabeled precursors, promises diagnostic gains (Prevost and Bannister, 1981).

Conventional antibody tests with unpurified antigens are useful in patients capable of producing normal immunoglobulin levels, but even then the immunodiffusion test is unable to discriminate antibodies stimulated by colonization of the gastrointestinal tract from those formed during systemic candidiasis. In a series of cancer patients, Gentry et al (1978) found that the agglutination of *Candida* antigen-coated latex particles was capable of monitoring fluctuations in antibody levels during therapy, but that results with latex agglutination and immunodiffusion tests were also positive in 26% to 39% of patients with no clinically detectable candidiasis. Of patients with disseminated candidiasis, 18% to 24% lacked detectable antibodies to *Candida*. Another series of 92 leukemia patients (Meckstroth et al, 1981) had even less encouraging results with unpurified antigens in latex agglutination and immunodiffusion tests. Thirty-four patients were colonized with *Candida* but without clinically recognized infection; of these 32% to 35% were serologically reactive, and nine of ten leukemics with disseminated candidiasis failed to produce detectable antibodies by these techniques.

A more meticulous kind of precipitin analysis, two-dimensional rocket immunoelectrophoresis has been applied to detect antibodies in humans to *Candida* species (Glew et al, 1978). Immunoprecipitins to mannan can be differentiated from those to candidal cytoplasmic proteins by placing the mannan-binding lectin Con A in an intermediate gel. Both types of Laurell rockets occurred in systemic candidiasis patients, but only four of 196 patients colonized with *Candida* species or normal controls had precipitins directed against candidal cytoplasmic proteins. This finding accords with the view that the increased phagocytosis and destruction of *Candida* during tissue invasion creates a different antigenic stimulus than colonization; one where cytoplasmic proteins are more apt to be involved. This series did not contain a majority of heavily immunosuppressed patients, those most likely to develop systemic candidiasis.

When using two-dimensional immunoelectrophoresis, the number of positive reactors among those colonized with *Candida* appears to be decreased, but because this test relies on the ability of patients to produce antibodies, and is relatively insensitive, its value in heavily immunosuppressed patients remains unproven. Positive tests in colonized patients and lack of reactivity in systemic candidiasis owing to the impairment of antibody synthesis are major drawbacks to precipitin tests as diagnostic adjuncts.

A more promising alternative immunoassay for systemic candidiasis is the detection of circulating antigen. Antibodies to mannan in humans are universal (Lehmann and Reiss, 1980b) and in leukemia anti-mannan IgG fluctuates with no exact relationship to disease activity (Meckstroth et al, 1981). The concentration of anti-mannan IgG in a group of leukemia patients was lower than that in normal persons. The proliferation of *Candida* species during tissue invasion in the face of a diminished ability to produce anti-mannan IgG upsets the clearance of this polysaccharide and leads to a steady state where mannan in the form of soluble immune complexes begins to circulate. Mannanemia has been observed by laboratories using diverse methods of detection in patients with CMC and in systemic candidiasis. A brief chronology of these findings of mannanemia is presented in Table 11.7 (de Repentigny and Reiss, 1984).

Mannanemia was first observed in a chronic mucocutaneous candidiasis patient by two-dimensional "rocket" immunoelectrophoresis with Con A in the intermediate gel (Axelsen and Kirkpatrick, 1973). Hemagglutination inhibition detected mannanemia in four of 14 systemic candidiasis patients (Weiner and Yount, 1976). The reagent for detection was novel: sheep erythrocytes coated with a subagglutinating concentration of Con A (Leon and Young, 1970) and supracoated with mannan. The agglutination of coated erythrocytes by a human reference antiserum containing antibodies to *Candida* could be inhibited by 300 ng/ml of *C albicans* A mannan. Treatment of sera with α-mannosidase resulted in a loss of mannan and a fourfold rise in antibodies to mannan, which provided evidence that mannan can occur as soluble immune complexes. Procedures that make no provision to detect or dissociate such complexes can be expected to have greatly reduced sensitivity. To provide effective coverage of clinical levels of mannanemia, a test should be capable of detecting mannan in the low ng/ml range, conditions satisfied by either radioimmunoassay or enzyme immunoassay.

Evidence for circulating immune complexes was found in sera of two of ten patients with invasive candidiasis and in a chronic mucocutaneous candidiasis patient (Burges et al, 1983). The immune complexes precipitated by polyethylene glycol 6000 were characterized further in the void volumes of ACA-34 and Sepharose 4B columns run in tandem. The immune complexes bound to staphylococcal protein A-agarose and contained C3. The presence of *Candida* antigen in the complexes was judged by counterimmunoelectrophoresis. More sensitive primary binding assays should increase the chances for detecting antigen in these complexes. Apparently the chromatographic steps to resolve complexes decreased the affinity of antibodies for the antigen, so that separate and harsher steps to dissociate antigen from the complexes were unnecessary. Surprisingly, Clq binding assays were ineffective in sequestering soluble immune complexes in this study.

Table 11.7. Summary of Published Studies on Detection of Serum Mannan for the Diagnosis of Candidiasis

	Assay Format	Dissociation Step	Minimum Sensitivity of Assay (ng/ml)	Source of Antiserum	Study Population (number with candidiasis)	Type of Candidiasis	Study Design	Sensitivity (percent)	Specificity (percent)
Axelsen and Kirkpatrick (1973)	Crossed IEP	No	nd	R	H (1)	CMCC	Retrospective	100	nd
Weiner and Yount (1976)	HAI	No	300	CPS	H (14)	Disseminated	Retrospective	32	99
Warren et al (1978)	Sandwich EIA	No	nd	R	H (3)	Disseminated	Retrospective	100	nd
Fischer et al (1978)	Inhibition of blastogenesis	No	nd	Nil	H (23)	CMCC	Retrospective	26	nd
Weiner and Coats-Stephen (1979)	RIA	Yes	0.125	R	H (11) R (29)	Disseminated Disseminated	Retrospective	47 52	96 100
Segal et al (1979)	Inhibition EIA	Yes	8–31	R	H (7)	Disseminated	Retrospective	100	100
Kerkering et al (1979)	CIE	No	nd	R	H (8)	Disseminated	Prospective	100	100
Poor et al (1979)	RIA	No	100	R	M (27)	Disseminated	Retrospective	70	100
Harding et al (1980)	Sandwich EIA	Yes	0.1–1.0	R	R (9)	Disseminated	Retrospective	100	100
Scheld et al (1980)	Sandwich EIA	Yes	1.0	R	R (16)	Endocarditis	Retrospective	75	100
Meunier-Carpentier et al (1981)	HAI	Yes	20	R	H (32)	Disseminated	Retrospective	50	100
Meckstroth et al (1981)	Inhibition EIA	Yes	4	H	H (10)	Disseminated	Prospective	70	92
Lew et al (1982)	Sandwich EIA	Yes	1.0	R	H (15) Rats (30)	Disseminated Disseminated	Retrospective	53 100	100 100
de Repentigny et al (1984)	Sandwich EIA	Yes	1.0	R	R (12)	Disseminated	Retrospective	100	100
de Repentigny et al (1985)	Sandwich EIA	Yes	1.0	R	H (16)	Disseminated	Retrospective	65	100

Abbreviations: IEP, immunoelectrophoresis; HAI, hemagglutination inhibition; EIA, enzyme immunoassay; RIA, radioimmunoassay; CIE, counterimmunoelectrophoresis; CPS, convalescent patient serum; R, rabbit; H, human; M, mouse; CMCC, chronic mucocutaneous candidiasis; nd, not determined.

A Farr-type RIA was devised in which mannan is detected by inhibition of precipitation of a standard amount of ^{125}I-mannan and reference antibodies in half-saturated ammonium sulfate (Weiner and Coats-Stephen, 1979). Preparation of radiolabeled mannan was accomplished by its activation with CNBr and coupling to tyramine before radioiodination. A key provision in the RIA was the dissociation of mannan-serum complexes with heat at acidic pH (pH 2.7, 96 °C), which increased the sensitivity of detection to 0.125 ng of *C albicans* mannan/ml serum. A majority of rabbits infected with *C albicans* died within 2 to 5 days; in this period mannanemia rose from 71 to 155 ng/ml and then a switch to antibody production occurred in the survivors. The same RIA was successful in detecting mananemia in five of 11 human candidiasis patients. Heavily immunosuppressed patients may be capable of only a feeble anti-*Candida* immunoglobulin response and mannanemia may continue to rise with the severity of infection until death, or remission, if antimycotic therapy is successful (Meckstroth et al, 1981).

Immunoperoxidase staining with anti-*C albicans* cell extract of lung specimens from patients with pulmonary candidiasis provided evidence that *Candida* species soluble antigens are present in the cytoplasm of alveolar macrophages, and in the bronchoalveolar fluid, with spillover into the systemic circulation (Humphrey and Weiner, 1983).

The observations of mannanemia stimulated development of enzyme immunoassays for its detection. Those tests that did not include a provision for dissociating complexes had no or limited success (Lehmann and Reiss, 1980a; Harding et al, 1980). In the development process, it was observed that high-titered antibodies to mannan could be raised in rabbits immunized with soluble mannan, provided it was in high enough molecular weight form. Thus peptidoglucomannan from *C albicans* cell walls is an effective immunogen, but mannan prepared by autoclaving whole cells followed by Fehling precipitation is not (Lehmann and Reiss, 1980a). A prototype indirect EIA-inhibition was developed that required enzyme-labeled, anti-human IgG as the indicator antibody (Segal et al, 1979). Dissociation of mannan-serum complexes was accomplished with heat and alkali digestion, followed by dialysis to neutralize and desalt the digested serum. Dissociated serum was incubated with unlabeled human reference anti-*Candida* serum. The mannan in dissociated serum competed for antibodies with mannan adsorbed to a polystyrene microtiter plate. The binding of peroxidase-labeled anti-human IgG was calculated and if the percentage inhibition exceeded 20%, corresponding to 40 ng/ml, a strong positive correlation with systemic candidiasis occurred. Mannanemia was evident in the 2 week period preceding dissemination and rose in proportion to disease severity (Meckstroth et al, 1981).

The variables in the slow and cumbersome prototype EIA have been analyzed (Reiss et al, 1982) with the conclusion that EIA-inhibition offered no advantages over double antibody sandwich EIA. When dissociation of complexes was accomplished by boiling sera in the presence of Na$_2$EDTA and mannan was detected by sandwich EIA, the time of performance was reduced to 2.7 hours. Definitive evaluations of the mannanemia sandwich EIA in a clinical setting began with Lew et al (1982). The appearance of mannanemia in rabbits immunosuppressed with cortisone acetate and injected with *C albicans* serotype A is shown in Figure 11.9. A switch occurs during dissemination away from antigenemia to a feeble antibody response. Serum mannose and arabinitol, a *Candida* species metabolite, are also elevated in the rabbit model. Mannose is presumed to derive from the degradation of

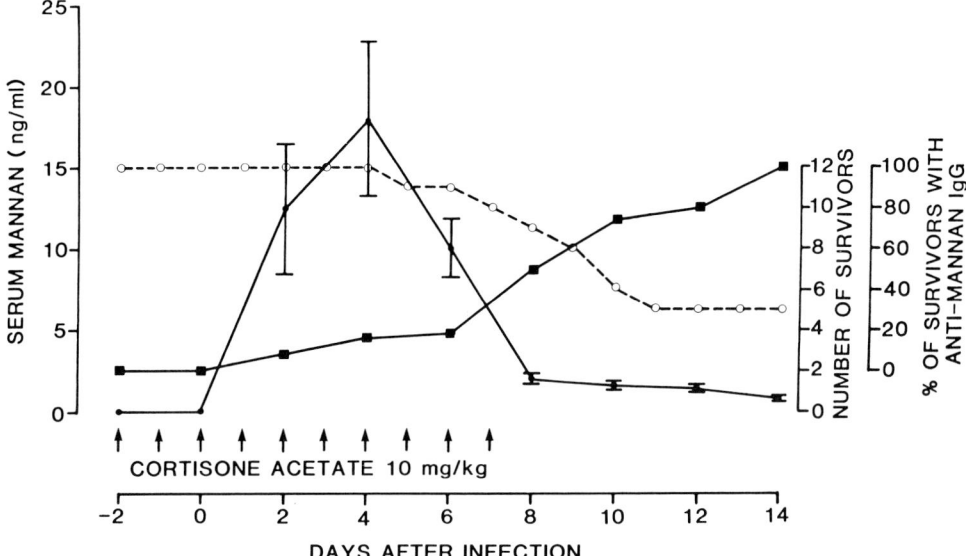

Figure 11.9. Twelve rabbits infected intravenously with *Candida albicans* 3181A on day 0. Open circles, number of survivors; closed circles, mean serum mannan ± SEM; closed rectangles, percent of survivors with anti-mannan IgG (de Repentigny et al, 1984). *Source:* reprinted with permission of the authors and the publisher, American Society for Microbiology.

mannan. Mannose and arabinitol also accumulate in renal insufficiency or cortisone therapy in the absence of invasive candidiasis. Further discussion of the pros and cons of using serum mannose or arabinitol as predictors of invasive candidiasis is found in Gold et al (1983); Monson and Wilkinson, 1981; and de Repentigny et al, 1984, 1985).

There is mounting evidence that mannan antigenemia detection overcomes the problem of discriminating mucocutaneous colonization from systemic candidiasis in heavily immunosuppressed patients, such as those with leukemia. The experience with methods to detect mannan is summarized in Table 11.7.

Some researchers have suggested that anti-mannan IgG concentrations measured by indirect EIA are useful in predicting the development of invasive candidiasis (Greenfield et al, 1983). Previously, it was shown that normal humans produce natural levels of anti-mannan IgG and that its production occurred independently of age, sex, or race (Lehmann and Reiss, 1980b). Meckstroth et al (1981) found that in leukemia patients anti-mannan IgG levels fluctuated without relation to disease severity and they could find no differences in anti-mannan IgG concentrations in sera obtained during the 2 week period before death from disseminated candidiasis and concentrations measured in the same patients over weeks and months beforehand. Thus the omnipresent concentrations of anti-mannan IgG complicate the interpretation of this parameter in disease.

Greenfield et al (1983) expressed the results of anti-mannan IgG concentrations as multiples of normal activity with concentrations in the upper 5% of the inpatient hospital population regarded as test-positive. Applying this criterion to ten patients with neoplastic disease and proven systemic candidiasis, four were test-positive, and in 22 patients with neoplasia who were colonized with *Candida* species, four (18.5%) were test-positive. The 40% sensitivity and 50% specificity achieved by measurement of anti-mannan IgG do not compare favorably with the measurement of mannan antigenemia by double antibody sandwich EIA (reviewed below). Measurement of IgM concentrations was not rewarding (Greenfield et al, 1983). If precautions are taken to exclude competition from IgG, measurement of IgM might provide additional insight about active disease if the humoral response to mannan displays the pattern observed with other microbial polysaccharides of short-term memory and a high potential to stimulate IgM.

The results obtained with the double antibody sandwich EIA are exemplified by de Repentigny et al (1985). The mean mannan concentration in normal blood donors was 0.04 ng/ml, and in leukemia patients without candidiasis, 0.08 ng/ml. The mean + 2 SD for serum mannan for leukemia patients without candidiasis of 0.46 ng/ml was used as the upper limit of normal. With this concentration as test-positive, 16 of 23 patients with invasive candidiasis had elevated serum mannan. The sensitivity was thus 70%, and the specificity, defined as the percentage of patients without invasive candidiasis whose serum mannan did not exceed 0.46 ng/ml, was 87%. If an upper limit of normal is taken as 1 ng/ml, the sensitivity of detection was 65% and the specificity, 100%.

11.4 GLUCAN

11.4.1 Chemistry

Less attention has been focused on this wall constituent than mannan, because glucan does not appear to be serologically active and its role in immune processes is more difficult to establish. D-glucan is a major mural constituent of *C albicans*, accounting for 47.0% of the yeast form mural dry weight as alkali-insoluble material with only 1% in alkali-soluble form according to one report (Chattaway et al, 1968). Estimates of glucan content may be low if no steps are taken to inactivate endogenous β-glucanase during preparation of walls. Soluble polymers of glucose and mannose were extracted from *C albicans* walls with cold dilute alkali as 10% of the wall dry weight, possibly as a glucomannan complex (Reiss et al, 1974). Anhydrous ethylenediamine was also shown to extract glucomannan complexes from yeast cell walls (Korn and Northcote, 1960).

Structural analysis of glucan solubilized from whole *C albicans* serotype B yeast with boiling dilute alkali (Yu et al, 1967) indicated a majority of β-1,6 linkages (67.4 mol%), many mannosyl residues branched through C-1, C-3, C-6 (16.6 mol%), and 6 mol% of 1,3 linkages. Earlier these investigators found negligible 1,3 linkages in glucan from serotype A and more frequent branch points (Bishop et al, 1960). The differences in these studies have not been clarified by more recent work. The alkali-soluble glucan of *C albicans* appears to be largely β-1,6-linked, and moderately branched. Ultracentrifugal analysis showed no evidence of different-sized components. Further hydrodynamic analysis and reinvestigation of sequence and linkages in *C albicans* D-glucans are necessary in view of the structural diversity discovered in *S cerevisiae* glucans.

11.4.2 Role in Infection

The *C albicans* cell wall fraction resistant to lysis in the host and responsible for granuloma formation was studied (Meister et al, 1977). Hepatic granulomas were induced in CBA mice by intravenous injection of heat-killed yeast, cell walls, or the mural glucan fraction. Remnants of walls were detected in Kupffer cells by electron microscopy for up to 20 days, even after no evidence of their presence could be found by light microscopy with immunofluorescent, Gram, or PAS techniques. The residual cell walls appeared swollen and more transparent as though hydration of fibrillar regions was underway. *Candida albicans* glucan was as active as whole cells or walls and was regarded as the factor responsible for this tissue reaction.

The presence of chitin in the *C albicans* glucan preparation needs to be clarified. It is unclear why glucan, with its predominant 1,6-linked structure, is not susceptible to periodate oxidation and is thus PAS-negative. This may be due to an extensive branching structure in glucan or may be due at least in part to chitin. Chitin is elevated in *C albicans* walls as compared with baker's yeast and is PAS-negative.

A positive correlation existed between the dose of glucan and the number of granulomas up to a dose of 3 mg per mouse, which was lethal if exceeded. Injection of baker's yeast glucan has been shown to stimulate the reticuloendothelial system at the site of bacterial, fungal, or tumoral implants (Williams et al, 1978). The intravenous injection of small amounts of baker's yeast glucan (<1 mg) also resulted in a granulomatous response with increased number and activity of macrophages and granulocytes. Injection thus appears to activate the innate phagocytic defense mechanisms. When mice were primed with baker's yeast glucan before injection with a lethal dose of *C albicans* yeast, there were reductions in mortality, weight loss, and kidney lesions.

Thus far separation of the overall stimulation of innate defense mechanisms from specific antigenic recognition of glucan or minor included determinants that may be shared by *C albicans* has not been possible. The time of administration of glucan, 7 days before infection, would allow recruitment of lymphocytes that recognize similar determinants on *Candida*. In the immunologically intact host D-glucan probably acts by both stimulation of innate phagocytosis and of specific antigen- responding clones that react with *Candida*. Three functions are therefore ascribed to glucan: (i) activation of the alternative complement pathway as the main constituent of zymosan; (ii) quantitative increase in monocytic and granulocytic function; and by implication, (iii) triggering of antigen-reactive lymphocytes.

11.5 PROTEIN ANTIGENS

Emphasis has been placed on the structural and antigenic properties of mannan of *Candida* species in order to gather many strands of data in the literature and to reinforce the idea that cell walls of fungi provide an antigenic stimulus distinct from that of procaryotes. Among the fungi, the theme and variations in mannans that explain crossreactions and the bases for immunospecificity are a major concern of this volume. Yet without knowledge of protein antigens a full appreciation of the fungal antigenic mosaic would be impossible. Microbial proteins can act as (i) determinants of pathogenicity, (ii) sources of specific antigens, (iii) stimulants of cell-mediated immunity, (iv) vaccines, (v) enzymes that have the potential to be detected in vivo as markers of infection.

The very large array of protein and glycoprotein antigens of *C albicans* has been enumerated by two-dimensional "rocket" immunoelectrophoresis (Axelsen, 1971). Seventy-eight antigenic factors were detected in candidal cytoplasmic solubles with serum from hyperimmunized rabbits (Figures 11.10 and 11.11). Further resolution

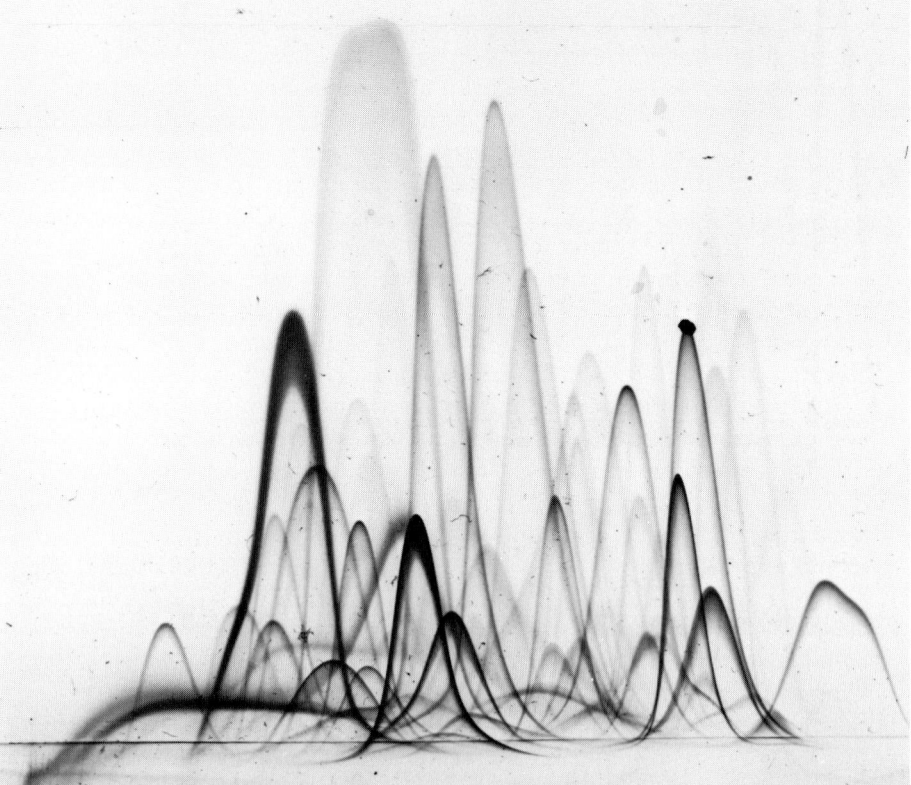

Figure 11.10. Two-dimensional crossed rocket immunoelectrophoresis of *Candida albicans* cytoplasmic proteins versus serum from a hyperimmunized rabbit. Anode to the right (first dimension) and top (second dimension). The dot indicates a rocket resulting from human albumin–anti-albumin spiked in as a reference precipitate. Stained with Coomassie blue R (Axelsen 1973). A maximum of 78 water-soluble antigens were enumerated, but not all could be detected in a single gel. The standard procedure was capable of resolving 54 precipitates and comparing their mobilities with respect to albumin–anti-albumin. *Source:* Reprinted with permission of the author and of the publisher, American Society for Microbiology.

of 168 candidal protein antigens was facilitated by two-dimensional acrylamide gel electrophoresis (O'Farrell technique) permitting simultaneous characterization of pI and molecular weight (Manning and Mitchell, 1980b) (Figure 11.12). The total range of possible antigens can be encompassed by these techniques which provide the tools for monitoring purification of individual factors. As the advances in enzyme immunoassays and monoclonal antibodies are adapted to the fungi, the means of producing large quantities of pure antigens will become available.

Candidal protein antigens have been described in various subcellular compartments. Exoenzymes have been recovered from the culture filtrate (Rüchel, 1981) or from the periplasmic space (Odds and Hierholzer, 1973). Mapping of antigens present in the cytoplasmic solubles of yeast and mycelial forms has been accomplished (Manning and Mitchell, 1980b), and a major cytoplasmic antigen was purified to near-homogeneity (Greenfield and Jones, 1981). Proteins capable of evoking delayed cutaneous hypersensitivity and in vitro correlates were solubilized from a mitochondria-membrane fraction (Domer and Moser, 1978).

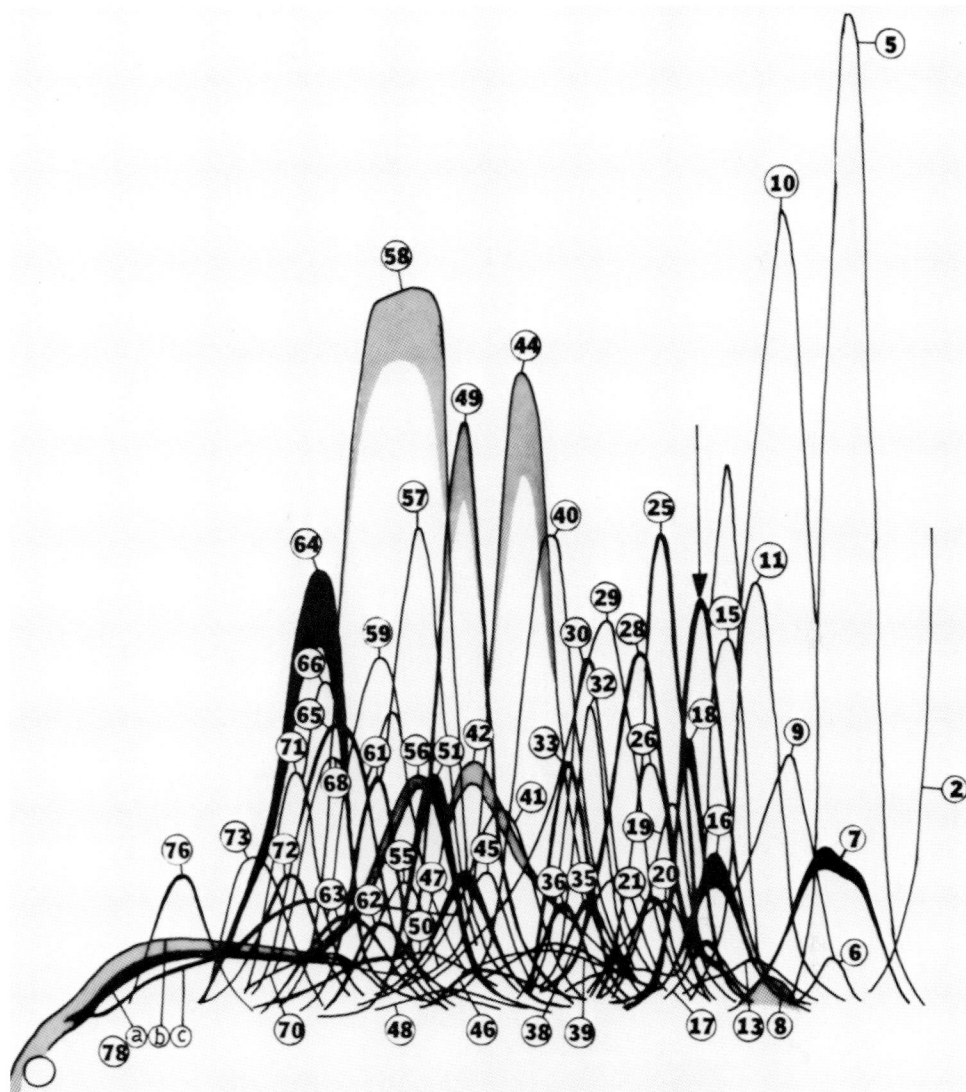

Figure 11.11. Numerical taxonomy of the rocket precipitates from the previous figure (Axelsen 1973). The arrow indicates the albumin–anti–albumin reference precipitate. *Source:* reprinted with permission of the author and of the publisher, American Society for Microbiology.

11.5.1 Acid Phosphatase

The occurrence of this enzyme in the periplasmic space of *C albicans* and its release after exposure of the yeast to DTT have already been discussed. Odds and Hierholzer (1973) characterized the enzyme from a whole cell homogenate as a glycoprotein having a 7:1 mannose to protein ratio and a molecular weight estimated at 124,000 (model E ultracentrifuge) or 136,000 (SDS-PAGE). The enzyme was purified eighty-sevenfold by a combination of gel permeation and ion exchange chroma-

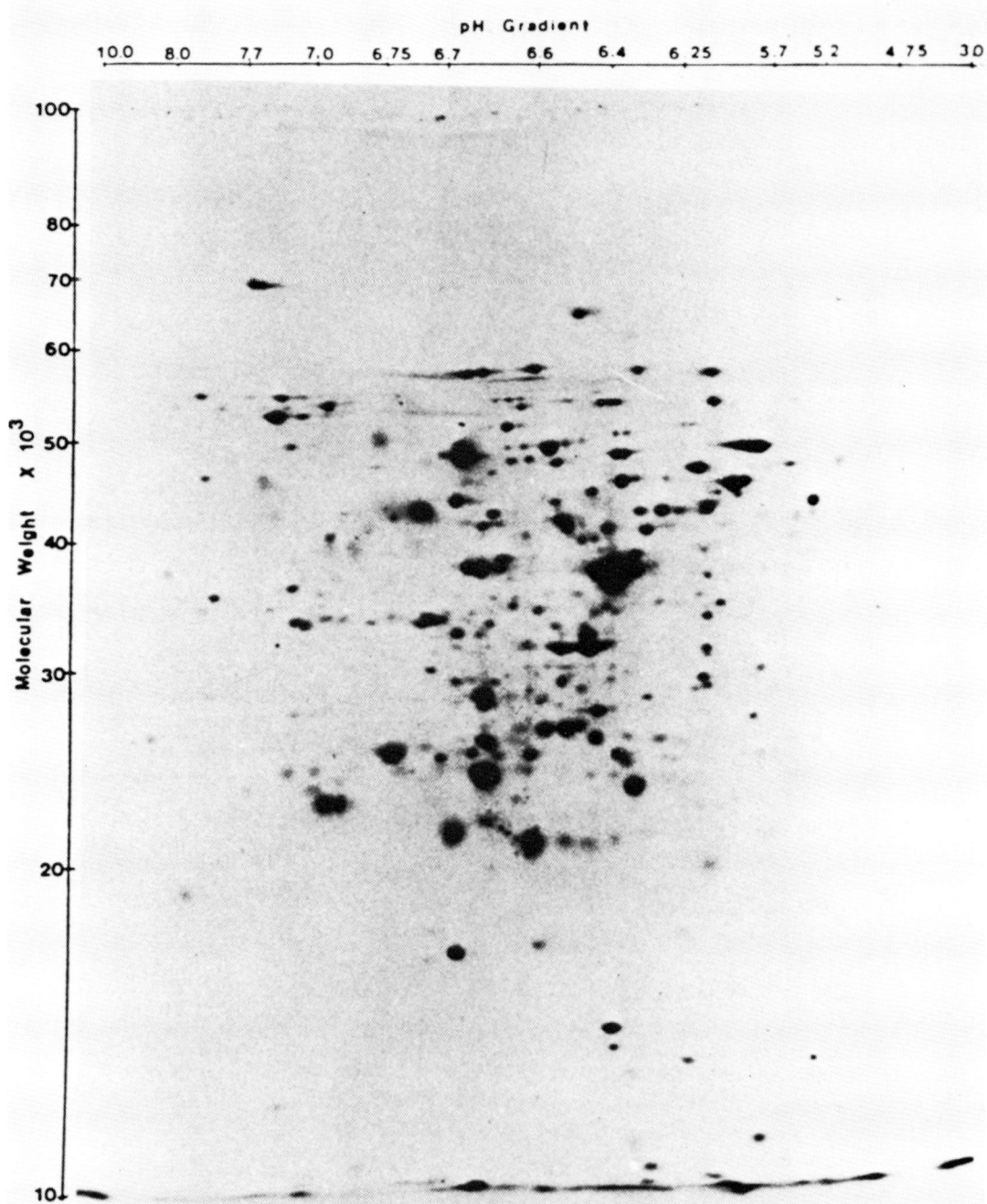

Figure 11.12. Two-dimensional polyacrylamide gel electrophoresis of the cytoplasmic proteins of *Candida albicans* mycelial forms after absorption with antiserum made against the yeast form of the same strain. This map technique does not imply that all of the proteins shown are specific for the mycelial form because absorption may have, with respect to some antigens, been incomplete (Manning and Mitchell, 1980). *Source:* reprinted with permission of the authors and of the publisher, American Society for Microbiology.

tography. Purified acid phosphatase was reactive in precipitin tests with sera from humans with candidiasis. This antigenic activity was heat-stable even after the enzymic function was inactivated. The growth of cultures in high phosphate medium to repress the enzyme did not remove antigenic activity of the extracts, taken as evidence that the glycosidic portion was immunodominant.

11.5.2 Acidic Carboxyl Proteinase

The fungistatic properties of serum, including humoral immunity, are not sufficient to kill *C albicans*, and instead germ tube formation and outgrowth occur. Some isolates of *C albicans* can utilize human serum albumin or bovine γ-globulin as a nitrogen source through induction of an exocellular acidic proteinase. Staib (1965) described the conditions for observing proteinase activity. Not unexpectedly, induction of the enzyme does not occur on the common medium, Sabouraud agar, because its glucose content is repressive. A simple basal salts medium with vitamins, containing 0.1% human serum albumin and adjusted to pH 5, was sufficient to induce proteinase in 75 of 100 *C albicans* isolates (Staib, 1965), and this result was confirmed (Chattaway et al, 1971) in 36 of 37 different isolates. Although it was suggested that only proteinase secretors caused widespread candidiasis in mice (Remold et al, 1968), others have found nonsecreting strains of high mouse virulence (Chattaway et al, 1971).

Exocellular acidic proteinase is induced on a simple basal salts medium plus human or bovine serum albumin. Maximum secretion occurred after 7 days in culture. The enzyme was purified one hundred fortyfold by combined gel permeation and ion exchange chromatography (Remold et al, 1968). A one-step purification was accomplished by adsorption onto dry DEAE-Sephadex and elution with dilute buffer (Macdonald and Odds, 1980b). The enzyme is subject to alkaline degradation; above pH 8.4 it aggregates into a dimer and undergoes autolysis. Purification was thus accomplished below pH 7 and by affinity chromatography on pepstatin immobilized on aminohexyl agarose beads (Rüchel, 1981). Good retention of proteinase activity was achieved by freezing in 10% glycerol or lyophilization from citrate buffer. Concentrated solutions are stabilized by the detergent Brij 35, which mildly stimulates proteolysis. Acid hydrolysis of the enzyme yields mannose and it was thus concluded to be a manno-enzyme (Macdonald and Odds, 1980b). The enzyme's molecular weight and pI were estimated by SDS-PAGE and isoelectric focusing (Table 11.8). Inhibition of proteolysis by an equimolar ratio of pepstatin helped to classify the enzyme as one of the carboxyl proteinases (EC 3.4.23.6).

The spectrum of activity in human serum is of particular interest if this enzyme is to be considered as a determinant of pathogenicity. It is not inactivated by two serum proteinase inhibitors, α-1-antitrypsin and α-2-macroglobulin. It is distinct among proteinases of human mucosal pathogens in its ability to digest both IgA myeloma protein subtypes IgA_1 and the structurally different IgA_2. Neutralization of the proteolytic activity is accomplished only by a several-fold excess of specific antibody (Rüchel, 1981).

Proof that the enzyme is expressed during infection was the bright immunofluorescence when sections of kidneys with multiple microabscesses from infected rabbits were stained with antiproteinase (Macdonald and Odds, 1980a). *C albicans* blastospores grown in Sabouraud medium showed only weak fluorescence possibly as a result of antigenic determinants shared by the proteinase and cell wall mannan.

Table 11.8. Properties of *Candida albicans* Acidic Carboxyl Proteinase EC 3.4.23.6

Molecular weight (da)	45,000
Isoelectric pH	4.4
Optimum pH	2.5–3.9
Range for activity	
Km at pH 3	7×10^{-5} M bovine serum albumin
labile	>pH 8.4; >45 °C
inhibitor	Pepstatin
glycoprotein	Contains mannose
clots milk	5 min at 35 °C; pH 5.5
Human IgA	Digested

Source: Macdonald and Odds, 1980; Rüchel, 1981.

The possibility that cytoplasmic extracts contain proteinase was ruled out by the lack of enzymatic activity and absence of serologic crossreactivity in two-dimensional crossed "rocket" immunoelectrophoresis. Candida proteinase was also localized by immunofluorescence within clogged blood vessels of a patient with acronecrosis of the fingers and toes as a consequence of systemic candidiasis (Rüchel, 1983b).

A quantitative CIE test detected levels of antibodies to proteinase in twelve candidiasis patients that were significantly higher than levels in sera from those colonized with *Candida*. No differentiation of these two groups of patients was possible with unpurified cytoplasmic antigen (Macdonald and Odds, 1980b). Purified acidic proteinase also detected antibodies in neat serum from persons with no evidence of candidiasis, thus requiring a clinical cut-off point for systemic disease of reactions occurring in a fourfold dilution of serum.

Taking care to appropriately dilute sera to exclude precipitin arcs to the proteinase in normal sera, precipitins to the proteinase occurred in a high percentage of post-surgical patients without a known history of candidiasis, as well as in systemic candidiasis patients (Macdonald and Odds, 1980b). Whether precipitins to the proteinase can be used as an indicator of infection in heavily immunosuppressed patients remains to be demonstrated. In a more sensitive format, like enzyme immunoassay, crossreactions between cell wall mannan and the glycosidic portion of the proteinase may interfere with antibody measurement. Removal of the mannosyl portion of the proteinase may provide a means of increasing its antigenic specificity. The proteinase may circulate in immunosuppressed patients, affording the prospect of its direct detection in body fluids. Another potential obstacle to the measurement of proteinase by immunoassays is the need to prevent autodigestion or hydrolysis of human Ig during the incubation.

Secretory carboxyl proteinase complexes with the plasma proteinase scavenger α-2-macroglobulin, but this does not completely occlude binding sites for antibody (Rüchel and Böning, 1983). By conducting a double antibody sandwich EIA at pH 6.1, the complex is cleaved, probably by the action of the trapped proteinase. Cleavage does not, however, lead to increased adsorption in the EIA because of the enzyme's capacity for autodigestion. Exposure of the enzyme to pH 9.8 (30 minutes, 22 °C) denatures it. After retitration to pH 6, the enzyme can be detected without the impediments of binding to α-2-macroglobulin or autodigestion.

The demonstration of proteinase in tissue sections of infected animals and the presence of specific antibodies in humans indicate that this enzyme is an aggressin, if not a toxin. Proteinases of *C albicans* are implicated in coagulation and vasoconstriction processes. The proteinases can activate coagulation factors IX and X, and they also display reninlike activation of the angiotensin vasoconstriction pathway (Rüchel, 1983a). A serine proteinase (EC 3.4.21) from *C albicans CBS2730* membranes had 20% of the ability to activate factor X as the coagulation standard, Russell's viper venom. Similarly, the secreted acid proteinase (EC 3.4.23) of the same *C albicans* strain had 1/10 the activity of pure renin in cleaving angiotensinogen, the first step in the generation of the vasoconstrictor peptide, angiotensin II. *Candida* proteinases are not able to convert prothrombin or to directly generate fibrin, yet their activities are of a scale that argues for involvement in the pathogenesis of candidiasis.

Further evidence favoring a role for acidic carboxyl proteinase as a determinant of pathogenicity is provided by the reduced mouse virulence of a nitrosoguanidine-induced proteinase-deficient *C albicans M12* (Macdonald and Odds, 1983). Isolating a mutant completely blocked in proteinase production was not possible, perhaps because of the diploid nature of *C albicans*.

Proteinase-secreting blastoconidia were less readily phagocytosed and killed than cells of the mutant indicating that the proteinase may be anti-opsonic. The mutant conformed in most respects, to the wild-type (ie, germ tube formation, rate of hyphal production), so that the tenfold increase in the 30 day LD_{50} could be attributed to the lack of proteinase alone.

Some strains of *C tropicalis* caused rapid proteolysis of bovine serum albumin into low molecular weight fragments (Rüchel et al, 1983). About a third of the *C tropicalis* strains studied were strongly proteolytic, whereas others were moderately or nonproteolytic, following a pattern similar to that of *C albicans*. Fewer than 10% of *C pseudotropicalis, C krusei*, and *Torulopsis (Candida) glabrata* strains were proteolytic. No attempt was made to determine if *C albicans* A or *C albicans* B (*stellatoidea*) assorted differently on the basis of proteinase expression. Rabbit antisera against proteinases of *C albicans* and *C tropicalis* reacted with the same but not the reciprocal species.

11.5.3 Major Cytoplasmic Proteins

A sensitive two-dimensional mapping method was used to survey changes in protein composition that may accompany dimorphism and thus lead to the identification of proteins unique to the mycelial or invasive form (Manning and Mitchell, 1980b) (Figure 11.12). Autoradiographs of ^{35}S-labeled cytoplasmic proteins revealed extremely complex electrophoregrams containing 168 antigens. High resolution is facilitated by two-dimensional SDS-PAGE and isoelectric focusing (O'Farrell technique). Adsorption of cytoplasmic solubles with hyperimmune rabbit serum produced by immunization with yeast forms or mycelial forms of *C albicans* was used in an attempt to identify form-specific proteins. Assurance of removing soluble immune complexes was increased by adsorption with protein A-containing staphylococci. The potency of the homologous hyperimmune rabbit serum was extraordinary in its ability to remove about 40% of the total trichloroacetic acid-precipitable radioactivity. Ten proteins were detected in mycelial forms that were absent in the map of yeast form proteins, but antiserum to the yeast forms absorbed five of these. Proteins with the same pI and molecular weight were detected in yeast form maps of another *Candida* isolate so that none was unique for mycelial morphology. Although

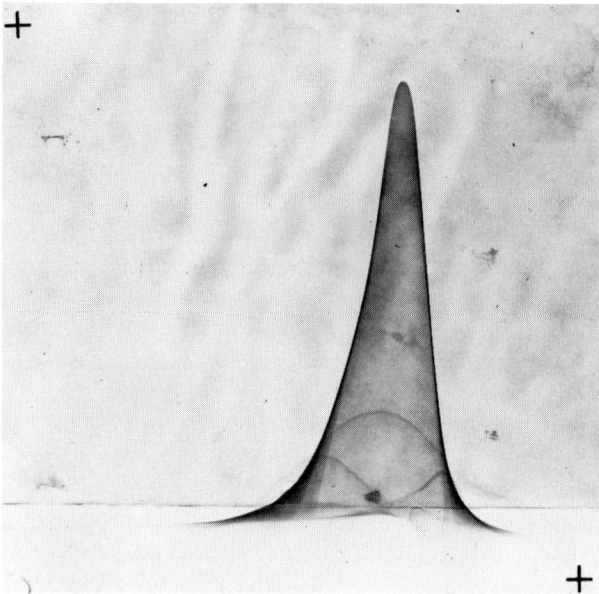

Figure 11.13. Purity of a major protein antigen prepared from *Candida albicans* cytoplasmic extract by diethylaminoethyl anion exchange and Con A-agarose column chromatography. The antiserum used was produced in rabbits by immunizing them with the purified antigen in complete Freund's adjuvant (Greenfield and Jones, 1981). *Source:* reprinted with permission of the authors and of the publisher, American Society for Microbiology.

this "shotgun" approach to delineating morphogenetic proteins may not be feasible, the resolving power of the O'Farrell method is unmatched, and its further development with partially purified fractions of *Candida* and other fungi will continue to provide physicochemical criteria for purity.

The ability of human antibodies to recognize candidal proteins has been amply demonstrated (Syverson et al, 1978; Jones, 1980; Ellsworth et al, 1977), and some researchers have suggested that internal cytoplasmic antigens represent the more profound antigenic stimulus of tissue invasion, as distinct from antibodies to mannan that occur naturally (Lehmann and Reiss, 1980b) and may increase during colonization (Meckstroth et al, 1981). Given the very large array of potentially antigenic proteins in the *Candida* cytosol (Manning and Mitchell, 1980b; Axelsen, 1971), it is surprising that a single factor could account for the major portion of antibody activity produced in response to *Candida* infection of rabbits and in a small series of humans with candidiasis (Jones, 1980). Such a cytoplasmic protein was purified from homogenate supernatants of *C albicans* (Greenfield and Jones, 1981). It was soluble in half-saturated ammonium sulfate and was purified to apparent homogeneity by adsorption on DEAE-cellulose. Removal of mannan was effected by adsorption on Con A-agarose. The purified protein was a single polypeptide chain having a molecular weight of 54,300 estimated by SDS-PAGE and <5% carbohydrate. It produced a strong rocket in two-dimensional immunoelectrophoresis and retained immunogenicity after purification (Figure 11.13). No enzymic activity could be ascribed to this protein after testing it against a battery of several likely substrates.

11.5.4 Membrane-Bound Antigens

The thrust of antigen research in *Candida* species includes the development of more specific probes of cell-mediated immunity in addition to serologic reagents. After screening a number of subcellular fractions prepared by differential centrifugation, attention was focused on a mural glycoprotein and a protein complex solubilized by hot buffer extraction (HEX) of mitochondria-membranes (Domer and Moser, 1978). The glycoprotein was isolated by ethylenediamine extraction of isolated cell walls and is a largely carbohydrate glucomannan complex. The HEX antigenic complex was superior to other fractions in evoking 24- and 48-hour delayed cutaneous reactions in infected mice. Peritoneal exudate cells from these mice, but not serum, could transfer HEX-responsiveness to naive recipients. A tendency toward early 4-hour cutaneous reactions was reduced after purifying proteins from HEX by ammonium sulfate precipitation (HEX: 75% protein, 25% carbohydrate, mannose : glucose = 1 : 2). The mural glucomannan–protein complex evoked large immediate cutaneous reactions in infected mice that could be prevented if the antigen was oxidized with periodate. Papain digestion of the antigen ablated all ability to evoke delayed cutaneous hypersensitivity.

11.6 IMMUNE RESPONSES TO INFECTION: THYMIC AND HUMORAL

11.6.1 Murine Infections

Murine candidiasis is a frequent model for dissecting the immune response to find which effector functions are important for host defense. Hyperimmune serum, but not lymphocytes, from donor mice was effective in protecting recipients from primary intramuscular (quadriceps) challenge, providing evidence in favor of innate and humoral immunity in defense against primary infection (Pearsall et al, 1978). Immunity to infection could also be demonstrated by a second intramuscular challenge resulting in smaller, more rapidly resolving lesions. The thigh lesion is a useful model deserving further investigation. For example, the antiserum for transfer was derived from hyperimmune mice, but lymphocytes were from those receiving a single dose of yeast, so that the transfer of these two immunostimulants do not provide an exact parallel. A useful control would be to transfer antiserum and lymphocytes from animals subject to the same antigenic stimulus. When that is done, successful adoptive transfer of immunity to *Candida* with lymphocytes from infected mice, but not with serum, can be demonstrated (Miyake et al, 1977). A modest but significant reduction in the number of *Candida* colonies recoverable from kidneys of recipients of immune lymphocytes was observed, accompanied by reduced mortality.

The value of footpad swelling as the sole criterion for transfer of delayed hypersensitivity is diminished if the cells transferred are not characterized as to T- or B-cell lineage and if in vitro responsiveness to antigenic stimulation is not assayed in donors or recipients. Consideration must also be given to inoculum size: if too large, it will mask efforts to show the effects of factors modulating immunity.

The view that humoral immunity is protective in resisting primary *C albicans* infection does not conflict with the value of cell-mediated immunity in effective development of immunologic memory that functions to limit infection after secondary challenge or in the late stages of primary infection. Most normal humans have positive delayed cutaneous hypersensitivity to *Candida* species, and thus the response of mice to primary infection does not find a parallel in human adults.

Evidence for the importance of cell-mediated immunity in the late stages of primary infection with *Candida* is the finding (Miyake et al, 1977) that after a systemic challenge, nu/nu mice become debilitated and *Candida* continue to proliferate in their kidneys, whereas euthymic littermates appear healthy and the renal growth of the fungus is restricted. This result qualifies the earlier report that nu/nu mice are more resistant to candidiasis (Cutler, 1976). Kidneys of euthymic mice show gross vegetations in the first few days after systemic challenge with high doses ($3 \times LD_{100}$) of yeast, in comparison with far less pathology in kidneys of athymic nu/nu counterparts. Innate resistance is important in the early stages of primary infection, and nu/nu mice may have an especially good phagocytic response, owing to their chronically activated macrophages and high incidence of natural killer lymphocytes. This response is exaggerated when the normal integument and mucosal barriers are short-circuited by systemic challenge. The early resistance of nu/nu mice gives way to wasting and death.

Immunologic memory is responsible for protection against a lethal systemic challenge with *C albicans*. The stimulus for this memory is provided by low doses of live yeast; for this purpose, heat-killed cells are poor substitutes. Protection of this kind can only be demonstrated if the systemic challenge dose is low enough so that host defenses are not overwhelmed (Giger et al, 1978a).

B-cell mice (neonatally thymectomized, lethally X-irradiated, and bone marrow reconstituted) and normal controls both develop lesions of comparable size and pathology after the first cutaneous infection. Upon a second systemic challenge, B-cell mice are more susceptible than controls. Although these mice have functional lymphocytes, they do not develop antibodies to a specific candidal immunostimulant, HEX, (candidal membrane-bound proteins) in contrast to a high incidence of such antibodies in immunologically intact challenged controls (Giger et al, 1978b). The ability to withstand a secondary systemic challenge correlates positively with intact T-lymphocytes, delayed cutaneous hypersensitivity, and T-dependent antibody production.

A further discrimination between delayed cutaneous hypersensitivity and antibodies as factors important in resistance is facilitated by cyclophosphamide, a DNA alkylating agent that, at an appropriate dose, inhibits the most rapidly dividing lymphocytes, B-cells (Moser and Domer, 1980). Cyclophosphamide treatment of mice has a paradoxical effect of increasing delayed cutaneous hypersensitivity to HEX antigens, but this may result from cyclophosphamide inhibition of T-suppressor cells.

Evidently, lymphocytes mediating delayed cutaneous hypersensitivity are not labile to low-dose cyclophosphamide treatment. Despite the presence of delayed cutaneous hypersensitivity, cyclophosphamide-treated mice survive poorly. This increased susceptibility is not related to low levels of PMN, macrophages, or delayed cutaneous hypersensitivity, but instead reflects poor development of antibodies. In the cyclophosphamide experiment T-dependent antibodies cannot be discriminated from T-independent antibodies, but other work with B-cell mice resulted in suppression of antibodies after challenge with *C albicans*, inferring a pivotal role for T-helper function in survival of mice after a secondary systemic challenge (Figure 11.14).

The observed cyclophosphamide-induced suppression of resistance to *C albicans* is disputed by the findings of Hurtrel et al (1981). This disparity is not explained by the dose of cyclophosphamide, because that was identical. In evaluating such experiments, the nature of the antigenic stimulus required for immunization, method of measuring cell-mediated immunity, and the size of the live challenge dose are critical parameters. Heat-killed *Candida* cells are poor stimulants for induction of cell-mediated immunity (Giger et al, 1978a, 1978b; Hurtrel et al, 1981) requiring high doses that may induce immunologic tolerance. Footpad testing may not be capable of measuring tolerance because the innate responses to whole *Candida* cells complicates measurement of delayed cutaneous hypersensitivity.

Whole killed *Candida* cells cannot be recommended as an antigen for assessment of delayed cutaneous hypersensitivity. The glucan portion of the cell wall is granulomagenic (Williams, 1978), whereas the particulate nature evokes a foreign body reaction. *Candida* glucomannan complexes are alternative pathway activators promoting an influx of PMN via chemotactic and immune adherence factors. Arthus reactions evoked by mannan may further cloud the picture. These competing inflammatory mechanisms make it important to measure in vitro correlates to validate that a state of T-cell-mediated immunity indeed exists.

If the live *C albicans* challenge dose is high, it will obscure efforts to measure factors modulating resistance. Life table techniques to measure survival, and determination of the LD_{50} dose for a given *C albicans* isolate are particularly important to resolve results obtained by different laboratories.

Figure 11.14. Analysis of deaths following intravenous challenge with 10^4 viable blastoconidia of *Candida albicans*. 0, not previously infected; 2, two previous cutaneous inoculations; SXB, sham operated; TXB, thymectomized, X-irradiated-bone marrow reconstituted 4-week-old male CBA/J mice (Giger et al, 1978). *Source:* Reprinted with permission of the authors and of the publisher, American Society for Microbiology.

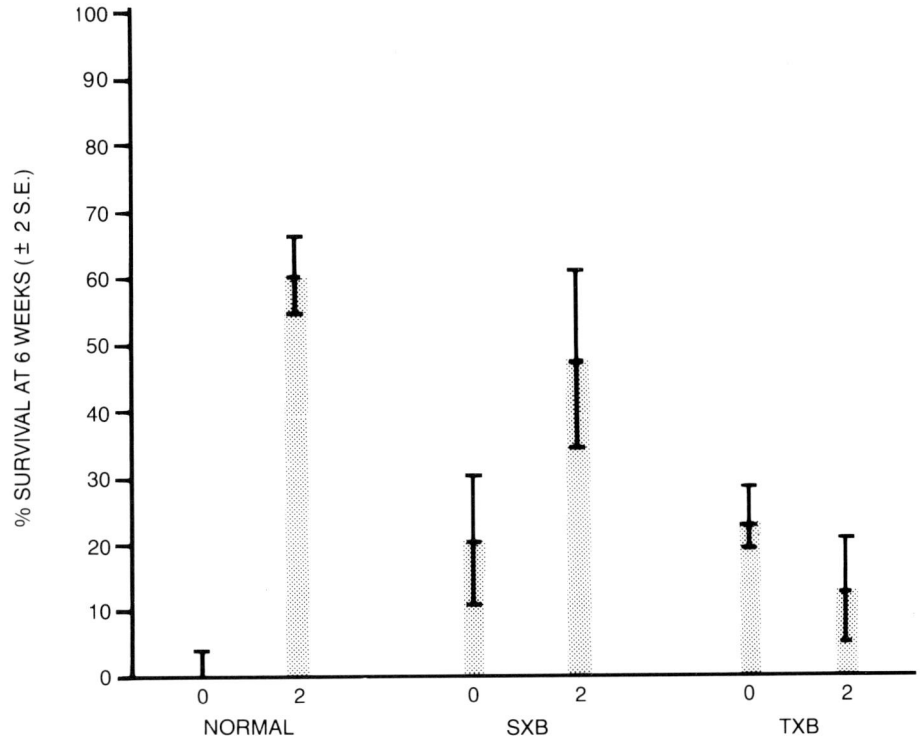

C5-deficient mice have significantly lower survival after *C albicans* challenge than wild type mice of the same CF1 strain (Morelli and Rosenberg, 1971a). The rate of in vitro phagocytosis of *Candida* by murine PMN is faster in the presence of serum from mice with intact complement, and the addition of specific antisera also stimulates phagocytosis. However, the presence of complement does not influence the rate of killing of ingested *Candida* (Morelli and Rosenberg, 1971b). Complement appears to function not as an isolated candidacidal mechanism but instead through opsonization via the alternative pathway. Complement components distal to C5 impart increased survival to candidal challenge in mice, but it is not yet clear whether this occurs through the anaphylatactic or chemotactic properties of C5a or whether the late components of the membrane attack complex play an adjunctive role in killing. Humans with deficiencies in C8 or C9 are not more prone to mycoses and only seem to be more susceptible to *Neisseria* infections.

Decomplementation of human plasma with zymosan results in significantly reduced in vitro killing of *C albicans* by leukocytes in whole human blood (Davies and Denning, 1972). This effect is seen when the concentration of *Candida* in the blood is low, (10 to 100 per ml). Adsorption of plasma with *Candida* cell walls, but not with cytoplasmic solubles, had a profound inhibitory effect on in vitro candidacidal capacity probably owing to the occurrence of natural levels of anti-mannan Ig opsonins. This kind of experiment makes no judgment about the nature of the effector cells, but the use of whole blood has the advantage of allowing cidal mechanisms to operate synergistically, so that potentiation of killing through complement-activation can be discerned.

Priming mice with mycobacteria (*M bovis* var *BCG*) before infection with *C albicans* enhances blastogenesis of splenocytes to candidal cytoplasmic solubles. At the same time the blastogenic response to PPD is suppressed (Rogers and Balish, 1978). This observation favors the view that candidiasis suppresses lymphocyte reactivity to a known immunostimulant, a finding that is consistent with findings in other mycoses (also see Chapter 17).

A study of humoral versus cell-mediated protection mechanisms in mice of five genetic backgrounds appeared to rule out differences in H-2 haplotype, antibody production, in vitro lymphocyte proliferation, and the ability to become immunized by cutaneous priming with live yeast cells (Hector et al, 1982). In vitro lymphocyte proliferation and circulating antibodies were both elevated in mice immunized by cutaneous priming, so that discerning which of these functions contributed to protection was not possible. C5 deficiency in the susceptible DBA/2J was the only immunologic parameter that separated it from the resistant BALB/cByJ. Susceptible DBA/2J mice opsonized blastoconidia via C3b so that the lack of C5 probably reduced chemotaxis and diminished the inflammatory response.

11.6.2 Chronic Mucocutaneous Candidiasis

Chronic mucocutaneous candidiasis occurs in the setting of a spectrum of immunodeficits and the infection varies from mild to severe. A genetic association is known in some families. The disease can thus be classified according to severity, genetic association, and type of immunodeficit. Attempts to group patients to account for all these observations have not been wholly successful, but progress has been made as more series are studied with tests of increased refinement. There is an appreciation of the variety of defects that contribute to CMC, but several pieces of the puzzle are missing. Although lines of evidence point to defects in the T-lymphocyte limb of the immune response, the critical T-cell functions required to keep *Candida albicans* as a commensal and eradicate lesions from cutaneous and mucosal foci are not known. Where familial genetic associations are known, HLA typing has not generally been performed and no histocompatibility antigens have been identified as risk factors.

The clinical features of CMC have been extensively reviewed (Kirkpatrick et al, 1971; Wells et al, 1972; Odds, 1979). This section is concerned with interpreting recent immunologic findings with the scheme for classifying CMC that has been proposed (Wells et al, 1972; Odds, 1979). The cutaneous infection is restricted to the nails and to the stratum corneum of the hands, arms, face, scalp, and less frequently the lower extremities. Hyperkeratosis is common, and in the extreme, results in cutaneous horn formation, the *Candida* granuloma. The basement membrane of the epidermis of most lesions is densely infiltrated by mononuclear cells. Mucosal infection occurs as oral or vaginal thrush, differing from their acute forms by a marked persistence. Sometimes the larynx or esophagus is infected, but in CMC patients the yeast rarely if ever invades the deep tissues.

Inborn errors in the central lymphoid tissues that affect thymic functions are known to lead to increased susceptibility to chronic dermatophytic, *Candida*, and viral infections. The potentially lethal defects, severe combined immunodeficiency, DiGeorge, or Nezelof-Allibone syndromes, are frequently complicated by CMC, but the *Candida* infection does not pose a threat to life compared to other infections that these patients are prone to develop. Candidiasis is also encountered in secondary immunodeficiency resulting from thymoma (Ruiz-Arguelles et al, 1983). Defects in innate phagocytic immunity, such as chronic granulomatous disease of childhood and defective myeloperoxidase, or chemotaxis, are also associated with candidiasis and these are mentioned below. In the narrow sense, the term CMC applies to those persons who do not have a lethal immunodeficiency, and of these, the majority become infected early in childhood.

11.6.2a Classification and Immune Defects

Analysis of the pedigrees of 46 CMC patients was the basis of a clinicogenetic classification into four groups (Wells et al, 1972). Immunologic findings in these patients were also reported (Valdimarsson et al, 1973). The number of patients fitting into this scheme was augmented by Odds, 1979. The most common immunologic finding in CMC is cutaneous anergy to *C albicans* and to other microbial antigens. The anergy is a reversible, specific defect that is probably secondary to infection because with favorable response to antimycotic drugs and a decreased antigenic load, conversion to positive delayed cutaneous hypersensitivity to *C albicans* occurs (Kirkpatrick and Smith, 1974). The defect resembles immunologic tolerance due to chronic antigen excess, which can be broken after removal of the antigen. The possible mechanisms involved in anergy are discussed below.

Group I contains patients with a mild form of candidiasis for whom there is evidence of an autosomal recessive mode of inheritance. Cutaneous anergy to *C albicans* extracts is typical, correlating with an inability of peripheral blood lymphocytes to produce MIF. A majority of patients in this category have lymphocytes that can be stimulated by *Candida* to undergo blastogenesis.

Group II encompasses severe CMC occurring in patients who represent sporadic cases with no consanguinuity or detectable familial inheritance of disease. The early onset and severity argue for a genetic basis, whether inherited or as a result of mutation. Such an impairment could, like some types of thymic aplasia, result from an insult during gestation that affects maturation of the immune system. Cutaneous anergy to *Candida* is the rule, as well as the inability to generate MIF after specific stimulation. Fewer patients fitting in Group II have lymphocytes that proliferate in vitro to *Candida* compared with other Groups. The more severely infected patients exhibit broader unresponsiveness to *Candida*, to other common microbes, and to contact sensitins. This anergy encompasses cutaneous and in vitro measures of T-cell functions.

Group III includes candidiasis-endocrinopathy. Analysis of pedigrees in these patients has led to the conclusion that susceptibility is inherited as an autosomal recessive trait (Wells et al, 1972; Arulanatham et al, 1979), with the extent of candidiasis being mild to moderate. Positive delayed cutaneous hypersensitivity to *Candida* is encountered in a higher proportion of these patients than in those of Groups I and II, but this difference is probably not significant (Odds, 1979). The inability to produce MIF upon *Candida* stimulation was marked in the relatively small number of patients studied. As seen in Groups I and II, blastogenic responses are dissociated from the delayed cutaneous and lymphokine functions in that significant proliferation is elicited by *C albicans* antigens in some of these patients. The endocrinopathies usually appear later than candidiasis (Kirkpatrick et al, 1971; Kirkpatrick and Smith, 1974) and are accompanied by organ-specific autoantibodies (Arulanatham et al, 1979) to thyroid, parathyroid, or adrenal glands resulting in endocrine insufficiency. Other diseases indicative of a lack of suppressor T-cell function have been detected in candidiasis-endocrinopathy including chronic active hepatitis with or without a detectable viral component, diabetes mellitus, ovarian failure (Arulanatham et al, 1979). The connection between these autoimmune phenomena and candidiasis is not yet understood. The derangement in immunomodulation is expressed in the production of autoantibodies, hypergammaglobulinemia and elevated IgE. Successful surveillance against *Candida* on the skin and mucosae seems to require a high order of "cross-talk" between regulator and effector cells.

Group IV patients are those in whom candidiasis has a late onset — 10 years old or older — and a genetic basis is less likely. Lesions are usually restricted to the mouth. In one series (Wells et al, 1972; Valdimarsson et al, 1973) delayed cutaneous hypersensitivity, MIF production, and blastogenesis were intact, leading to the speculation that their candidiasis may result from nutritional deficiency or less effective innate phagocytic immunity. Other cases of late onset reviewed by Odds (1979) displayed cutaneous anergy to *Candida*, and in over half of these blastogenic responses were suppressed. This disparity underscores the inadequacy of artificial categories to accommodate an opportunist like *Candida* that takes advantage of various predisposing factors. Crossing over from Group III to Group IV has been suggested to occur either spontaneously or as a result of therapy (Valdimarsson et al, 1973).

Children with CMC may improve as they reach adulthood but not at sites such as the nails or vagina (Hay et al, 1980). The advent of orally administered, less toxic, antimycotic agents such as ketoconazole provides a maintenance therapy that reduces the symptoms of CMC to a low level or even cures the disease, but the drug dosage necessary to prevent relapses has not yet been determined (Hay et al, 1980).

The inability to produce MIF has been taken as a logical explanation for the absence of delayed cutaneous hypersensitivity (Valdimarsson et al, 1973) but a CMC patient with cutaneous anergy to *C albicans* and intact MIF production has been described (Bice et al, 1974). Delayed cutaneous hypersensitivity to *C albicans* in healthy adults correlates more closely with blastogenic responses than with the migration index. The diameter of *C albicans*-elicited delayed cutaneous hypersensitivity correlated with blastogenic indices in another group of healthy adults ($r = 0.73$ P less than 0.001) (Ferguson et al, 1977), so that an induration of 5 mm or more corresponded to a blastogenic index equal to or greater than 5.

It is noteworthy that lymphocytes producing MIF are not actively dividing cells compared to cells proliferating in response to *C albicans* and thus susceptible to "suicide" by uptake of bromodeoxyuridine and light (Rocklin, 1976). The impression in Groups I–III, and to a lesser extent in Group IV, is that in CMC delayed cutaneous hypersensitivity is dissociated from blastogenic responses. Inhibition of blastogenesis has been related directly to the presence of serum blocking factor and indirectly to the consequences of complement and IgE-mediated release of histamine from mast cells and basophils.

11.6.2b Mechanisms of Blocking Blastogenesis

The inability of lymphocytes from many CMC patients to proliferate in vitro when stimulated with *Candida albicans* is related to the presence of blocking factor in serum or plasma. Serum containing blocking factor will inhibit the lymphocyte responses to *Candida* of active cultures obtained from normal donors. This property was taken advantage of to demonstrate blocking factor in sera from seven CMC patients, the majority of whom were placed in Group III (Valdimarsson et al, 1973). The series studied by Fischer et al (1978, 1982) detected six CMC patients with unresponsive lymphocytes whose sera contained blocking factor. They characterized the factor as mannan or mannan-Ig complexes (also see section 11.3.8). The blocking factor activity is a reversible, specific defect that wanes during periods of remission. This property of acting in a concentration-dependent manner may help to explain the variability in CMC patients' proliferative responses.

Interference by mannan with *C albicans*-induced proliferation also occurred in lymphocytes from normal donors. Fischer et al (1982) speculated that defective metabolism of mannan by macrophages from some CMC patients is a lesion that contributes to causing this disease.

Lymphocytes or serum from women with recurrent vaginitis blocked *C albicans*-induced blastogenesis (Hollister-Stier antigen) of lymphocytes from normal controls (Witkin et al, 1983). This reaction was antigen-specific because responses to phytomitogens were unaffected. Recurrent vaginitis may be a suitable clinical entity with which to explore the possibility that faulty regulation of T-suppressor cells contributes to disease.

The evolution of immune responses in chronic dermatophytosis is interpreted by some (see Chapter 9) to proceed through a stage where immediate cutaneous hypersensitivity to dermatophyte extracts supercedes delayed hypersensitivity, resulting in eventual anergy. Experimental evidence in guinea pigs has shown that histamine inhibits the expression of delayed cutaneous hypersensitivity and in vitro correlates, particularly MIF production (Rocklin, 1976). This suppression is blocked by treatment with H-2 histamine receptor antagonists. A group of four adult family members with CMC who were anergic to *C albicans* in cutaneous, leukocyte MIF, and blastogenesis tests were treated with the H-2 histamine receptor antagonist cimetidine (Jorizzo et al, 1980). After 4 weeks, positive delayed cutaneous hypersensitivity to *Candida* was restored in all four and leukocyte MIF production commenced in three of four, but blastogenesis remained suppressed. Discontinuing the drug caused these positive immune responses to decline, but resumption of cimetidine reversed this trend again. The important implications of this experiment are that the cutaneous anergy in CMC may be controlled by lymphocytes with H-2 histamine receptors and that positive delayed cutaneous hypersensitivity can be dissociated in this disease from in vitro blastogenesis to C albicans.

These experiments show that the two in vitro parameters relied on to characterize the immunologic status of CMC patients are subject to regulation by different and perhaps exclusive means: regulation of blastogenesis by blocking factor, and MIF production subject to control by histamine-sensitive lymphocytes.

Primary thymic defects may coexist with, and perhaps cause CMC (Ruiz-Arguelles et al, 1983). A patient with late-onset CMC was found to have malignant thymoma, a rare concurrence, but one that has been previously reported. In this instance the immunologic defect was characterized as a reduced number of T-helper cells and postthymic precursor T-cells. Declines were observed in several T-helper functions, including a decreased response to PHA and to soluble antigens, and a decrease, measured in vitro, of Ig-secreting cells. The latter response was reversible upon addition of normal T-cells or PHA-stimulated cell-supernates.

11.6.2c Hyper-IgE-Recurrent Infection Syndrome

Progress has been made in the immunology of chronic pustular dermatitis and CMC. These patients frequently have symptoms of atopy, hyperproduction of IgE, and depressed leukocyte chemotaxis (van Scoy et al, 1975; Berger et al, 1980). Specific IgE was demonstrated by adsorptions with whole killed staphylococci (protein A-negative) and C albicans, but the total elevated IgE was not reduced. Chemotaxis to Escherichia coli filtrates was less than 70% of normal in one series (Berger et al, 1980). The hyper-IgE is considered to be a primary defect and not a result of chronic antigenic stimulation. These antibodies are regarded as pathogenetic, reacting with mast cells and basophils to cause release of histamine and pharmacologic mediators that decrease chemotactic responses and inhibit delayed cutaneous hypersensitivity. In view of the familial mode of inheritance that has been described for this disease and the common finding of CMC, including a determination of IgE in the battery of tests for evaluating CMC patients would seem desirable.

11.6.2d Lesions of CMC

Cross-sections of the lesions in CMC show C albicans growing in the stratum corneum of the epidermis with inflammatory cells in the upper dermis and along the basement membrane (Sohnle et al, 1976a). No differences have been detected in the inflammatory lesions in persons who display delayed cutaneous hypersensitivity to C albicans and in those who do not. In both there is hyperkeratosis with an intense mononuclear cell infiltrate. These lesions can be modeled in guinea pigs by rubbing C albicans into the skin (Sohnle et al, 1976a, 1976b). Small papules develop, followed by sloughing of scales containing the yeast. Animals immunized cutaneously with live yeast a month before challenge undergo a stronger eruption and rapid confluent scaling. Sometimes the Candida are sealed and sloughed off alive in the upper epidermis before a noticeable inflammatory infiltrate can occur. The rapidity of the scaling process is a function of cell-mediated immunity and can be adoptively transferred to naive recipients by peritoneal exudate cells from immune donors. The added stress created by occlusive dressings of plastic film superimposed on infected skin results in conditions more favorable to C albicans proliferation and a stronger antigenic stimulus.

Immunofluorescent probes have localized humoral factors in biopsied human skin lesions and in those induced by *C albicans* under occlusive dressings in guinea pigs. In animals, C3 (but not C4 or Ig) was detected at the stratum corneum and coating the yeast cells. Properdin factor B and C3 were detected along the basement membrane of human lesions, but again no C4 or Ig could be found. These findings support a role for the alternative pathway in vivo in the guinea pig.

The release of chemotactic and immune adherence factors changes the character of the inflammatory infiltrate from mononuclear to PMN. Normal and immune animals respond alike to infection under occlusive dressings: in both groups crusts containing PMN and many *Candida* are shed. Two distinct but mutually supporting immune responses are involved in the cutaneous lesions, one innate and the other dependent on immunologic memory: (i) complement-mediated accumulations of PMN represent a potent innate primary response. Further evidence of direct complement activation has been obtained in vitro using *Candida* yeast or extract in C2-deficient serum. (ii) The memory-dependent effector function of cell-mediated immunity appears to be in increasing the rate of mitosis and hence the effluvial current of the upper epidermis, possibly as a result of secretion of an unknown lymphokine. Both of these processes may be independent of antibody or the classical complement pathway. Failure of CMC patients to eradicate *Candida* from the upper epidermis even in the presence of a marked mononuclear infiltrate argues for a defect in T-effector function at the level of a keratinization-stimulating lymphokine, but there is no direct evidence for the existence of such a molecule. The resulting inflammatory infiltrates cannot rid hosts of the infection, but they may prevent its dissemination.

11.6.2e Immunotherapy

The effect of transfer factor alone or in combination with antimycotic therapy illustrates the experience gained with this treatment of CMC (Kirkpatrick and Smith, 1974). Transfer factor is dialyzable extract of leukocytes from donors expressing positive delayed cutaneous hypersensitivity to the antigen. A "dose" has been reported as the extract from 6×10^8 (Kirkpatrick and Smith, 1974) or 1×10^9 (Ballow and Hyman, 1977) lymphocytes. The potency of TF is not standardized because the active principle and mechanism are not yet known. Amphotericin B therapy alone resulted in complete remissions for over a year in three of four patients with CMC of late onset; in the fourth person mucosal candidiasis persisted. Two of those receiving this treatment converted to giving positive delayed cutaneous hypersensitivity. Five other CMC patients who acquired the disease in infancy were treated with TF alone in multiple doses without any improvement in their condition. Two patients with extensive CMC since infancy and marked suppression of cell-mediated immunity received amphotericin B until the lesions were cleared, then multiple doses of TF were started and tapered to one dose every 4 months. This regimen was successful in sustaining remissions of cutaneous candidiasis for 2 years, but mild mucosal candidiasis persisted. Positive delayed cutaneous hypersensitivity and MIF production have been maintained, although lymphocyte transformation to *Candida* was not significantly improved.

Blockages in the development of central lymphoid tissues are unlikely to respond to TF alone, and at the other end of the spectrum late-onset candidiasis may only require antimycotic therapy to sustain prolonged remission. The regimen offering the most promise for early-onset CMC appears to be to reduce the antigenic load with systemic antimycotic agents, before multiple doses of TF. The length of remissions, if any, are gauged. Fetal thymus grafts combined with vigorous TF therapy provides a more intensive follow up that has produced remissions where TF alone has not, and this remission has been prolonged (Figure 11.15).

In another CMC patient where TF had no effect on the disease, a fetal thymus graft without additional TF also failed to exert any clinical improvement, although gains were recorded in in vitro cell-mediated immune responses and increased cellularity of the paracortical areas of a regional lymph node (Kirkpatrick et al, 1976). Fetal thymus grafts and TF may act synergistically, but more instances of this immunotherapy have to be accumulated. Evidence of chimerism from the engrafted thymus has not been detected even when successful remission of candidiasis was achieved (Ballow and Hyman, 1977). The implants may function by providing a favorable microenvironment and thymic hormones to stimulate maturation of Candida-specific T-cell responses. Signs of responsiveness are acquisition of positive delayed cutaneous hypersensitivity, increased MIF production and blastogenesis to Candida and to other common microbial antigens, and the ability of contact sensitins to induce delayed cutaneous hypersensitivity. The long-term remission achieved by combined fetal thymus grafts and TF indicate that significant immunologic reconstitution can be achieved.

11.7 PHAGOCYTIC CANDIDACIDAL MECHANISMS

Candida albicans yeast forms are readily phagocytosed in vitro by normal human neutrophils and monocytes only when serum is present (Lehrer and Cline, 1969; Leijh et al, 1977). Heat-stable opsonins suffice to promote phagocytosis, but native serum is a better source so that heat-labile opsonins, ie, complement components, are necessary for optimal phagocytosis of yeast forms (Lehrer and Cline, 1969). About half of the ingested *Candida* are killed within one hour (Leijh et al, 1977). The primary mechanism for killing *Candida* yeast is oxidative, operating via the myeloperoxidase-H_2O_2-halide pathway (Roos, 1980; Cheson et al, 1977). Upon ingestion of *Candida*, a respiratory burst occurs accompanied by chemiluminescence, generation of H_2O_2, superoxide anion O_2^-, and iodination of candidal proteins (Diamond and Krzesicki, 1978). Some evidence for the participation of singlet $^1O_2$ and hypochlorous acid also has been found (Diamond and Krzesicki, 1978). Pseudohyphal forms, too large for endophagocytosis, are still subject to damage and probable killing by neutrophils. Adherence is followed by spreading of the phagocyte along the hypha and discharge of granules. Cytopathic effects on the target cell are detected by inhibition of ^{14}C-cytosine uptake and the dissolution of the cell wall and cytoplasmic organelles (Diamond et al, 1978; Diamond and Krzesicki, 1978; Diamond and Haudenschild, 1981). Damage to pseudohyphae can occur in the absence of serum but is potentiated by natural levels of anti-*Candida*-IgG. The adherence of neutrophils occurs preferentially to viable *Candida* and specifically involves a chymotrypsin-sensitive receptor on the *Candida* surface, presumably an outer coat protein. A similar means was invoked for the adherence of *C albicans* to vaginal epithelium (King et al, 1980). The mechanism of central importance for damage to pseudohyphal forms is also the MPO-H_2O_2-halide pathway.

Iodination of candidal proteins, on the other hand, has been characterized as a "noncausal concomitant" of killing *Candida* (Lehrer, 1975) on grounds that inhibitors of iodination did not impair monocytes' ability to kill *C albicans*. Others (Diamond and Krzesicki, 1978; Diamond and Haudenschild, 1981) have emphasized the

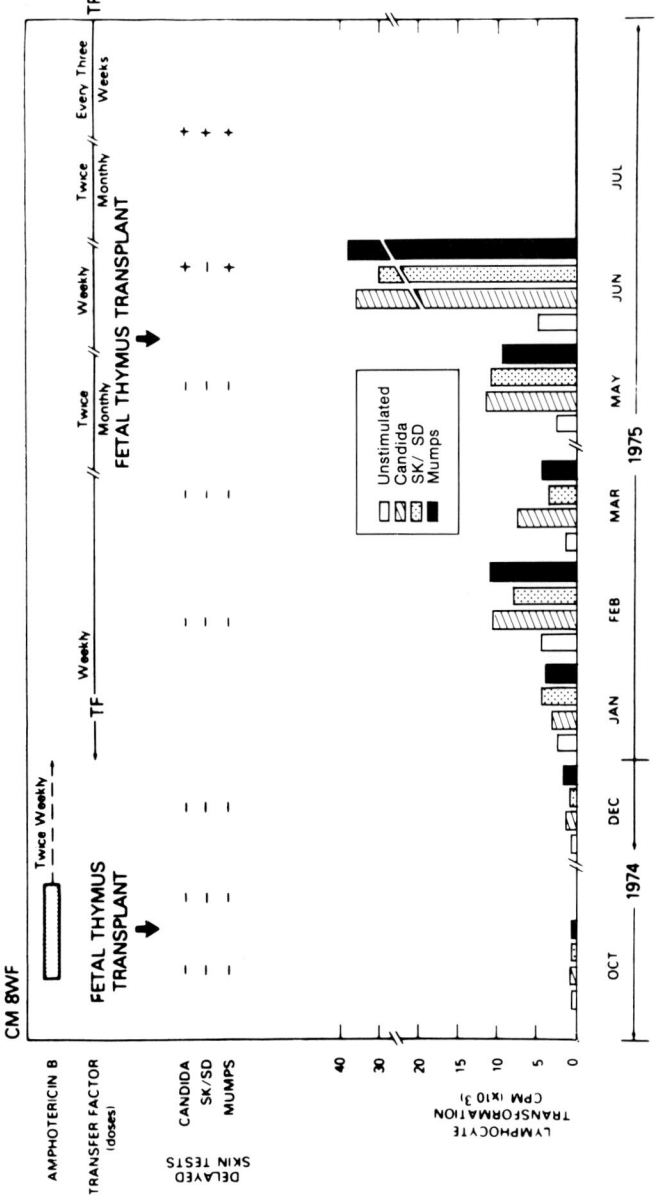

Figure 11.15. Fluctuations in lymphocyte blastogenesis and skin tests in an 8-year-old girl with chronic mucocutaneous candidiasis who received combined therapy with amphotericin B, transfer factor (TF), and a fetal thymus graft. SK/SD, streptokinase-streptodornase; "candida", dermatophytin O, a product of Hollister-Stier Laboratories (Ballow and Hyman, 1977). *Source:* Reprinted with permission of the authors and of the publisher, Academic Press.

ability of neutrophils and monocytes to iodinate *Candida* pseudohyphae and moreover, that isolated neutrophil granules inflict damage on *Candida* only in the presence of halide ions. Normal chemotaxis, adherence, and ingestion of *C albicans*, but defective killing is observed in two different and rare inherited defects in innate phagocytic immunity: myeloperoxidase deficiency and chronic granulomatous disease (CGD) of childhood. Persons with a selective absence of peroxidase in the azurophil granules of their neutrophils and monocytes lack MPO and are more susceptible to *Candida albicans* or *Staphylococcus aureus* (Lehrer and Cline, 1969a; Lehrer, 1975; Cheson et al, 1977; Roos, 1982).

Myeloperoxidase-deficient phagocytes can inhibit the intracellular germination of *C albicans* but cannot kill the yeast (Lehrer, 1970). This underscores the need for integrity of the oxidative microbicidal pathway to provide optimal resistance to *Candida*. In spite of these in vitro findings many of the persons with MPO deficiency are generally free of serious infection because of intact compensatory mechanisms. A normal or supranormal respiratory burst occurs in MPO-deficient cells and both H_2O_2 and superoxide are produced. Monocytes from an MPO-deficient person had a greater ability to kill *C parapsilosis* and *tropicalis* than did those from normal donors (Lehrer, 1975). The secondary reactions dependent on MPO that are subnormal or absent in individuals deficient in this enzyme are (i) chemiluminescence, (ii) iodination of microbial proteins, (iii) generation of hypochlorous acid and singlet oxygen (Cheson et al, 1977; Roos, 1980). Quenchers of hypochlorous acid and singlet $^1O_2$ were found to inhibit damage by monocytes to *C albicans* pseudohyphae (Diamond and Haudenschild, 1981). Beside the supranormal respiratory burst in MPO-deficient phagocytes, the evidence for a nonoxidative, supplemental *Candida*-killing mechanism is considered below.

Children with CGD are blocked earlier in the oxidative pathway than those with MPO-deficiency. Their neutrophil granules have a full complement of enzymes, including MPO, but no respiratory burst occurs in CGD granulocytes or monocytes upon addition of activators, and hence, no H_2O_2 or superoxide are produced. Bacteria that produce their own H_2O_2 are readily killed by CGD phagocytes (Roos, 1980), but *C albicans* and other *Candida* species are not killed, nor is germ tube formation and pseudohyphal growth retarded. Consequently, *Candida* species are among those pathogens that cause infections in children with CGD (Oh et al, 1969; Lehrer, 1970; Cheson et al, 1977).

Damage to *C albicans* by nonoxidative processes has been detected. Monocytes from MPO-deficient humans inhibited uptake of ^{14}C-cytosine by *C albicans* pseudohyphae, and a similar effect was observed with monocytes from one of four patients with CGD (Diamond and Haudenschild, 1981). Neither whole human monocyte lysates or purified "granules" were able to damage *C albicans* pseudohyphae under anaerobic conditions, so the mechanism of nonoxidative damage is still not known. Others (Lehrer, 1975) have reported that killing of phagocytosed *Candida* species yeast was very poor in monocytes from CGD patients, even against those species of *Candida* (ie, *parapsilosis* and *pseudotropicalis*) that were killed by monocytes from MPO-deficient persons. Granule-rich fractions of a lysate of human MPO-deficient neutrophils readily killed *C parapsilosis* yeast. A cationic esterase that was cidal for *C parapsilosis* was resolved by PAGE of soluble granule proteins, but *C albicans* was more difficult to kill (Lehrer et al, 1975). Further evidence for potent microbicidal activity of cationic proteins was obtained from lysates of rabbit alveolar macrophages (Patterson-Delafield et al, 1980). Active cationic protein-containing fractions were banded by density gradient zonal centrifugation. Markers of known lysosomal enzymes did not band at a similar density, so proof of a lysosomal origin is still elusive. These proteins, which are more cationic than lysozyme, are particularly cidal against *Candida* species, less effective against Gram-positive bacteria, and Gram-negative bacteria were least susceptible (Figure 11.16). Assays of cidal effect of macrophage cationic protein were not carried out under anaerobic conditions so the contribution of oxidative processes, if any, was not determined.

11.8 SUMMARY

The outer layers of the *C albicans* cell wall contain protein and possibly mannan adhesins. The wall also provides a niche for periplasmic enzymes such as acid phosphatase and N-acetylglucosaminidase. Structural protein helps to maintain the cell wall integrity as shown by the beneficial effect of sulfhydryl reagents on protoplast formation.

Readily soluble mannan is an outer coat and also occurs as a deep mural layer. The major fibrillar component of the cell wall matrix is β-1,3-glucan. Chitin is localized in bud scars and as a thin microcrystalline sleeve on the lateral walls. Hyphal forms of *C albicans* have more chitin than yeast forms.

Mannan is a glycoprotein with major polymannose strands linked via chitobiosylasparatamido bonds, and short oligomannosides linked via base-labile hydroxyamino acid esters, to the peptide moiety. Three domains characterize the general plan of yeast mannans: the outer chain containing the major antigenic sites, the inner core, and the base-labile oligomannosides. The linear strands in the α-1,6-mannan backbone are heavily substituted with a majority of α-1,2 linkages in the antigenic side chains.

Serotype A is more abundant in clinical isolates. The mannan of serotype A contains the determinants present in type B plus the A-specific antigen, factor "6." A B-specific factor is suspected on chemical and serologic grounds. In addition to serologic differences, mannan preparations may differ according to the harshness of extraction methods which affect the degree of polymerization, charge interactions derived from O-phosphonomannan, and the presence of glucomannan complexes. The fractionation of mannan with borate cetyltrimethylammonium bromide conserves the molecular weight of mannan.

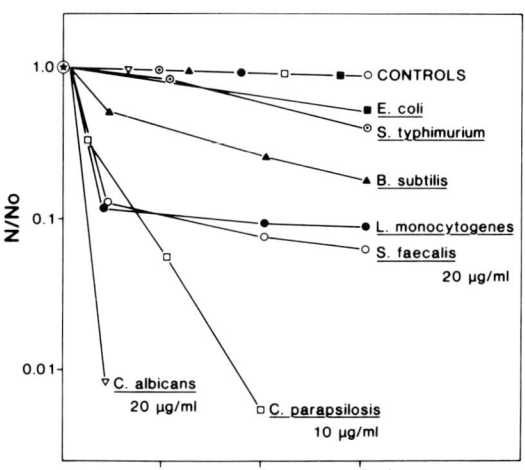

Figure 11.16. Microbicidal activity of cationic proteins from rabbit alveolar macrophages. Each incubation mixture contained 10^6 organisms per ml and 20 μg of cationic proteins. Controls are each organism resuspended in saline. N/N_0, number of viable organisms, initial, divided by the number of viable organisms at the time of measurement (Patterson-Delafield et al, 1980). *Source:* Reprinted with permission of the authors and of the publisher the American Society for Microbiology.

Methylation-fragmentation-GLC analysis of *C albicans* mannan shows a greater abundance of 1,2 linkages in serotype A, that accords with hapten-inhibition analysis for longer oligomannoside chains in type A mannan. The A-specific determinants in mannan are the sparse terminal nonreducing 1,3 linkages. There is agreement based on PMR spectral evidence that rare β-linked mannose residues occur exclusively in serotype A mannan but the glycosidic bond arrangement and participation as an antigenic determinant are not fully established.

Controlled partial acetolysis cleaves the linear α-1,6-mannan backbone releasing antigenic oligomannosides that after gel permeation chromatography provide a unique fingerprint for a given mannan.

Considering the results of acetolysis, methylation-fragmentation, and hapten inhibition, type specificity resides in the mannohexaose fragment in these mannans.

The sequence and linkage assignments of the mannohexaose haptens of serotypes A and B are still incomplete. In type A, two possible structures are proposed. The simplest structure is a straight chain α-1,2-pentaose with an α-1,3 nonreducing terminal. A more complex epitope for serotype A may consist of a β-1,2-mannosyl terminal and a single mannosyl branch point linked $1 \rightarrow 6$ to the oligosaccharide (Figure 11.1). Type B mannohexaose appears to lack a 1,3 terminal, to have α-1,2 linkages and one mannose branch point linked α-1,3 to the linear portion of the determinant. The shape of the molecule appears to be treelike rather than comblike with type-specific residues representing a minority of the total serologically active side chains.

O-phosphonomannan is a structural feature common to mannans of many yeasts including *C albicans*. About one third of the bulk mannan of serotype B is eluted at high ionic strength from anion exchangers indicating *O*-phosphonomannan. Indirect evidence suggests a positive correlation between the degree of phosphorylation and antigenicity. A highly phosphorylated fraction of baker's yeast mannan is a small portion by weight, but is enriched in glucomannan complexes. A function of phosphodiester mannan may thus be to serve as the "missing link" between these two wall polymers.

Exo-α-mannan hydrolase can be used to differentiate between *C albicans* mannan of type A (resistant) and type B (susceptible).

β-D-glucan accounts for almost half of the mural dry weight of *C albicans*, partly as an alkali-soluble polymer with an increase in β-1,6 linkages but mainly as a β-1,3 fibrillar glucan. The fibrillar mural skeleton activates complement directly, promotes formation of granulomas, and confers protection against infection.

A large array of 168 candidal soluble cytoplasmic proteins have been mapped as antigens by two-dimensional PAGE and autoradiography. Proteins or glycoproteins that are especially capable of evoking cell-mediated responses have been solubilized from a mitochondrial-membrane fraction, the "HEX" antigen.

Acid phosphatase from the periplasmic space of *C albicans* is a mannoprotein of about 130,000 da with immunodominant epitopes in the mannan portion of the molecule. Acidic carboxyl proteinase is an inducible enzyme in many *C albicans* isolates whose secretion is repressed by the peptones or glucose in Sabouraud agar. After purification on immobilized pepstatin this enzyme is characterized by a pH optimum of 2.5 to 3.9 and a molecular weight of 45,000 da. Acidic carboxyl proteinase is expressed during infection and acts on IgA subtypes 1 and 2. Antibodies to proteinase occur in persons colonized or infected with *C albicans*. This enzyme, too, appears to be a mannoprotein.

Blastogenic responses of rabbit lymphocytes to whole formalin-killed *Candida* are brisk in endocarditis, but a trend towards anergy develops when multisystem dissemination occurs. Proliferative responses of B-lymphocytes to a glycopeptide complex of *C albicans* requires the presence of T-cells or supernates of T-cell cultures. Prolonged in vitro exposure of human peripheral blood lymphocytes to *C albicans* glycopeptides generates T-suppressor cells.

Soluble complexes in sera of chronic mucocutaneous candidiasis patients containing *C albicans* mannan are capable of blocking in vitro blastogenic responses to candidal culture filtrates. The accumulation of mannan or soluble immune complexes containing mannan in chronic mucocutaneous candidiasis patients' sera results from a defect in their monocytes, which are inefficient in the metabolism of this polysaccharide.

Several effects of *C albicans* mannan are traceable to the direct activation of complement by the alternative pathway. They include production of anaphylatoxin, chemotactic factor, and triggering of mast cell degranulation. Symptoms of anaphylactic shock in rodents have followed injection of mannan.

Biopsies of human cutaneous lesions show evidence of the alternative complement pathway component, properdin factor B, and C3, coating yeast forms and also depositing at the base of the stratum corneum.

Patients being treated with corticosteroids or cytotoxic drugs for management of neoplastic, autoimmune disease, or for maintenance of renal allografts are at risk of developing invasive candidiasis. In that event, the granulocytopenia and compromise of humoral immunity results in a steady state between sloughing of soluble mannan into the plasma and its metabolism to mannose and eventual excretion. Mannan can be detected by radioimmunoassay or enzyme immunoassay in the serum of these patients in the low ng/mL range, provided that soluble immune complexes are first dissociated. Mannanemia is a transient phenomenon and is replaced by antibodies in the serum even when the disease progresses.

In the murine model of candidiasis, adoptive transfer of hyperimmune serum or sensitized lymphocytes have given increments of protection. Innate resistance is important in the early stages of primary infection. Phagocytosis and destruction of *C albicans* by PMN and by macrophages remain intact in athymic mice, so that initially the infection is restricted. Later, *Candida* proliferates in the kidneys, but in euthymic mice such growth is limited. Similarly, secondary challenge with live candidae in thymectomized mice finds them more susceptible than euthymic controls, thus arguing for a role for immunologic memory, specifically T-cell help in producing antibodies to candidal proteins.

Genetic deletion of C5 renders mice more susceptible to candidiasis, but a role for the membrane-attack phase of the complement cascade is unproven as a candidacidal mechanism.

Cutaneous anergy in chronic mucocutaneous candidiasis is a reversible, specific defect secondary to the infection, which is removed by favorable response to antimycotic therapy and a decreased antigenic load. Soluble immune complexes containing mannan are presumptive factors blocking lymphocyte blastogenesis. Migration inhibitory factor production is controlled by a subset of histamine-sensitive lymphocytes. In healthy adults, delayed cutaneous hypersensitivity to *C albicans* correlates more closely with the blastogenic index than with the leukocyte migration index.

A clinicogenetic classification is useful in demonstrating that CMC is a spectrum of diseases. Group I patients have genetically determined mild disease. They typically lack cutaneous and MIF responses, but with blastogenesis remaining intact.

Patients with severe disease and no detectable genetic component are in Group II. Broad cutaneous anergy occurs, expressed as the lack of both MIF and blastogenesis. A genetic component and endocrinopathy are found in Group III. Some of these patients' lymphocytes undergo blastogenesis but are otherwise anergic to *C albicans*. Endocrinopathy occurs later in the disease and may be the result of organ-specific autoantibodies. Other immune defects in this group including hypergammaglobulinemia and hyper-IgE production suggest a failure of T-suppressor cells. Late onset CMC patients with mild oral candidiasis are placed in Group IV. Several of these patients have intact MIF and blastogenic responses to *Candida*. This category contains a mixed group of patients with mild defects in phagocytic immunity or immunoregulation.

The rapidity of scaling in the epidermal layers of guinea pigs infected with *C albicans* is a function of cell-mediated immunity and can be adoptively transferred with lymphocytes. The effluvial current of the upper epidermis may be regulated by a keratinization-stimulating lymphokine.

The strategy for immunotherapy in CMC that has the most promise of success is to first reduce the antigenic load by systemic antimycotic therapy. Then multiple doses of transfer factor have been given and the length of remissions, if any, were gauged. Fetal thymus grafts and transfer factor appeared to be synergistic, achieving long remissions in a few cases.

Phagocytosis of *C albicans* yeast forms occurs readily when they are opsonized by natural antibodies and complement. Pseudohyphal forms are too large for endophagocytosis, but adherence to phagocytes permits the close contact sufficient for deposition of lysosomal enzymes. Intracellular killing is efficiently accomplished by the myeloperoxidase-H_2O_2-halide pathway accompanied by iodination of candidal proteins.

Myeloperoxidase-deficient neutrophils and monocytes can ingest and inhibit germination of candidae in vitro but cannot kill the yeast. Yet myeloperoxidase-deficient persons are generally free of serious candidiasis because of in vivo compensatory mechanisms, ie, a supranormal respiratory burst and nonoxidative killing by cationic proteins. Children with chronic granulomatous disease fail to exert a respiratory burst and thus no H_2O_2 or superoxide are generated, leaving them vulnerable to *C albicans*.

REFERENCES

Ahearn DG, Roth FJ Jr, Meyers SP, 1968. Ecology and characterization of yeasts from aquatic regions of South Florida. Marine Biology—Inter J on Life in Oceans and Coastal Waters 1:291–309.

Armstrong D, Chmel H, Singer C, Tapper M, Rosen PP, 1975. Non-bacterial infections associated with neoplastic disease. Eur J Cancer Clin Oncol 11 (Suppl):79–94.

Arulanatham K, Dwyer JM, Genel M, 1979. Evidence for defective immunoregulation in syndrome of familial candidiasis endocrinopathy. N Engl J Med 300:164–168.

Auger P, Dumas C, Joly J, 1979. A study of 666 strains of *Candida albicans*: correlation between serotype and susceptibility to 5-fluorocytosine. J Infect Dis 139:590–594.

Axelsen NH, 1971. Antigen-antibody crossed electrophoresis (Laurell) applied to the study of the antigenic structure of *Candida albicans*. Infect Immun 4:525–527.

Axelsen NH, 1973. Quantitative immunoelectrophoretic methods as tools for a polyvalent approach to standardization in the immunochemistry of *Candida albicans*. Infect Immun 7:949–960.

Axelsen NH, Kirkpatrick CH, 1973. Simultaneous characterization of free *Candida* antigens and *Candida* precipitins in a patient's serum by means of crossed immunoelectrophoresis with intermediate gel. J Immunol Methods 2:245–249.

Baker RD, 1962. Leukopenia and therapy in leukemia as factors predisposing to fatal mycoses. Am J Clin Pathol 37:358–373.

Ballou C, 1976. Structure and biosynthesis of the mannan component of the yeast cell envelope. Adv Microb Physiol 14:93–158.

Ballou CE, 1980. Genetics of yeast mannoprotein biosynthesis. pp. 1–14. In Fungal Polysaccharides, Sandford PA, and Matsuda K, eds. Washington, DC: American Chemical Society.

Ballow M, Hyman LR, 1977. Combination immunotherapy in chronic mucocutaneous candidiasis. Synergism between transfer factor and fetal thymus tissue. Clin Immunol Immunopathol 8:504–512.

Barrett JT, 1983. Textbook of Immunology, 4th edition. 520 pp. St. Louis, MO: The C. V. Mosby Company.

Bellanti JA, 1985. Immunology III. 598 pp. Philadelphia, PA: W.B. Saunders.

Berger M, Kirkpatrick CH, Goldsmith PK, Gallin JI, 1980. IgE antibodies to *Staphylococcus aureus* and *Candida albicans* in patients with the syndrome of hyperimmunoglobulin E and recurrent infections. J Immunol 125:2437–2443.

Bice DE, Lopez M, Rothchild H, Salvaggio J, 1974. Comparison of *Candida* delayed hypersensitivity skin test size with lymphocyte transformation, migration inhibitory factor production and antibody titer. Inter Arch Allergy Appl Immunol 47:54–62.

Bishop CT, Blank F, Gardner PE, 1960. The cell wall polysaccharides of *Candida albicans*: glucan, mannan and chitin. Can J Chem 38:869–881.

Bodey GP, 1966. Fungal infections complicating acute leukemia. J Chronic Dis 19:667–687.

Burges G, Holley HP, Virella G, 1983. Circulating immune complexes in patients with *Candida albicans* infections. Clin Exp Immunol 53:165–174.

Cabib E, Bowers B, 1971. Chitin and yeast budding. Localization of chitin in yeast bud scars. J Biol Chem 246:152–159.

Cassone A, Mattia E, Boldrini L, 1978. Agglutination of blastospores of *Candida albicans* by concanavalin A and its relationship with the distribution of mannan polymers and the ultrastructure of the cell wall. J Gen Microbiol 105:263–273.

Centeno A, Davis CP, Cohen MS, Warren MM, 1983. Modulation of *Candida albicans* attachment to human epithelial cells by bacteria and carbohydrates. Infect Immun 39:1354–1360.

Chandler FW, Kaplan W, Ajello L, 1980. A Colour Atlas and Textbook of the Histopathology of Mycotic Diseases. 333 pp. London: Wolfe Medical Publications, Ltd.

Chattaway FW, Holmes MR, Barlow AJE, 1968. Cell wall composition of the mycelial and blastospore forms of *Candida albicans*. J Gen Microbiol 51:367–376.

Chattaway FW, Odds FC, Barlow AJE, 1971. An examination of production of hydrolytic enzymes and toxins by pathogenic strains of *Candida albicans*. J Gen Microbiol 67:255–263.

Chattaway FW, Shenolikar S, Barlow AJE, 1974. Release of acid phosphatase and polysaccharide- and protein-containing components from the surface of dimorphic forms of *Candida albicans* by treatment with dithiothreitol. J Gen Microbiol 83:423–425.

Chattaway FW, Shenolikar S, O'Reilly J, 1976. Changes in the cell surface of the dimorphic forms of *Candida albicans* by treatment with hydrolytic enzymes. J Gen Microbiol 95:335–347.

Cherniak R, Reiss E, Slodki ME, Plattner RD, Blumer SO, 1980. Structure and antigenic activity of the capsular polysaccharides of *Cryptococcus neoformans* serotype A. Mol Immunol 17:1025–1032.

Cheson BD, Curnette JT, Babior BM, 1977. The oxidative killing mechanism of the neutrophil. Prog Clin Immunol 3:1–66.

Colonna WJ, Lampen JO, 1974a. Structure of the mannan from *Saccharomyces* strain FH4C, a mutant constituitive for invertase biosynthesis. I. Significance of phosphate to the structure and refractoriness of the molecule. Biochemistry 13:2741–2748.

Colonna WJ, Lampen JO, 1974b. Structure of the mannan from *Saccharomyces* strain FH4C, a mutant constituitive for invertase biosynthesis. II. Protein moiety and components of the carbohydrate-peptide bonds. Biochemistry 13:2748–2753.

Craig JM, Farber S, 1953. The development of disseminated, visceral mycosis during therapy for acute leukemia. Am J Pathol 29:601 (abstract).

Cutler JE, 1976. Acute systemic candidiasis in normal and congenitally thymus-deficient (nude) mice. J Reticuloendothel Soc 19:121–124.

Davies RR, Denning RJV, 1972. *Candida albicans* and the fungicidal activity of the blood. Sabouraudia 10:301–312.

de Repentigny L, Kuykendall RJ, Chandler FW, Broderson JR, Reiss E, 1984. Comparison of serum mannan, arabinitol, and mannose in experimental disseminated candidiasis. J Clin Microbiol 19:804–812.

de Repentigny L, Marr LD, Keller JW, Carter AW, Kuykendall RJ, Kaufman L, Reiss E, 1985. Comparison of enzyme immunoassay and gas-liquid chromatography for the rapid diagnosis of invasive candidiasis in cancer patients. J Clin Microbiol 21:972–979.

de Repentigny L, Reiss E, 1984. Current trends in immunodiagnosis of candidiasis and aspergillosis. Rev Infect Dis 6:301–312.

Diamond RD, Haudenschild CC, 1981. Monocyte-mediated serum-independent damage to hyphal and pseudohyphal forms of *Candida albicans* in vitro. J Clin Invest 67:173–183.

Diamond RD, Krzesicki R, 1978. Mechanisms of attachment of neutrophils to *Candida albicans* pseudohyphae in the absence of serum, and of subsequent damage to pseudohyphae by microbicidal processes of neutrophils in vitro. J Clin Invest 61:360–369.

Diamond RD, Krzesicki R, Jao W, 1978. Damage to pseudohyphal forms of *Candida albicans* by neutrophils in the absence of serum in vitro. J Clin Invest 61:349–358.

Djaczenko W, Cassone A, 1971. Visualization of new ultrastructural components in the cell wall of *Candida albicans* with fixatives containing TAPO. J Cell Biol 52:186–190.

Domer JE, Moser SA, 1978. Experimental murine candidiasis–cell mediated immunity after cutaneous challenge. Infect Immun 20:88–98.

Edwards JE, Foos RY, Montgomerie JZ, Guze LB, 1974. Ocular manifestations of candida septicemia: review of seventy-six cases of hematogenous *Candida* endophthalmitis. Medicine 53:47–75.

Ellsworth JH, Reiss E, Bradley RL, Chmel H, Armstrong D, 1977. Comparative serological and cutaneous reactivity of candidal cytoplasmic proteins and mannan separated by affinity for concanavalin A. J Clin Microbiol 5:91–99.

Elorza MV, Rico H, Gozalbo D, Sentandreu R, 1983. Cell wall composition and protoplast regeneration in *Candida albicans*. Antonie van Leeuwenhoek J 49:457–469.

Farkas V, 1979. Biosynthesis of cell walls of fungi. Microbiol Rev 43:117–144.

Ferguson AC, Kershnar HE, Collin WK, Stiehm ER, 1977. Correlation of cutaneous hypersensitivity with lymphocytic response to *Candida albicans*. Am J Clin Pathol 68:499–504.

Fieser LF, 1957. Experiments in Organic Chemistry, 3rd edition, rev. 360 pp. Boston: D.C. Heath and Co.

Fischer A, Ballet J-J, Griscelli C, 1978. Specific inhibition of in vitro *Candida* induced lymphocyte proliferation by polysaccharidic antigen present in the serum of patients with chronic mucocutaneous candidiasis. J Clin Invest 62:1005–1013.

Fischer A, Pichat L, Audinot M, Griscelli C, 1982. Defective handling of mannan by monocytes in patients with chronic mucocutaneous candidiasis resulting in a specific cellular unresponsiveness. Clin Exp Immunol 47:653–660.

Friis J, Ottolenghi P, 1970. The genetically determined binding of Alcian blue by a minor fraction of yeast cell walls. Compt Rend Trav Lab Carlsberg 37:327–341.

Fukazawa Y, Nishikawa A, Suzuki M, Shinoda T, 1980. Immunochemical basis of serologic specificity of the yeast: immunochemical determinants of several antigenic factors of yeasts. Zentralbl Bakteriol Mikrobiol [Suppl] 8:127–136.

Funayama M, Nishikawa A, Shinoda T, Fukazawa Y, 1983. Immunochemical determinant of *Candida parapsilosis*. Carbohydr Res 117:229–239.

Gentry LO, McNitt TR, Kaufman L, 1978. Use and value of serologic tests for the diagnosis of systemic candidiasis in cancer patients: a prospective study of 146 patients. Curr Microbiol 1:239–242.

Giger DK, Domer JE, McQuitty JT Jr, 1978a. Experimental murine candidiasis: pathological and immune responses to cutaneous inoculation with *Candida albicans*. Infect Immun 19:496–509.

Giger DK, Domer JE, Moser S, McQuitty JT Jr, 1978b. Experimental murine candidiasis: pathological and immune responses in T lymphocyte depleted mice. Infect Immun 21:729–737.

Glew RH, Buckley HR, Rosen HM, Moellering RC Jr, Fischer JE, 1978. Serologic tests in the diagnosis of systemic candidiasis. Enhanced diagnostic accuracy with crossed immunoelectrophoresis. Am J Med 64:586–591.

Gold JWM, Wong B, Bernard EM, Kiehn TE, Armstrong D, 1983. Serum arabinitol concentrations and arabinitol/creatinine ratios in invasive candidiasis. J Infect Dis 147:504–513.

Gorin PAJ, 1973. The position of phosphate groups in the phosphonomannan of *Hansenula capsulata* as determined by carbon 13 magnetic resonance spectroscopy. Can J Chem 51:2105–2109.

Gorin PAJ, Spencer JFT, 1968. Galactomannans of *Trichosporon fermentans* and other yeasts. Proton magnetic resonance and chemical studies. Can J Chem 46:2299–2304.

Greenfield RA, Jones JM, 1981. Purification and characterization of a major cytoplasmic antigen of *Candida albicans*. Infect Immun 34:469–477.

Greenfield RA, Stephens JL, Bussey MJ, Jones JM, 1983. Quantitation of antibody to *Candida* mannan by enzyme linked immunosorbent assay. J Lab Clin Med 101:758–771.

Harding SA, Brody JP, Normansell DE, 1980. Antigenemia detected by enzyme-linked immunosorbent assay in rabbits with systemic candidiasis. J Lab Clin Med 95:959–966.

Hasenclever HF, Mitchell WO, 1961. Antigenic studies of *Candida* I. Observation of two antigenic groups in *Candida albicans*. J Bacteriol 82:570–573.

Hasenclever HF, Mitchell WO, 1964. Immunochemical studies on polysaccharides of yeasts. J Immunol 93:763–771.

Hasenclever HF, McAtee FJ, 1977. Antigenic relationships of *Candida albicans*, *Saccharomyces telluris* and *Saccharomyces cerevisiae*, pp. 126–137. In Host–Parasite Relationships in Systemic Mycoses, Proceedings of the 21st OHOLO Biological Conference, Ma'alot, Israel, A.M. Beemer, et al, eds. Basel: S. Karger.

Hay RJ, Wells RS, Clayton YM, Wingfield HJ, 1980. Treatment of chronic mucocutaneous candidosis with ketoconazole: a study of 12 cases. Rev Infect Dis 2:600–605.

Hector RF, Domer JE, Carrow EW, 1982. Immune responses to *Candida albicans* in genetically distinct mice. Infect Immun 38:1020–1028.

Horisberger M, Rosset J, 1976. Localization of wheat germ agglutinin receptor sites on yeast cells by scanning electron microscopy. Experientia (Basel) 32:998–1000.

Humphrey DM, Weiner MH, 1983. Candidal antigen detection in pulmonary candidiasis. Am J Med 74:630–640.

Hurley DL, Barlow AJE, Fauci AS, 1975. Experimental disseminated candidiasis II. Administration of glucocorticosteroids, susceptibility to infection, and immunity. J Infect Dis 132:393–398.

Hurtrel B, Langrange PH, Michel JC, 1981. Absence of correlation between delayed-type hypersensitivity and protection in experimental systemic candidiasis in immunized mice. Infect Immun 31:95–101.

Jones GH, Ballou CE, 1969a. Studies on the structure of yeast mannan. I. Purification and some properties of an α-mannosidase from an *Arthrobacter* species. J Biol Chem 244:1043–1051.

Jones GH, Ballou CE, 1969b. Studies on the structure of yeast mannan. II. Mode of action of the *Arthrobacter* α-mannosidase on yeast mannan. J Biol Chem 244:1052–1059.

Jones JM, 1980. Quantitation of antibody against cell wall mannan and a major cytoplasmic antigen of *Candida* in rabbits, mice and humans. Infect Immun 30:78–89.

Jorizzo JL, Sams WM, Jegasothy B, Olansky AJ, 1980. Cimetidine as an immunomodulator: chronic mucocutaneous candidiasis as a model. Ann Intern Med 92:192–195.

Kerkering TM, Espinell-Ingroff A, Shadomy S, 1979. Detection of *Candida* antigenemia by counterimmunoelectrophoresis in patients with invasive candidiasis. J Infect Dis 140:659–664.

Kind LS, Kaushal PK, Drury P, 1972. Fatal anaphylaxis-like reaction induced by yeast mannans in non-sensitized mice. Infect Immun 5:180–182.

King RD, Lee JC, Morris AL, 1980. Adherence of *Candida albicans* and other *Candida* species to mucosal epithelial cells. Infect Immun 27:667–674.

Kirkpatrick CH, Ottenson EA, Smith TK, Wells SA, Burdick JF, 1976. Reconstitution of defective cellular immunity with fetal thymus and dialysable transfer factor. Long term studies in a patient with chronic mucocutaneous candidiasis. Clin Exp Immun 23:414–428.

Kirkpatrick CH, Rich RR, Bennett JE, 1971. Chronic mucocutaneous candidiasis: model building in cellular immunity. Ann Intern Med 74:955–978.

Kirkpatrick CH, Smith TK, 1974. Chronic mucocutaneous candidiasis: immunologic and antibiotic therapy. Ann Intern Med 80:310–320.

Kocourek J, Ballou CE, 1969. Method for fingerprinting yeast cell wall mannans. J Bacteriol 100:1175–1181.

Korn ED, Northcote DH, 1960. Physical and chemical properties of polysaccharides and glycoproteins of the yeast cell wall. Biochem J 75:12–17.

Lechevalier MP, 1972. Description of a new species, *Oerskovia xanthineolytica*, and emendation of *Oerskovia* Prauser et al. Int J Systematic Bacteriol 22:260–264.

Lee JC, King RD, 1983. Characterization of *Candida albicans* adherence to human vaginal epithelial cells in vitro. Infect Immun 41:1024–1030.

Lee KL, Buckley HR, Campbell CC, 1975. An amino acid synthetic medium for the development of mycelial and yeast forms of *Candida albicans*. Sabouraudia 13:148–153.

Lehmann PF, Reiss E, 1980a. Detection of *Candida albicans* mannan by immunodiffusion, counterimmunoelectrophoresis and enzyme-linked immunoassay. Mycopathol 70:83–88.

Lehmann PF, Reiss E, 1980b. Comparison of serum anti-*Candida albicans* mannan IgG levels in ELISA tests of a normal population and diseased patients. Mycopathol 70:89–93.

Lehrer RI, 1970. Measurement of candidacidal activity of specific leukocyte types in mixed cell populations I. Normal, myeloperoxidase-deficient, and chronic granulomatous disease neutrophils. Infect Immun 2:42–47.

Lehrer RI, 1975. The fungicidal mechanisms of human monocytes I. Evidence for myeloperoxidase-linked and myeloperoxidase independent candidacidal mechanisms. J Clin Invest 55:338–346.

Lehrer RI, Cline MJ, 1969a. Leukocyte myeloperoxidase deficiency and disseminated candidiasis: the role of myeloperoxidase in resistance to *Candida* infection. J Clin Invest 48:1478–1488.

Lehrer RI, Cline MJ, 1969b. Interaction of *Candida albicans* with human leukocytes and serum. J Bacteriol 98:996–1004.

Lehrer RI, Ladra KM, Hake RB, 1975. Nonoxidative fungicidal mechanisms of mammalian granulocytes. Demonstration of components with candidacidal activity in human, rabbit, and guinea pig leukocytes. Infect Immun 11:1226–1234.

Leijh PCJ, van den Barselaar MT, van Furth R, 1977. Kinetics of phagocytosis and intracellular killing of *Candida albicans* by human granulocytes and monocytes. Infect Immun 17:313–315.

Leon MA, Young NM, 1970. Concanavalin A: a reagent for the sensitization of erythrocytes with glycoproteins and polysaccharides. J Immunol 104:1556–1557.

Lew MA, Siber GR, Donahue DM, Maiorca F, 1982. Enhanced detection with an enzyme-linked immunosorbent assay of *Candida* mannan in antibody-containing serum after heat extraction. J Infect Dis 145:45–56.

Lloyd KO, 1970. Isolation, characterization and partial structure of peptido galactomannans from the yeast form of *Cladosporium werneckii*. Biochemistry 9:3446–3470.

Macdonald F, Odds FC, 1980a. Purified *Candida albicans* proteinase in the serological diagnosis of systemic candidosis. JAMA 243:2409–2411.

Macdonald F, Odds FC, 1980b. Inducible proteinase of *Candida albicans* in diagnostic serology and in the pathogenesis of systemic candidosis. J Med Microbiol 13:423–435.

Macdonald F, Odds FC, 1983. Virulence for mice of a proteinase-secreting strain of *Candida albicans* and a proteinase-deficient mutant. J Gen Microbiol 129:431–438.

Manning M, Mitchell TG, 1980a. Strain variation and morphogenesis of yeast and mycelial phase *Candida albicans* in low sulfate, synthetic medium. J Bacteriol 142:714–719.

Manning M, Mitchell TG, 1980b. Analysis of cytoplasmic antigens of the yeast and mycelial phases of *Candida albicans* by two-dimensional electrophoresis. Infect Immun 30:484–495.

Meckstroth KL, Reiss E, Keller JW, Kaufman L, 1981. Detection of antibodies and antigenemia in leukemic patients with candidiasis by enzyme-linked immunosorbent assay. J Infect Dis 144:24–32.

Meister H, Heymer B, Schäfer H, Haferkamp O, 1977. Role of *Candida albicans* in granulomatous tissue reactions II. In vivo degradation of *C albicans* in hepatic macrophages of mice. J Infect Dis 135:235–242.

Meunier-Carpentier F, Armstrong D, 1981. *Candida* antigenemia, as detected by passive hemagglutination inhibition, in patients with disseminated candidiasis or *Candida* colonization. J Clin Microbiol 13:10–14.

Meunier-Carpentier F, Kiehn TE, Armstrong D, 1981. Fungemia in the immunocompromised host: changing patterns, antigenemia, high mortality. Am J Med 71:363–370.

Miyake T, Takeya K, Nomoto K, Muraoka J, 1977. Cellular elements in the resistance to *Candida* infection in mice I. Contribution of T lymphocytes and phagocytes at various stages of infection. Microbiol Immun 21:703–725.

Monson TP, Wilkinson KP, 1981. Mannose in body fluids as an indicator of invasive candidiasis. J Clin Microbiol 14:557–562.

Morelli R, Rosenberg LT, 1971a. Role of complement during experimental *Candida* infection in mice. Infect Immun 3:521–523.

Morelli R, Rosenberg LT, 1971b. The role of complement in the phagocytosis of *Candida albicans* by mouse peripheral blood leukocytes. J Immunol 7:476–480.

Moser SA, Domer JE, 1980. Effects of cyclophosphamide on murine candidiasis. Infect Immun 27:376–386.

Myerowitz RL, Pazin GJ, Allen CM, 1977. Disseminated candidiasis. Changes in incidence, underlying diseases, and pathology. Am J Clin Pathol 68:29–38.

Nakajima H, Itoh N, Kawasaki T, Yamashina I, 1979. Reaction of antimannan antibodies with oligomannosides and glycopeptides. J Biochem (Tokyo) 85:209–216.

Nakajima T, Ballou CE, 1974a. Characterization of the carbohydrate fragments obtained from *Saccharomyces cerevisiae* mannan by alkaline degradation. J Biol Chem 249:7679–7684.

Nakajima T, Ballou CE, 1974b. Structure of the linkage region between the polysaccharide and protein parts of *Saccharomyces cerevisiae* mannan. J Biol Chem 249:7685–7694.

Nishikawa T, Harada S, Hatano H, Fukazawa Y, Tsuchiya T, 1970. Biological and serological properties of *Candida* isolated from cutaneous candidiasis. Jpn J Med Mycol 11:120–124.

Nosal R, Novotny J, Sikl D, 1974. The effect of glycoprotein from *Candida albicans* on isolated rat mast cells. Toxicon 12:103–108.

Odds FC, 1979. *Candida* and Candidosis. 382 pp. Baltimore, MD: University Park Press.

Odds FC, Hierholzer JC, 1973. Purification and properties of a glycoprotein acid phosphatase from *Candida albicans*. J Bacteriol 114:257–266.

Oh MK, Rodey GE, Good RA, Chilgren RA, Quie PG, 1969. Defective candidacidal capacity of polymorphonuclear leukocytes in chronic granulomatous disease of childhood. J Pediatr 75:300–302.

Okubo Y, Honma Y, Suzuki S, 1979. Relationship between phosphate content and serological activities of the mannans of *Candida albicans* strains NIH A-207, NIH B792, and J-1012. J Bacteriol 137:677–680.

Okubo Y, Ichikawa T, Suzuki S, 1980. Immunochemistry of *Candida albicans* mannan. pp.

96–111. In Fungal Polysaccharides. Sandford PA, Matsuda K, eds. Washington DC: American Chemical Society.

Patterson-Delafield J, Martinez RJ, Lehrer RI, 1980. Microbicidal cationic proteins in rabbit alveolar macrophages: a potential host defense mechanism. Infect Immun 30:180–192.

Pearsall NN, Adams BL, Bunni R, 1978. Immunologic responses to *Candida albicans*. III. Effects of passive transfer of lymphoid cells or serum on murine candidiasis. J Immunol 120:1176–1180.

Peat S, Whelan WJ, Edwards TE, 1961. Polysaccharides of baker's yeast IV. Mannan. J Chem Soc (London) 1:29–34.

Piccolella E, Lombardi G, Morelli R, 1980. Human lymphocyte-activating properties of a purified polysaccharide from *Candida albicans*. B and T cell cooperation in the mitogenic response. J Immunol 125:2082–2088.

Piccolella E, Lombardi G, Morelli R, 1981a. Generation of suppressor cells in the response of human lymphocytes to a polysaccharide from *Candida albicans*. J Immunol 126:2151–2155.

Piccolella E, Lombardi G, Morelli R, 1981b. Mitogenic response of human peripheral blood lymphocytes to a purified *C albicans* polysaccharide fraction: lack of helper activities is responsible for the in vitro unresponsiveness to a second antigenic challenge. J Immunol 126:2156–2160.

Poor AH, Cutler JE, 1979. Partially purified antibodies used in a solid phase radioimmunoassay for detecting candidal antigenemia. J Clin Microbiol 9:362–368.

Poulain D, Tronchin G, Dubremetz JF, Biguet J, 1978. Ultrastructure of the cell wall of *Candida albicans* blastospores: study of its constitutive layers by use of a cytochemical technique revealing polysaccharides. Annales de Microbiologie (Institut Pasteur) 129:141–153.

Prevost E, Bannister E, 1981. Yeast septicemia: detection by biphasic and radiometric methods. J Clin Microbiol 13:655–660.

Pugh D, Cawson RA, 1978. The surface layer of *Candida albicans*. Microbios 23:19–23.

Raschke WC, Ballou CE, 1972. Characterization of a yeast mannan containing N-acetyl-D-glucosamine as an immunochemical determinant. Biochemistry 11:3807–3816.

Ray TL, Hanson A, Ray LF, Wuepper KD, 1979. Purification of a mannan from *Candida albicans* which activates serum complement. J Invest Dermatol 73:269–274.

Reiss E, Patterson DG, Yert LW, Holler JS, Ibrahim BK, 1981. Structural analysis of mannans from *Candida albicans* serotypes A and B and from *Torulopsis glabrata* by methylation gas chromatography mass spectrometry and exo-α-mannanase. Biomed Mass Spectrom 8:252–255.

Reiss E, Stockman L, Kuykendall RJ, Smith SJ, 1982. Dissociation of mannan-serum complexes and detection of *Candida albicans* mannan by enzyme immunoassay variations. Clin Chem 28:306–310.

Reiss E, Stone SH, Hasenclever HF, 1974. Serological and cellular immune activity of peptidoglucomannan fractions of *Candida albicans* cell walls. Infect Immun 9:881–890.

Remold H, Fasold H, Staib F, 1968. Purification and characterization of a proteolytic enzyme from *Candida albicans*. Biochim Biophys Acta 167:399–406.

Rippon JW, 1982. Medical Mycology—The Pathogenic Fungi and the Pathogenic Actinomycetes 2nd edition. Philadelphia, PA: W.B. Saunders Co.

Roberts GD, Horstmeier C, Hall M, Washington JA II, 1975. Recovery of yeast from vented blood culture bottles. J Clin Microbiol 2:18–20.

Rocklin RE, 1976. Modulation of cellular immune responses in vivo and in vitro by histamine receptor-bearing lymphocytes. J Clin Invest 57:1051–1058.

Rogers TJ, Balish E, 1978. Immunity to experimental renal candidiasis in rats. Infect Immun 19:737–740.

Roos D, 1980. The metabolic response to phagocytosis. pp 338–385. In The Cell Biology of Inflammation, Vol. 2 G. Weissmann, ed. Amsterdam: Elsevier-North Holland.

Rüchel R, 1981. Properties of a purified proteinase from the yeast *Candida albicans*. Biochim Biophys Acta 659:99–113.

Rüchel R, 1983a. On the renin-like activity of *Candida* proteinases and activation of blood coagulation in vitro. Zentralbl Bakteriol Mikrobiol Hyg (A) 255:368–379.

Rüchel R, 1983b. On the role of proteinases from *Candida albicans* in the pathogenesis of acronecrosis. Zentralbl Bakteriol Mikrobiol Hyg (A) 255:524–536.

Rüchel R, Böning B, 1983. Detection of *Candida* proteinase by enzyme immunoassay and interaction of the enzyme with α-2-macroglobulin. J Immunol Meth 61:107–116.

Rüchel R, Uhlemann K, Böning B, 1983. Secretion of acid proteinases by different species of the genus *Candida*. Zentralbl Bakteriol Mikrobiol Hyg (A) 255:537–548.

Ruiz-Arguelles A, Jett JR, Ritts RE Jr, 1983. Impaired generation of helper T cells in a patient with chronic mucocutaneous candidiasis and malignant thymoma. J Clin Lab Immunol 10:165–169.

Scheld WM, Brown RS Jr, Harding SA, Sande MA, 1980. Detection of circulating antigen in experimental *Candida albicans* endocarditis by an enzyme-linked immunosorbent assay. J Clin Microbiol 12:679–683.

Schumacher HR, Ginns DA, Warren WJ, 1964. Fungus infection complicating leukemia. Am J Med Sci 247:313–323.

Scott JH, Schekman R, 1980. Lyticase: endoglucanase and protease activities that act together in yeast cell lysis. J Bacteriol 142:414–423.

Segal E, Berg RA, Pizzo PA, Bennett JE, 1979. Detection of *Candida* antigen in sera of patients with candidiasis by an enzyme linked immunosorbent assay inhibition technique. J Clin Microbiol 10:116–118.

Seymour FR, Slodki ME, Plattner RD, Stodola RM, 1976. Methylation and acetolysis of extracellular D-mannans from yeast. Carbohydr Res 48:225–237.

Skerl KG, Scheld WM, Allegro GA, Calderone RA, 1980. Lymphocyte blastogenesis during experimental endocarditis caused by *Candida albicans*. J Reticuloendothel Soc 28:495–505.

Slodki ME, 1980. Structural aspects of exocellular yeast polysaccharides. pp. 183–196. In Fungal Polysaccharides American Chemical Society Symposium Series no. 126. Sandford PA, Matsuda K, eds. Washington, DC: American Chemical Society.

Sohnle PG, Frank MM, Kirkpatrick CH, 1976a. Deposition of complement components in the cutaneous lesions of chronic mucocutaneous candidiasis. Clin Immunol Immunopathol 5:340–350.

Sohnle PG, Frank MM, Kirkpatrick CH, 1976b. Mechanisms involved in elimination of organisms from experimental cutaneous *Candida albicans* infections in guinea pigs. J Immunol 117:523–530.

Staib F, 1965. Serum-proteins as nitrogen source for yeast-like fungi. Sabouraudia 4:187–193.

Stewart TS, Ballou CE, 1968. A comparison of yeast mannans and phosphomannans by acetolysis. Biochemistry 7:1855–1863.

Stewart TS, Mendershausen PB, Ballou CE, 1968. Preparation of a mannopentaose, mannohexaose and mannoheptaose from *Saccharomyces cerevisiae* mannan. Biochemistry 7:1843–1854.

Summers DF, Grollman AP, Hasenclever HF, 1964. Polysaccharide antigens of the *Candida* cell wall. J Immunol 92:491–499.

Sunayama H, 1970. Studies on the antigenic activities of yeasts. IV. Analysis of the antigenic determinant groups of the mannan of *Candida albicans* serotype A. Jpn J Microbiol 14:27–39.

Sunayama H, Suzuki S, 1970. Studies on the antigenic activities of yeasts. VI. Analysis of the antigenic determinants of the mannan of *Candida albicans* serotype B-792. Jpn J Microbiol 14:371–379.

Suzuki M, Fukazawa Y, 1982. Immunochemical characterization of *Candida albicans* cell wall antigens: specific determinant of *Candida albicans* serotype A mannan. Microbiol Immunol 26:387–402.

Suzuki S, Suzuki M, Yokota K, Sunayama H, Sakaguchi O, 1967. On the immunochemical and biochemical studies of fungi XI. Cross reactions of the polysaccharides of *Aspergillus fumigatus, Candida albicans, Saccharomyces cerevisiae,* and *Trichophyton rubrum* against *Candida albicans* and *Saccharomyces cerevisiae* antisera. Jpn J Microbiol 11:269–273.

Svec P, 1974. On the mechanism of action of glycoprotein from *Candida albicans*. J Hyg Epidemiol Microbiol Immun (Prague) 18:373–376.

Syverson RE, Buckley HR, Gibian JR, 1978. Increasing the predictive value positive of the precipitin test for the diagnosis of deep-seated candidiasis. Am J Clin Path 70:826–831.

Tabeta H, Mikami Y, Abe F, Ommura Y, Arai T, 1984. Studies on defense mechanisms against *Candida albicans* infection in congenitally athymic nude (nu/nu) mice. Mycopathol 84:107–113.

Takeda N, Okubo Y, Ichikawa T, Suzuki S, 1979. Cross-reaction between the mycelial galactomannans of three *Hormodendrum* strains and the mannans of two *Candida albicans* strains of different serotypes, A and B. Infect Immun 23:146–149.

Thieme TR, Ballou CE, 1972. Subunit structure of the phosphomannan from *Kloeckera brevis* yeast cell wall. Biochemistry 11:1115–1120.

Thiery JP, Rambourg A, 1974. Cytochimie des polysaccharides. J Micr (Paris) 21:225–232.

Torres-Bauza LJ, Riggsby WS, 1980. Protoplasts from yeast and mycelial forms of *Candida albicans*. J Gen Microbiol 119:341–349.

Trinci APJ, 1978. Wall and hyphal growth Sci Prog Oxford 65:75–99.

Tsuchiya T, Fukazawa Y, Taguchi M, Nakase M, Shinoda T, 1974. Serologic aspects of yeast classification. Mycopathol 53:77–91.

Valdimarsson H, Higgs J, Wells R, Yamamura M, Hobbs J, Holt P, 1973. Immune abnormalities associated with chronic mucocutaneous candidiasis. Cell Immunol 6:348–361.

van Scoy RE, Hill HR, Ritts RE Jr, Quie PG, 1975. Familial neutrophil chemotaxis defect, recurrent bacterial infections, mucocutaneous candidiasis and hyperimmunoglobulinemia E. Ann Intern Med 82:766–771.

Warren RC, Bartlett A, Bidwell DE, Richardson MD, Voller A, White LO, 1977. Diagnosis of invasive candidosis by enzyme immunoassay of serum antigen. Br Med J 1:1183–1185.

Weiner MH, Coats-Stephen M, 1979. Immunodiagnosis of systemic candidiasis: mannanemia detected by radioimmunoassay in experimental and human infections. J Infect Dis 140:989–993.

Weiner MH, Yount WJ, 1976. Mannan antigenemia in the diagnosis of invasive *Candida* infections. J Clin Invest 58:1045–1053.

Wells RS, Higgs JM, MacDonald A, 1972. Familial chronic mucocutaneous candidiasis. J Med Genet 9:302–310.

Williams D, Cook J, Hoffman E, DiLuzio N, 1978. Protective effect of glucan in experimentally induced candidiasis. J Reticuloendothel Soc 23:479–490.

Wilton JMA, Lehner T, 1980. Immunology of candidiasis. Comprehensive Immunol 8:525–559.

Wingard JR, Merz WG, Saral R, 1979. *Candida tropicalis*: a major pathogen in immunocompromised patients. Ann Intern Med 91:539–543.

Witkin SS, Yu IR, Ledger WJ, 1983. Inhibition of *Candida albicans*-induced lymphocyte proliferation by lymphocytes and sera from women with recurrent vaginitis. Am J Obstet Gynecol 147:809–811.

Yu RJ, Bishop CT, Cooper FP, Hasenclever HF, Blank F, 1967a. Structural studies of mannans from *Candida albicans* (serotypes A and B), *Candida parapsilosis, Candida stellatoidea* and *Candida tropicalis*. Can J Chem 45:2205–2211.

Yu RJ, Bishop CT, Cooper FP, Blank F, Hasenclever HF, 1967b. Glucans from *Candida albicans* (serotype B) and from *Candida parapsilosis*. Can J Chem 45:2264–2267.

Zhang WJ, Ballou CE, 1981. *Saccharomyces kluyveri* cell wall mannoprotein. J Biol Chem 256:10073–10079.

Chapter 12

Cryptococcus neoformans

12.1 INTRODUCTION

Cryptococcus neoformans is an encapsulated yeast that reproduces by multipolar budding. It is world-wide in distribution in the nests and accumulated droppings of pigeons. The yeast cells are typically 4 to 10 μm in diameter, but the outside diameter is dependent on the thickness of the capsule which varies from barely detectable to greater than 4 μm. Kwon-Chung (1975) discovered that compatible mating types complete a basidiomycetous sexual cycle. The perfect state *Filobasidiella neoformans* var *neoformans* is produced from A and D serotype isolates, whereas mating pairs of B and C serotypes form *F neoformans* var *bacillispora*. The corresponding asexual stages are referred to as *C neoformans* and *C neoformans* var *gattii* (Kwon-Chung et al, 1982).

Cryptocccus neoformans infections occur after inhalation of dust containing viable cells. Although there are many opportunities for exposure, clinical illness is uncommon. The incidence of cryptococcal meningitis in the USA for the year 1976 was 338 human cases reported to the Centers for Disease Control (Kaufman and Blumer, 1977). In the United Kingdom and Eire the incidence is lower, about eight cases per year reported to the London School of Hygiene (Hay et al, 1980) but this figure probably underestimates the true incidence.

After inhalation of the yeast, a local, and usually transitory, pulmonary infection develops which heals leaving a cryptococcal granuloma (= cryptococcoma) or heals without a residue. Sometimes the pulmonary phase may develop into a pneumonia, but if dissemination occurs the central nervous system is usually involved, and less frequently the joints, bones, eyes, and skin. The yeast spreads hematogenously, crosses the blood-brain barrier, and grows in the spinal meninges, the subarachnoid space, or forms cystic masses in the brain.

Cryptococcosis is an opportunistic mycosis and about half of those infected have an underlying condition that compromises their immune system. Meningitis caused by *C neoformans* has been reported in patients receiving systemic prednisone therapy for management of renal allografts (Schroeter et al, 1976). Hematologic malignancy, Hodgkin's lymphoma, acquired immunodeficiency disease syndrome (AIDS), and autoimmune disease such as lupus erythematosus and autoimmune hemolytic anemia are some of the other predisposing conditions (Schupbach et al, 1976; Diamond, 1981).

The tissue reaction of the host varies as a function of the anatomical site involved, the degree of encapsulation of the isolate, and the host's immune status. The immune response in the central nervous system tends to be hyporeactive, which is consistent with the view that the brain is an immunoprivileged site. Masses of cryptococci are surrounded by a minimal inflammation consisting of moderate perivascular cuffing with mononuclear cells and small foci of mixed inflammatory cells (Chandler et al, 1980). In some cases a granulomatous reaction is present, as evidence of a functioning immune system, occurring particularly when the infecting isolate is lightly encapsulated and more often in the lungs than the brain. Cryptococcomas contain intracellular cryptococci within multinucleate giant cells. Macrophages and variable numbers of polymorphonuclear neutrophils are also present.

12.2 THE CAPSULE AS AN AGGRESSIN

There is much evidence supporting the idea that the acidic polysaccharide capsule is a determinant of pathogenicity in *C neoformans*. It is a physical and chemical barrier to phagocytosis. The capsular polysaccharide persists in body fluids and is recalcitrant to digestion by host enzymes. Immunologically, the major capsular antigen is T-independent and subject to tolerance. Stable, acapsular mutants are poorly pathogenic. There are probably other virulence factors for *C neoformans*, but the capsule is so conspicuous in this yeast, as is the case in many bacterial agents of meningitis, that it has been the subject of many investigations.

12.2.1 Physical Barrier

The capsule is a viscous, acidic polysaccharide that varies in diameter depending on the isolate from barely detectable to greater than 4 μm (Staib et al, 1977). Cell diameters of large capsule forms have been reported in the range of 14 to 70μm (Aronson and Kletter, 1973). Capsules of large size (greater than 4 μm) physically inhibit endophagocytosis.

12.2.2 Persistence in Body Fluids and Tissues

The detection of cryptococcal polysaccharide in cerebrospinal fluid with antibody-coated latex particles is a reliable and standard method for diagnosing *C neoformans* meningitis (Palmer et al, 1977; Kaufman and Blumer, 1977). Even though clinical isolates may appear poorly encapsulated or as dry variants when first cultured, of 632 cryptococcal meningitis patients followed through therapy, capsular polysaccharide appeared in the cerebrospinal fluid in 99% of the cases (Kaufman and Blumer, 1977). This circulating antigen persists in direct proportion to disease severity and its replacement by free antibody is a sign of favorable response to therapy (Diamond, 1981). In normal mice, cryptococcal polysaccharide is cleared from the sera in 1 or 2 weeks and is lodged in the kidney tubular cells as granules detectable by immunofluorescence for up to 14 weeks (Kozel et al, 1977). High molecular weight and apparent resistance to host glycosidases are impediments to its excretion.

12.2.3 Rodent Virulence

Encapsulated cryptococci are virulent for mice, and many isolates can be recovered from the brain after intravenous inoculation. *Cryptococcus neoformans 602* is a spontaneous and stable mutant derived from an encapsulated serotype D isolate (Kozel and Cazin, 1971). The original clinical isolate had a 60 day LD_{50} dose of 4900 yeast, yet the acapsular mutant failed to kill mice in 60 days even at doses of 10^7. No high molecular weight viscous polysaccharide was isolated from cultures of the mutant, but a small amount of lower molecular weight polysaccharide was found (Kozel and Cazin, 1971). This second polysaccharide is probably galactoxylomannan, chemically and antigenically distinct from the major polymer and unable to organize a capsule (Cherniak et al, 1982).

Acapsular variants have been selected by ultraviolet irradiation of wild type cultures (Bulmer et al, 1967). Frequently these dry variants are unstable and revert spontaneously, or because of selection pressures of the host. In the few instances where mutants remained stable during the experimental period, their pathogenicity was greatly reduced in comparison with the wild type.

Treatment of *C neoformans*-infected rats with intravenous hyaluronidase during the acute stage of the infection resulted in a marked change in the histology of cerebral and extracerebral lesions (Radhakrishnan et al, 1982). In untreated, infected control rats there was a minimal host inflammatory response, whereas a macrophage infiltrate and granuloma formation occurred in rats treated with hyaluronidase. This response was interpreted as enzymatic depolymerization of the capsule leading to immune recognition. If it is confirmed, further credence must be given to the idea that even in immunologically intact hosts, the presence of the capsule foils the normal T-cell-mediated immune response. Farmer and Komorowski (1973) have drawn attention to a weakly encapsulated clinical isolate of *C neoformans* that produced a suppurative granulomatous pneumonia. This isolate, inoculated intracerebrally in mice, also evoked a strong inflammatory response, after which the mice survived.

Pathogenicity is not a direct function of capsule diameter. Staib et al (1977) assorted 31 isolates into dry, medium, or large capsule forms and found that these categories could not predict the organ distribution after mouse infection. Some cultures tended to disseminate widely, others had a low ability to penetrate the central nervous system, and a third group had a cerebral focus reminiscent of human infection. Of the six brain-specific cultures, four had medium size capsules, one a small capsule, and one had no capsule detectable by India ink. The pathogenesis of these isolates showed early dissemination to other organs, which was controlled in 4 weeks. The yeasts persisted in the brain for longer periods in small clusters at scattered sites, termed cryptococcomas. There was little tissue reaction but when inflammatory responses finally occurred, cryptococcomas tended to heal spontaneously.

12.2.4 Tolerance

Evidence that the major viscous polysaccharide is a tolerogen is both direct and indirect. Priming mice with a greater than optimally immunogenic dose resulted in (i) suppression of splenic immunoplaque forming cells (Murphy and Cozad, 1972), (ii) suppression of circulating antibodies (Kozel et al, 1977), and (iii) increased death rates after challenge with live cryptococci (Bennett and Hasenclever, 1965). The capsular polysaccharide is a T-independent antigen resulting in IgM production (Cauley and Murphy, 1979), but IgM levels are probably regulated by suppressor T-cells (Breen and Combee, 1980).

In one report, mice primed with 1 mg of cryptococcal polysaccharide before challenge with 10^5 live *C neoformans* had higher death rates than controls (Bennett and Hasenclever, 1965). This observation could not be consistently verified and has not been reproduced. Direct splenic plaque forming cells (PFC) were suppressed in mice primed with >5 µg of polysaccharide, compared with the optimal immunogenic dose of 0.5 µg (Murphy and Cozad, 1972). This model differs from the classical low dose tolerance to microbial polysaccharides: immune paralysis to type III pneumococcal polysaccharide (SIII) (Baker, 1975). The direct PFC response to SIII is much greater and the tolerant state is durable, lasting many weeks. In the cryptococcal model thus far reported (Murphy and Cozad, 1972), higher priming doses are required for tolerance induction and termination of tolerance occurred after 7 days (Figure 12.1). The distribution of SIII in mice was in the macrophages, fibroblasts, and capillary endothelium in contrast to the lodgment of cryptococcal polysaccharide in the renal tubular cells.

Direct PFC responses to cetyltrimethylammonium bromide purified cryptococcal polysaccharide was increased in mice treated with anti-thy 1, implying that regulation by a T-cell subset, probably T-suppressor cells, operated in the *C neoformans* model as has also been shown in response to pneumococcal SIII (Breen and Combee, 1980; Baker, 1975). Suppression also occurs at the level of circulating antibodies (Kozel et al, 1977). Partial suppression of circulating anti-*C neoformans* polysaccharide immunoglobulin in mice occurred when the priming dose exceeded 50 µg, and complete suppression, lasting for more than a month, followed doses greater than 400 µg.

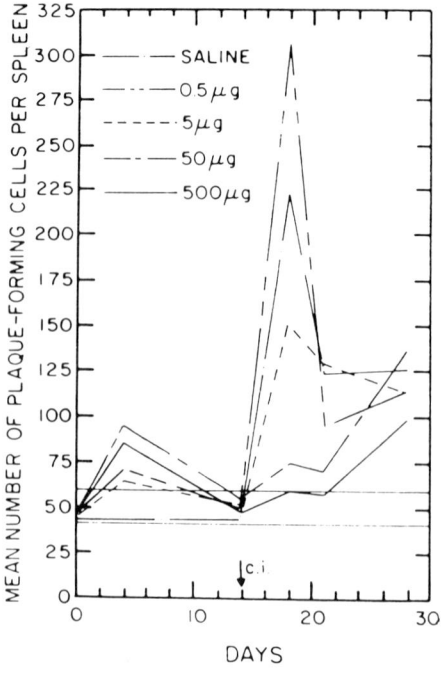

Figure 12.1. Mean numbers of hemolytic immunoplaque forming cells (PFC) in CBA/J mice primed with various amounts of cryptococcal polysaccharides on day 0. Each point is the mean of six mice per group. The horizontal continuous line represents the background PFC in saline-treated control mice. The arrow indicates challenge immunization with 0.125 µg of polysaccharides in incomplete Freund's adjuvant (Murphy and Cozad, 1972). Note the suppressive effect of priming with >5 µg of antigens. *Source:* Reprinted with permission of the authors and of the publisher, American Society for Microbiology.

T-independent IgM agglutinins reactive with capsular polysaccharide are produced in congenitally athymic nude (nu/nu) mice (Cauley and Murphy, 1979). The observed lack of inflammatory response surrounding microfoci in the brain provided indirect evidence for the capsule as tolerogen (Chandler et al, 1980).

In many of these studies no attempt was made to purify the major acidic polysaccharide, glucuronoxylomannan. It is now known that a second, immunochemically distinct polysaccharide, galactoxylomannan, is also produced, and that the responses reported are probably composite responses to at least two different antigens. For that reason, extension of investigations to measure tolerogenicity of individual exopolysaccharides will be necessary.

12.2.5 Antiphagocytic Nature

The antiphagocytic nature of the *C neoformans* capsule was described when capsular polysaccharide was added in vitro to a mixture of human PMN, autologous serum, and a stable acapsular mutant yeast (Bulmer and Sans, 1968). Phagocytosis was decreased from 78% to 28% in comparison with 18% phagocytosis of encapsulated forms. Inhibition of the adherence stage occurred because extraneous capsular polysaccharide was bound to specific receptors in the *C neoformans* cell wall (Kozel, 1977). Small amounts of polysaccharide, 1 to 10 μg per 10^6 cryptococci, are inhibitory (Kozel and Mastroianni, 1976). Phagocytosis of cryptococci by guinea pig alveolar macrophages is optimal in the presence of autologous serum, and killing of the yeast is dependent on a serum factor (Bulmer and Tacker, 1975).

Tacker et al (1972) reported that human PMN did not require serum for phagocytosis and killing of cryptococci, but that opsonization with IgG and complement increased the rate of phagocytosis. The inverse relation between capsule diameter and phagocytosis by macrophages was observed by Mitchell and Friedman (1972). Only 10 to 30 heavily encapsulated yeast were ingested by 100 rat peritoneal macrophages after 2 hours at a yeast : macrophage ratio of 5–10:1, whereas 270 to 560 of the small capsule yeasts were phagocytosed. If the exposure time was lengthened to 6 hours, phagocytosis of large capsule forms increased to a level comparable to that achieved with small capsule forms after 2 hours. Normal rat serum was essential for macrophages to phagocytose living or killed cryptococci (Mitchell and Friedman, 1972). Complement was implicated by the lability of the serum factor after incubation at 56 °C, but the possibility of alternative or classical pathway activation operating through low levels of natural cytophilic antibody was not explored. The susceptibility of ingested yeast to killing was characteristic for a particular isolate, independent of capsular size, suggesting that properties apart from capsular diameter protect cryptococci. This is consistent with the finding that organ specificity is independent of capsular size (Staib et al, 1977).

Formation of macrophage rosettes around large capsule forms, distinct from endophagocytosis, has been described in vitro and in peritoneal membrane mounts obtained after injection of viable cryptococci (Figure 12.2) (Kalina et al, 1971). The relevance of this mechanism in natural infection is unknown, but it illustrates how the capsule size hinders effective phagocytosis. Macrophage rosette formation occurred in mice inoculated intraperitoneally with cryptococci only if immune serum was also injected. The injection of cryptococci into the peritoneal cavity of rabbits and guinea pigs was followed by prompt migration of PMN which adhered to the yeasts. After 24 to 48 hours, PMN were replaced by macrophages. Immune rabbit

256 *Cryptococcus neoformans*

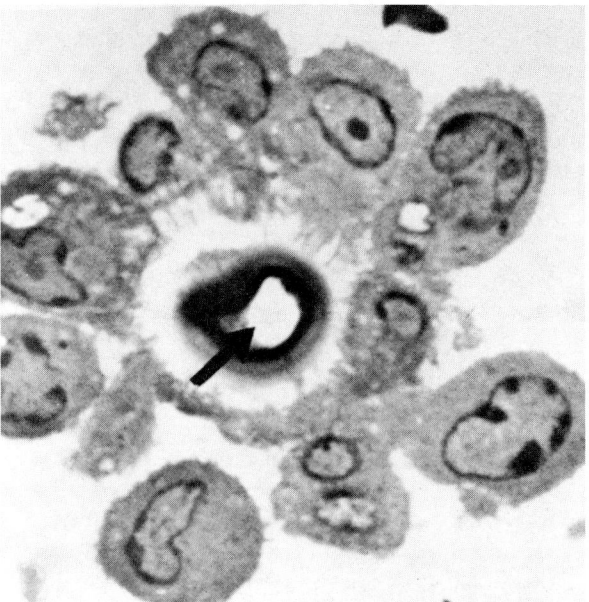

Figure 12.2. A rosette of rabbit phagocytes adhering to a large capsule form of *Cryptococcus neoformans* in vitro in the presence of 40% isologous serum. The cryptococci were heat-killed before reconstitution with leukocytes in order to demonstrate localization of acid phosphatase as a marker for lysosomal enzyme release. The dark band interior to the cell wall is the reaction product. The rosette is composed of both monocytes and polymorphonuclear neutrophils (Kalina et al, 1971).

serum was not required in these animals. In guinea pigs no further changes were observed, but in rabbits the macrophages fused into giant cells and granulomata developed. Yeasts were killed within 18 to 24 hours in rabbits, perhaps because (i) their normal body temperature (39.5 °C) was restrictive, and (ii) normal rabbit serum was inhibitory to cryptococcal replication (Aronson and Kletter, 1973). Cryptococci survived for longer times in guinea pigs, and indeed, cavies are a suitable model for studying chronic cryptococcosis (Diamond, 1977).

Adherence of macrophages to cryptococci progresses by penetration of pseudopodia through the capsule to contact the plasmalemma, with resulting dissolution of yeast intracellular organization (Aronson and Kletter, 1973). A series of studies of the interactions between murine peritoneal macrophages and acapsular mutant yeast have shown that addition of capsular polysaccharide inhibits the adherence stage of phagocytosis (Kozel and Mastroianni, 1976).

12.3 CAPSULAR AND CELL WALL HETEROGLYCANS OF *CRYPTOCOCCUS* AND *TREMELLA* SPECIES

Glucuronoxylomannan has been described as the major viscous, acidic, capsular polysaccharide of *C neoformans* (Drouhet et al, 1950; Evans and Mehl, 1951; Rebers et al, 1958; Blandamer and Danishefsky, 1966; Bhattacharjee et al, 1978; Cherniak et al, 1980) of the related saprophyte, *Cryptococcus laurentii* (Abercrombie et al, 1960; Slodki et al, 1966) and in the dimorphic yeastlike fungus *Tremella mesenterica* (Fraser et al, 1973a, 1973b). Thirty years elapsed from the composition analysis of *C neoformans* capsular material to the first definitive assignments of linkage and sequence. In the intervening years the presence of a second distinct heteroglycan was a matter of conjecture.

An early attempt was made to fractionate a "contaminating galactan" from the viscous, acidic glucuronoxylomannan of serotype A by immunoprecipitation with equine anti-*Streptococcus pneumoniae* type XIV as a probe for β-1,4-galactose (Rebers et al., 1958; Lindberg et al, 1977). The anti-type XIV precipitated only a minor portion of the bulk polysaccharide but the product was enriched in galactose. Anti-type II used as a probe for glucuronic acid was able to precipitate the cryptococcal polysaccharide, but no separation of glucuronosyl and galactosyl components was effected. It was concluded that the galactose polymer was a minor component and also contained uronic acid. This perceptive observation received no corroboration for almost 25 years. Galactose was detected in the major capsular polysaccharide of *C neoformans* serotype B (Blandamer and Danishefsky, 1966), but this was not confirmed (Bhattacharjee et al, 1980). Selective precipitation of the major, viscous glucuronoxylomannan with CTAB left a soluble galactose-containing polymer in the supernate that was an antigenic galactoxylomannan (Cherniak et al, 1982). Previous reports on the chemical and immunologic properties of unfractionated "cryptococcal polysaccharide" were thus actually composites of responses to at least two antigens.

Structural analyses of polysaccharides from morphologically similar saprophytic fungi can be useful models for the frank pathogens. This is true in the instance of *C neoformans*. Slodki et al (1966) reported that the major, viscous glucuronoxylomannan from *C laurentii* NRRL-1401 had molar ratios of xylose : mannose : glucuronic acid : O-acetyl = 1 : 4 : 1 : 1.5. The corresponding ratios in *Tremella mesenterica* were = 7 : 5 : 1 : 0.7. (Fraser et al, 1973a, 1973b). In *T mesenterica* the main chain linkage contains a 1,3-mannan with glucuronic acid residues suspended singly, and xylose occurring as disaccharide side chains. The majority of xylose residues were stripped from the polymer by heating with anionic exchange resin (Amberlite IR 120), with only minor release of glucuronic acid or mannose. As de-xylosylation proceeded, the specific rotation increased from $-26°$ to $116°$ indicating xylosyl units were β-linked.

Equine anti-*S pneumoniae* type II globulins with narrow specificity for glucuronic acid could not precipitate the *T mesenterica* glucuronoxylomannan, except after the polysaccharide was subjected to Smith degradation. In that event, periodate oxidation of xylose occurred but the 1,3-mannan backbone was resistant. Reduction and mild acid hydrolysis did not destroy glucuronic acid residues because they were protected by O-acetylation at C-3.

12.3.1 Structure of Glucuronoxylomannan

The major capsular polysaccharide of *C neoformans* is glucuronoxylomannan. Proof that serotype specificity resides in the capsule depends on the observed loss of serotype in acapsular mutants (Kozel and Cazin, 1971; Jacobson et al, 1982), but admittedly this is indirect evidence. Glucuronoxylomannans are viscous, high molecular weight (Table 12.1) essentially nitrogen-free polysaccharides with the main structural features being a linear α-1,3-mannan backbone heavily substituted with xylosyl or glucuronosyl residues. In serotypes B and C (Figure 12.3), the glucuronoxylomannans are completely substituted and every third mannose is doubly substituted at O-4 with xylosyl and at O-2 with glucuronic acid or vice versa (Bhattacharjee et al, 1979b). Serotypes A and D have some unbranched mannose residues in the backbone and no tetrasubstituted ones (Figure 12.4). For most serotypes the

258 Cryptococcus neoformans

Table 12.1. Composition of the Acidic, Capsular Glucuronoxylomannan of *C neoformans*

Serotype	Mannose	Xylose	Glucuronic Acid	O-Acetyl (%)	Reference
A	3	2	1	nd	Merrifield and Stephen, 1980
A	5	2	1	6.5	Cherniak et al, 1980
A	3.3	1.5	1	nd	Bhattacharjee et al, 1981
B	3	3	1	10.5	Bhattacharjee et al, 1980
C	3	4	1	3.0	Bhattacharjee et al, 1978, 1979b
D	3	1	1	10.3	Bhattacharjee et al, 1979a
D	4.4	1.6	1	nd	Bhattacharjee et al, 1981

Source: Reprinted from Cherniak and Reiss, 1985, with permission from the authors and from the publisher, Springer Verlag.
Abbreviation: nd, not determined.

glucuronoxylomannans were isolated from 1-week-old Sabouraud broth cultures and purified by DEAE-chromatography. *Cryptococcus neoformans* serotype A was grown on semisynthetic medium for 72 hours at 35 °C and the glucuronoxylomannan was isolated by precipitation with CTAB (Cherniak et al, 1980). The material sloughed off into the medium during growth is assumed to be identical to the capsular polysaccharide that remains adherent to the yeast cells.

Figure 12.3. Proposed repeat unit structures for glucuronoxylomannan of *Cryptococcus neoformans* var *gattii* (serotypes B and C). Bhattacharjee et al (1980) found evidence for tetrasubstituted mannose residues. *Abbreviations:* Man, mannose; Xyl, xylose; Glc A, glucuronic acid; p, pyranose. *Source:* Reprinted from Cherniak and Reiss (1986) with permission of the publisher, Springer Verlag.

SEROTYPE

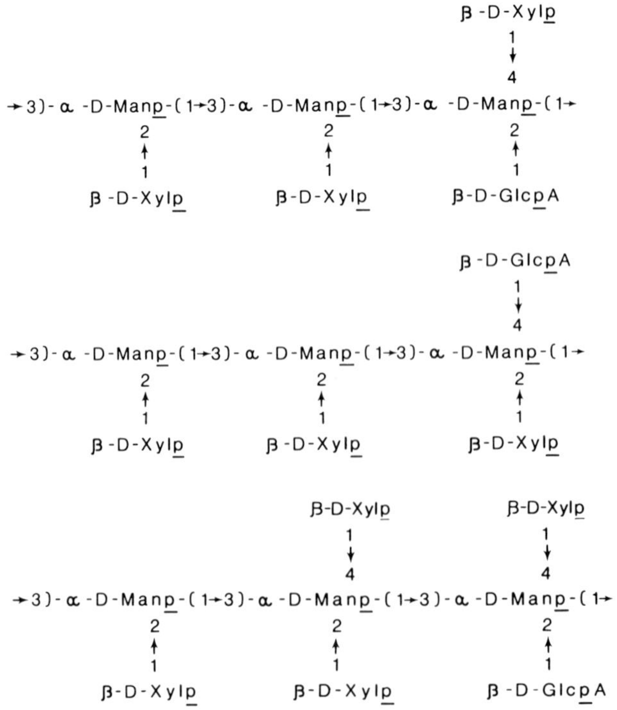

B1:
β-D-Xylp
1
↓
4
→3)-α-D-Manp-(1→3)-α-D-Manp-(1→3)-α-D-Manp-(1→
     2   2   2
     ↑   ↑   ↑
     1   1   1
   β-D-Xylp   β-D-Xylp   β-D-GlcpA

B2:
β-D-GlcpA
1
↓
4
→3)-α-D-Manp-(1→3)-α-D-Manp-(1→3)-α-D-Manp-(1→
     2   2   2
     ↑   ↑   ↑
     1   1   1
   β-D-Xylp   β-D-Xylp   β-D-Xylp

C:
             β-D-Xylp   β-D-Xylp
               1   1
               ↓   ↓
               4   4
→3)-α-D-Manp-(1→3)-α-D-Manp-(1→3)-α-D-Manp-(1→
     2   2   2
     ↑   ↑   ↑
     1   1   1
   β-D-Xylp   β-D-Xylp   β-D-GlcpA

12.3 CAPSULAR AND CELL WALL HETEROGLYCANS

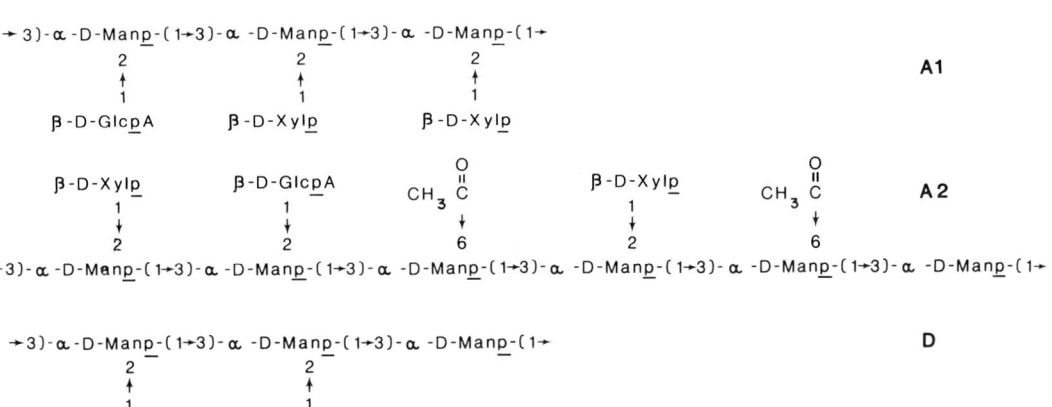

Figure 12.4. Models for the repeat unit structure of *Cryptococcus neoformans* var *neoformans* (serotypes A and D) glucuronoxylomannan. A1: Bhattacharjee et al (1981) show few, if any, unsubstituted mannosyl residues in the α-1,3-linked backbone. The structure shown as A2 is that proposed by Cherniak et al (1980) based on methylation-fragmentation analysis augmented more recently with carbon 13 nuclear magnetic resonance spectrometry (Cherniak and Reiss, 1986) that permitted assignment of O-acetyl groups to the O-6 position of mannose residues. The serotype D structure of Bhattacharjee et al (1979) differs from A only in that the backbone is less highly branched. *Source:* Reprinted from Cherniak and Reiss (1986) with permission of the publisher, Springer Verlag.

The β configuration in the side substituents was shown by the lability of xylosyl to CrO_3 oxidation (Bhattacharjee et al, 1979b) and by isolation of an acid-resistant mannobiuronic acid that was susceptible to β-glucuronidase. The α-1,3-mannan backbone was deduced from its periodate resistance, methylation at O-6, and optical rotation of the mannan resistant to Smith degradation. The A and C serotype mannans produced by Smith degradation were water-insoluble. A minor feature in the glucuronoxylomannan of serotype D was that one of fifteen mannose residues had a mannose side branch at O-2 (Bhattacharjee et al., 1979a).

Two clones were isolated from a stock culture of serotype A isolate, NIH371 (Bhattacharjee et al, 1981). This isolate had been carried in culture for 15 years and was originally isolated from cuckoo bird droppings in Thailand. The clone 371-a had a xylose : mannose : glucuronic acid ratio of 1.6 : 5.5 : 1 with 3 of 8 unbranched mannose residues in the linear α-1,3 backbone whereas for clone 371-3 the xylose : mannose : glucuronic acid was in a ratio of 1.5 : 3.3 : 1, and one in ten mannosyl residues were unsubstituted. Some differences were seen in the ^{13}C-NMR spectra of the two glucuronoxylomannans in the region $\delta \sim 102$, but because of the complexity of the signals, resonances were not assigned to individual anomeric carbons. In other respects, such as optical rotation and reciprocal immunodiffusion reactions, the polysaccharides from these two clones were indistinguishable. The authors considered clone 371-a to represent serotype D, and 371-3 to represent serotype A, although there was no discernible difference in their serologic activity and the characteristic of unsubstituted mannosyl residues — up to then considered to be a property of serotype D — were present in the A serotype. To reconcile these results, the authors suggested that a spectrum of variations in the glucuronoxylomannans exists in *C neoformans*, ranging from A to AD to D. A small quantitative difference in the amount of unsubstituted mannan is not likely to be sufficient to determine serotype specificity, especially in view of the microheterogeneity that occurs in polysaccharides resulting from substrate limitations and the possibility of action by endogenous glycosidases. If serotype is to be designated on this basis, then it should be supported by some immunologic assay, the most obvious of which would be quantitative precipitin analysis. Future studies should include recent isolates from clinical sources.

12.3.2 Structural Features of the Galactoxylomannan

At least three polysaccharides are released in the broth cultures of *C neoformans* (Cherniak et al, 1982; Turner et al, 1984). These are resolved by CTAB into the precipitable glucuronoxylomannan (88%) and the nonprecipitable galactoxylomannan complex (12%) (Table 12.2). Galactose is present only in the galactoxylomannan, thus explaining the sporadic reports of this monosaccharide in "cryptococcal polysaccharide" going back to Rebers et al (1958). The galactoxylomannan contains 2% O-acetyl and molar ratios of xylose : mannose : glucuronic acid = 1 : 1.8 : 1.9 : 0.2. The glucuronoxylomannan contains more glucuronic acid and O-acetyl and is much higher in molecular weight than galactoxylomannan. The galactoxylomannan is susceptible to periodate oxidation indicating fewer 1,3 linkages or C-2, C-4 disubstitutions. The glucuronoxylomannan and galactoxylomannan are direct opposites in their viscosity, Con A affinity, and complexation with CTAB. They are also different serologically, as seen in counterimmunoelectrophoresis and enzyme immunoassay (see below).

Fine structural analysis of the galactoxylomannan complex has not yet been accomplished and it is now recognized that Con A affinity can separate a galactoxylomannan-enriched fraction from a mannan that binds to the lectin (Turner et al,

Table 12.2. Summary Comparison of Properties of Exopolysaccharides of *C neoformans* CDC B2500 Serotype A[a]

Property	GalXM	GXM
Percent of total exopolysaccharide	12	88
Molar ratios		
Xylose	1	2
Mannose	1.8	5
Glucuronic acid	0.2	1
Galactose	1.9	0
O-Acetyl	2	6.5
Phosphate	0.4	nd[b]
Peptide	3%	1.6%
Molecular weight	$275{,}000 \pm 25{,}000$[c]	$\sim 8 \times 10^{5}$[d]
Periodate susceptibility (Smith degradation)	90% susceptible	38%
Viscosity	low	high
Con A affinity	reactive	nonreactive
Complexation with CTAB	no	yes

Source: Reprinted with permission of the authors and of Elsevier Science Publishing Co.
[a] Cherniak et al, 1980, 1982.
[b] nd, not determined.
[c] Determined by gel-permeation chromatography on Sepharose CL-6B.
[d] Determined by Bhattacharjee et al, 1979, for GXM serotype D.

1984). The mannan is enriched in phosphate (mannose to P molar ratio of 4.5 : 1) and contains 17% covalent peptide with serine and threonine predominating. The galactoxylomannan complex is also secreted by acapsular mutants (ie, cap 67). The source and function of the galactoxylomannan are matters for further research. It is possible that, although the galactoxylomannan is incapable of organizing a capsule by itself, it intercalates with the major capsular glucuronoxylomannan. Galactoxylomannan may represent the readily soluble outer wall layer, or it may be a polysaccharide like the peptidophosphogalactomannan of *Penicillium charlesii*, which is actively secreted and appears to stabilize periplasmic enzymes (Gander et al, 1980).

A major cell wall glycoprotein of *C laurentii* also contains mannose, galactose, and xylose (Raizada et al, 1975). Oligosaccharide portions of the glycoprotein were synthesized in vitro from microsomal glycoside transferases and radiolabeled sugar nucleotide precursors. The structure of the radiolabeled glycoprotein was analyzed and evidence was obtained for three separate oligosaccharide chains of different composition (Figure 12.5) connected by base-labile O-glycosidic hydroxyamino acid esters to protein. One was an oligogalactoside sequence composed of (6-α-alpha-galactosyl)$_{10}$ β-galactosyl-mannose. A mannotetraose with β-xylose linked to the reducing terminal mannose and an 1,2-α-mannotriose were also delineated. This glycoprotein is largely periodate-sensitive and binds to Con A via terminal mannosyl residues. In these respects the *C laurentii* glycoprotein resembles the galactoxylomannan of *C neoformans*.

12.3.3 Serologic Heterogeneity in the *Cryptococcus neoformans* Capsule

Three serotypes of *C neoformans*, A, B, C, were first described (Evans, 1950) on the basis of agglutinin-adsorption and tube agglutinin tests. Later, a fourth D serotype was discovered (Wilson et al, 1968). Serotypes A and D occur in soil, type A being the most frequent clinical isolate in the USA. The ecologic niche of B and C types is not known. Compatible mating types of serotypes A and D complete the basidiomyce-

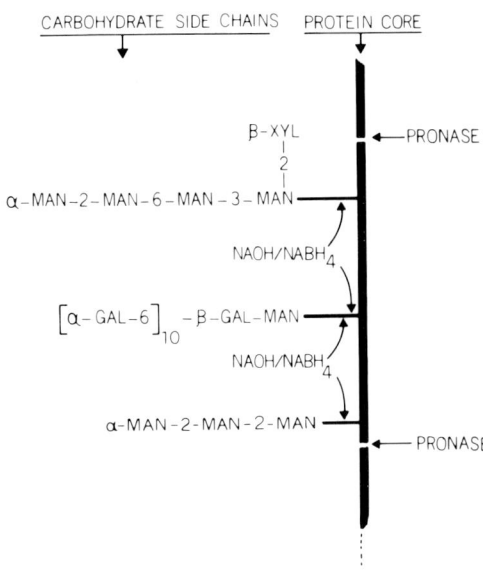

Figure 12.5. Schematic representation of glycoprotein characterized from cell extracts of *Cryptococcus laurentii*. Three types of O-glycosidically linked carbohydrate side chains are susceptible to alkali cleavage (Raizada et al, 1975). The major value of this structure is to compare it to the galactoxylo polymer and the mannoprotein of *C neoformans* (Turner et al, 1984).

tous perfect state *Filobasidiella neoformans* (Kwon-Chung, 1975). Later, Kwon-Chung et al (1978) refined these types on biochemical, serologic, and genetic grounds as *Cryptococcus neoformans* A and D and *Cryptococcus bacillisporus* serotypes B and C. This decision was reversed when it was discovered that *C bacillisporus* was synonymous with *C neoformans* var *gattii* (Kwon-Chung et al, 1982).

The array of determinants recognized on the *C neoformans* cell surface was confirmed by fluorescent antibody cross-staining and adsorption studies with encapsulated cryptococci (Kaplan et al, 1981). Antisera to each of the four serotypes recognizes a common determinant. An additional determinant is shared by the B-C group and one shared by A and D. Within these groupings there is a serotype-specific determinant. The occurrence of panspecific, A + D and B + C group-specific, and type-specific factors has been confirmed by Ikeda et al (1982) who assigned numbers to these factors. They reported a fifth serotype, AD.

Heterogeneity exists among the glucuronoxylomannans of four serotypes and also in the galactoxylomannan produced by the yeast that has not yet been related to serotypes. Epitopes directing serotype specificity are not yet delineated but B and C types share an unusual structural feature: tetrasubstituted sugars in the mannan backbone of the major glucuronoxylomannan (Bhattacharjee et al, 1978, 1979b, 1980). Definitive relationships among the serotypes have not been confirmed in the few reports using purified glucuronoxylomannans in immunodiffusion or quantitative precipitin reactions. Preliminary results indicate (Cherniak et al, 1980) that although glucuronoxylomannans appear chemically and immunoelectrophoretically homogeneous, the glucuronoxylomannan of serotype A can be fractionated on an anti-*C neoformans* D IgG immunoadsorbent column. An effluent fraction reacted in immunodiffusion with anti-A and not with anti-D. A fraction remaining bound to the column had a higher affinity for antibodies to D.

The serologic activity of chemically altered glucuronoxylomannans was tested by Bhattacharjee et al (1980). Smith-degraded glucuronoxylomannan was devoid of activity, indicating that mannose itself is not immunodominant. Carboxy-reduction did not affect the homologous immunodiffusion reaction of the glucuronoxylomannan of serotype B, but ablated the reaction of anti-*C neoformans* C with the glucuronoxylomannan of serotype B, showing that this anti-C serum recognized glucuronic acid. De-O-acetylation of the glucuronoxylomannan of serotype D did not inhibit precipitin arc formation in immunodiffusion.

Agglutination of anti-*C neoformans* serotype A-coated latex particles permitted an estimate of the activity of chemically altered glucuronoxylomannan of the same serotype (Cherniak et al, 1980). De-O-acetylation resulted in a 50% reduction; and the reduction of glucuronic acid to glucose resulted in a 75% loss of activity. Autohydrolysis in Dowex 50 resin preferentially removed xylosyl residues and completely inhibited the latex agglutination reaction. The results of these two studies agree that glucuronic acid is a determinant in the glucuronoxylomannan of serotypes A and C. Evidence for O-acetyl haptens is reported only for serotype A. Both studies imply that xylosyl is an important hapten. It is possible that adjacent xylosyl and glucuronic acid residues form a conformational determinant.

A minor portion of the total anti-*C neoformans* type D IgG bound to a β-glucuronic acid affinity column, and an even smaller fraction of the total antibodies was eluted from a column of immobilized type D-glucuronoxylomannan by β-methyl xyloside (Bhattacharjee et al, 1983). The majority of the IgG that is precipitable with intact glucuronoxylomannan, but is not specific for either xylose or glucuronic acid, is considered to interact with both determinants and part of a mannose residue. This view is buttressed by hard sphere exo-anomeric effect equations that calculated the distance between adjacent glucuronic acid and xylose residues as 1.6 nm, which is compatible with the capacity of the combining site of IgG.

The glucuronoxylomannan and the "minor component," galactoxylomannan, are serologically distinct. The high molecular weight of glucuronoxylomannan retards its diffusion through agar gel, whereas the galactoxylomannan migrates well forming a distinct immunoprecipitate (Cherniak et al, 1982). The serologic difference between these two heteroglycans is clearly demonstrated in counterimmunoelectrophoresis. The ethanol-precipitated culture filtrate resulted in a single arc because the high concentration of glucuronoxylomannan obscures detection of the galactoxylomannan. After CTAB fractionation, galactoxylomannan produces two arcs, one corresponding to residual glucuronoxylomannan. Only after chromatography on DEAE-Sephadex or Sepharose CL6B was the galactoxylomannan devoid of the major heteroglycan.

The galactoxylomannan reacted with anti-type XIV pneumococcal polysaccharide and with the antiserum to an artificial branched mannotriose conjugated to bovine serum albumin (Reiss et al, 1984):

$$\text{mannose-}\alpha\text{-1} \rightarrow \text{2-mannose-1} \rightarrow \text{BSA}$$
$$\overset{6\uparrow}{\underset{1\uparrow}{}}$$
$$\alpha\text{-mannose}$$

It should be noted that the linkages shown are not proven to occur in the galactoxylomannan. This reaction occurred in indirect EIA with galactoxylomannan adsorbed to the polystyrene solid phase. Type XIV pneumococcal polysaccharide contains galactose determinants (Lindberg et al, 1977). These observations provide evidence for both galactosyl and mannosyl epitopes in the galactoxylomannan complex. The galactoxylomannan of *C neoformans* evokes serologic reactions in humans as estimated by EIA measurement of specific IgG and IgM (Reiss et al, 1984). Immunoglobulin G was present in sera of some normal individuals and in candidiasis patients, making it impossible to discriminate these reactions from those in cryptococcosis cases even when sera were first adsorbed with *Candida albicans* cell walls. Immunoglobulin M reactions to galactoxylomannan were not detected in normal subjects, the frequency of crossreactions with candidiasis patients was much lower, and these could be removed by adsorption with *C albicans* cell walls. About 23% of the 55 cryptococcosis cases tested had IgM antibodies to galactoxylomannan that were considered significant at a serum dilution of 1/16. These reactions were not spuriously due to the presence of rheumatoid factor. Greater than 90% of the serologic activity of the galactoxylomannan complex resided in the mannan fraction purified by adsorption to Con A when measured against rabbit antiserum. Humans, however, may have an increased capacity to recognize the galactoxylo-moiety. The IgM antibodies seem to show increased affinity for the determinants specific for *C neoformans*. This affinity is fortunate because this class appears early in the primary immune response. The measurement of the IgM to the galactoxylomannan complex of *C neoformans* has implications for early diagnosis when no circulating antigen can be detected. This is especially relevant because the IgG or IgM response to the major glucuronoxylomannan is weak, which reinforces the concept of the capsule as a tolerogenic molecule (Reiss et al, 1984).

12.4 OPSONINS

Chemotaxis and adherence of phagocytes to cryptococci are mediated by heat-labile and heat-stable opsonins, but even in the presence of these opsonins in vitro phagocytosis of encapsulated cryptococci remains low. If normal human serum is adsorbed with cryptococci the serum loses most of its ability to support phagocytosis (Diamond et al, 1974). Acapsular mutant *C neoformans* are phagocytosed without added opsonins, but at a lower rate, indicating an as yet unknown recognition mechanism (Kozel and McGaw, 1979). Delineation of the role of opsonins with and without the presence of capsular polysaccharide was facilitated by the availability of acapsular mutants.

Both acapsular and encapsulated *C neoformans* can activate complement by the alternative pathway, but capsular polysaccharides isolated from any of the four known serotypes were unable to deplete properdin factor B from human serum (Laxalt and Kozel, 1979). However, Diamond et al (1974) reported that isolated *C neoformans* capsular material fixed late complement components in normal and C4-deficient serum. This disparity probably derives from impurities present in some polysaccharide preparations, necessitating the fractionation of the ethanol-precipitated culture filtrate into the major viscous, acidic glucuronoxylomannan and separating it from other mannose-based heteroglycans (Cherniak et al, 1980, 1982; Turner et al, 1984).

Chemotaxis of rabbit PMN in vitro towards cryptococci as the attractant occurs only after the yeasts have been opsonized by alternative pathway activation. The capsular polysaccharide appears neither to enhance nor inhibit direct complement activation and does not interfere with chemotaxis. Encapsulated cryptococci that are opsonized by complement components, monitored by measurement of factor B depletion, are still resistant to phagocytosis. The capsular polysaccharide may physically present a barrier between the cell wall-bound complement fragments, thus preventing immune adherence.

IgG in normal human serum accounts for the total heat-stable opsonic activity. About 0.1% to 0.2% of ^{125}I-labeled normal human IgG, at limiting dilution, was bound to acapsular mutant cryptococci. Encapsulated cryptococci also bind IgG present in normal human serum, but this opsonization does not remove the inhibition of phagocytosis by murine macrophages (Kozel and McGaw, 1979; McGaw and Kozel, 1979). Because the IgG in normal serum is directed against the cell wall of *C neoformans* and the Fc portion remains masked, recognition by macrophage Fc receptors is inhibited. The evidence for this recognition is that (i) antimacrophage IgG inhibited the adherence of opsonized acapsular cryptococci but not the ingestion of cryptococci once adherence occurred, and (ii) heavy chain-specific anti-IgG is unable to agglutinate encapsulated cryptococci previously opsonized with radioiodinated normal human IgG. Normal spinal fluid does not opsonize cryptococci, and this may contribute to the immunoprotected niche that enables meningitis to ensue. The stimulus for production of low levels of natural antibodies reactive with the cryptococcal cell wall, but not with the capsule itself, is still unknown. Some evidence exists that *C albicans* and *C neoformans* are crossreactive and that antibodies from candidiasis patients bind to the galactoxylomannan from *C neoformans*. This crossreactivity is removed by adsorption of sera with *C albicans* cell walls (Reiss et al, 1984).

Murine macrophages take up soluble immune complexes containing cryptococcal polysaccharide (Griffin, 1981). In so doing, the membrane Fc receptors are blocked and hence are unable to phagocytose cryptococci that have been opsonized by IgG. Macrophages contain C3b receptors, but these can only be activated by a lymphokine, the production of which is stimulated by immune complexes. Addition of this lymphokine to macrophages that have ingested cryptococci renders them capable of phagocytosing C3b-coated cryptococci, thus providing a potentially critical link between the humoral and cell-mediated limbs of the immune response to *C neoformans*.

Patients with cryptococcal sepsis had depressed levels of total hemolytic complement, low C3, and split products of properdin factor B, indicating alternative pathway activation (Macher et al, 1978). Polymorphonuclear neutrophils from such patients phagocytosed cryptococci poorly when reconstituted with autologous serum. Normal complement levels and phagocytic indices were observed in cryptococcal meningitis patients without sepsis (Diamond et al, 1972; Macher et al, 1978) except in one report (Mohr et al, 1974) where phagocytic indices were depressed.

Guinea pigs injected with high doses (1×10^9) of *C neoformans* also exhibit complement depletion and depressed phagocytosis (Macher et al, 1978). When heat-killed cryptococci were suspended in normal serum and washed, C3b was detected by RIA. The addition of up to 250 μg/ml of cryptococcal polysaccharide resulted in a slight lowering of total complement, but phagocytic indices remained normal with this serum source, thus providing more evidence that capsular polysaccharide itself does not directly activate complement via the alternative pathway (Macher et al, 1978).

12.4.1 Genetic Control of Humoral Immunity

Genetic control is illustrated in the tenfold greater susceptibility of mice that lack hemolytic complement including the widely used A/J strain (Rhodes et al, 1980). Genetic susceptibility was linked to the Hc° allele coding for the C5 protein on chromosome 2. Resistance was inherited as a single dominant trait apparently independent of the H-2 haplotype. This central defect in complement blocks activation of late complement components and deletes the anaphylatoxin and chemotactic properties of the C5a fragment. Opsonization of cryptococci can still occur in deficient mice because that property is served by Ig or C3b. The increased susceptibility seems to depend on a failure to attract sufficient phagocytes or to a dual need for phagocytic and complement-mediated damage to effect lysis of the target yeast. Thus the adhered phagocyte's lysosomal enzymes may plasticize the cell wall, letting late complement components attack the *C neoformans* plasmalemma.

12.5 PROTEIN ANTIGENS

The role of the cryptococcal capsule as an aggressin is now generally appreciated. Studies on the molecular and cellular level continue to unfold the tolerogenic and chemical properties of the capsule that protects the yeast from host surveillance. The relationship of cryptococcal proteins in the immune response is still largely an unexplored issue.

Acid phosphatase has been cytochemically detected in the cell wall and throughout the capsule, hinting that far from being a passive structure, the capsule serves as a diffusion medium through which gradients of enzymes are secreted by the host (Mahvi et al, 1974).

Although auxanographic methods have not revealed proteolytic activity by C neoformans, a more sensitive immunoelectrophoretic technique showed split products of human fibrinogen in extracts of plasma-agar inoculated with either of several C neoformans isolates (Müller and Sethi, 1972). A potential role for fibrinolytic enzymes in the pathogenicity was suggested, but this aspect has not been reinvestigated.

12.5.1 Cryptococcin

Few efforts have been systematically made to assess the extent of subclinical exposure to C neoformans in the community as reflected in positive delayed cutaneous hypersensitivity. The limiting factor has been the lack of commercial skin test preparations. Recent developments, reviewed below, have increased optimism about the practicability of improved C neoformans skin test antigens. The conventional wisdom is that because the yeast is ubiquitous in rural and urban areas, the extent of subclinical exposure and hence positive skin tests should be high. Available data do not bear that out; on the contrary, few persons in the general population manifest positive cutaneous responses and those who are test-positive may have more contact with pigeons. Positive delayed cutaneous hypersensitivity responses were recorded in 32.9% of pigeon fanciers compared to 4.2% of controls (Newberry Jr. et al, 1967).

Lacaz and Melhem (1978) found 6% positive reactors among 198 soldiers in São Paolo, Brazil. Schimpff and Bennett (1975) found a high incidence of positive skin tests in mycology laboratory workers, but none in a small sample of student volunteers. Unlike H capsulatum, mere residence in an area known to contain culture-positive sites (eg, pigeon roosts) is insufficient to induce delayed type hypersensitivity.

The potential to acquire a heightened state of immunity (in the absence of disease) is present, as indicated by the reactions among laboratory workers. Further systematic appraisal of the cutaneous and in vitro correlates of delayed hypersensitivity will help to estimate the advisability of vaccines if groups at higher risk could benefit from such prophylaxis.

A C neoformans skin test antigen was made from a small capsule serotype A culture grown on yeast extract dialyzate medium (Atkinson and Bennett, 1968; Bennett, 1981). Yeast cells were resuspended in borate buffered 11.6M urea and stirred at 4 °C for 72 hours. After dialysis and removal of the precipitate, the supernate was filter sterilized. Yields using this method are low, 3.5 mg protein/g of wet packed cells. Although approximately 75% of the proteins that were extracted were dialyzable, this fraction did not evoke delayed cutaneous hypersensitivity in guinea pigs sensitized with killed cryptococci. All reactivity resides in the nondialyzable portion and is precipitable with half-saturated ammonium sulfate. Results of gel permeation and analytical ultracentrifugation agree that the proteins extracted by this method that are active in evoking delayed cutaneous hypersensitivity are low in molecular weight, ie, approximately 10,000 da but not dialyzable.

12.5.2 Homogenate Fractions

The occurrence of large amounts of polysaccharide in culture supernatants of *C neoformans* has held back development of lysate antigens. Encapsulated yeast are more difficult to disrupt, but a postmitochondrial supernate obtained from small capsule isolate was found to elicit specific footpad swelling and migration inhibitory factor production in *C neoformans*-infected mice (Hay and Reiss, 1978).

Differential centrifugation of a homogenate from an acapsular *C neoformans* mutant resulted in a 105,000 × g supernate and a microsomal fraction (Jones et al, 1981). Both evoked specific delayed cutaneous hypersensitivity and blastogenesis of peripheral blood lymphocytes and peritoneal exudate cells from guinea pigs infected with encapsulated cryptococci. The proteins of each fraction were resolved by sodium dodecyl sulfate-polyacrylamide gels. An obstacle to their development as skin test antigens was their immunogenicity, ie, primed normal animals responded in recall tests. This same property is noteworthy, considering the difficulty experienced in vaccinating animals with whole killed, encapsulated cryptococci (Abrahams and Gilleran, 1960). Protection experiments with subcellular fractions of acapsular cryptococci have not been done but deserve further attention.

Blastogenic responses elicited by subcellular fractions (Jones et al, 1981) showed large variations among animals that probably result from single samplings of lymphocytes at one month postinfection. Multiple samplings are necessary to chart the activation of lymphocytes and the predicted decline towards anergy as dissemination progresses.

The importance of subcellular antigens from acapsular mutants is underscored. Avoidance of capsular polysaccharide is assured, permitting a more well-defined antigenic stimulus. The abundance of protein in the extracts and its potency in the 1 to 10 μg/ml range will be useful in studies to manipulate factors that stimulate or suppress immunity (Table 12.3).

A variety of antigens from *C neoformans* 145 (type A) were tested in mice infected in the thigh. Responses of spleen cells were flat, but inguinal node cells responded to the soluble cytoplasmic substances and to "B-HEX". The latter antigen is prepared

Table 12.3. Blastogenesis of Lymphocytes from Guinea Pigs Infected with *C neoformans* B551-A (Small Capsule) or 371-A (Large capsule)[a]

		Small Capsule, s/c[b]			Large Capsule, s/c		
Addition	Cell Type	Animal 1	Animal 2	Animal 3	Animal 4	Animal 5	Animal 6
ms	PBL	6.6	1.0	31.7	2.8	35.4	1.3
105K[c]	PBL	1.4	1.1	15.1	1.4	17.4	1.0
conA	PBL	35.1	70.8	82.9	61.2	89.1	19.9
ms	PEC	4.0	0.7	1.3	7.1	7.2	2.0
105K	PEC	5.7	0.7	2.6	2.0	10.1	5.1
conA	PEC	15.4	20.9	4.4	11.7	4.7	2.6

Source: Reprinted from Mycopathologia with permission from Dr. W. Junk, publisher.

Abbreviations: ms, microsomal fraction, 1 μg/ml; conA, concanavalin A, 25 μg/ml; PBL, peripheral blood lymphocytes; PEC, peritoneal exudate cells.
[a] Jones et al, 1981.
[b] Ratio counts/min ³H-thymidine: stimulated culture/control culture.
[c] 105,000 g supernatant, 1 μg/ml.

(as was previously done by Domer in the instance of *C albicans*) starting with a mitochondria-membranes fraction extracted first with *n*-butanol and then with hot phosphate buffer. The solubilized proteins were then recovered by precipitation with ammonium sulfate. Whole killed cryptococci were ineffective stimulants of blastogenesis (^{3}HTdR incorporation) and culture filtrate products were weakly reactive. Taken in the context of previous research with cryptococcin, and with the microsomal and 105,000 × g supernate antigens, the promising avenue for development of *C neoformans* antigens as stimulants of in vitro lymphocyte blastogenesis is clear (Table 12.4). The objectives are immunochemical purity and retention of activity. The use of acapsular mutant *C neoformans* excludes the glucuronoxylomannan and increases the ease of disrupting cryptococci for maximum antigenic preservation. With the discovery of a second antigenic polysaccharide complex, galactoxylomannan, it would appear worthwhile to remove the mannoprotein from cell extracts by Con A affinity chromatography. The solubilization of mitochondria-membranes, microsomal, and ribosomal proteins along the lines of Domer et al (1983) is promising. More knowledge of the polyacrylamide gel profile of these proteins will be necessary.

12.6 HOST DEFENSES

The replacement of antigenemia with antibodies reactive with whole cryptococci signals a favorable response to therapy in human cryptococcal meningitis (Diamond and Bennett, 1974). The capsule is known to mask heat-labile and heat-stable opsonins that bind to the *C neoformans* cell wall, preventing effective phagocytosis.

Table 12.4. Summary of *C neoformans* Antigens and the Immune Responses they Evoke

Antigen	Immune Response	References
Killed *C neoformans*	Protection	Abrahams and Gilleran, 1960
Live *C neoformans*	Protection by adoptive transfer of T-lymphocytes	Graybill and Mitchell, 1979
Cell wall (acapsular mutant)	Complement activation	Kozel and McGaw, 1979
Capsular polysaccharides (glucuronoxylomannan and galactoxylomannan complex)	Antiphagocytic, tolerogen	Kozel and Mastroianni, 1976; Murphy and Cozad, 1972
Galactoxylomannan complex	IgM	Reiss et al, 1984
Cryptococcin, 10,000 da protein	Delayed cutaneous hypersensitivity, leukocyte-MIF, lymphocyte blastogenesis	Bennett, 1981
Postmitochondrial supernatant, microsomal fraction 105,000 g supernatant	Delayed cutaneous hypersensitivity, MIF, lymphocyte blastogenesis	Hay and Reiss, 1978; Jones et al, 1981
Soluble cytoplasmic substances "B-HEX"; a butanol and hot water extract of mitochondria–membranes precipitated with saturated ammonium sulfate	Delayed cutaneous hypersensitivity, lymphocyte blastogenesis	Domer et al, 1983

The capsular polysaccharide itself is not an alternative complement pathway activator. Antibodies to the capsule should be important in neutralizing the antiphagocytic property and in mediating antibody-dependent cell-mediated cytotoxicity, thus providing a rationale for determining ways to increase levels of these antibodies in animal experiments.

Natural antibodies of the IgG class are the major heat-stable opsonins of cryptococci present in normal human serum, but it is by no means clear that passive immunity would protect mice against challenge with live cryptococci. Such protection occurred only under narrow conditions: when mice received anti-*C neoformans* IgG by the same route and simultaneously with the infective dose (Graybill et al, 1981). Both BALB/c and athymic nu/nu mice resisted the challenge dose but the latter breed succumbed in the late stages of primary infection at a time when cell-mediated immunity would ordinarily mature. The passively transferred antibodies were opsonic but not lethal for cryptococci.

Mouse antibodies reactive with ethanol-precipitated culture filtrate rose from day 9 to day 21 after intracerebral infection with *C neoformans* and then abruptly fell as cryptococcal polysaccharide in the serum and urine gradually increased to peak levels (Scott et al, 1981) (Figure 12.6). When antibodies could no longer be detected by indirect EIA, the mice began to die. Early in the primary immune response (<15 days) IgM predominated; afterwards, IgG became evident. Further dissection of the humoral responses in mice would be fruitful, for example, to chart the development of IgM against galactoxylomannan.

Low-zone tolerance can be overcome to some extent by the production of glycoconjugates and their injection in adjuvant. Complexation of cryptococcal polysaccharide with methylated bovine serum albumin or γ-globulin was effective in inducing antibodies in mice when administered with incomplete adjuvant. The

Figure 12.6. Appearance of antibodies in *Cryptococcus neoformans*-infected Swiss-Webster mice. Circles are indirect EIA titers; Rectangles denote the whole cell agglutinin titers; Bars indicate mortality. Each point is the median for 3 animals/group (Scott et al, 1981). *Source:* Reprinted from Sabouraudia with permission from the copyright holder, the International Society for Human and Animal Mycology.

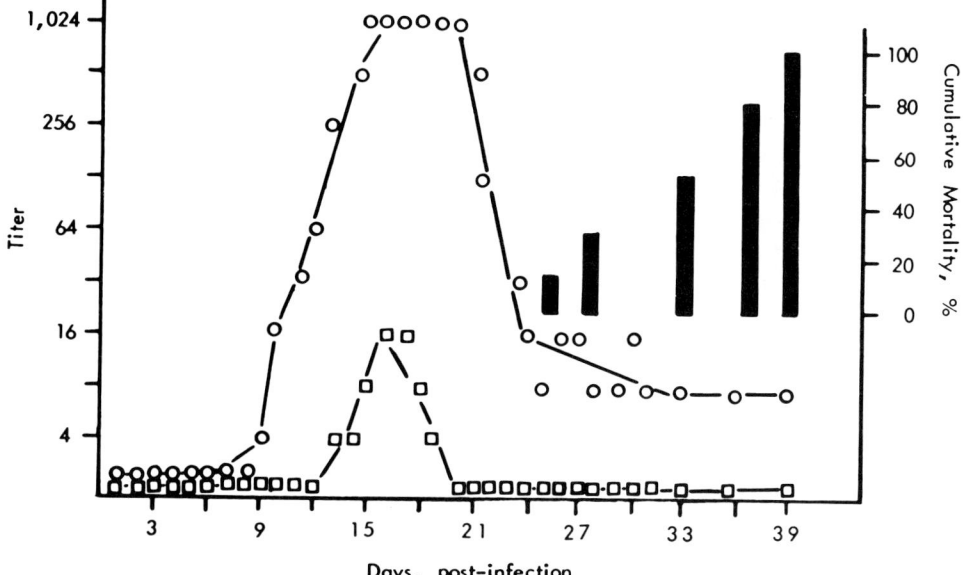

antibodies were detected by hemagglutination of polysaccharide-coated erythrocytes (Kozel and Cazin, 1974).

Conjugation of cryptococcal polysaccharide to bovine gammaglobulin via azocarbanilate led to glycoconjugates with high levels of carrier substitution (protein : polysaccharide = 2.3 : 1) and good retention of O-acetyl groups, which are believed to be important for antigenicity (Goren and Middlebrook, 1967). This glycoconjugate injected in complete Freund adjuvant evoked high agglutinin titers. Perhaps the best way to overcome the tolerogenic potential of the intact, viscous glucuronoxylomannans would be to produce haptenic oligosaccharides and conjugate them to protein carriers. Unfortunately, an effective means of producing oligosaccharides from *C neoformans* heteroglycans is not known. Surprisingly, there are no reports where animals were primed to produce moderate levels of capsular antibodies and then challenged with live cryptococci.

12.6.1 Acquired Immunity

Early workers were perplexed by the difficulty of protecting mice with whole killed yeast or polysaccharide vaccines. Effective resistance to lethal challenges was achieved in one study by immunizing mice with formalin-killed small capsule forms, resulting in increased survival time with reduced multiplication of yeast in liver, spleen, and to a lesser extent in brain (Abrahams and Gilleran, 1960). Immunization with high doses of *C neoformans* culture filtrate or capsular polysaccharide did not protect mice from lethal challenge nor did it induce anti-capsular Ig. Removal of readily soluble polysaccharide by several cycles of mechanical abrasion of the yeast, insufficient to disrupt cells, liberated a soluble fraction proximal to the cell wall that, weight for weight, was an effective immunogen and stimulant of antibodies capable of agglutinating erythrocytes coated with capsular polysaccharide (Gadebusch, 1963). The failure to protect mice by priming them with high doses of capsular polysaccharide or with large capsule forms is understood as an immunologic tolerance phenomenon (Murphy and Cozad, 1972; Kozel et al, 1977; Goren and Middlebrook, 1967; Kozel and Cazin, 1974), but the possibility that some subfraction of the capsule or cell wall is a more potent immunogen has not been investigated.

12.6.2 Cell-Mediated Cytotoxicity

Antibody and complement were not found to kill cryptococci in vitro but provided the means for phagocytes to recognize the opsonized yeast (Diamond and Allison, 1976). The promotion of yeast growth in the central nervous system and the lack of tissue response often observed may be partially explained by the absence of opsonins in cerebrospinal fluid (Diamond, 1981). The killing of encapsulated cryptococci by PMN or by macrophages is not efficient, particularly with large capsule forms, and a more efficient nonphagocytic killing mechanism operating by antibody-dependent cell-mediated cytotoxicity was described (Diamond and Allison, 1976). Human peripheral blood leukocytes required rabbit anti-*C neoformans* to kill small capsule forms. Purified T-lymphocytes were not capable of killing cryptococci. Granulocytes killed 47.2% of the target cells and killing by lymphocytes amounted to 63.6%. The most efficient killing (75.9%) occurred with a mixture of monocytes and lymphocytes that adhered to, but did not phagocytose the cryptococci. This finding recalls earlier studies of rosette formation by macrophages around large capsule forms (Aronson and Kletter, 1973). Lysosomal degranulation was not a consistent finding in the antibody-dependent cell-mediated cytotoxicity and the mechanism by which killing is effected is unknown.

Polymorphonuclear neutrophils from normal human donors reconstituted in vitro with rabbit anti-cryptococcal Ig effectively killed greater than 80% of small capsule *C neoformans* within 4 hours, whereas mononuclear cells were moderately less effective (Miller and Kohl, 1983). The reactions were carried out in heat-inactivated serum to exclude complement. The discrepancies between these results and those of Diamond and Allison are explained by the (i) higher concentration of rabbit antibodies used by Miller and Kohl; (ii) a different mechanism because the PMN-mediated killing was intracellular and implicated a myeloperoxidase-H_2O_2-cidal pathway.

Normal *CBA/N* mice have high levels of natural killer lymphocytes, a cell-type whose lineage is different from T and B-cells, lacking both thy 1 and surface Ig markers. The nylon wool nonadherent portion of splenic lymphocytes from these mice significantly inhibited growth (40% to 85% inhibition) of cryptococci, independent of serotype and degree of encapsulation after an overnight incubation at an effector:target ratio of 500:1 (Murphy and McDaniel, 1982). Negative selection, by treatment of splenic lymphocytes with anti-thy 1, anti-mouse Ig or anti-Ia plus complement did not reverse in vitro killing. The only treatment that ablated killer activity was anti-asialo GM-1, an antibody specific for a cell membrane glycoprotein found at high density on natural killer lymphocytes. Other manipulations were also consistent with the identity of the effector as a natural killer cell. Factors known to stimulate natural killer function, priming mice with the synthetic polynucleotide poly I-C, or *Corynebacterium parvum*, potentiated in vitro killing of cryptococci. The prediction that inbred mice with low levels of natural killer lymphocytes, *A.TH* or old *CBA/N*, would display reduced in vitro killing proved correct.

These observations lead us to postulate a varied role for innate immunity in surveillance against cryptococci. The paradoxical resistance of athymic nude mice in the days immediately after cryptococcal infection may be understood in terms of the high levels of both activated macrophages and natural killer lymphocytes present in these mice (Cheers and Waller, 1975; Herberman, 1978).

12.6.3 T-Lymphocyte Functions

12.6.3a Mice

Congenitally athymic nude (nu/nu) mice are not able to control cryptococcal infection as well as thymus-bearing nu/+heterozygotes (Cauley and Murphy, 1979). At first, the rate of multiplication of cryptococci is slower in nu/nu organs of the reticuloendothelial system, liver and spleen, probably due to more active macrophages and natural killer (null) lymphocytes in these mice. At the time that heterozygotes develop delayed type hypersensitivity, 14 days postinfection, the extent of infection begins to plateau, whereas in nu/nu mice dissemination increases in spite of the production of IgM agglutinins. The difference in host resistance that limits or reverses cryptococcal dissemination is likely a granulomatous mononuclear inflammatory response. The possibility exists that T-helper function may be required for production of antibodies to yeast determinants in the cell wall that are T-dependent and important in mediating target cell lysis via complement activation or antibody-dependent cell-mediated cytotoxicity, but this is only speculation.

Adoptive transfer of splenic lymphocytes from BALB/c mice immunized with live cryptococci significantly prolonged survival after challenge with a lethal dose of *C neoformans* (Graybill and Mitchell, 1979). Fifty percent survival at 30 days was observed in contrast to less than 10% survival of mice receiving lymphocytes from naive donors. Nylon wool passaged lymphocytes were more potent in the adoptive transfer of immunity (eg, 2×10^6 T-enriched lymphocytes conferred greater protection than 5×10^7 unfractionated cells).

Mice primed intravenously with a culture filtrate of *C neoformans* or with antigenemic serum from infected mice gave rise to a suppressor cell population in peripheral and mesenteric lymph nodes that was absent in the spleen (Murphy and Moorhead, 1982). Adoptive transfer of these lymph node cells inhibited the ability (i) of mice to mount a delayed cutaneous hypersensitivity response to the same antigen and (ii) of spleen cells to inhibit in vitro growth of cryptococci. The suppression was antigen-specific and operated at the level of induction of the immune responses (ie, adoptive transfer could not abrogate an already established state of immunity). The suppressor population was identified as T-lymphocytes by both negative and positive selection: suppressive ability was susceptible to anti-thy-1 and complement and occurred in the effluent cell population from a nylon wool column.

The phenotype of the T-suppressor cells was lyt-1^+, Ia$^+$ (I-J$^+$), (Murphy et al, 1983). Adoptive transfer of these cells or supernates of the freeze-thawed suppressor cells induces a second (efferent) suppressor cell population with the same phenotype. The occurrence of lyt-1^+ suppressor cells is not a contradiction if they are regarded as suppressor-inducer cells.

The antigenic complex contained cryptococcal polysaccharides and protein, thus the nature of the antigen inducing the T-suppressor cells is uncertain. The major, viscous, capsular polysaccharide (glucuronoxylomannan) is a T-independent antigen because IgM agglutinins can be raised in athymic nude mice. The T-suppressor cell can still function to regulate production of such antibodies even if their induction is independent of the need for T-cell help. This has been amply demonstrated in the case of the splenic plaque forming cell response to type III pneumococcal polysaccharide (Baker, 1975). This mechanism can explain the inhibition of in vitro growth of cryptococci by spleen cells induced to produce anti-capsular Ig. The in vitro inhibition assay contained a source of complement (Murphy and Moorhead, 1982) and the mechnism may act via antibody-dependent cell-mediated cytotoxicity, as demonstrated by Diamond and Allison (1976) or Miller and Kohl (1983). However, T-suppressor cell inhibition of delayed hypersensitivity must act via a different pathway because the capsular polysaccharide, being essentially nitrogen-free, does not evoke classical type IV cutaneous hypersensitivity. Clearly, as these assays become capable of more sensitive measurements of cell–cell interactions, a more rigorous sorting out of the antigens providing the stimuli will be essential, in order to deduce the mechanism of action.

T-Lymphocytes from infected mice cultivated in vitro with unpurified cell walls of *C neoformans* gave rise to cell supernates that contain a suppressor factor (Morgan et al, 1983). The functional assay for suppression was reduction in phagocytosis of acapsular cryptococci by murine macrophages. Neither walls of *Saccharomyces cerevisiae*, nor lymphocytes from noninfected mice could spare the specific requirements for production of the suppressor factor. The effect was genetically restricted so that the T-cells and macrophages must share a histocompatible H-2 haplotype. Suppressor factor induced by *C neoformans* cell walls in T-cells of *C neoformans*-infected mice can also suppress phagocytosis by macrophages of *S cerevisiae*. This *C neoformans* factor differs from the soluble immune response suppressor factor (see Chapter 17) in that its induction is antigen-specific and its effect is genetically restricted.

12.6.3b Humans

The T-lymphocyte reactivity of a series of 13 cryptococcosis patients was studied, half of whom had primary diseases that compromised their immune systems (Graybill and Alford, 1974). All patients showed mild reduction in total circulating T-lymphocytes and a lower frequency and amplitude of cutaneous hypersensitivity to *Candida* species and to histoplasmin in comparison with normal controls. Whole killed cryptococci stimulated blastogenic responses in normal persons, more vigorously in those under forty years old. This natural immunity is not surprising given the many opportunities for environmental exposure to the yeast. Lymphocyte reactivity is not usually expressed as delayed cutaneous hypersensitivity in normal persons when urea extracts of sonified cryptococci, termed cryptococcin (see below), are used as the antigen. It is unclear if the disparity between cutaneous and in vitro activity is a result of some deficiency in the antigen preparation or represents a dissociation between recognition and effector cell functions. Half of the meningitis patients were cutaneously anergic to cryptococcin; two with primary immunosuppressive conditions were also anergic at the level of blastogenic responses to *C neoformans*. Some patients with no known underlying disease also had blastogenic indices to cryptococcin of less than two: furthermore, no differences in the magnitude of blastogenic indices were apparent between those with or without known immunocompromised conditions. Generalizing from the T-cell functions of only a limited number of patients is problematic, but anergy seems to occur more often at the time of diagnosis and early in therapy. Blastogenic responses return first during convalescence, and then delayed cutaneous hypersensitivity returns. However, potent cutaneous and in vitro reactions were observed in a disseminated cryptococcosis patient who was also subjected to thoracic duct drainage resulting from rejection of a renal allograft (Graybill and Alford, 1974).

Fifteen cryptococcosis patients with no known preexisting immunosuppressed condition and who were cured of the yeast infection for at least 1 year were assessed for cell-mediated immune functions in comparison with mycology laboratory workers and young persons with no known exposure to *C neoformans* (Schimpff and Bennett, 1975). Cryptococcin skin tests were negative in "unexposed" controls but positive in 85% of laboratory workers. Only one third of the "normal" cured cryptococcosis patients responded with delayed cutaneous hypersensitivity to cryptococcin. Cutaneous responses to streptokinase and mumps were also less frequent than in normal controls. Laboratory workers had brisk direct leukocyte migration inhibitory factor responses to whole killed cryptococci, but cured "normal" patients had a flat mean response with much individual variation. Cryptococcin was less potent in evoking MIF but more active in inducing blastogenesis of lymphocytes from laboratory workers. The variation in blastogenic indices among the cured "normal" patients was high, but the mean response did not differ from that of positive controls. Blastogenic responses to other microbial antigens gave no evidence of defects in the cured "normal" patients. Cryptococcal meningitis patients with known immunocompromised states showed profound *C neoformans*-specific and general anergy. Cutaneous anergy coexisted with positive lymphocyte blastogenesis in cured "normal" patients giving evidence of a block in delayed cutaneous hypersensitivity effector function, and raising the possibility that such a defect may antedate the infection, perhaps increasing susceptibility to cryptococcosis.

The possibility that defects in *C neoformans*-specific cell-mediated immune responses are a result of infection has not been ruled out. Few experimental studies bear on this point, but Hay and Reiss (1978) showed a decrease in murine MIF responses to a *C neoformans* postmitochondrial supernate corresponding to the extent of dissemination.

12.7 SUMMARY

The polysaccharide capsule of *C neoformans* is a physical and chemical barrier to phagocytosis and persists in cerebrospinal fluid and plasma in proportion to disease severity. Encapsulated cryptococci are more virulent for mice than stable acapsular mutants. Pathogenicity is not, however, a direct function of capsule diameter, and other virulence factors are probably involved. The major viscous acidic glucuronoxylomannan capsular polysaccharide is a tolerogen because when the optimally immunogenic dose is exceeded, in vitro splenic immunoplaques and circulating antibodies are suppressed. Glucuronoxylomannan is a T-independent elicitor of IgM, but the antibody level is regulated by T-suppressor cells. The adherence stage of phagocytosis of acapsular cryptococci is inhibited by small amounts of glucuronoxylomannan. When the capsule diameter is too large for endophagocytosis, rosettes of macrophages can encircle cryptococci, fuse, and their pseudopodia penetrate to the plasmalemma and damage the yeast.

The glucuronoxylomannan consists of a linear α-1,3-mannan backbone, with single β-1,2-glucuronic acid and xylose residues linked β-1,2 in serotypes A and D, or in serotypes B and C one finds β-1,2- and β-1,4-linked xylose. The ratios of component sugars and O-acetyl substitution differ according to the serotype. In serotypes B and C glucuronoxylomannan the mannan backbone is completely substituted, and every third mannose is disubstituted with xylose and glucuronic acid.

Twelve percent of the exopolysaccharides recovered from *C neoformans* cultures is not precipitated by CTAB. This fraction represents the galactoxylomannan, with xylose : mannose : galactose : glucuronic acid molar ratios of 1 : 1.8 : 1.9 : 0.2. The galactoxylomannan is unable to organize a capsule and is lower in both viscosity and molecular weight than the glucuronoxylomannan. Galactoxylomannan can be fractionated by affinity for Con A into a mannan which is bound to the lectin and a galactoxylo-enriched fragment that appears in the effluent.

Type specificity is considered to reside in the major capsular glucuronoxylomannan, but the epitopes responsible for serotypes are unknown. Structural features that contribute to antigenicity are O-acetyl, glucuronic acid, xylose, and mannose.

The antibody response in humans against glucuronoxylomannan is weak, probably because of immunologic tolerance, but the galactoxylomannan is more serologically active. Some determinants in the galactoxylomannan are shared with *C albicans* mannan, but IgM reactions with *C neoformans* are a specific indication of disease that occurs in about 20% of cryptococcosis patients. However, the most unequivocal evidence of infection is capsular polysaccharide circulating in the cerebrospinal fluid or serum.

The cell wall, but not the glucuronoxylomannan, is a direct activator of complement by the alternative pathway. Cryptococci opsonized by normal serum induce chemotaxis of macrophages, but these encapsulated cells still resist phagocytosis because the glucuronoxylomannan masks the wall-bound complement components. Low levels of IgG in normal serum are reactive with the galactoxylomannan of the cell wall, not with the capsule.

Cryptococcal sepsis in humans and in guinea pigs is accompanied by complement depletion and results in a reduction of the phagocytic index. C5-Deficient mice have increased susceptibility to *C neoformans*. Although these mice can still effect opsonization via C3b or by immunoglobulins, the result of deleting the late components suggests that the chemotactic properties of C5a or the membrane attack phase contribute an important increment of protection.

Killing of cryptococci, especially large capsule forms, by PMN and macrophages is not very efficient, but two other cell-mediated cytotoxic mechanisms have been demonstrated in vitro: Antibody-dependent cytotoxicity mediated by monocytes and lymphocytes (not T-cells) and natural killing in nonimmunized animals mediated by thy 1^-, surface Ig^-, Ia^-, and asialo-GM-positive natural killer lymphocytes.

Euthymic mice begin to control cryptococcal infection at a time when T-cell-mediated immunity matures, but athymic mice cannot control dissemination despite the production of IgM agglutinins. Similarly, T-enriched lymphocytes from immune donors show enhanced ability to confer protection in adoptive transfer experiments. Regulation of immunity to *C neoformans* occurs at the level of T-suppressor cells which can be induced in mice by priming them with *C neoformans* antigens.

T-cell-mediated immunity probably functions in different elements of the immune response to *C neoformans:* (i) regulating levels of antibodies to the capsule, (ii) marshalling a granulomatous response to eradicate cryptococcal foci, (iii) providing help in recognition of T-dependent protein antigens. How these diverse functions may be orchestrated, and which ones are critical to host defense, are matters for further research.

Three specific levels of T-cell regulation of immunity have been identified in experimental murine cryptococcosis: (i) secretion of a lymphokine in response to soluble immune complexes that enables macrophages to phagocytose opsonized cryptococcosis via C3b receptors; (ii) induction of afferent and efferent T-suppressor cells that suppress delayed cutaneous hypersensitivity; (iii) production by T-cells from infected mice of a soluble suppressor factor that down-regulates macrophage phagocytosis. If all three of these functions are to be exerted effectively, an intricate control mechanism akin to a computer program is needed.

In humans cutaneous anergy is observed in cryptococcal meningitis, and during convalescence blastogenic responses are reconstituted before the appearance of a positive delayed skin test. The level of cutaneous hypersensitivity in the general population is probably low, but is high among laboratory workers with opportunities for exposure.

A hypothetical sequence of immune responses to *C neoformans* is initiated when alveolar macrophages adhere to and phagocytose cryptococci once they are opsonized by the alternative and classical pathways of complement activation (see Figure 17.6). Natural killer cells mediate nonphagocytic killing of the target cells. Natural immunoglobulins crossreactive with *C albicans* are important opsonins early in the immune response. B-Lymphocyte clones stimulated by the major capsular polysaccharide expand and produce IgM. In the presence of an intact immune response, the infection is confined to the lung. This depends on concerted killing by macrophages, natural killer lymphocytes, and PMN. Antibody-dependent cell-mediated cytotoxicity may also have a role once the latent period is past for production of antibody against the glucuronoxylomannan. When there is a failure to opsonize, or when killing is defective, the antigenic stimulus of the capsular glucuronoxylomannan exceeds the optimally immunogenic dose and immunologic tolerance occurs, shutting down secretion of IgM agglutinins. This in turn decreases opsonization and chances for successful cytotoxicity, setting the stage for hematogenous spread to the central nervous system. T-Lymphocyte-mediated immunity is needed to activate a mononuclear inflammatory response; if not, cryptococcomas will not be eradicated and meningitis will result.

REFERENCES

Abercrombie MJ, Jones JKN, Lock MV, Perry MB, Stoodley RJ, 1960. The polysaccharides of *Cryptococcus laurentii* (NRRL Y-1401) Part I. Can J Chem 38:1617–1624.

Abrahams I, Gilleran TG, 1960. Studies on actively acquired resistance to experimental cryptococcosis in mice. J Immunol 85:629–635.

Aronson M, Kletter J, 1973. Aspects of the defense against a large-sized parasite, the yeast, *Cryptococcus neoformans*. pp. 132–162. In Dynamic Aspects of Host–Parasite Relationships. Zuckerman A, Weiss DW, eds. New York: Academic Press.

Atkinson AJ, Bennett JE, 1968. Experience with a new skin test antigen prepared from *Cryptococcus neoformans*. Am Rev Respir Dis 97:637–643.

Baker PJ, 1975. Homeostatic control of antibody responses: a model based on the recognition of cell-associated antibody by regulatory T-cells. Transplantation Rev 26:3–20.

Bennett JE, 1981. Cryptococcal skin test antigen: preparation variables and characterization. Infect Immun 32:373–380.

Bennett JE, Hasenclever HF, 1965. *Cryptococcus neoformans* polysaccharide: studies on serological properties and role in infection. J Immunol 94:916–920.

Bhattacharjee AK, Bennett JE, Bundle DR, Glaudemans CPJ, 1983. Anticryptococcal type D antibodies raised in rabbits. Mol Immunol 20:351–359.

Bhattacharjee AK, Kwon-Chung KJ, Glaudemans CPJ, 1978. On the structure of the capsular polysaccharide from *Cryptococcus neoformans* serotype C. Immunochem 15:673–679.

Bhattacharjee AK, Kwon-Chung KJ, Glaudemans CPJ, 1979a. On the structure of the capsular polysaccharide from *Cryptococcus neoformans* serotype D. Carbohydr Res 73:183–192.

Bhattacharjee AK, Kwon-Chung KJ, Glaudemans CPJ, 1979b. On the structure of the capsular polysaccharide from *Cryptococcus neoformans* serotype C. II. Mol Immunol 16:531–532.

Bhattacharjee AK, Kwon-Chung KJ, Glaudemans CPJ, 1980. Structural studies on the major capsular polysaccharide from *Cryptococcus bacillisporus* serotype B. Carbohydr Res 82:103–111.

Bhattacharjee AK, Kwon-Chung KJ, Glaudemans CPJ, 1981. Capsular polysaccharides from a parent strain and from a possible mutant strain of *Cryptococcus neoformans* serotype A. Carbohydr Res 95:237–247.

Blandamer A, Danishefsky I, 1966. Investigations on the structure of the capsular polysaccharides from *Cryptococcus neoformans* type B. Biochim Biophys Acta 117:305–313.

Breen JF, Combee CL, 1980. Immunogenicity of cryptococcal capsular polysaccharide. Abstr Ann Mtg Amer Soc Microbiol, no F3. p. 320.

Bulmer GS, Sans MD, 1968. *Cryptococcus neoformans* III. Inhibition of phagocytosis. J Bacteriol 95:5–8.

Bulmer GS, Sans MD, Gunn CM, 1967. *Cryptococcus neoformans* I. Nonencapsulated mutants. J Bacteriol 94:1475–1479.

Bulmer GS, Tacker JR, 1975. Phagocytosis of *Cryptococcus neoformans* by alveolar macrophages. Infect Immun 11:73–79.

Cauley LK, Murphy JW, 1979. Response of congenitally athymic nude and phenotypically normal mice to *Cryptococcus neoformans* infection. Infect Immun 23:644–651.

Chandler FW, Kaplan W, Ajello L, 1980. p. 56. In A Colour Atlas and Textbook of Histopathology of Mycotic Diseases. London: Wolfe Medical Publications.

Cheers C, Waller RP, 1975. Activated macrophages in congenitally athymic "nude" mice and in lethally irradiated mice. J Immunol 115:844–847.

Cherniak R, Reiss E, 1986. The chemistry of soluble polysaccharides of *Cryptococcus neoformans* and their interactions with the immune system. Curr Top Med Mycol 2: submitted for publication.

Cherniak R, Reiss E, Slodki ME, Plattner RD, Blumer SO, 1980. Structure and antigenic activity of the capsular polysaccharide of *Cryptococcus neoformans* serotype A. Mol Immunol 17:1025–1032.

Cherniak R, Reiss E, Turner SH, 1982. A galactoxylomannan antigen of *Cryptococcus neoformans* serotype A. Carbohydr Res 103:239–250.

Diamond RD, 1977. Effects of stimulation and suppression of cell-mediated immunity on experimental cryptococcosis. Infect Immun 17:187–194.

Diamond RD, 1981. Immunology of invasive fungal infections. Comprehensive Immunol 8:585–633.

Diamond RD, Allison AC, 1976. Nature of the effector cells responsible for antibody-dependent cell-mediated killing of *Cryptococcus neoformans*. Infect Immun 14:716–720.

Diamond RD, Bennett JE, 1974. Prognostic factors in cryptococcal meningitis. Ann Intern Med 80:176–181.

Diamond RD, May JE, Kane MA, Frank MM, Bennett JE, 1974. The role of the classical and alternate complement pathways in host defenses against *Cryptococcus neoformans* infection. J Immunol 112:2260–2270.

Diamond RD, Root RK, Bennett JE, 1972. Factors influencing killing of *Cryptococcus neoformans* by human leukocytes in vitro. J Infect Dis 125:367–376.

Domer JE, Lyon FL, Murphy JW, 1983. Cellular immunity in a cutaneous model of cryptococcosis. Infect Immun 40:1052–1059.

Drouhet E, Segretain G, Aubert JP, 1950. Polyoside capsulaire d'un champignon pathogene, *Torulopsis neoformans*. Relation avec la virulence. Ann Inst Pasteur 79:891–900.

Evans EE, 1950. The antigenic composition of *Cryptococcus neoformans*. I. A serologic classification by means of the capsular and agglutination reactions. J Immunol 64:423–430.

Evans EE, Mehl JW, 1951. A qualitative analysis of capsular polysaccharides from *Cryptococcus neoformans* by filter paper chromatography. Science 114:10–11.

Farmer SG, Komorowski RA, 1973. Histologic response to capsule-deficient *Cryptococcus neoformans*. Arch Pathol 96:383–387.

Fraser CG, Jennings HJ, Moyna P, 1973a. Structural analysis of an acidic polysaccharide from *Tremella mesenterica* NRRL Y-6158. Can J Biochem 51:219–224.

Fraser CG, Jennings HJ, Moyna P, 1973b. Structural features inhibiting the cross-reaction of the acidic polysaccharide from *Tremella mesenterica* with type II anti-pneumococcal serum. Can J Biochem 51:225–230.

Gadebusch HH, 1963. Immunization against *Cryptococcus neoformans* by capsular polysaccharide. Nature 199:710.

Gander JE, Beachy J, Unkefer CJ, Tonn SJ, 1980. Toward understanding the structure, biosynthesis and function of a membrane-bound fungal glycopeptide—structural studies. pp. 49–79. In Fungal Polysaccharides, Sandford PA, Matsuda K, eds. Washington, DC: American Chemical Society.

Goren MB, Middlebrook, GM, 1967. Protein conjugates of polysaccharide from *Cryptococcus neoformans*. J Immunol 98:901–913.

Graybill JR, and Alford RH, 1974. Cell-mediated immunity in cryptococcosis. Cell Immunol 14:12–21.

Graybill JR, Hague RM, Drutz, DJ, 1981. Passive immunization in murine cryptococcosis. Sabouraudia 19:237–244.

Graybill JR, Mitchell L, 1979. Host defense in cryptococcosis III. In vivo alteration of immunity. Mycopathol 69:171–178.

Griffin FM, Jr, 1981. Roles of macrophage Fc and C3b receptors in phagocytosis of immunologically coated *Cryptococcus neoformans.* Proc Nat Acad Sci (USA) 78:3853–3857.

Hay RJ, Reiss E, 1978. Delayed-type hypersensitivity responses in infected mice elicited by cytoplasmic fractions of *Cryptococcus neoformans.* Infect Immun 22:72–79.

Hay RJ, MacKenzie DWR, Campbell CK, Philpot CM, 1980. Cryptococcosis in the United Kingdom: an analysis of 69 cases. J Infect 2:13–22.

Herberman RB, 1978. Natural cell-mediated cytotoxicity in nude mice. p. 135. In The Nude Mouse in Experimental and Clinical Research, Fogh J, Giovanella GC, eds. New York: Academic Press.

Ikeda R, Shinoda T, Fukazawa Y, Kaufman L, 1982. Antigenic characterization of *Cryptococcus neoformans* serotypes and its application to serotyping of clinical isolates. J Clin Microbiol 16:22–29.

Jacobson ES, Ayers, DJ, Harrell AC, Nicholas CC, 1982. Genetic and phenotypic characterization of capsule mutants of *Cryptococcus neoformans.* J Bacteriol 150:1292–1296.

Jones AE, Reiss E, Spira T, 1981. A microsomal fraction of *Cryptococcus neoformans* elicits in vitro blastogenesis of lymphocytes from infected guinea pigs. Mycopathol 75:129–138.

Kalina M, Kletter Y, Shahar A, Aronson M, 1971. Acid phosphatase release from intact phagocytic cells surrounding a large-sized parasite. Proc Soc Exptl Biol Med 136:407–410.

Kaplan W, Bragg SL, Crane S, Ahearn DG, 1981. Serotyping *Cryptococcus neoformans* by immunofluorescence. J Clin Microbiol 14:313–317.

Kaufman L, Blumer S, 1977. Cryptococcosis: the awakening giant. pp. 176–182. In Proc Fourth Intern Conf Mycoses. The Black and White Yeasts—Brasilia. Washington, DC: Pan American Health Organization.

Kozel TR, 1977. Non-encapsulated variant of *Cryptococcus neoformans* II. Surface receptors for cryptococcal polysaccharide and their role in inhibition of phagocytosis by polysaccharide. Infect Immun 16:99–106.

Kozel TR, Cazin J, Jr, 1971. Non-encapsulated variant of *Cryptococcus neoformans* I. Virulence studies and characterization of soluble polysaccharide. Infect Immun 3:287–294.

Kozel TR, Cazin J, Jr, 1974. Induction of humoral antibody response by soluble polysaccharide of *Cryptococcus neoformans.* Mycopathol Mycol Appl 54:21–30.

Kozel TR, Gulley WF, Cazin J, Jr, 1977. Immune response to *Cryptococcus neoformans* soluble polysaccharide: immunological unresponsiveness. Infect Immun 18:701–707.

Kozel TR, Mastroianni RP, 1976. Inhibition of phagocytosis by cryptococcal polysaccharide: dissociation of the attachment and ingestion phases of phagocytosis. Infect Immun 14:62–67.

Kozel TR, McGaw TG, 1979. Opsonization of *Cryptococcus neoformans* by human immunoglobulin G: role of immunoglobulin G in phagocytosis by macrophages. Infect Immun 25:255–261.

Kwon-Chung KJ, 1975. A new genus *Filobasidiella,* the perfect state of *Cryptococcus neoformans.* Mycologia 67:1197–1200.

Kwon-Chung KJ, Bennett JE, Theodore TS, 1978. *Cryptococcus bacillisporus* sp. nov: serotype B-C of *Cryptococcus neoformans.* Int J Syst Bacteriol 28:616–620.

Kwon-Chung KJ, Polacheck I, Bennett JE, 1982. Improved diagnostic medium for separation of *Cryptococcus neoformans* var. *neoformans* (serotypes A and D) and *Cryptococcus neoformans* var. *gattii* (serotypes B and C). J Clin Microbiol 15:535–537.

Lacaz C da S, Melhem M de SC, 1978. Immunoallergic assay with cryptococcin in police recruits in the city of São Paulo (Brazil). Rev Hosp Clin Fac Med Univ São Paulo 33:48–51.

Laxalt KA, Kozel TR, 1979. Chemotaxigenesis and activation of the alternative complement pathway by encapsulated and nonencapsulated *Cryptococcus neoformans.* Infect Immun 26:435–440.

Lindberg B, Lonngren J, Powell, DA, 1977. Structural studies on the specific type-14 pneumococcal polysaccharide. Carbohydr Res 58:177–186.

Macher AM, Bennett JE, Gadek JE, Frank MM, 1978. Complement depletion in cryptococcal sepsis. J Immunol 120:1686–1690.

Mahvi TA, Spicer SS, Wright NJ, 1974. Cytochemistry of acid mucosubstance and acid phosphatase in *Cryptococcus neoformans*. Can J Microbiol 20:833–838.

McGaw TG, Kozel TR, 1979. Opsonization of *Cryptococcus neoformans* by human immunoglobulin G: masking of immunoglobulin G by cryptococcal polysaccharide. Infect Immun 25:262–267.

Merrifield EH, Stephen AM, 1980. Structural investigations of two capsular polysaccharides from *Cryptococcus neoformans*. Carbohydr Res 86:69–76.

Miller GP, Kohl S, 1983. Antibody-dependent leukocyte killing of *Cryptococcus neoformans*. J Immunol 131:1455–1459.

Mitchell TG, Friedman L, 1972. In vitro phagocytosis and intracellular fate of variously encapsulated strains of *Cryptococcus neoformans*. Infect Immun 5:491–498.

Mohr JA, Muchmore HG, Tacker R, 1974. Stimulation of phagocytosis of *Cryptococcus neoformans* in human cryptococcal meningitis. J Reticuloendothal Soc 15:149–154.

Morgan MA, Blackstock RA, Bulmer GS, Hall NK, 1983. Modification of macrophage phagocytosis in murine cryptococcosis. Infect Immun 40:493–500.

Müller HE, Sethi KK, 1972. Proteolytic activity of *Cryptococcus neoformans* against human plasma proteins. Med Microbiol Immunol 158:129–134.

Murphy JW, Cozad GC, 1972. Immunological unresponsiveness induced by cryptococcal capsular polysaccharide assayed by the hemolytic plaque technique. Infect Immun 5:896–901.

Murphy JW, McDaniel DO, 1982. In vitro reactivity of natural killer (NK) cells against *Cryptococcus neoformans*. J Immunol 128:1577–1583.

Murphy JW, Moorhead JW, 1982. Regulation of cell mediated immunity in cryptococcosis I. Induction of specific afferent T suppressor cells by cryptococcal antigen. J Immunol 128:276–283.

Murphy JW, Mosley RL, Moorhead JW, 1983. Regulation of cell-mediated immunity in cryptococcosis II. Characterization of first-order T suppressor cells (Ts1) and induction of second-order suppressor cells. J Immunol 130:2876–2881.

Newberry WM Jr, Walter JE, Chandler JW, Tosh FE, 1967. Epidemiologic study of *Cryptococcus neoformans*. Ann Intern Med 67:724–732.

Palmer DF, Kaufman L, Kaplan W, Cavallaro JJ, 1977. Serodiagnosis of Mycotic Diseases 191 pp. Springfield, IL: C. C. Thomas.

Radhakrishnan VV, Mathai A, Shanmugham J, Mathews GJ, 1982. The role of hyaluronidase in experimental cryptococcal infections. Surg Neurol 17:239–244.

Raizada MK, Schutzbach JS, Ankel H, 1975. *Cryptococcus laurentii* cell envelope glycoprotein. Evidence for separate oligosaccharide chains of different composition and structure. J Biol Chem 250:3310–3315.

Rebers PA, Barker SA, Heidelberger M, Dische Z, Evans EE, 1958. Precipitation of the specific polysaccharide of *Cryptococcus neoformans* A by types II and XIV antipneumococcal sera. J Amer Chem Soc 80:1135–1137.

Reiss E, Cherniak R, Eby R, Kaufman L, 1984. Enzyme immunoassay detection of IgM to galactoxylomannan of *Cryptococcus neoformans*. Diagn Immunol 2:109–115.

Rhodes JC, Wicker LS, Urba WJ, 1980. Genetic control of susceptibility to *Cryptococcus neoformans* in mice. Infect Immun 29:494–499.

Schimpff SC, Bennett JE, 1975. Abnormalities in cell mediated immunity in patients with *Cryptococcus neoformans* infection. J Allergy Clin Immunol 55:430–441.

Schroeter GPJ, Temple DR, Husberg BS, Weil R, III, Starze TE, 1976. Cryptococcosis after renal transplantation: report of ten cases. Surgery 79:268–277.

Schupbach CW, Wheeler CE, Briggman RA, Warner NA, Kanof EP, 1976. Cutaneous manifestations of disseminated cryptococcosis. Arch Dermatol 112:1734–1738.

Scott EN, Muchmore HG, Felton FG, 1981. Enzyme-linked immunosorbent assays in murine cryptococcosis. Sabouraudia 19:257–265.

Slodki ME, Wickerham LJ, Bandoni RJ, 1966. Extracellular heteropolysaccharides from *Cryptococcus* and *Tremella:* a possible taxonomic relationship. Can J Microbiol 12:489–494.

Staib F, Mishra SK, Grosse G, Abel R, 1977. Pathogenesis and therapy of cryptococcosis in animal experiments. pp. 48–59. In Host–Parasite Relationships in Systemic Mycoses. Part I. Proc of 21st Annual OHOLO Biol Conf, Ma'alot, Israel. Beemer AM, Ben-David A, Klingberg MA, Kuttin ES, eds. Basel: S. Karger.

Tacker JR, Farhi F, Bulmer GS, 1972. Intracellular fate of *Cryptococcus neoformans*. Infect Immun 6:162–167.

Turner SH, Cherniak R, Reiss E, 1984. Fractionation and characterization of galactoxylomannan from *Cryptococcus neoformans*. Carbohydr Res 125:343–349.

Wilson DE, Bennett JE, Bailey JW, 1968. Serologic grouping of *Cryptococcus neoformans*. Proc Soc Exp Biol Med 127:820–823.

Chapter 13

Nocardiae

13.1 INTRODUCTION

The principal pathogenic nocardiae are *Nocardia asteroides*, the major species of medical importance, *Nocardia brasiliensis*, more frequent in the tropics, and rarely *Nocardia caviae*. They are soil-dwelling bacteria that grow as long, branching filaments, 0.5 to 1.0 μm in diameter. Older filaments fragment into coccoid and bacillary forms in liquid medium. Sporulation occurs on solid medium when aerial hyphae fragment into arthrospores. All three species give a beaded Gram-positive reaction in tissue. Nocardiae are partially acid-fast, and this reaction is labile to acid–alcohol decolorization, unlike that obtained with the mycobacteria. In abscesses, nocardial filaments are surrounded by polymorphonuclear neutrophils or are ingested by macrophages. Morphologic and tinctorial features are reviewed by Robboy and Vickery (1970) and the histopathology by Chandler et al (1980).

Infection may occur by inhalation or by traumatic implantation in the skin. Human-to-human transmission is not known to occur. In the USA, 100 to 1000 new cases are recognized annually (Beaman et al, 1976; Curry, 1980); of these 85% are serious pulmonary or systemic infections. *Nocardia asteroides* is also pathogenic for dogs and causes severe mastitis in cattle, but no evidence exists of transmission from animals to humans.

The unit lesion of pulmonary nocardiosis is a focal pneumonitis that progresses either as necrotizing lobar pneumonia, segmental infiltrates, single or multiple nodules, cavities, or pleural effusions. Chest radiographs are nonspecific and variable. The disease remains pulmonary in 70% of cases and is usually chronic, but may be rapid in debilitated patients. When dissemination occurs, cerebral abscesses are notable (Byrne et al, 1979) and may occur even in the absence of a detectable pulmonary phase. Persons with no known immunologic dysfunction and not receiving immunosuppressive therapy may still develop this disease. The outlook for patients receiving immunosuppression who have pulmonary nocardiosis is bleak at present. The aggressive use of immunosuppressive agents for neoplastic, or autoimmune processes and for management of renal allografts should heighten the index of suspicion of nocardiosis among these patients. (Table 13.1) Compromised host defenses due to underlying primary disease were cited in 50% to 75% of nocardiosis cases in four series of patients, although the majority do not suffer from cancer (Curry, 1980).

Difficulties exist in the current methods of diagnosis of pulmonary or multisystem nocardiosis. Chest radiographs are nonspecific and variable. Nocardiae are slow-growing and often are mixed with other rapidly growing bacteria that obscure detection. The differential diagnosis must exclude tuberculosis because antituberculous therapy is ineffective against nocardiosis. Therapeutic choices of proven efficacy are limited at present to sulfa drugs so that successful empirical antibiotic therapy is difficult. Sputum cultures may be negative, requiring more invasive procedures to obtain a specimen. For these reasons immunodiagnosis and monitoring of the immune responses during therapy could provide a valuable clinical adjunct. Progress in this area has been slow, primarily because of antigenic crossreactions with mycobacteria. The immunosuppressed status of some patients may further limit diagnostic procedures that rely on the detection of antibodies.

- Actinomycotic mycetoma. Localized extrapulmonary nocardiosis in North America has been reported as leg ulcers, or following compound fractures (Goodman and Koenig, 1970). Chronic infections in the tropics or subtropics result from traumatic implantation in the lower extremities during barefoot labor or on the back from carrying produce in burlap sacks. These conditions may persist for long periods before treatment is sought. The lesions present as large swellings with multiple draining sinus tracts sometimes involving adjacent bone, but usually remaining localized. The drainage contains granules, which are masses of nocardiae. The triad of tumefaction, sinus formation, and granules is present in actinomycotic mycetoma. The species involved are *N brasiliensis, N asteroides, Actinomadura madurae, Actinomadura pelletieri,* and *Streptomyces somaliensis.*

13.2 MURAL COVALENT SKELETON

Structural homology exists between cell walls of mycobacteria and nocardiae. The covalent skeleton is the wall residue after extraction of free lipids, peptides that are susceptible to proteinases, pigment, and minor neutral polysaccharides, eg, glucan (Petit and Lederer, 1978). The covalent skeleton is defined as the insoluble matrix composed of three elements: nocardomycolic acids, arabinogalactan, and peptidoglycan. Interest in the covalent skeleton is due to its adjuvant and mitogenic activi-

Table 13.1. Predisposing Factors for Nocardiosis

Steroid therapy including maintenance of renal allografts (Goodman and Koenig, 1970, Young et al, 1971)

Cancer, especially the leukemias, and immunosuppressive cancer therapy (Scully et al, 1980; Young et al, 1971)

Pulmonary alveolar proteinosis (Andriole et al, 1964)

Chronic granulomatous disease of childhood

Autoimmune diseases, lupus erythematosus (Gorevic et al, 1980)

Ulcerative colitis

Cushing's disease (Geisler and Anderson 1979)

Alcoholism (Byrne et al, 1979)

ties. The minimal subunit retaining the adjuvant activity of the mycobacterial or nocardial component of complete Freund adjuvant is a muramyl dipeptide, N-acetylmuramyl-L-alanyl-D-isoglutamine. A large subunit of some nocardial walls is a B-lymphocyte mitogen (see 13.7).

13.2.1 Nocardomycolic Acids

Nocardomycolic acids are α-branched, β-hydroxy fatty acids with skeletons of about 50 to 70 carbon atoms. The α-substituent is a C_{10} or C_{12} side chain (Azuma et al, 1973) or C_{10} to C_{16} (Lechevalier, 1977), or C_{16}, C_{18} (Beaman, 1975) whereas the main chain length varies from C_{38} to C_{60} (Lechevalier et al, 1973). Vacheron et al (1972) characterized the main β-hydroxy fatty acid as a tetra-unsaturated $C_{62}H_{116}O_6$ nocardic acid. The nocardomycolic acids are taxonomic acids, useful in differentiating nocardiae from mycobacteria and corynebacteria (Lechevalier et al, 1973).

The distribution of nocardomycolic acids varies in the wild type N asteroides and in L-form revertants (see also 13.6.1). Mycolic acids make up between 15% to 20% of the total wall mass (Beaman et al, 1981). Mycolic acid with a total chain length of C_{54} was a major constituent in the cell walls of all revertants. One revertant had longer chain mycolic acids ($>C_{54}$) in higher concentration. During pyrolysis at temperatures above 300 °C, fragments of the α-chain and an aldehyde from the β-chain are produced. These pyrolysis products are characteristic of the genus and allow discrimination between corynebacteria, mycobacteria and nocardiae. The diversity in mycolic acid composition among L-form revertants may explain their different colonial morphology (Beaman and Bourgeois, 1981).

13.2.2 Arabinogalactan and Arabinomannan

Arabinogalactan is a branched polysaccharide with a proposed unit structure shown in Figure 13.1 (ara:galactose = 2.33–2.87:1). Arabinogalactan consists of a β-1,4-galactose disaccharide alternating with α-1,5-arabinofuranose disaccharides. Branching with arabinofuranosides of a degree of polymerization less than 6 occurs through C-3 of arabinose in the main chain. The structural features are constant for diverse species including *Mycobacterium tuberculosis*, *Mycobacterium bovis*, *Mycobacterium phlei*, *Mycobacterium smegmatis* (Yamamura et al, 1972; Misaki et al, 1974) *N asteroides*, and *Corynebacterium diphtheriae* (Yamamura et al, 1972; Azuma et al, 1973; Beaman, 1975) (Table 13.2). About one in ten arabinofuranosyl residues is esterified to mycolic acid via a 5'-hydroxyl of the terminal arabinose. Proof of this

Figure 13.1. Proposed repeat unit structure of mycobacterial (*M bovis* BCG) D-arabino-D-galactan. ara, arabinose; gal, galactose; f, furanose; p, pyranose. Molecular weight, 31,000 da, molar ratio of ara:gal is 5:2 (Misaki et al, 1974).

$$5\,\alpha\,\text{ara}\,f\,1 \rightarrow 5\,\alpha\,\text{ara}\,f\,1 \rightarrow 4\,\beta\,\text{gal}\,p\,1 \rightarrow 4\,\beta\,\text{gal}\,p\,1 \rightarrow 4\,\beta\,\text{gal}\,1$$
$$\uparrow_1^3$$
$$2\,\alpha\,\text{ara}\,f$$
$$\uparrow_1^3$$
$$[5\,\alpha\,\text{ara}\,f]_{n>3}$$
$$\uparrow_1$$
$$\alpha\,\text{ara}\,f$$

Table 13.2. D-Arabino-D-Galactans of Mycobacteria and Nocardiae

Source	Molecular Weight	$[\alpha]_D$	Arabinose : Galactose	Reference
Mycobacterium tuberculosis var *hominis*	31,000	23.5–25.9°	5:2	Misaki et al, 1974
M bovis BCG, *M phlei, M smegmatis*	31,000	23.5–25.9°	5:2	Misaki et al, 1974
Nocardia asteroides 131	nd	22.8°	2.7:1	Azuma et al, 1973
N asteroides 14759	nd	nd	2:1	Beaman, 1975
N brasiliensis	nd	48°	3:1	Estrada-Parra et al, 1965

Abbreviation: nd, not determined.

linkage was provided by isolation of arabinose-5-O-mycolate after mild hydrolysis of bacilli or of wax D (Petit and Lederer, 1978). Structural analysis of mycobacterial arabinogalactan was aided by the action of "M-2" endo-glycosidase of a soil bacterium (Kotani et al, 1971; Yamamura et al, 1972; Misaki et al, 1974) that liberated oligoarabinosyl side chains, including a hexasaccharide containing α-1,5- and α-1,2-arabinofuranosides in a 3:2 ratio. This hapten was a potent inhibitor of immune precipitation of the intact polysaccharide from *M bovis* var BCG cell walls. Another enzyme, β-galactosidase of *Aspergillus niger*, hydrolyzed most galactoside linkages in the arabinogalactan leaving an arabinose-rich core (Misaki et al, 1974). The molecular weight of arabinogalactan from *M bovis* BCG walls was 3.1×10^4.

The lectin concanavalin A binds sugar residues with a 3,4,6-arabino-α-D-glycopyranosyl or α-D-arabinofuranosyl configuration thus accurately predicting that the arabinogalactans react well with Con A (Misaki et al, 1974). Arabinogalactan is crosslinked to peptidoglycan by phosphodiester bridges between the C-6 hydroxyl of one of ten muramic acid residues and arabinose. Both muramic acid-6-phosphate and arabinose phosphate were isolated (Petit and Lederer, 1978).

Arabinomannan is formed more in the cytoplasm of mycobacteria and nocardiae (Yamamura et al, 1972). It is isolated from defatted cytoplasm with dilute alkali and purified on diethylaminoethyl-cellulose. A mannan fraction is further purified by Cu^{++} complexation with Fehling solution. The structural features of *M smegmatis* arabinomannan are a mannan region comprised of a linear α-1,6-mannan with single mannose or mannosyl disaccharides linked to the backbone through C-2. The arabinose segment of the molecule is similar to that found in that arabinogalactan.

Arabinogalactans of the mycobacteria listed above and that from *N asteroides* form single immunoprecipitin arcs that fuse completely with each other. They form spurs, indicating partial identity, with arcs containing arabinomannan (Yamamura et al, 1972).

An early report that *N asteroides* contains a species-specific arabinogalactomannan (Zamora et al, 1963) has not been verified. More likely it was a mixture of arabinogalactan and arabinomannan. Careful analysis of the mycobacterial polysaccharides in an immunoelectrophoresis system (Janicki et al, 1971; Beck, 1972) showed that of the 13 reference immunoprecipitates formed with antigens of *M tuberculosis* var *hominis* only one, number 2, identified as arabinomannan, was held in common among 12 other mycobacterial species tested. The search for antigens specific for *N asteroides* proceeds on the assumption that the presence of arabinogalactan and arabinomannan in antigenic complexes are impediments to the achievement of specificity and their removal is desirable.

13.2.3 Peptidoglycan

The general plan of bacterial peptidoglycans consists of strands of the repeating subunit β-1,4-N-acetylglucosaminyl-N-acetylmuramic acid linked to pentapeptide. An interlocking cagelike structure is formed by peptide bridges joining the glycopeptide subunits. This plan is varied in some respects in mycobacterial and nocardial peptidoglycan. Muramic acid occurs as the N-glycolyl derivative. A tetrapeptide is present, L-alanine-D-α-glutamic acid-(L)-diaminopimelic acid-(D)-NH$_2$-(L)-D-alanine, with the α-carboxyl of glutamic acid and the carboxyl on the D-carbon of meso-diaminopimelic acid amidated. The bridge units in mycobacterial peptidoglycan are short, consisting of either D-alanine or *meso*-diaminopimelic acid. The latter seems restricted to mycobacteria (Petit and Lederer, 1978) and D-alanine serves that function in *Nocardia kirovani* (Vacheron et al, 1972).

Morphogenesis and cell wall changes occur during the growth cycle of *N asteroides* in vitro. Midlog phase cells are long branching filaments that gradually fragment into short rods and cocci. The wall has a trilaminar profile in the log phase composed of two dense layers separated by a lucent layer (Beaman, 1975; Beadles et al, 1980). Consideration of the antigenic structure cannot proceed rationally without an understanding of the morphogenesis of the organism.

Growth of *N asteroides* 14759 on brain-heart infusion broth at 34 °C is characterized by changes in form with the age of the culture that correlate with changes in the cell wall composition (Beaman, 1975) (Figure 13.2). The unicellular inoculum undergoes a lag phase of 7 hours during which time the bacteria elongate. Long, branching filaments are present in the logarithmic phase, at 16 hours, and the stationary phase is reached by 30 hours. After that, the filaments begin to fragment into short rods and cocci. Cross-sections of the cell wall in logarithmic phase reveal (i) an outer layer of lipid that declines in the stationary phase; (ii) A dense middle layer composed of carbohydrate and lipid; and (iii) An inner osmophilic layer of peptidoglycan and polysaccharide. The inner layer remains intact during the stationary phase and is a major component accounting for 40% of the wall weight. With age, the lipid composition declines and shifts towards relatively more tuberculostearic and longer chain fatty acids. The proportion of bound lipids not removable with ethanol: diethyl ether, or with chloroform, also increases. The cytoskeleton in the stationary phase is thus composed mostly of peptidoglycan-arabinogalactan, with a small but significant bound-lipid fraction. This change in form and composition with culture age is accompanied by a major decline in virulence (Beaman and Maslan, 1978) (Table 13.3).

13.3 ANTIGENIC COMPLEXES

13.3.1 Serologic Analysis

The study of antigens of the nocardiae has been pursued intermittently for over 50 years (Drake et al, 1943). Some studies have been devoted to taxonomic differentiation of related aerobic actinomycetes and others to the diagnosis of infection.

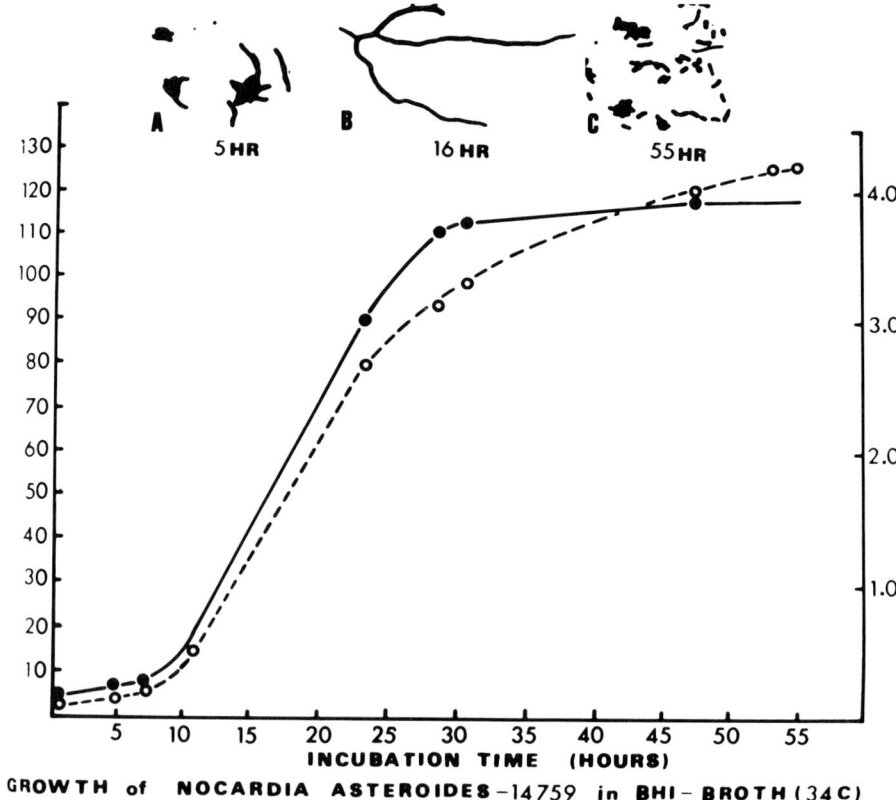

Figure 13.2. Growth curve and morphology of *Nocardia asteroides* ATCC 14759 on brain-heart infusion broth at 34 °C on a gyratory shaker, 150 rev/min. Letters denote morphologic changes: A, coccoid or rod shapes of the initial inoculum increase in diameter and filaments begin to form; B, long branching filaments observed during mid-logarithmic growth; C, total fragmentation into short rods and cocci at 55 hours. Filled circles, colony-forming units ($\times\ 10^7$); open circles, A_{580nm} (Beaman 1975). *Source:* Reprinted with permission of the author and of the publisher, American Society for Microbiology.

Table 13.3. Cell Wall Composition of *N asteroides* ATCC 14759

	Percent of Wall Dry Weight		
Constituent	Logarithmic Phase, 16 hours	Stationary Phase, 55 hours	After 1 week
Peptidoglycan	25.1	36.6	43.3
Amino acids	17.1	20.6	16.9
Arabinose	17.0	17.0	18.2
Galactose	11.1	9.9	9.4
Glucose	3.4	3.4	3.7
Lipid	21.0	11.0	7.0
Total recovered	94.7	98.5	98.5

Source: Beaman, 1979.

Complex antigenic structures have been demonstrated among bacteria of the genera *Nocardia* and *Mycobacterium*. Antigenic differences also occur within the species of *Nocardia*. The antigens used for evaluation have been unfractionated culture filtrates or cytoplasmic solubles prepared by mechanical disruption or ultrasonic treatment. Differences in antigen preparations aside, a variety of serologic techniques have been used to assess antigenic variations including agglutination, hemagglutination, complement fixation, and precipitin analyses by diffusion in agar gels. In some of these reports evidence for genus-specific as well as species-specific antigens was presented, but none has been corroborated by others or has been used extensively, probably because of the low degree of sensitivity, specificity, or both that can be obtained with unpurified antigenic complexes. Shortcomings due to lack of sensitivity may be overcome by using enzyme- or radiolabeled antibodies. In the setting of immune compromise, antigenemia or soluble immune complexes may be present.

13.3.1a Agglutinins

Schneidau and Shaffer (1959) observed crossreactivity between species of *Nocardia* and *Mycobacterium tuberculosis*. In 1962, Cummins analyzed the antigenic structure of cell walls of these two groups and found a common antigenic component, which had arabinose and galactose as the principal sugars.

Erythrocytes sensitized with antigens from culture filtrates of *N asteroides* reacted with antisera prepared against nocardiae and some mycobacteria, but specificity could be discerned by a combination of hemagglutination and hemolytic inhibition tests (Thurston et al, 1968).

13.3.1b Complement Fixation

Kwapinski and Seeliger (1965) found crossreactivities in the cell wall components from different nocardial and mycobacterial species. Pier and Keeler (1965) used *N asteroides* culture filtrate antigens in complement-fixation tests to detect antibodies in bovine nocardiosis and compared sera from cattle with nocardiosis or mycobacteriosis (Pier et al, 1968). The mycobacterial complement-fixing antigens did not differentiate between sera from cattle infected with either *Nocardia* or *Mycobacterium* species, but nocardial antigens did not crossreact with sera of cattle infected with mycobacteria. Using *N asteroides* culture filtrate antigens, Shainhouse et al (1978) detected antibodies in human nocardiosis as well as in sera from human tuberculosis or leprosy patients.

13.3.1c Immunodiffusion

The most extensive studies on serologic relationships between *Nocardia* and *Mycobacterium* species have been performed with immunodiffusion methods. All mycobacterial and nocardial strains possess at least one antigen or hapten in common (Castelnuovo et al, 1968; Kwapinski et al, 1973).

A battery of mycobacterial immunoprecipitation reference antisera produced in sheep were compared in immunodiffusion to determine the extent of crossreactions with nocardiae (Figure 13.3). Four- to six-week-old cultures grown on chemically defined Sauton's medium at 37 °C were the source of culture filtrate and homogenate-supernate antigens. No differences were detected in either the exocellular or cytoplasmic soluble antigens. Ten isolates of *N asteroides* were tested; of these, eight had one or two antigens that crossreacted with anti-*M bovis* BCG, the remaining two

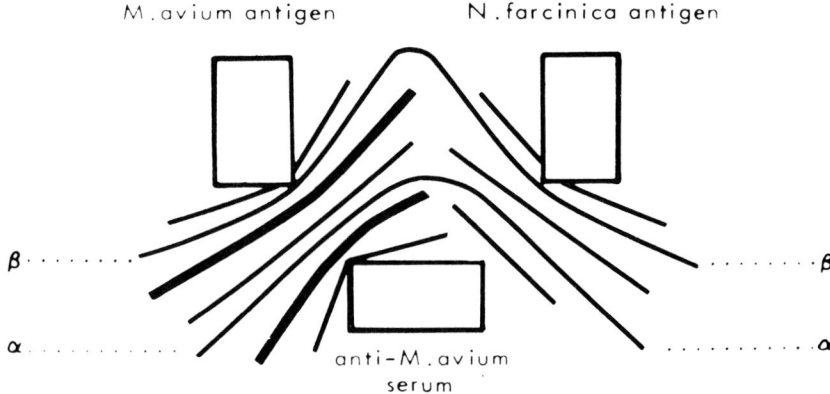

Figure 13.3. Tracing of mycobacterial antigen-antibody reference immunoprecipitation system (*Mycobacterium avium*) illustrates crossreactions with *Nocardia farcinica*. Common factors α and β are shown. Factor α activity was present in all ten *Nocardia asteroides* tested and factor β was present in two of them. Culture supernates and cell extract antigens were found by Ridell and Norlin (1973) to be interchangeable. *Source:* Reprinted with permission of the authors and of the publisher, American Society for Microbiology.

reacted with anti-*Mycobacterium kansasii* and with other species. This similarity was even shown when these systems were evaluated with ribosomal preparations (Ridell et al, 1979).

Culture filtrates of two isolates of *N brasiliensis* were found to crossreact extensively with antimycobacterial sera. The reaction of *M bovis* BCG with its homologous antiserum produced a multiplicity of twelve arcs, thus suggesting that crossreactions with nocardiae are possibly the result of one or two antigens. If that assumption has merit, the task of removing crossreactivity does not appear very difficult. A heat-stable polysaccharide factor, β, common to the mycobacteria, was also detected in extracts of two of ten *N asteroides* cultures (Ridell and Norlin, 1973).

Homogenate-supernate antigens from many isolates of related *Nocardia* species have been screened for serologic crossreactions in immunodiffusion versus antisera produced against *N asteroides*, and *N caviae* reference cultures (Ridell, 1981). The reference antigens contained six or seven factors when reacted with homologous antisera. Several *N asteroides* isolates formed a cluster, sharing four to six common antigens, but others did not show such close antigenic relationships. *Nocardia brasiliensis* isolates contained fewer common antigens reactive with antisera to *N asteroides*. A clustering of *N caviae* isolates occurred with up to five to seven shared factors.

Serologic relationships among the nocardiae, mycobacteria, and fungi were studied by immunodiffusion using homogenate-supernates of cultures as antigens (Kwapinski et al, 1973). Pathogenic and saprophytic nocardiae shared a common antigen. Some differences observed between *N asteroides* and *N brasiliensis* support the view that species-specific factors occur. Cytoplasmic antigens of *N brasiliensis* crossreacted with pathogenic, but not with saprophytic mycobacteria. Nocardiae containing an antigen designated as "A" (including *N asteroides*) reacted with anti-*Candida albicans* sera, an observation that bears remembering as more sensitive immunoassay formats are adapted to nocardiosis serology.

Nocardia asteroides cells were sonically disrupted for seven hours to prepare antigens for an evaluation of nocardiosis patients' sera in immunodiffusion (Humphreys et al, 1975). The need for such prolonged sonication was not explained. A relatively low percentage of nocardiosis patients (nine of 20) were positive for precipitins and significant crossreactions with tuberculosis patients occurred. Somewhat greater sensitivity was achieved when culture filtrate antigens of N asteroides were used to detect complement-fixing antibodies in 13 of 16 nocardiosis patients.

Three N asteroides isolates were selected because they elicited more precipitins in rabbits than other isolates (Blumer and Kaufman, 1979). Filtrates of 2-week-old cultures grown on peptone dialyzate medium at 36 °C were pooled from these three isolates and a homogenate-supernate from one of these, N asteroides B1042, also was prepared. Immunodiffusion patterns of sera from 71 nocardiosis patients were obtained alongside a multivalent reference antiserum from a patient infected with N brasiliensis. Forty-nine percent of patients reacted with the homogenate antigen, most showing a line of identity with the control antiserum. Higher sensitivity, 70%, was achieved with the culture filtrate pool. The culture filtrate antigen was more prone to crossreact with mycotic, actinomycotic, or tuberculosis antisera. Crossreactions produced lines of identity with the nocardiosis reference antiserum in only a small fraction of cases. Thus the device of using a human reference antiserum increased the test's reliability. When both homogenates and culture filtrate antigens were used, an increment in sensitivity reaching 79% was achieved. Most crossreactions occurred with sera from tuberculosis patients. If a line of identity with the human reference antiserum was taken as the criterion, only 47% of nocardiosis patients could be detected as test positive.

13.3.1d Serotypes

Pier and Fichtner (1971, 1981), in reports concerned mostly with bovine nocardiosis, identified four antigens in filtrates of 2-week-old cultures grown on neopeptone dialyzate medium at 36 °C. Antisera specific for each factor were produced by immunizing rabbits with one of four N asteroides cultures selected because each secreted only one of the serotypic antigens (Table 13.4). Results of screening many isolates of N asteroides showed that factor II was produced alone or with factor IV; factor III also occurred alone or with factor IV; but factors I, II, III were mutually exclusive, were restricted to N asteroides, and were not found in N brasiliensis, N caviae, Actinomadura pelletieri, Mycobacterium fortuitum, or M bovis BCG. Factor IV also was produced by N brasiliensis. Whether sera from persons infected with M tuberculosis or other mycobacteria would react with these N asteroides factors has not been determined. Although the antisera have been known for some time, the biochemical nature of the antigens involved is unknown. A large number of N asteroides isolates of diverse clinical, environmental, and geographic origins were serotyped: types 1 and 4 were the most frequent clinically encountered, type 3 was intermediate in occurrence, and serotype 2 was comparatively rare (Pier and Fichtner, 1981).

Table 13.4. *N asteroides* Cultures Representative of Four Serotypes

Isolate	Factor(s) Secreted
40 N	I
47 N	II
India 10	III
36 N	IV

Source: Pier and Fichtner, 1971.

The heterogeneity and potential for type-specific factors among *N asteroides* were also demonstrated by Kurup and coworkers (1981, 1983) although they found seven immunotypes, each with its own distinct antigenic determinant. They characterized immunoabsorbed antigens in immunodiffusion, two-dimensional polyacrylamide gel electrophoresis, and crossed "rocket" immunoelectrophoresis. The approaches taken in the more recent studies increase the probability that factors specific for *N asteroides* are capable of purification.

13.3.1e Mycobacterial Reference Precipitins

Thirteen mycobacterial antigens were classified in an immunoelectrophoresis reference system, the components of which are available for distribution (Janicki et al, 1971; Beck, 1972). The system is of interest because (i) the polysaccharide antigens present in cell extracts are known to crossreact with the nocardiae and (ii) attempts towards standardization of nocardial antigens should be encouraged. Although 13 antigenic factors were identified, the additional resolving power of two-dimensional crossed "rocket" immunoelectrophoresis detected 36 factors in the same antigenic complex (Beck, 1972). Furthermore, other antigens detected in primary binding assays may not precipitate in agar gel. Most of the antigens present in *M tuberculosis* var *hominis* are present in *M bovis* BCG, but in twelve other mycobacterial species only arabinomannan (band 2) was held in common. This illustrates the antigenic diversity among the medically important aerobic actinomycetes. Bands "1" and "3" have been identified as arabinogalactan (Seibert's polysaccharide I) and a glucan (Seibert's "II").

The limitations of using unpurified *N asteroides* antigens to detect specific antibodies in patients are amply demonstrated by these studies. No attempt at fractionation of these complex mixtures has been reported, although the identity of arabinogalactan as a common antigen among nocardiae and mycobacteria is well established. Furthermore, the lability of arabinogalactan to periodate oxidation and its affinity for Con A have also been reported (Estrada-Parra et al, 1965; Azuma et al, 1973; Misaki et al, 1974).

13.3.2 Delayed Hypersensitivity

Cytoplasmic extract antigens were made from 3-week-old cultures of *N asteroides* IP766 and *N brasiliensis* UPHG24. (Ortiz-Ortiz et al, 1972c). The bacteria were defatted with ethanol-diethyl ether, dried, and resuspended in Tris buffer before being disrupted in a Ribi cell fractionator. The 144,000 g supernates were dialyzed versus water and lyophilized. The ability of these antigens to elicit specific delayed cutaneous hypersensitivity and direct migration inhibitory factor production was assessed in guinea pigs infected with the two *Nocardia* species or with *M bovis* BCG. Skin tests showed that for *N brasiliensis* and *N asteroides*, relative specificity was achieved expressed as a *specific difference* of 7 according to Magnusson (1961):

$$\text{specific difference} = (A_a + B_b) - (A_b + B_a)$$

Capital letters denote skin reactions in animals sensitized with either organism a or b (lower case letters). The source of the antigen is from either organism A or B. A_a is an homologous reaction, B_b is an homologous reaction, A_b is the reaction when antigen A is injected into animals sensitized with organism b and vice versa for B_a. When the specificity difference is greater than 0, then the homologous reaction is larger than the heterologous one. The specific differences to *Nocardia* antigens in BCG-sensitized animals were 12–13 indicating the homologous reaction is far greater than the heterologous one. Surprisingly, an even greater degree of specificity was seen at the level of percent inhibition of migration in the direct MIF assay. Only peritoneal exudates from guinea pigs sensitized with the homologous antigen were able to respond in vitro.

Delayed cutaneous hypersensitivity to a homogenate-supernate of defatted *N brasiliensis* or *N asteroides* was estimated in patients with mycetoma or those with tuberculosis or leprosy or in healthy individuals (Ortiz-Ortiz and Bojalil, 1972). Responses to 5 μg of protein in the cytoplasmic extract were measured 48 hours after skin tests. In addition to protein the extract contained 27% carbohydrate. All ten of the *N brasiliensis* mycetoma patients displayed positive delayed cutaneous hypersensitivity with the homologous antigen; reactions with similarly prepared *N asteroides* antigens were fewer and lower in diameter. Low, if any, reactivity was observed in tuberculosis or leprosy patients or in healthy persons residing outside the endemic area. The tropical state of Morelos in Mexico, where sugar cane cultivation is the main agricultural occupation, is endemic for *N brasiliensis* mycetoma and 500 colonies of the organism were recovered per 0.1 g of soil. Two thirds of the 15 healthy residents tested reacted with the *N brasiliensis* antigens. The cytoplasmic extract appears to be adequate to survey exposure in the community. A more extensive evaluation is needed including persons with positive tuberculin tests.

13.4 FRACTIONATION OF NOCARDIAL EXTRACTS

13.4.1 *Nocardia*-Active Polypeptides

Nocardia asteroides-active polypeptides are produced by a method similar to that devised to make mycobacterial-active peptides (see below). *Nocardia*-active polypeptides are specific for the genus *Nocardia* in cutaneous (Kingsbury and Slack, 1967, 1972) and lymphocyte blastogenesis tests (Hiramine et al, 1981) showing little or no crossreactivity with the mycobacterial component of Freund adjuvant. The extract is prepared from defatted whole nocardiae by digestion with dilute HCl and precipitation of the liberated polypeptides by picrate complexation. Delayed-type hypersensitivity resides in the dialyzate of *Nocardia*-active polypeptides (Kingsbury and Slack, 1967, 1972) and others (Hiramine et al, 1981) observed that these antigens partitioned at equivalent specific activity in both the ultrafiltrate and retentate after filtration through a membrane having a nominal weight limit of 25,000 da (Amicon CF 25).

Six guinea pigs were sensitized with different *N asteroides* isolates and four responded positively to *Nocardia*-active polypeptides pooled from three different *N asteroides* cultures. Specificity was narrow, and *N brasiliensis*-sensitized animals showed no crossreactions. These preparations have potential as epidemiologic or even diagnostic adjuncts; moreover, they are not toxic in animals and do not induce delayed cutaneous hypersensitivity after repeated injections (Kingsbury and Slack, 1967).

Self-limited subcutaneous infection of guinea pigs in the nuchal area provides a model to measure the kinetics of T-lymphocyte activation in the draining cervical nodes, and to relate histopathology of the lesion to in vitro cell-mediated immune functions evoked by *Nocardia*-active polypeptides (Figure 13.4). In the first week

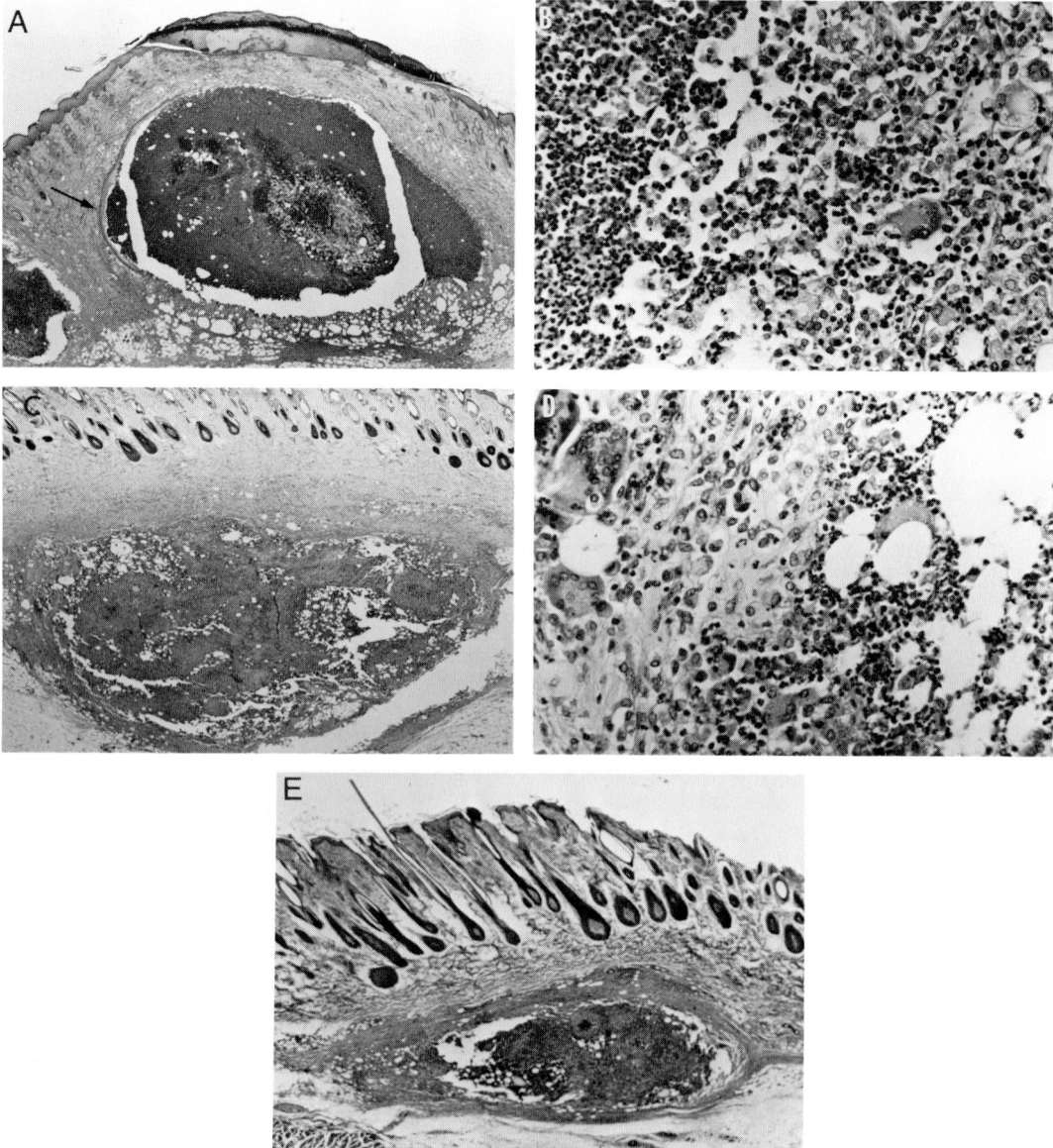

Figure 13.4. Histologic sequence of lesions in the nuchal area of a guinea pig infected subcutaneously with *Nocardia asterioides*. (a) A low power view of the inoculated site 7 days after infection. An abscess lies in the subcutis to dermis. To the left side the squamous epithelium is deep in the dermis (arrow) sharply delimiting the abscess cavity from normal tissue. Crust formation has occurred in the overlying skin accompanied by thin regenerating epithelium. (b) A high power view of the inoculated site 2 weeks after infection. On the left side, is a boundary zone separating the collection of PMN from the surrounding chronic inflammation composed of epithelioid and giant cells (hematoxylin-eosin, H and E, $\times$ 170). (c) The infected site after 3 weeks. The lesion is a large mass derived from replacement of the central zone of suppurative necrosis by actively proliferating granulomatous tissue (H and E, $\times$ 10). (d) Higher magnification of panel c showing an individual granuloma with a central area of degenerating PMN surrounded by granulation tissue containing epithelioid and giant cells on the right side (H and E, $\times$ 170). (e) The inoculated site after 4 weeks shows a small cutaneous lesion that is granulomatous and completely encapsulated with thick fibrous tissue (H and E $\times$ 15) (Hiramine et al, 1981). *Source:* Reprinted from the International Archives of Allergy and Applied Immunology with permission from the authors and the publisher, S. Karger AG, Basel.

after nuchal infection, an abscess perforating the skin formed, becoming encapsulated by a broad granulomatous zone at 2 weeks. The lesion contained active macrophages, epithelioid cells, and Langhans giant cells. Thereafter, the granulomas became circumscribed and completely resolved 6 weeks after infection.

Dissemination from the subcutaneous focus to the draining cervical nodes occurred, and nocardiae were also observed in the spleen and lungs, inducing hyperplasia of lymphoid follicles in these organs. Skin tests with *Nocardia*-active polypeptides (NAP) were positive for delayed hypersensitivity from 1 week postinjection, revealing a mononuclear infiltrate with few PMN.

Blastogenic responses of cervical lymph node cells showed a sharp peak of in vitro activity two weeks postinjection coinciding with the appearance of a granulomatous type IV reaction in the infected subcutaneous focus. Blastogenic indices were high: a blastogenic index of 80 at a NAP concentration of 20 μg/ml.

T-cell help and macrophage processing of NAP are required to drive blastogenesis of purified lymphocytes from *N asteroides* infected guinea pigs (Hiramine et al, 1981). Collaboration between mitomycin-treated T-cells and B-cells was necessary for B-lymphocyte blastogenesis. T-cells proliferated in the presence of NAP only when macrophages were present. The magnitude of T-cell responses to NAP was much higher, with a blastogenic index equal to 17.9, than B-cell blastogenesis, with a blastogenic index of 2.1, even in the presence of accessory cells.

13.4.2 Superoxide Dismutase

Nocardia asteroides resists killing by PMN despite the oxidative burst (Filice, 1983). The rate of iodination of *N asteroides* was 1/10 that of zymosan and a ten to one hundredfold higher concentration of H_2O_2 and superoxide was required to kill the bacteria than *Staphylococcus aureus*. The resistance cannot solely be explained on the basis of catalase because even after neutralization of catalase with aminotriazole, nocardiae were not as susceptible as staphylococci to killing by H_2O_2 or by lactoperoxidase-H_2O_2-iodide, or by neutrophils.

Nocardia asteroides induces an oxidative burst in human PMN and monocytes, yet this organism is not killed by the production of H_2O_2 or superoxide radicals (Beaman et al, 1983). Superoxide dismutase (SOD) is the predominant protein secreted into the medium during all phases of *N asteroides* growth and represents over 75% of the total secreted protein when the culture reaches stationary phase. The enzyme is associated with the outer cell envelope as shown by immunofluorescence. It can be localized on polyacrylamide gels by the photochemical reduction of nitroblue tetrazolium to the colored formazan. *Nocardia asteroides* SOD from a virulent strain was unique and species-specific. It has a molecular weight of 100,000 da consisting of four equal subunits held together noncovalently. It also contained 2 g atoms Mg^{++}, Fe^{++} and Zn^{++} with trace amounts of Cu^{++}, Co^{++} and Ni^{++}, which was unusual because other bacterial SOD have only one or two metals. This SOD was different from the trimer of *Mycobacterium tuberculosis* SOD which has a molecular weight of 65,000 da and contains only Fe^{++}.

13.4.3 Enzyme-Linked Immunoelectro Transfer Blot

Proteins in the culture filtrate and homogenate of *N asteroides* can be separated analytically on pH gradient polyacrylamide gels (Rajki et al, 1982; El-Zaatari et al, 1986). The proteins present in filtrates of 4-week-old cultures grown on chemically defined Sauton's medium have isoelectric points in the pH 4.0 to 5.2 range. The proteins were electrotransferred onto nitrocellulose paper and the strips cut from the "blotted" proteins act as solid-phase adsorbed antigens suitable for enzyme immunoassays (El-Zaatari et al, 1986). At least 20 of these proteins are antigenic as judged by their reactivity with antisera from rabbits immunized with culture filtrates in incomplete Freund adjuvant. Only quantitative differences were seen between the antigens present in the supernate of homogenized *N asteroides* and in the culture filtrate. Seventy-five percent of 20 human nocardiosis cases reacted in the immunoblot test. The pattern of recognized antigens was variable (Figure 13.5) but a specific antigen focusing at a pI of 4.68 was reactive with most nocardiosis case sera, but not with any sera of 17 tuberculosis patients. The nocardial antigens react with sera from patients infected with *N asteroides*, *N brasiliensis*, and *N caviae*, but not with four patients from whom *Rhodococcus* species were isolated. This technique permits the identification of specific antigenic factors. The enzyme immunoelectrotransfer blot (EITB) method can be recommended as a prototype immunodiagnostic technique because, once formed, the blot strips are stable and provide the dual advantage of identifying individual protein reactions coupled with the sensitivity afforded by a primary binding assay, in this instance an enzyme immunoassay.

13.4.4 Polysaccharides

The accepted roles for nocardial polysaccharides are (i) as arabinogalactan precipitinogens that crossreact with similar if not identical components of mycobacteria and corynebacteria and (ii) as portions of the cytoskeleton that are mitogens (see below). One report, however, claims that glycans from nocardiae evoke delayed cutaneous hypersensitivity and MIF production in infected guinea pigs. The conventional view is that purified polysaccharides are T-independent stimulants of IgM. Some fungal polysaccharides such as mannans from baker's yeast, and proba-

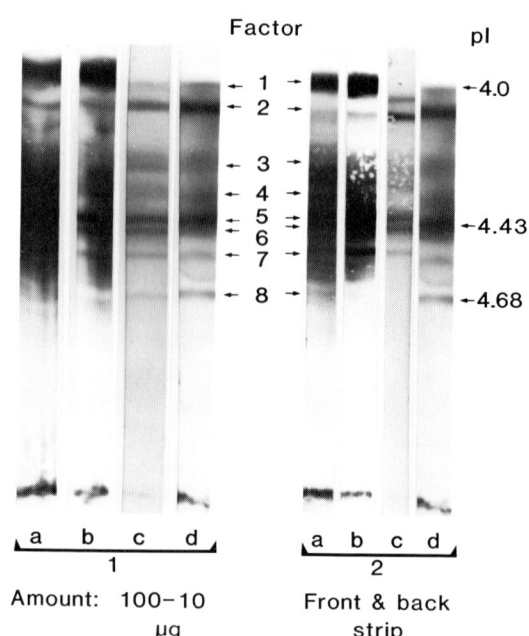

Figure 13.5. Enzyme-linked immunoelectro transfer blot method with isoelectrically focused *Nocardia asteroides* culture filtrate proteins showing the effect of different antigen concentrations on the appearance of the nitrocellulose blot strips. All major reactive antigenic factors in nocardin are shown. The blot strips were developed against serum from an immunized rabbit (a and b) or from a nocardiosis patient (c and d). Panel 1. The antigen protein concentrations in µg / lane were: a, 100; b, 10; c, 10; d, 100. Panel 2. The antigen concentration in all 4 lanes was 30 µg /lane; in addition the appearance of the front of the nitrocellulose strips (lanes a and d) and back (lanes b and c) are shown. Reducing the antigen concentration helps to clarify the pattern. Factor 1 with an isoelectric point of 4.0 penetrates to the back of the strip and into a second nitrocellulose sheet. *Source:* El-Zaatari et al, 1986.

bly from *C albicans*, contain covalent structural protein. The requirements for T-cell help in antibody responses to such complex molecules may not follow the dogma about polysaccharides. The most cautious approach to take is that until the purity and sequence analysis of a polysaccharide are known, reports of its ability to evoke delayed cutaneous hypersensitivity and its in vitro correlates should be received with skepticism. In the case of the nocardiae, some polysaccharides were extracted with a KCl solution from mechanically disrupted *N asteroides* and *N brasiliensis* containing varying amounts of L-arabinose, galactose, mannose, and less than 1% nitrogen. This carbohydrate complex was not characterized chromatographically or by sequence analysis, but at pH 10 one component "poly I", a presumptive arabinogalactan was soluble, whereas another component, "poly II," insoluble at pH 10, was soluble at pH 5 (Zamora et al, 1963). "Poly II" was serologically specific with no reciprocal crossreactivity between these antigens extracted from the two nocardial species when tested against antisera produced in rabbits immunized with defatted whole bacteria. No further studies with "poly II" complex have been reported where the sources of serum were infected animals or humans. The crossreaction of "poly II" with sera produced against mycobacteria is likewise unknown.

In later studies (Ortiz-Ortiz et al, 1972a) proteinases and organic solvents were used to reduce the nitrogen content of nocardial polysaccharide complexes to <0.04% No positive steps were taken, however, to characterize the glycans hydrodynamically or with respect to carbohydrate composition. The relation of this preparation to the former "poly I" and "poly II" is thus unclear. Guinea pigs primed with live *Nocardia* were skin tested with nitrogen-deficient polysaccharides and the histologic picture obtained was of an Arthus reaction being replaced after 24 hours by a predominantly lymphocytic and macrophage infiltrate with some PMN. Direct MIF production was observed when peritoneal exudate cells from infected guinea pigs were reconstituted in vitro with nocardial polysaccharides. Controls for cytophilic antibody were not included. Because the polysaccharides were not characterized, nonspecific stimulation of lymphocytes by nocardial mitogen cannot be ruled out.

13.4.5 Ribosomal Proteins

The same set of culture conditions employed to produce a homogenate-supernate antigen of *Nocardia* (Ortiz-Ortiz and Bojalil, 1972) was used to prepare ribosomal proteins. Ribosomes were precipitated with ammonium sulfate, redissolved, and sedimented at 165,000 g. Ribosomal proteins were separated after 2-chloroethanol precipitation of RNA (Ortiz-Ortiz et al, 1972b; 1976). Guinea pigs were sensitized by a method that has been standardized by these investigators: 1 mg wet weight of live *N brasiliensis* in incomplete Freund adjuvant was injected weekly for 3 weeks and animals were skin tested 3 weeks later. Whether these conditions are optimal for sensitization or simply arbitrary is not known. The antigen contained no RNA, DNA, or carbohydrate, and evoked positive delayed cutaneous hypersensitivity at a dose of 1 μg protein. Ribosomal proteins from *N asteroides* crossreacted with *N brasiliensis*, but reciprocal intradermal tests indicated significant specific differences between these antigens according to Magnusson's formula (See section 13.3.2). In the instance of ribosomal proteins from *N brasiliensis* and *N asteroides* a specificity difference of 9 indicated a good ability to differentiate animals sensitized with either *Nocardia* species.

Antigenic activity of the ribosomal proteins for delayed cutaneous hypersensitivity was abolished by pronase digestion but not by boiling. The T-cell-mediated reactivity of *Nocardia* ribosomal proteins was also assayed by their ability to elicit MIF in the direct MIF assay of infected guinea pigs. Significant inhibition of migration was stimulated by 50 µg of ribosomal proteins from either *Nocardia* species, and homologous and heterologous reactions were significantly different.

13.4.6 Glycolipid

Agar gel precipitin reactions between mannophosphoinositides of *N asteroides* and sera from infected guinea pigs were reported (Trana et al, 1978).

13.5 INTERACTION WITH MACROPHAGES

The interaction of nocardiae with alveolar macrophages is an early event in infection that provides insight into the pathogenic potential of *N asteroides* isolates. In an in vitro assay, *N asteroides* 10905 (low virulence) was phagocytosed and killed by normal rabbit alveolar macrophages (Beaman, 1977). The virulent isolate *GUH-2* elicited rapid migration of macrophages followed by aggregation and fusion into Langhans giant cells. The virulent strain appeared to send a chemotactic, activating signal to the macrophages. Nocardiae inside giant cells were killed in 24 hours but in macrophages that did not fuse, bacteria produced strongly acid-fast filaments that grew out of the phagocyte. Macrophage fusion into giant cells increases their phagocytic power, enabling them to kill virulent nocardiae.

Several different microbial agents known to activate macrophages or to generate nonspecific resistance have been tested to assess their effects on phagocytosis of nocardiae. The rationale for this approach with respect to *N brasiliensis* is that mycetomas are resistant to inactivation by the host, and without specific therapy lesions progress through subcutaneous tissue and bone. Thus immunization with agents known to activate particular immune mechanisms provides a probe with which to study the modulation of resistance to *Nocardia*.

Activated macrophages from mice infected with *Toxoplasma gondii* or injected with *Corynebacterium parvum* inhibited the growth of *N asteroides* in vitro, causing them to remain in the coccobacillary form and killed 50% of the targets in 6 hours (Filice et al, 1980). Nocardiae in control macrophages from normal mice grew as filaments and destroyed the macrophage monolayer.

Mycobacterium bovis BCG is known to generate nonspecific protection towards other bacterial agents and has antigens that crossreact with nocardiae. Immunization with live BCG only slightly reduced the ability of *N brasiliensis* to establish mycetoma after footpad injection (Ximénez et al, 1980). Delayed cutaneous hypersensitivity to nocardial cytoplasmic extract measured in the contralateral footpad was only slightly increased in BCG-infected mice compared with controls. In contrast to the effect of BCG, immunization with killed *N brasiliensis* inhibited the induction of mycetoma and increased specific delayed footpad swelling to *Nocardia* extract. Mice immunized with killed nocardiae so that mycetoma was prevented always had higher delayed cutaneous hypersensitivity than those with mycetoma. The failure of BCG to induce resistance to *N brasiliensis* is consistent with the occurrence in humans of dual infection with *Nocardia* and *M tuberculosis*. The crossreactive antigens shared by mycobacteria and nocardiae do not appear to be those that are responsible for protection.

The reverse of this experiment, where mice were first infected with *Nocardia brasiliensis* and then at intervals infected with another intracellular bacterium, *Listeria monocytogenes*, has also been reported (Melendro et al, 1978). Resistance to *Listeria* in the liver and spleen was significantly higher in *Nocardia*-infected mice at day 15 after injection, coinciding with the onset of positive delayed cutaneous hypersensitivity to *Nocardia* cytoplasmic extract. The index of resistance increased to a peak 25 days after infection with *Nocardia* and then declined in parallel with decreasing delayed cutaneous hypersensitivity. Peritoneal exudate cells removed from *N brasiliensis*-infected mice at their peak of resistance to *Listeria* were pulsed in vitro with *Nocardia* extract and showed significantly greater killing of phagocytosed *Listeria* than macrophages from normal mice with or without a pulse of *Nocardia* extract. Thus the maturation of T-cell-mediated immunity to *N brasiliensis* is linked with enhanced cidal power of their macrophages.

13.6 EXPERIMENTAL MURINE NOCARDIOSIS

Immunosuppressive therapy to control a primary neoplastic, autoimmune, or host-versus-graft process is an important factor increasing the risk of pulmonary nocardiosis. Yet few reports exist that attempt to clarify the effect of immunosuppressants on alterations in the host response in experimental nocardiosis. Swiss-Webster mice were primed with cyclophosphamide, and then infected with one of three isolates of *N asteroides* differing in virulence and target organ distribution (Beaman and Maslan, 1977). The drug, 2 mg per mouse, was injected 72 hours before intravenous infection. The most virulent isolate, *GUH-2*, having a tropism for mouse kidney, was ten times more virulent in the drug-primed group as measured by the LD_{50} dose. Intermediate virulence isolate *14759* produced multiple heart and lung abscesses. Cyclophosphamide priming before infection with this strain increased susceptibility one hundredfold. The low virulence isolate, *10905*, which grew only in the lungs, was 40 times more virulent in the drug-primed group than in controls receiving no cyclophosphamide. Although drug-primed mice were less able to halt nocardial growth, the organ tropisms did not change. Immunocompromise created much greater susceptibility in mice infected with strains of low or intermediate virulence. These strains are preferred in studies to dissect factors modulating murine resistance in order to avoid rapidly progressing acute infection with a highly virulent isolate. The conclusion was that humoral immunity protects mice from *Nocardia* because cyclophosphamide has a more profound toxicity for B-lymphocytes. This hypothesis needs more direct support by monitoring antibodies stimulated by *N asteroides* directly or as a function of their role as opsonins.

The congenitally athymic nu/nu mouse was adapted as a suitable living test tube to assay the need for T-cell-mediated immunity in host resistance to nocardiae (Beaman et al, 1978a, 1978b). Possibly as a genetic compensatory mechanism, these mice have chronically activated macrophages (Emmerling et al, 1977). The results obtained with three routes of infection (intravenous, footpad, and intranasal) in nu/nu and nu/+ thymus-bearing heterozygotes provide evidence that the macrophage barrier is not sufficient to eradicate nocardiae. In the absence of a secondary T-lymphocyte directed granulomatous response, infection leads to dissemination and death. In nu/+ mice, a smoldering infection produced type IV hypersensitivity that augments bactericidal capacity. The nu/+ heterozygotes were more prone to die in the first week after intranasal infection than nu/nu mice because of a fulminant pulmonary infection, but nu/+ survivors healed the infection, whereas in nu/nu mice dissemination occurred to the kidney according to the tropism of the *GUH-2* isolate.

Athymic mice are more susceptible to *N asteroides* than euthymic heterozygotes and immunity can be transferred to nu/nu mice by *N asteroides* primed T-cells (Deem et al, 1982). T-lymphocytes from immunized mice kill *N asteroides* in vitro in the absence of macrophages. Neither normal mouse lymphocytes nor B-cells purified from immune mice are effector cells (Deem et al, 1983). Proof that effector cells are T-cells is provided by the enrichment in cytotoxic cells in a T-rich population, sensitivity to anti-thy 1 and complement, and indirect immunofluorescence that localized T-cells in contact with the lysing target cells. Furthermore natural killer lymphocyte activity against YAC-1 targets was low in the lymphocytes enriched in *Nocardia*-specific killing. Enhancement of cytotoxicity against *Nocardia* is accomplished if immune lymphocytes are incubated with *N asteroides* cell wall fragments 12 to 48 hours before reconstitution with live nocardiae. Killing is highly specific for *N asteroides* because other related actinomycetes are not effectively killed.

Injection of *N brasiliensis* into the footpads of mice initiates a mycetomalike process with the production of actinomycotic granules and eventual spontaneous loss of the infected leg (Rico et al, 1982). The importance of cell-mediated versus humoral immunity was surveyed in this model. Congenitally athymic nude mice were more susceptible to *Nocardia* infection than euthymic mice, but in this inbred strain one cannot evaluate the relative contributions of delayed type hypersensitivity and of T-cell help in antibody production. Mice were therefore thymectomized, lethally irradiated, and reconstituted with syngeneic bone marrow. These "B-mice" had an enhanced susceptibility to infection measured as the spontaneous loss of the infected limb, but it was also observed that B-mice could not produce antibodies that reacted with erythrocytes coated with nocardial cytoplasmic extract. The passive transfer of antibodies from convalescent mice, intravenously or by first opsonizing live nocardiae, resulted in both cases in more severe disease. Next, the spleen cells of immune mice were depleted of *Nocardia* antigen-specific B-cells by rosetting them with *Nocardia*-antigen-coated erythrocytes before removal by gradient centrifugation. When the nocardial-depleted spleen cells were adoptively transferred to irradiated recipients, they responded to their challenge infections much better, despite the fact that they were unable to make detectable antibodies to nocardiae. These findings argue that the humoral response in nocardiosis is detrimental, and that the antibodies produced are of the blocking type that shield the organisms from recognition by T-cells. Moreover, the development of delayed-type hypersensitivity is mandatory for successful resistance to nocardiae.

Specific antibodies and C3 are localized in *N brasiliensis*-containing granules in the footpads of experimentally infected mice 2 weeks after infection but are not found in tissues surrounding the actinomycotic granules or in the plasma until 30 days later (Conde et al, 1983). The concentration of antibodies plus complement in the granules does not lead to their destruction, but rather the infected legs are sloughed off. The humoral response may thus be harmful if it blocks recognition by T-lymphocytes. Failure to detect antibodies in the plasma does not, however, rule out the possibility of circulating immune complexes.

13.6.1 L-Forms

Despite the fact that nocardiae are excellent adjuvants and produce a T-independent B-cell mitogen, the humoral response in infected mice was lower than expected and detrimental to the host. This model is thus suitable for further evaluation of the regulation of humoral responses to an intracellular pathogen. Does the presence of antibodies hasten the conversion of nocardiae to become L-forms so that they are better able to avoid immune surveillance?

Wall-deficient L-forms of *N asteroides* have been induced by cultivation in cultured mouse peritoneal macrophages or in medium containing glycine and cycloserine (Bourgeois and Beaman, 1974, 1976). Revertants of L-forms of *N asteroides* 10905 showed a divergence in form and physiology from the parent culture (Beaman and Bourgeois, 1981). These mutations affected colonial and microscopic morphology, acid-fastness, fermentation patterns, and shifts in the chain length of mycolic acids. The ease with which these changes occur underscores the variability among isolates of this actinomycete, complicating efforts to formulate numerical taxonomic schemes. It would be of great interest to know how such changes affect the antigenic structure.

The slow, progressive mycetomas caused by *N brasiliensis*, *N caviae*, and *N asteroides* are most often seen in subtropical or tropical countries where traumatic implantation of nocardial spores is an occupational hazard of barefoot laborers. In the USA localized extrapulmonary nocardiosis has been reported as leg ulcers, or infections following compound fracture of the humerus (Goodman and Koenig, 1970). The triad of tumefaction, sinus formation, and granules composed of bacterial aggregates is always involved in mycetoma. Lesions enlarge by direct extension involving muscle and bone. No satisfactory interpretation has been offered about how these growths can evade defenses in the apparently immunocompetent host. Finely detailed studies in mice infected with *N asteroides* (Bourgeois and Beaman, 1974) or with *N caviae* (Beaman and Scates, 1981) now suggest that bacteria in these granules are wall-defective L-forms and that occult foci in the brain, eyes, spinal cord, marrow, and kidney also contain L-forms (Figure 13.6). Mycetomas became visible in mice 6 months post intravenous infection with *N caviae* and were usually situated at the base of the skull or at the posterior of the spine. Sometimes these

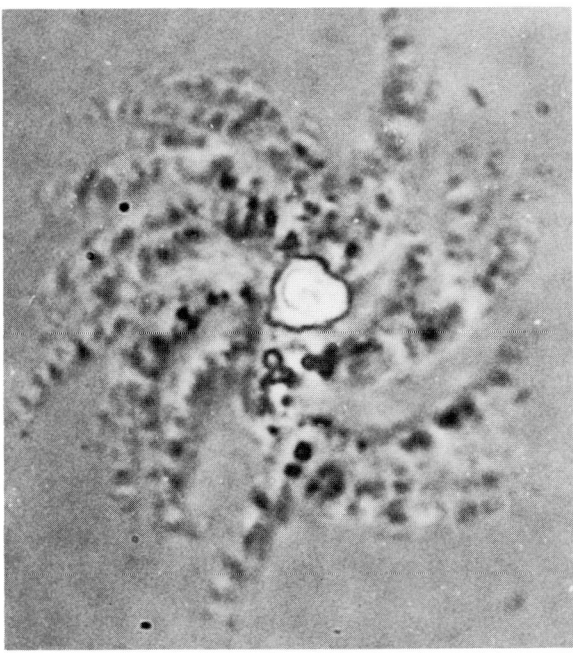

Figure 13.6. L-Form variant of *Nocardia caviae* incubated on Barile-Yarguchi-Eveland agar at 37 °C. L-Forms were obtained from the spinal cord homogenate of a mouse infected for 6 months. The large refractile central body with radiating spiral arms is the L-form colony. The colony diameter is approximately 100 μm (Beaman and Scates, 1981). *Source:* Reprinted with permission of the authors and the publisher the American Society for Microbiology.

tumefactions weighed up to 50 g! Mycetomas do not develop in athymic nu/nu or hereditarily asplenic mice because they could not effectively suppress the intact mural form of *N caviae* which flourished without restriction. Immunocompetent mice, however, stripped the walls off the intact bacteria, but the resulting L-forms were refractory to the milieu of degradative enzymes. L-Forms may play a major role in forming the bacterial granule. The isolation of L-form variants in humans (Beaman and Scates, 1981) taken together with findings in murine nocardiosis predict that conversion to L-forms occurs in clinical nocardiosis and that antigenic stimulation is provided by proteins of the cytoplasmic membrane and cytosol. The identification of L-forms as *N caviae* was made in part because of their immunofluorescence after incubation with anti-*N caviae* serum but not with anti-*N asteroides*.

In contrast to the obvious lesion of mycetoma, the eyes of mice blinded by *N caviae* did not show abscesses or granulomas. Only L-forms were isolated from the eye and a role for immunopathology in these foci remote from the mycetoma lesion has not been studied.

The tropism for the central nervous system observed in murine *N caviae* infection is also reflected in human nocardiosis where cerebral abscesses are a notable extrapulmonary site of dissemination.

13.7 NOCARDIAL IMMUNOSTIMULANTS

The progress in mycobacterial immunostimulant research has a parallel in the nocardiae. Even in the early days Jules Freund described how heat-killed nocardiae could substitute for the mycobacterial component of Freund adjuvant (Freund, 1956). The cellular reaction caused by *N asteroides* in water-in-oil emulsion is almost identical to that provoked by killed mycobacteria in the same emulsion. The range of immunostimulatory effects of these bacteria includes, adjuvanticity, mitogenicity, nonspecific protection against bacterial infection, antitumor activity, and stimulation of interferon (Barot-Ciorbaru et al, 1979). Many of these effects have been found to reside in the mycobacterial or nocardial cell wall (Azuma et al, 1976; Adam et al, 1973). Research on immunostimulatory subfractions of the mycobacterial or nocardial cell wall have diverged into studies of cord factor (6,6'-dimycolate of α,α-D-trehalose) or the peptidoglycan. The properties of cord factors have been studied in the mycobacteria, but it should be borne in mind that nocardiae contain lower molecular weight homologues (Lederer, 1977). The toxic, granulomagenic and adjuvant activity of cord factors are discussed elsewhere (Lederer, 1980). Soon after it became know that adjuvant activity was concentrated in the cell wall, lysozyme digests of defatted walls were found to liberate a soluble fragment—water-soluble adjuvant from *Nocardia* and from mycobacteria (Adam et al, 1973).

The material was a complex of arabinogalactan and peptidoglycan. The advantages of the water-soluble adjuvant are that unlike whole mycobacteria, the water-soluble adjuvant induces neither hypersensitivity to itself nor polyarthritis. The minimal unit that can replace the mycobacterial or nocardial component of Freund adjuvant is *N*-acetyl-muramyl-L-alanyl-D-isoglutamine (muramyl dipeptide, MDP) (Lederer, 1980). The availability of a water soluble adjuvant that is nonantigenic and has low or no toxicity for laboratory animals (rodents, cavies, or lagomorphs) should clarify many experiments that up to now have relied on the use of entire bacteria in a water-in-oil emulsion. The mural unit capable of evoking B-cell mitogenicity is somewhat larger than MDP (Figure 13.7).

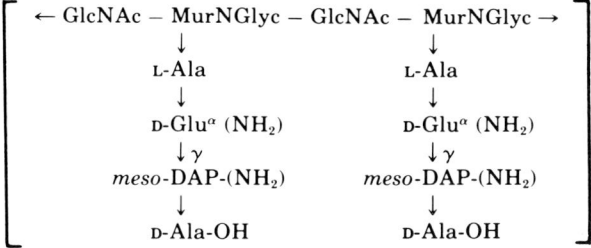

Figure 13.7. Tentative structure of a soluble mitogenic peptidoglycan fragment obtained by the action of *Streptomyces albus* G peptidases on *Nocardia rubra* walls. Mechanically disrupted cells were centrifuged and the walls were digested with DNAse, trypsin, and chymotrypsin. They were then subjected to a five-stage delipidation with organic solvents. Acid hydrolysis occurred in 0.1M HCl for 12 hours at 60 °C. Most of the arabinogalactan was solubilized and nocardomycolic acids became soluble in chloroform. GlcNAc, N-acetylglucosamine; MurNGlyc, N-glycolylmuramic acid (Ciorbaru et al, 1976). *Source:* Reprinted with permission of the authors and the publisher, American Society for Microbiology.

Although MDP appears competent to stimulate both humoral and delayed cutaneous hypersensitivity reactions, the adjuvant properties were found to give more stimulation of T-cell-mediated immunity after chemical conjugation with nocardomycolic acids (Saiki et al, 1980). Insertion of nocardomycolic acid from *N asteroides* was accomplished at either the 6-O position of N-acetylmuramic acid or the carboxyl group of D-isoglutamine via lysine. The conjugate had significantly greater activity than MDP alone to promote cellular cytotoxicity of splenic lymphocytes from allogeneic recipient mice transplanted with mastocytoma *P815-X2*. The 6-O-nocardomycolyl MDP conjugate was a potent suppressant of the growth of *Meth-A* tumor cells in BALB/c mice, but MDP alone was ineffective.

13.7.1 B-Lymphocyte Mitogen

Mitogenic activity resides in the cell wall peptidoglycan and also in a cytoplasmic fraction devoid of peptidoglycan. Cell walls were produced from mechanically disrupted nocardiae (*Nocardia opaca* syn *Mycobacterium rhodochrous, Nocardia rubra* syn *Nocardia corynebacterioides*), were digested with proteinases and defatted with a series of five organic solvents (Ciorbaru et al, 1977). Mild acid hydrolysis then was used to solubilize arabinogalactan and deacylate nocardomycolic acids. At this stage the cell wall preparation was enriched in peptidoglycan, as both free and bound lipid and arabinogalactan were removed. Two other enzymes were used to further fractionate the peptidoglycan: lysozyme or D,D-carboxypeptidase from *Streptomyces albus G*. The latter enzyme cleaves D-alanine-*meso* diaminopimelic acid peptide bonds (Petit and Lederer, 1978; see nocardial peptidoglycan, 13.2.3) thus fragmenting susceptible peptidoglycans by scission of interglycan peptide bridges. These enzymes profoundly affected *N rubra* cell walls, each resulting in 80% to 90% solubilization (Figure 13.7). Cell wall peptidoglycan digested with lysozyme retains adjuvanticity but is not mitogenic. The products solubilized by D,D-carboxypeptidase from acid-treated, defatted cell walls of *N rubra* (syn *N corynebacterioides*) and from *N opaca* = *M rhodochrous* were strongly mitogenic for splenocytes from AKR and athymic nu/nu mice (Ciorbaru et al, 1976). The alternative method of preparing cytoplasmic mitogen from lysozyme-digested, defatted nocardiae (including *N brasiliensis*, Ortiz-Ortiz et al, 1979) has been extensively used (Petit and Lederer, 1978;

Bona et al, 1979; Bona, 1979). The supernate after lysozyme digestion is filtered through Sephadex G 75 and the mitogenic material is localized in the void volume. Centrifugation at high speed yields a residue that is mitogenic although the supernate is not (Ciorbaru et al, 1977). Cell wall fragments were not present as judged by the absence of the marker, diaminopimelic acid, in hydrolyzates (Ciorbaru et al, 1976). Thus the term nocardial water-soluble mitogenic (NWSM) is ambiguous in that it refers to two distinct components of the nocardial cell: (i) a high speed sedimented fraction of lysozyme digests and (ii) a fragment of nocardial peptidoglycan solubilized by D,D-carboxypeptidase of *Streptomyces albus* G. A more precise nomenclature of the mitogenic fractions is needed, and the high speed sedimented fraction should be characterized.

Polyclonal B-lymphocyte activators, or mitogens, in the strict sense, must produce a general increase in immunoglobulin synthesis — not restricted to a single antigen — in the absence of T-lymphocytes. Conforming to this view, several B-cell mitogens are known from work with mice, but none are effective as such in humans except for the NWSM. The spectrum of species in which nocardial mitogen is active is broad, including human, rabbit, mouse, rat, and avian B-lymphocytes (Bona et al, 1974a,b, 1975; Bona et al, 1978; Bona, 1979).

Proof that NWSM meets the criteria of a B-cell mitogen in humans requires that the following variables are controlled: (i) known crossreactions between tuberculin and nocardin may result in humans primed to respond to *Nocardia* extracts; (ii) natural sensitization to saprophytic or pathogenic nocardiae may account for part of the human proliferative response to NWSM; and (iii) contamination with lipopolysaccharide endotoxin.

Direct stimulation of human cord blood lymphocytes with NWSM from *N opaca* showed that prior sensitization is not required for the mitogenic effect (Brochier et al, 1976). Peripheral blood lymphocytes from PPD-negative human donors respond to NWSM, and the magnitude of responses of tuberculin-positive donors to PPD did not correlate with blastogenic indices to NWSM (Brochier et al, 1976). The presence of endotoxin was ruled out by the absence of the marker substance, keto-deoxyoctanoic acid (KDO) and positive responses to NWSM in C_3H/HeJ mice, which are genetic nonresponders to lipopolysaccharide endotoxin, LPS (Bona et al, 1974a).

Nocardial water-soluble mitogen stimulated proliferation and immunoglobulin synthesis of lymphocytes from a majority of human donors tested (Brochier et al, 1976; Lethibichthuy et al, 1978) and blastogenic indices increased when T-lymphocytes were depleted as sRBC rosetting cells. T-Lymphocytes from a few persons were stimulated by NWSM, possibly because of previous exposure to *Nocardia* species (Brochier et al, 1976). Other enrichment procedures, including purification of B-lymphocytes on anti-Fab columns (Brochier et al, 1976) or depletion of human T-cells with anti-thymocyte serum plus complement did not ablate NWSM-induced mitogenesis (Brochier et al, 1976). Human B-lymphocytes exposed to NWSM are driven to increased polyclonal immunoglobulin synthesis that is detected as plaque forming cells against heterologous erythrocytes, trinitrophenyl hapten, tetanus toxoid, and keyhole limpet hemocyanin (Bona et al, 1979). T-Lymphocytes are not required to help activate human B-cells to proliferate in response to NWSM as is the case with pokeweed mitogen, but T-lymphocytes enhance immunoglobulin production by nocardial water-soluble mitogen-stimulated B-cells (Bona et al, 1979).

The subset of B-lymphocytes stimulated by NWSM has been characterized. They are (i) highly cortisone sensitive (Bona et al, 1974a,b, 1978); (ii) and are a different subset than is stimulated by other mitogens, as shown by bromodeoxyuridine (BUdR) suicide experiments in which lymphocytes proliferating in response to pokeweed mitogen or lipopolysaccharide incorporate the nucleotide and are killed by light. The remaining lymphocytes are still able to respond to NWSM (Bona et al, 1979; Bona, 1979).

Surface markers on NWSM-responding cells were assayed in selective lysis experiments with (i) complement and anti-mouse Ig and with (ii) murine alloantisera directed against the Ia antigens coded for by the I region of the major histocompatibility complex (Bona et al, 1978). Alloantisera to the gene products of the I-A subregion inhibited the response to NWSM. Partial inhibition resulted from incubation with anti-Ia^7(I-E/C) serum. A two-stage lysis experiment showed that a lymphocyte subset with no (or low density) immunoglobulin receptors and not expressing I-E/C gene products was capable of being stimulated by NWSM. This information is of value in defining the specificity of probes for B-lymphocytes, like NWSM. Such probes are potentially useful in differentiating immunodeficiency states involving B-cell dysfunction from those due to failure of T-lymphocyte regulation.

The B-lymphocyte mitogenic activity of soluble extracts of cell walls of saprophytic nocardiae has been known for some time, but the relevance of this finding to the pathogenic nocardiae has only recently received experimental support (Ortiz-Ortiz et al, 1979). Defatted *N brasiliensis* were exposed to lysozyme and the filtrate was lyophilized. This material did not contain endotoxin when tested by the *Limulus* lysate assay, and, as a further precaution C$_3$H/HeJ mice, which are nonresponders to LPS were operated as a control source of splenocytes. The *N brasiliensis* extract evoked in vitro polyclonal activation of splenocytes from conventional, nu/nu and C$_3$H/HeJ mice and from splenocytes of conventional mice treated with antithymocyte serum + complement. No mitogenesis of murine thymocytes occurred, verifying that *N brasiliensis* extract is a source of a soluble B-cell mitogen. Genetic control of mitogen responsiveness to *N brasiliensis* extract was observed to the extent that mice bearing the H-2^K genotype (AKR, C$_3$H/St, B10.BR) were consistently higher responders than other H-2 genotypes. *N brasiliensis* extract stimulated polyclonal activation and direct plaque forming cell responses in murine splenocytes to trinitrophenyl, human γ-globulin, and sheep erythrocytes. The ability of nocardial extracts to activate polyclonal humoral responses may play an immunopathological role in infection if some of these antibodies were directed against the self.

13.8 SUMMARY

Pulmonary nocardiosis must be differentiated from tuberculosis because the therapy for tuberculosis is not effective against the nocardiae. *Nocardia asteroides* is a primary pathogen where no underlying disease is known, but also has emerged as an opportunist associated with systemic steroid therapy, neoplasia, autoimmune disease, and other primary conditions. Actinomycotic mycetoma induced by traumatic implantation is a distinct clinical entity presenting on the lower legs or back. The triad of tumefaction, sinus formation, and granules is always present.

Cell walls of nocardiae, in agreement with those of mycobacteria, contain mycolic acids, arabinogalactan, and peptidoglycan. The arabinogalactan consists of β-1,4-galactose disaccharides alternating with α-1,5-arabinofuranose disaccharides, Antigenic α-1,5-arabinofuranosides with a dp less than 6 are suspended from the main chain at arabinofuranose branch points. One of ten arabinofuranose units is esterified to mycolic acid by a 5'-hydroxyl on the terminal arabinofuranose. The arabinogalactan is crosslinked by phosphodiester bridges through C-6 hydroxyl groups of muramic acid. Arabinomannan is a distinct polysaccharide associated more with the cytoplasm, and forms an insoluble Cu^{++} complex with Fehling solution. The peptidoglycan of nocardiae differs from the general plan of bacterial peptidoglycan by the presence of the N-glycolyl derivative of muramic acid and a tetrapeptide containing diaminopimelic acid and D-alanine units.

During the *N asteroides* cell cycle, branching filaments fragment into short rods and cocci. With age, the lipid content declines and shifts more towards tuberculostearic and tightly bound lipids. The arabinogalactan and peptidoglycan are the major components of the cell wall in the stationary phase.

Three factors present in filtrates of *N asteroides* cultures are mutually exclusive and do not appear in other pathogenic nocardiae or related organisms. Two antigens of *N asteroides* react in precipitin tests with anti-BCG. About half of nocardiosis cases can be detected by immunodiffusion if a line of identity with a human reference antiserum is taken as the criterion for a specific immune response. The positions of arcs containing polysaccharide antigens have been located in a standard mycobacterial reference immunoelectrophoretogram. Band 1 and band 3 contain arabinogalactan, and band 2 is the arabinomannan.

Reciprocal skin test measurements in guinea pigs with cytoplasmic extracts of nocardiae and BCG show that crossreactions occur, but that homologous reactions are stronger. In humans, homogenate antigens of *N brasiliensis* evoke delayed cutaneous hypersensitivity in mycetoma patients but weak or negative reactions occur in patients with tuberculosis or in healthy controls from outside the endemic area.

Nocardia-active polypeptides are present in the dialyzable, acid extract of defatted bacteria. They evoke species-specific skin test responses and high blastogenic indices in sensitized guinea pigs. Proteins present in homogenates and filtrates of *N asteroides* cultures are acidic, with pIs in the 4.0 to 5.2 range. The isofocused proteins electro transferred to nitrocellulose paper are useful for enzyme immunoassays. One antigen with a pI of 4.6 to 4.7 was reactive with nocardiosis patients but not with tuberculosis patients' sera. Ribosomal proteins from *N asteroides* and *N brasiliensis* evoked delayed cutaneous hypersensitivity and MIF production.

Wall-deficient L-forms of *N asteroides* and *N brasiliensis* probably enable the bacteria to evade host defenses and favor chronic infection in mice. These infections disseminate to occult foci in the brain, eyes, marrow and kidneys. The transition of *N asteroides* to L-forms may also potentiate pathogenicity in humans and further complicates primary isolation of the organism from clinical samples.

Immunization with *Mycobacterium bovis* BCG does not provide cross-protection against *N brasiliensis*. Cyclophosphamide treatment before infection results in a lower LD_{50} dose of *N asteroides* for mice.

Athymic nude mice have increased susceptibility to disseminated nocardiosis, independent of the route of infection. Thus the macrophage surveillance barrier is not, by itself, a sufficient host defense. Euthymic mice are prone to die in the first week after intranasal infection because of the intensity of the pulmonary inflammation. One function of T-cell-mediated immunity in murine nocardiosis is to provide help for T-dependent antigens. Without T-cells, thymectomized, X-irradiated, and bone marrow reconstituted mice produce negligible *Nocardia*-specific antibodies. The production of antibodies does not, however, appear to be the reason for enhanced resistance. In fact, removal of *Nocardia*-specific antibody-producing B-cells before adoptive transfer increases the degree of protection conferred to recipient mice. These findings are paradoxical within the limits of our current knowledge, because nocardiae produce a T-independent mitogen that stimulates polyclonal B-cell blastogenesis and immunoglobulin production. It is possible that the nocardial mitogen is not expressed in vivo by *N asteroides,* or that clones secreting immunoglobulins specific for nocardial proteins are not stimulated by the mitogen. The conversion of *N asteroides* to L-forms may be accompanied by a loss of mitogenicity and of other surface markers with consequent failure of immune recognition.

Conjugation of muramyl dipeptide with nocardomycolic acids enhances the cytotoxic activity of lymphocytes against tumor cells. T-independent mitogenic activity for human B-lymphocytes resides in the mural peptidoglycan and in the cytoplasm of *N opaca, N corynebacterioides,* and *N brasiliensis.* The nocardial water-soluble mitogen is heterogeneous because the term refers to (i) the sedimented fraction of lysozyme digests containing no cell wall markers, and (ii) a soluble peptidoglycan released by D,D-carboxypeptidase cleavage of D-alanine-*meso*-diaminopimelic acid. Human B-lymphocytes are directly stimulated by nocardial mitogen to polyclonal immunoglobulin synthesis but the response is augmented by T-cells. B-Lymphocytes stimulated by nocardial mitogen differ from those that are transformed by other mitogens as determined in "BUdR suicide" experiments. Surface markers on murine B-lymphocytes subsets responding to nocardial mitogen are encoded in the I-A and I-E/C region of the murine H-2 major histocompatibility complex.

A hypothetical sequence of immune responses important for defense against nocardiae consistent with the present state of research may be constructed. Inhaled bacilli are rapidly phagocytosed by alveolar macrophages, which fuse into multinucleate giant cells for increased bacteriolytic power. Components of the nocardiae evoke nonspecific immunostimulation. The peptidoglycan exerts an adjuvant and polyclonal B-cell mitogen effect. Cord factor evokes granuloma formation. Some isolates are more virulent and can survive in macrophages. Maturation of a type IV delayed hypersensitivity response coincides with circumscribed and shrinking lesions. When T-cell-mediated immunity is suppressed, bacilli can proliferate in the lung. Humoral responses occur to species-specific factors and to factors that crossreact with mycobacteria, including the common cell wall arabinogalactan. These antibodies opsonize the bacteria and activate complement. There is no concrete evidence that humoral immunity is important for host defense and some evidence in mice that it is deleterious. Nocardiae can transform into L-forms, evade immune recognition and disseminate from the pulmonary focus. L-Forms may also protect nocardiae from eradication in actinomycotic mycetoma.

REFERENCES

Adam A, Ciorbaru R, Lederer E, Petit JF, Chedid L, Lamensans A, Parant F, Rosselet JP, Berger FM, 1973. Preparation and biological properties of water soluble adjuvant fractions from delipidated cells of *Mycobacterium smegmatis* and *Nocardia opaca.* Infect Immun 7:855–861.

Andriole VT, Ballas M, Wilson GL, 1964. The association of nocardiosis and pulmonary alveolar proteinosis. A case study. Ann Intern Med 60:266–275.

Azuma I, Kanetsuna F, Tanaka Y, Mera M, Yanagihara Y, Mifuci I, Yamamura Y, 1973. Partial chemical characterization of cell wall of *Nocardia asteroides* 131. Jpn J Microbiol 17:154–159.

Azuma I, Taniyama T, Sugimura K, Aladin AA, Yamamura Y, 1976. Mitogenic activity of cell walls of mycobacteria, nocardia, corynebacteria, and anaerobic coryneforms. Jpn J Microbiol 20:263–271.

Barot-Ciorbaru R, Yokota Y, Petit JF, Chedid L, Atanasiu P, 1979. Induction de la synthèse d'interféron chez le hamster par des fractions de Nocardia: essais de protection contre la rage par la NWSM. Annales d'Microbiologie Inst Pasteur 130B:263–269.

Beadles TA, Land GA, Knezek DJ, 1980. An ultrastructural comparison of the cell envelopes of selected strains of Nocardia asteroides and Nocardia brasiliensis. Mycopathol 70:25–32.

Beaman BL, 1975. Structural and biochemical alterations of Nocardia asteroides cell walls during its growth cycle. J Bacteriol 123:1235–1253.

Beaman BL, 1977. In vitro response of rabbit alveolar macrophages to infection with Nocardia asteroides. Infect Immun 15:925–937.

Beaman BL, Bourgeois AL, 1981. Variations in properties of Nocardia asteroides resulting from growth in the cell wall deficient state. J Clin Microbiol 14:574–578.

Beaman BL, Bourgeois AL, Moring SE, 1981. Cell wall modification resulting from in vitro induction of L-phase variants of Nocardia asteroides. J Bacteriol 148:600–609.

Beaman BL, Burnside J, Edwards B, Causey W, 1976. Nocardial infections in the US. 1972–1974. J Infect Dis 134:286–289.

Beaman BL, Gershwin ME, Maslan S, 1978a. Infectious agents in immunodeficient murine models: pathogenicity of Nocardia asteroides in congenitally athymic (nude) and hereditarily asplenic (Dh/+) mice. Infect Immun 20:381–387.

Beaman BL, Goldstein E, Gershwin ME, Maslan S, Lippert W, 1978b. Lung response of congenitally athymic (nude), heterozygous, and swiss webster mice to aerogenic and intranasal infection by Nocardia asteroides. Infect Immun 22:867–877.

Beaman BL, Maslan S, 1977. Effect of cyclosphosphamide on experimental Nocardia asteroides infection in mice. Infect Immun 16:995–1004.

Beaman BL, Maslan S, 1978. Virulence of Nocardia asteroides during its growth cycle. Infect Immun 20:290–295.

Beaman BL, Scates SM, 1981. Role of L-forms of Nocardia caviae in the development of chronic mycetomas in normal and immunodeficient murine models. Infect Immun 33:893–907.

Beaman BL, Scates SM, Moring SE, Deem R, Misra HP, 1983. Purification and properties of a unique superoxide dismutase from Nocardia asteroides. J Biol Chem 258:91–96.

Beck ES, 1972. Mycobacterial antigen workshop. An evaluation of a reference serum for antigens of Mycobacterium tuberculosis. Amer Rev Respir Dis 106:142–147.

Blumer SO, Kaufman L, 1979. Microimmunodiffusion test for nocardiosis. J Clin Microbiol 10:308–312.

Bona C, 1979. Modulation of immune responses by Nocardia immunostimulants. Progress Allergy 26:97–136.

Bona C, Broder S, Dimitriu A, Waldmann TA, 1979. Polyclonal activation of human B lymphocytes by Nocardia water soluble mitogen (NWSM). Immunol Rev 45:69–92.

Bona C, Chedid L, Damais C, Ciorbaru R, Shek PN, Dubiski S, Cinader B, 1975. Blast transformation of rabbit B-derived lymphocytes by a mitogen extract from Nocardia. J Immunol 114:348–353.

Bona C, Damais C, Chedid L, 1974a. Blastic transformation of mouse spleen lymphocytes by a water-soluble mitogen extracted from Nocardia. Proc Nat Acad Sci 71:1602–1606.

Bona C, Damais C, Dimitriu A, Chedid L, Ciorbaru R, Adam A, Petit JF, Lederer E, Rosselet JP, 1974b. Mitogenic effect of a water-soluble extract of Nocardia opaca: a comparative study with some bacterial adjuvants on spleen and peripheral lymphocytes of four mammalian species. J Immunol 112:2028–2035.

Bona C, Yano A, Dimitriu A, Miller RG, 1978. Mitogenic analysis of murine B-cell heterogeneity. J Exp Med 148:136–147.

Bourgeois L, Beaman BL, 1974. Probable L-forms of Nocardia asteroides induced in cultured mouse peritoneal macrophages. Infect Immun 9:576–590.

Bourgeois L, Beaman BL, 1976. In vitro spheroplast and L-form induction within the pathogenic nocardiae. J Bacteriol 127:584–594.

Brochier J, Bona C, Ciorbaru R, Revillard J-P, Chedid L, 1976. A human T-independent B-lymphocyte mitogen extracted from *Nocardia opaca*. J Immunol 117:1434–1439.

Byrne E, Brophy BP, Perrett LV, 1979. *Nocardia* cerebral abscess: new concepts in diagnosis, management and prognosis. J Neurol Neurosurgery and Psychiatry 42:1038–1045.

Castelnuovo G, Bellezza G, Giuliani HJ, Asselineau J, 1968. Relations chimiques et immunologiques chez les *Actinomycetales*. Ann Inst Pasteur 114:139–147.

Chandler FW, Kaplan W, Ajello L, 1980. A Colour Atlas and Textbook of Histopathology of Mycotic Diseases. 333pp. London: Wolfe Medical Publications.

Ciorbaru R, Petit JF, Lederer E, Bona C, Chedid L, 1977. Isolement et caractérisation de mitogènes de *Nocardia*. Annales d'Immunologie Inst Pasteur 128C:41–45.

Ciorbaru R, Petit JF, Lederer E, Zissman E, Bona C, Chedid L, 1976. Presence and subcellular localization of two distinct mitogenic fractions in the cells of *Nocardia rubra* and *Nocardia opaca*: preparation of soluble mitogenic peptidoglycan fractions. Infect Immun 13:1084–1090.

Cisar JO, Barsumian EL, Curl SH, Vatter AE, Sandberg AL, Siraganian RP, 1981. Detection and localization of a lectin on *Actinomyces viscosus* T14V by monoclonal antibodies. J Immunol 127:1318–1322.

Conde C, Mancilla R, Fresan M, Ortiz-Ortiz L, 1983. Immunoglobulin and complement in tissues of mice infected with *Nocardia brasiliensis*. Infect Immun 40:1218–1222.

Cummins GS, 1962. La composition chimique des parois cellulaires d'actinomycetes et son application taxonomique. Ann Inst Pasteur 103:385–391.

Curry WA, 1980. Human nocardiosis. A clinical review with selected case reports. Arch Intern Med 140:818–826.

Deem RL, Beaman BL, Gershwin ME, 1982. Adoptive transfer of immunity to *Nocardia asteroides* in nude mice. Infect Immun 38:914–920.

Deem RL, Doughty FA, Beaman BL, 1983. Immunologically specific direct T-lymphocyte mediated killing of *Nocardia asteroides*. J Immunol 130:2401–2406.

Drake CH, Henrici AT, 1943. *Nocardia asteroides*. Its pathogenicity and allergic properties. Amer Rev Tuberc 48:184–198.

El-Zaatari FA, Reiss E, Yakrus MA, Bragg SL, Kaufman L, 1986. Monoclonal antibodies against isoelectrically focused *Nocardia asteroides* proteins characterized by the enzyme-linked immunoelectro-transfer blot method. Diagn Immunol: in press

Emmerling P, Finger H, Hof H, 1977. Cell-mediated resistance to infection with *Listeria monocytogenes* in nude mice. Infect Immun 15:382–385.

Estrada-Parra S, Zamora A, Bojalil L, 1965. Immunochemistry of the group-specific polysaccharide of *Nocardia brasiliensis*. J Bacteriol 92:571–574.

Filice GA, 1983. Resistance of *Nocardia asteroides* to oxygen-dependent killing by neutrophils. J Infect Dis 148:861–867.

Filice GA, Beaman BL, Remington JS, 1980. Effects of activated macrophages on *Nocardia asteroides*. Infect Immun 27:643–649.

Freund J, 1956. The mode of action of immunologic adjuvants. Adv Tuberc Res 7:130–148.

Geisler PJ, Andersen BR, 1979. Results of therapy in systemic nocardiosis. Amer J Med Sci 278:188–194.

Goodman JS, Koenig MG, 1970. *Nocardia* infections in a general hospital. Ann N Y Acad Sci 174:552–567.

Gorevic PD, Katler EI, Agus B, 1980. Pulmonary nocardiosis. Occurrence in men with systemic lupus erythematosus. Arch Intern Med 140:361–363.

Hiramine CK, Hojo K, Yano I, 1981. Development of cell-mediated hypersensitivity in guinea pigs following infection with *Nocardia asteroides*. Int Arch Allergy Appl Immunol 65:220–234.

Humphreys DW, Crowder JG, White A, 1975. Serological reactions to *Nocardia* antigens. Amer J Med Sci 269:323–326.

Janicki BW, Chaparas SD, Daniel TM, Kubica GP, Wright GL Jr, Yee GS, 1971. A reference system for antigens of *Mycobacterium tuberculosis*. Amer Rev Respir Dis 104:602–604.

Kingsbury EW, Slack JM, 1967. A polypeptide skin-test antigen from *Nocardia asteroides*. I. Production, chemical and biologic characterization. Amer Rev Respir Dis 95:827–832.

Kingsbury EW, Slack JM 1972. A polypeptide skin-test antigen from *Nocardia asteroides*. II. Further studies on the specificity of a nocardin active polypeptide. Sabouraudia 10:85–89.

Kotani S, Kato T, Matsuda T, Kata K, Misaki A, 1971. Chemical structure of the antigenic determinants of cell wall polysaccharide of *Mycobacterium tuberculosis*. Biken J 14:379–387.

Kurup VP, Scribner GH, 1981. Antigenic relationship among *Nocardia asteroides* immunotypes. Microbios 31:25–30.

Kurup VP, Piechura JE, Ting EY, Orlowski JA, 1983. Immunochemical characterization of *Nocardia asteroides* antigens: support for a single species concept. Can J Microbiol 29:425–432.

Kwapinski JB, Kwapinski EH, Dowler J, Hosman G, 1973. The phyloantigenic position of nocardiae revealed by examination of cytoplasmic antigens. Can J Microbiol 19:955–964.

Kwapinski JB, Seeliger HPR, 1965. Investigations on the antigenic structure of actinomycetales. IX. Serological classification of the nocardiae with the polysaccharide fractions of their cell walls. Mycopathol 25:173–182.

Lechevalier MP, 1977. Lipids in bacterial taxonomy. A taxonomist's view. CRC Crit Rev Microbiol 15:109–210.

Lechevalier MP, Lechevalier H, Horan AC, 1973. Chemical characteristics and classification of nocardiae. Can J Microbiol 19:965–972.

Lederer E, 1977. Natural and synthetic immunostimulants related to the mycobacterial cell wall. Medicinal Chem 5:257–279.

Lederer E, 1980. Cord factor and related synthetic trehalose diesters. pp. 95–110. In Immunostimulation, Chedid L, Miescher PA, Mueller-Eberhard HJ, eds. Berlin: Springer Verlag.

Lethibichthuy, Ciorbaru R, Brochier J, 1978. Human B-cell differentiation I. Immunoglobulin synthesis induced by *Nocardia* mitogen. Eur J Immunol 8:119–123.

Magnusson M, 1961. Specificity of mycobacterial sensitins. Amer Rev Respir Dis 83:57–68.

Melendro EI, Contreras MF, Ximénez C, García-Maynez AM, Ortiz-Ortiz L, 1978. Changes in host resistance caused by *Nocardia brasiliensis* in mice: cross-protection against *Listeria monocytogenes*. Int Arch Allergy Appl Immunol 57:74–81.

Misaki A, Seto N, Azuma I, 1974. Structure and immunological properties of D-arabino-D-galactans isolated from cell walls of *Mycobacterium* species. J Biochem 76:15–27.

Ortiz-Ortiz L, Bojalil LF, 1972. Delayed skin reactions to cytoplasmic extracts of *Nocardia* organisms as a means of diagnosis and epidemiological study of *Nocardia* infection. Clin Exp Immunol 12:225–229.

Ortiz-Ortiz L, Bojalil LF, Contreras MF, 1972a. Delayed hypersensitivity to polysaccharides from *Nocardia*. J Immunol 108:1409–1413.

Ortiz-Ortiz L, Contreras MF, Bojalil LF, 1972b. The assay of delayed hypersensitivity to ribosomal proteins from *Nocardia*. Sabouraudia 10:147–151.

Ortiz-Ortiz L, Contreras MF, Bojalil LF, 1972c. Cytoplasmic antigens from *Nocardia* eliciting a specific delayed hypersensitivity. Infect Immun 5:879–882.

Ortiz-Ortiz L, Contreras MF, Bojalil LF, 1976. Delayed hypersensitivity to *Nocardia* antigens. pp. 418–428. In Biology of the Nocardiae, Goodfellow M, Brownell GH, Serrano JA, eds. London: Academic Press.

Ortiz-Ortiz L, Parks DE, Lopez JS, Weigle WO, 1979. B-lymphocyte activation with an extract of *Nocardia brasiliensis*. Infect Immun 25:627–634.

Petit JF, Lederer E, 1978. Structure and immunostimulant properties of mycobacterial cell walls. Symp Soc Gen Microbiol 28:177–199.

Pier AC, Fichtner RE, 1971. Serologic typing of *Nocardia asteroides* by immunodiffusion. Amer Rev Respir Dis 103:698–707.

Pier AC, Fichtner RE, 1981. Distribution of serotypes of *Nocardia asteroides* from animal, human and environmental sources. J Clin Microbiol 134:548–553.

Pier AC, Keeler RF, 1965. Extracellular antigens of *Nocardia asteroides:* production and immunologic characterization. Amer Rev Respir Dis 91:391–399.

Pier AC, Thurston JR, Larsen AB, 1968. A diagnostic antigen for nocardiosis: comparative tests in cattle with nocardiosis and mycobacteriosis. Amer J Vet Res 29:397–403.

Rajki K, Brehmer W, Hammer H-J, Fischer W, Daus H, Mauch H, 1982. Analysis of soluble cytoplasmic components of mycobacteria and *Nocardia* by crossed immunoelectrofocusing. Zbl Bakt Hyg I Abt Orig A 251:389–398.

Rico G, Ochoa R, Oliva A, Gonzalez-Mendoza A, Walker SM, Ortiz-Ortiz L, 1982. Enhanced resistance to *Nocardia brasiliensis* infection in mice depleted of antigen specific B cells. J Immunol 129:1688–1693.

Ridell M, 1981. Immunodiffusion studies of some *Nocardia* strains. J Gen Microbiol 123:69–74.

Ridell M, Baker R, Lind A, Ouchterlony O, 1979. Immunodiffusion studies of ribosomes in classification of mycobacteria and related taxa. Int Arch Allergy Appl Immunol 59:162–172.

Ridell M, Norlin M, 1973. Serological studies of *Nocardia* using mycobacterial precipitation reference systems. J Bacteriol 113:1–7.

Robboy SJ, Vickery AL, 1970. Tinctorial and morphologic properties distinguishing actinomycosis and nocardiosis. N Engl J Med 282:593–596.

Saiki I, Uemiya M, Kusama T, Yamamura Y, Azuma I, 1980. Adjuvant activity of nocardomycolyl derivatives of N-acetylmuramyldipeptide in mice and guinea pigs. Microbiol Immunol 24:265–269.

Schneidau JD Jr, Shaffer MF, 1959. Studies on *Nocardia* and other actinomycetales II. Antigenic relationships shown by slide agglutination tests. Amer Rev Respir Dis 82:64–76.

Scully RE, Galdabini JJ, McNeely BU, 1980. Case records of the Massachusetts General Hospital. Weekly clinicopathological exercises. Case 20-1980. New Engl J Med 302:1194–1199.

Shainhouse JZ, Pier AC, Stevens DA, 1978. Complement fixation antibody tests for human nocardiosis. J Clin Microbiol 8:516–519.

Thurston JR, Phillips M, Pier AC, 1968. Extracellular antigens of *Nocardia asteroides* III. Immunologic relationships demonstrated by erythrocyte-sensitizing antigens. Amer Rev Respir Dis 97:240–247.

Trana AK, Sehgal SC, Khuller GK, 1978. Antibodies to phospholipids in experimental nocardiosis. Antonie van Leeuwenhoek J Microbiol Serol 44:391–394.

Vacheron MJ, Guinard M, Michel G, Ghuysen JM, 1972. Structural investigations on the cell walls of *Nocardia* sp: the wall lipid and peptidoglycan moieties of *Nocardia kirovani.* Eur J Biochem 29:156–166.

Vilkas E, Amar C, Markovits J, Vliegenthart JFG, Kamerling JP, 1973. Occurrence of a galactofuranose disaccharide in immunoadjuvant fractions of *Mycobacterium tuberculosis* (cell walls and wax D). Biochim Biophys Acta 297:423–435.

Ximénez C, Melendro EI, Gonzalez-Mendoza A, Garcia AM, Martinez A, Ortiz-Ortiz L, 1980. Resistance to *Nocardia brasiliensis* infection in mice immunized with either *Nocardia* or BCG. Mycopathol 70:117–122.

Yamamura Y, Misaki A, Azuma I, 1972. Chemical and immunological studies on polysaccharide antigens of mycobacteria, *Nocardia,* and corynebacteria. Bull Intern Union Against Tuberculosis 47:181–191.

Young LS, Armstrong D, Blevins A, Lieberman P, 1971. *Nocardia asteroides* infection complicating neoplastic disease. Amer J Med 50:356–367.

Zamora A, Bojalil LF, Bastarrachea F, 1963. Immunologically active polysaccharides from *Nocardia asteroides* and *Nocardia brasiliensis.* J Bacteriol 85:549–555.

Chapter 14

Microaerophilic Actinomycetes

14.1 INTRODUCTION

14.1.1 Actinomycetes as Agents of Periodontal Disease

The microaerophilic actinomycetes, *Actinomyces viscosus*, *Actinomyces naeslundii*, *Actinomyces israelii* and *Rothia dentocariosa* are common oral inhabitants and constituents of dental plaque. One or more species of oral actinomycetes can be isolated from saliva in the smooth surface dental plaque of almost all normal human adults (Ellen, 1976). When normal subjects with good oral hygiene suspended tooth brushing for 3 weeks, heavy plaque and gingivitis developed in all, yielding cultures dominated by *Actinomyces* species. There is a shift from *Streptomyces mutans* to *Actinomyces* species correlating with the onset of gingivitis (Syed and Loesche, 1978). Actinomycetes have been isolated from root surface caries (Jordan and Hammond, 1972) and from lesions of advanced periodontal disease. Infection of germ-free rats and hamsters with human isolates of *A viscosus* and *A naeslundii* results in the full spectrum of plaque, gingivitis, root surface caries, and bone destruction fulfilling Koch's postulates (Socransky et al, 1970; Jordan and Keyes, 1964; Jordan et al, 1972; Brecher et al, 1978).

No toxins have been identified as involved in periodontal inflammation or bone resorption but immunologic reactions induced by intact actinomycetes, their exocellular polysaccharides, and cytoplasmic fractions account for the chronic inflammation and pathology. Whole dental plaque elicits blastogenesis of lymphocytes from periodontitis patients at higher levels than in normal subjects (Ivanyi and Lehner, 1970; Baker et al, 1976). Levan, an exocellular storage polysaccharide which is a poly-2,1- and 2,6-linked fructosan, is induced in several oral actinomycetes by sucrose in the medium. Levan activates complement by the classical and alternative pathways; this activation promotes in vitro toxicity against human gingival fibroblasts (Lesher and Gerencser, 1977).

14.2 POLYSACCHARIDES

An exocellular heteroglycan of *A viscosus* composed largely of *N*-acetylglucosamine (van der Hoeven, 1974) is a polyclonal B-cell activator in rats and mice (Burckhardt et al, 1977). Soluble cytoplasmic substances from a homogenate of *A viscosus* also display a potent B-cell mitogenic activity (Burckhardt et al, 1977) that, in the presence of T-cells, results in polyclonal immunoglobulin production (Mangan and Lopatin, 1981). Surface fimbriae mediate (i) the attachment of *A viscosus* to teeth and (ii) coaggregation of *A viscosus* with oral streptococci (Cisar et al, 1979).

Peptidoglycan of *A viscosus* is a more potent activator of the alternative complement pathway than that of *Streptococcus pyogenes* group A (Baker and Billy, 1983). The latter was previously regarded as the most potent activator. The two peptidoglycans were purified by hot formamide extraction of cell walls and differ only in that the peptide bridge of *S pyogenes* contains glycine$_5$, whereas that from *A viscosus* consists of glutamic acid and ornithine. These various immunostimulatory and lectin activities contribute to chronic inflammation in the host.

14.2.1 Levan

When grown in the presence of sucrose, *Actinomyces viscosus* and *Rothia* elaborate a capsular polysaccharide, levan, a homopolymer of fructose (Howell and Jordan, 1967; Lesher and Gerencser, 1977). Levan serves as a storage polysaccharide for this organism because *A viscosus* produces a levan hydrolase (β-fructofuranosidase) in the late stationary phase of growth that depolymerizes levan to fructose (Miller et al 1975).

Gram quantities of levan substantially free of other macromolecules are obtained by resuspending washed *A viscosus* cells in 5% sucrose and penicillin (Miller et al, 1975). After 48 hours at 37 °C the levan is precipitated from the culture supernate with ethanol. The product contains 100% ketohexose, and hydrolysis in 0.2M H_2SO_4 gave fructose as the only monosaccharide. β-Fructofuranosidase activity was detected in culture supernates and was also cell-associated (Miller and Somers, 1978). Murine myeloma proteins with specificity for β-2,6- and β-2,1-linked fructose residues are structural probes for the levan of *A viscosus*. Immunoprecipitin arcs formed with both myeloma proteins provided evidence that both glycosidic linkages were present in *A viscosus* levan.

The role of levan as a determinant of pathogenicity is unclear. Its presence as a sticky tenacious capsule may promote adherence of plaque to the teeth. The adherence of *A naeslundii* to human buccal epithelial cells was inhibited by IgA (Williams and Gibbons, 1972), but the ligand is probably protein fimbriae on the cell surface. Interbacterial aggregation of *A naeslundii* and oral streptococci occurs, facilitating surface accumulation, but this phenomenon also appears to be mediated by lactosyl-specific fimbriae (Cisar et al, 1979, 1983). The presence of a slimy capsule renders *A viscosus* less subject to phagocytosis as is also the case for *Cryptococcus neoformans*. The proven immunologic activity of levan is its ability to activate complement. In an in vitro system, human gingival fibroblasts serving as target cells were incubated with levan. This resulted in clumping but not killing of the targets (Lesher and Gerencser, 1977). The addition of complement reduced the fibroblast viability from 93.5% to 54.5%. No antibody was present so that the destruction of fibroblasts proceeded through activation of complement by the alternative pathway. Sheep erythrocytes passively coated with levan could also trigger the classical pathway leading to hemolysis.

14.2.2 6-Deoxy-L-talose Polymer

Before it was known that the density of fimbriae was responsible for the difference between the T14V and T14AV strains of *A viscosus* a polysaccharide antigen isolated from T14V was implicated as the factor affecting pathogenicity (Hammond et al, 1976). A hot water extract of T14V cell walls was purified by anion exchange chromatography and the polysaccharide was eluted with 0.5M NaCl. The antigen was largely carbohydrate, 80% to 90%, and the predominant sugar, 6-deoxy-L-talose, was most unusual, accounting for 67% of the dry weight. This antigen displayed anodic mobility in immunoelectrophoresis using an antiserum from animals infected with T14V. The identity of 6-deoxy-L-talose was made on the basis of melting point, infrared spectrum, and hapten inhibition in which this monosaccharide was the only one capable of inhibiting the immune precipitation of the antigen. About 10% of the antigen was protein and both muramic acid and N-acetylglucosamine indicated the presence of peptidoglycan, underscoring the relationship of this polymer to the cell wall. The 6-deoxy-L-talose polymer bound tightly to the anion exchanger probably because of the peptide moiety.

Cell walls of *A viscosus* T14V contained 15.8% 6-deoxytalose whereas there was 10.8% 6-deoxytalose in walls of the T14AV strain (Brown et al, 1980). Obviously, incremental differences such as these cannot explain differences in virulence. The same monosaccharide was a constituent of *A naeslundii*, *A odontolytica*, and *Arachnia propionica*.

14.2.3 Viscous Exocellular N-Acetylglucosamine Polymer

Levan is not required to initiate periodontitis in gnotobiotic rats and hamsters infected with *A naeslundii* or *A viscosus* because the disease progresses even without a high sucrose diet (Rosan and Hammond, 1974). Slime polysaccharides other than levan have been reported in *A viscosus*. An exocellular polysaccharide was produced by *A viscosus* when cultures were grown on a sucrose-free medium with a low glucose content. The culture supernate was precipitated with acetone and treated with trichloroacetic acid to remove peptones from the growth medium. A yield of 108 mg of polysaccharide per liter of culture supernate was obtained. Nucleic acid and protein were below the limits of detection. The composition, as a percent of the dry weight, was as follows: N-acetylglucosamine, 45%; hexose, 16%; uronic acid, 3.4%; and phosphate, 5.3%. Once lyophilized the polymer was not completely soluble. It was not clear if more than one type of molecule was present because no separation techniques were applied. The isolated polymer was not serologically active when tested against antiserum to formalin-killed *A viscosus*. This finding may be explained if the glycan is not firmly bound to the bacteria but diffuses into the medium. The glycan is similar in its high N-acetylglucosamine content to a polymer from *A viscosus* reported by van der Hoeven (1974). That preparation had more protein, less hexose, no uronic acid, and was mitogenic for rodent B-cells.

The virulent *A viscosus* T14V is subject to a high rate of phagocytosis and lysosomal release by polymorphonuclear neutrophils from normal humans (Taichman et al, 1978). A polysaccharide slime layer on the surface of the avirulent variant T14AV inhibits both phagocytosis and lysosomal release. This inhibition is reversed by the opsonizing activity of normal serum. The slime polysaccharide is a viscous gel rich in N-acetylglucosamine (45%) with small amounts of glycerol and neutral sugars but no 6-deoxy-L-talose. More recently, this latter sugar has been identified as a marker for the cell wall, not the exocellular viscous slime. Others have described a polysaccharide from a different isolate of *A viscosus* (Wicken et al, 1978) the bulk of which is composed of glucose, galactose, mannose, N-acetylgalactosamine and 16.8% of O-esterified fatty acids. This antigen is amphipathic and attaches readily to erythrocytes, which facilitates the measurement of hemagglutinins.

Further separations to purify the viscous amino sugar-containing slime polysaccharide were conducted by Imai and Kuramitsu (1983). This polysaccharide is increased in the avirulent T14AV mutant and was isolated from chemically defined broth medium after 48 hours static incubation of the mutant. Chromatographic purifications were accomplished by gel permeation on Sepharose 4B and by ion exchange on DEAE-Biogel A. The major viscous component appeared in the first effluent peak from Sepharose 4B. The fraction was resolved into two viscous components on diethylaminoethyl. The minor A-1 component in the breakthrough volume contained glucose, 53 mol%, and N-acetylglucosamine, 23.7 mol%; whereas the major A-2 component eluted with 0.05M NaCl and contained glucose, 15.2 mol%; galactose, 31.7 mol%; and N-acetylglucosamine, 29.4 mol%. Two nonviscous components were resolved from the bulk polysaccharide by gel permeation, both lower in molecular weight than component A. Fraction B contained L-rhamnose and glucose as the major constituent sugars. Fraction C is of interest because it contains 6-deoxy-L-talose (13.5 mol%) with galactosamine as the major constituent. In contrast to Taichman et al (1978), 6-deoxy-L-talose was detected in cell walls and in the exocellular polysaccharides of both T14V and T14AV by Imai and Kuramitsu (1983).

14.3 FIMBRIAE AND VIRULENCE ASSOCIATION

The first step in periodontal disease is adherence to and colonization of the gingival crevice between the teeth. The laboratory-derived *A viscosus* T14AV mutant lacks this ability as shown in rodent experiments (Brecher et al, 1978). Fine surface fimbriae present on the virulent T14V, and to a lesser extent on the avirulent T14AV isolate are sensitive to chymotrypsin. Digestion with this enzyme inhibits the in vitro adherence of T14V cells to saliva-coated hydroxyapatite, a tooth enamel analog (Wheeler et al, 1979). Antigenic sites were localized on fimbriae of *A viscosus* T14V by immunoelectron microscopy (Cisar et al, 1978) with rabbit antibodies to T14V adsorbed with T14AV cell walls and also with monoclonal antibodies (Cisar et al, 1981). Fimbriae were demonstrated with the same labeling technique in human dental plaque.

Soluble antigen was prepared by digesting T14V cell walls with lysozyme. This antigen reacted with both the anti-T14V and anti-T14AV in immunoelectrophoresis suggesting that only a quantitative difference existed between the two cultures. The weak affinity of T14AV cells for hydroxyapatite and for human teeth may be related to a large amount of glucosamine-containing polysaccharide that masks the fimbriae, inhibiting adherence (Brecher et al, 1978) but later it was discovered that the T14AV contains a greatly reduced density of fimbriae. The addition of extraneous glucosamine-polysaccharide from the avirulent isolate to the virulent cells impaired the latter's ability to adhere to hydroxyapatite. Here, then, is a paradox, because the slime polysaccharide contributes to intercellular plaque material but inhibits adherence of the bacteria to the gingival crevice.

Monoclonal antibodies to *A viscosus* agglutinated whole bacteria and could also aid in demonstrating the lectinlike activity of the fimbriae (Cisar et al, 1981). The lectin activity is responsible for coaggregation of *A viscosus* with certain strains of *Streptococcus sanguis* and is reversible with lactose or β-methyl galactoside. Purified fimbrial preparations (from cells subjected to mild sonication) form soluble immune complexes with these monoclonal antibodies that agglutinate neuraminidase-treated erythrocytes. This is analogous to the lactose-reversible coaggregation of *S sanguis* with *A viscosus*. Solubilized fimbriae react in two-dimensional crossed "rocket" immunoelectrophoresis demonstrating two major precipitin arcs with rabbit antiserum. All nine monoclonal antibodies reacted with a single fimbrial antigen, the one responsible for the lectin activity. The function of the second type of fimbriae is to mediate adherence of the bacteria to saliva-coated tooth enamel. The functional differences between two types of surface fimbriae were further delineated with type-specific monoclonal antibodies, and a coaggregation-deficient *A viscosus* mutant (Cisar et al, 1983). Type 1 fimbriae are responsible for adherence to saliva-coated hydroxyapatite or tooth enamel. A spontaneous coaggregation-deficient mutant was selected by cultivating the T14V strain on medium containing *Streptococcus sanguis*. The resulting mutant lacked type 2 fimbriae. Proof of this was provided by correlating the lack of lactose-reversible coaggregation with immunoperoxidase staining that revealed positive staining with MAb specific for type 1 and the absence of fimbriae reactive with MAb specific for type 2.

Quantitative binding of ^{14}C-labeled MAbs corroborated these findings and also proved that the T14AV mutant contains greatly reduced concentrations of fimbriae of both types. Soluble fimbrial antigen from the coaggregation-deficient mutant was compared to a similar preparation from the T14V parent strain by two-dimensional crossed line immunoelectrophoresis with antiserum capable of binding fimbriae of both types (Figure 14.1). This experiment further established the presence of type 1 and the absence of type 2 fimbriae in the coaggregation-deficient mutant.

14.4 B-CELL MITOGEN OF *ACTINOMYCES VISCOSUS*

A B-cell mitogen for rodents was prepared from *A viscosus* cells by mechanical disruption and low-speed centrifugation. The lyophilized supernate is referred to as AVIS (Engel et al, 1976). Mitogenic activity was localized in a 20,000 $\times$ g supernate of sonicated *A viscosus* cells and also in an extracellular heteroglycan (Burckhardt et al, 1977). Lymphocyte blastogenesis measured by the uptake of radiolabeled nucleic acid precursors, uridine and thymidine, was maximal in splenocytes from germ-free rats after 48 hours contact with 10 μg of mitogen. The high speed supernate had a higher specific activity than the heteroglycan. Splenocytes from athymic nude mice were stimulated by the heteroglycan and supernate at levels that exceeded the blastogenic response to lipopolysaccharide endotoxin.

14.4.1 Cellular Requirements for Activation

AVIS is mitogenic in the 10 to 100 μg/ml range for spleen cell cultures from C57Bl and athymic nude mice (Engel et al, 1977). Splenocytes from genetic low responders to lipopolysaccharide, the C3H/HeJ mice, are also stimulated by AVIS. This rules

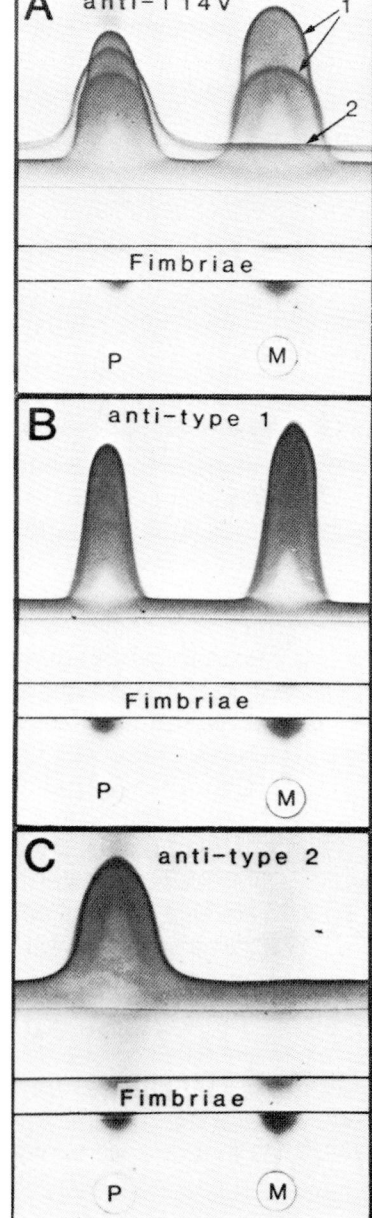

Figure 14.1. Comparison of fimbriae isolated from the coaggregation-positive parent strain, *Actinomyces viscosus* T14V, with those isolated from the coaggregation-negative mutant strain, by rocket-line immunoelectrophoresis. The wells contained 5 μg of fimbrial protein from the parent (P) strain or mutant (M) strain. The trough labeled "Fimbriae" was filled with fimbrial antigens of *A viscosus* T14V to provide reference lines for the comparison of parent and mutant antigens. The upper portion of the agarose gel on different plates contained antiserum against either whole *A viscosus* T14V bacteria, panel (A); against type 1 fimbriae, panel (B); or against type 2 fimbriae, panel (C). The arrows in panel (A) identify type 1 and type 2 fimbrial antigens. The anode for electrophoresis is towards the top. Type 2 fimbriae mediate coaggregation of *A viscosus* with *Streptococcus sanguis*; type 1 fimbriae are the adhesin that causes *A viscosus* to bind to saliva-coated hydroxyapatite (Cisar et al, 1983). *Source:* Reprinted with the permission of the authors and the publisher, American Society for Microbiology.

out the possibility of endotoxin contamination, which is unlikely because *A viscosus* is a Gram-positive organism. A requirement for T-cells was excluded because positive responses were obtained in athymic nude mice, and in splenocytes from euthymic mice depleted of T-cells by treatment with rabbit anti-rat brain and complement. The polyclonal activation of spleen cells is illustrated by the 13% to 25% of AVIS-treated splenocytes that converted to blast forms and were driven to IgM synthesis.

Inactivation experiments attempted to define the nature of the mitogen in AVIS. Activity was reduced by periodate oxidation, by digestion with bacterial proteinases, and was abolished by alkaline digestion. Neither carbohydrate nor protein determinants were excluded as components of the active mitogen. Mice sensitized with whole killed *A viscosus* in incomplete Freund adjuvant were compared with normal mice for their footpad reactions to 800 μg of AVIS. Normal mice, C57Bl/6J, produced an infiltrate of polymorphonuclear neutrophils but in immunized mice a predominantly mononuclear cell infiltrate was observed that persisted for 48 hours, presumptive evidence for delayed cutaneous hypersensitivity (Engel et al, 1976). Chemotaxis of human neutrophils was elicited by AVIS in the presence of fresh serum. Heat inactivated serum (56 °C) could not support chemotaxis so that complement activation is likely involved.

The effect of AVIS on fibroblast cultures was tested because human dental plaque extracts have been shown to be toxic for several cultured mammalian cell lines (Engel et al, 1978). The fibroblasts bound AVIS, and phagosomes containing the antigen were observed in the cytoplasm. The cells remained viable because significant ^{51}Cr release did not occur, and collagen synthesis was normal. The adherence of AVIS to fibroblasts is an active process inhibited by azide and may be harmful to the host if antibody-dependent complement activation were to occur, as observed in the interaction between levan and human gingival fibroblasts (Lesher and Gerencser, 1977). AVIS-coated mouse gingival fibroblasts are recognized in vitro as targets and killed by lymphocytes from AVIS-immunized mice.

Lymphocyte blastogenesis to *A viscosus* antigens is found in gingivitis and periodontal disease but not in peripheral blood lymphocytes from normal individuals. Under certain conditions polyclonal B-cell activation in normal peripheral blood lymphocytes by sonicated *A viscosus* cell extracts can be demonstrated (Mangan and Lopatin, 1981). The factors influencing in vitro B-cell activation and polyclonal immunoglobulin synthesis are the presence of T-cell help, removal of the antigen after pulsing cultures for two days, and depletion of monocytes. Immunoglobulin M appeared on day 5 followed by IgA and IgG on day 7. Although a sensitive enzyme immunoassay was used, no antibodies specific for the stimulating antigen were detected. The composition of the antigen was not studied, but extracts of the GA strain and the familiar T14V were both competent to induce polyclonal immunoglobulin synthesis. The infiltration of B-lymphocytes in periodontal lesions may be explained by the occurrence of the polyclonal B-cell-activating property of *A viscosus* and the resulting immunoglobulin production could contribute to (immunopathology) through complement-mediated damage.

14.4.2 Blastogenesis Correlates with Periodontitis

Ivanyi and Lehner (1970) reported a direct relation between the severity of periodontal disease and the degree of blastogenesis evoked by plaque in peripheral blood lymphocytes in vitro. If the blastogenic response indicates the degree of inflammation, then the microorganism that increases blastogenesis ought to be the cause of periodontal disease. Baker et al, 1976, prepared cell walls of common plaque organisms to test this possibility. Six patients with periodontal disease responded to solubilized dental plaque (mean blastogenic index = 9.8), and the blastogenic index of *A viscosus* cell wall antigen was 11.5. The blastogenic indices of normal subjects were flat. *A naeslundii* was the most potent antigen tested and, in general, the actinomycetes were more potent activators of blastogenesis than were the oral streptococci. *Actinomyces viscosus* is now known to contain a T-dependent B-cell mitogen that is active in normal persons. The reaction in periodontal disease may be qualitatively different because B-cells from normal subjects are driven to polyclonal immunoglobulin synthesis in the absence of appreciable uptake of DNA precursors (Mangan and Lopatin, 1981).

14.5 SUMMARY

Actinomyces viscosus and related microaerophilic actinomycetes are major cultivatable organisms in mature and calcifyinig plaque. They induce neutrophil chemotaxis and lysosomal enzyme release indirectly through complement activation. A mitogen in homogenates of *A viscosus* triggers B-cell activation and T-dependent polyclonal immunoglobulin synthesis in lymphocytes from normal persons. Higher blastogenic responses are recorded when gingivitis or periodontitis is present. Two kinds of surface fimbriae of *A viscosus* mediate adherence of the bacteria to teeth and participate in coaggregation with *Streptococcus sanguis*. Exopolysaccharides are produced by these microaerophils including levan and an N-acetylglucosamine polymer that contribute to the matrix of dental plaque. These immunostimulatory properties of oral actinomycetes set in train the events leading to chronic inflammation.

REFERENCES

Baker JJ, Billy SA, 1983. Activation of the alternate complement pathway by peptidoglycan of *Actinomyces viscosus*, a potentially pathogenic oral bacterium. Arch Oral Biol 28:1073–1075.

Baker JJ, Chan SP, Socransky SS, Oppenheim JJ, Mergenhagen SE, 1976. Importance of *Actinomyces* and certain Gram-negative anaerobic organisms in the transformation of lymphocytes from patients with periodontal disease. Infect Immun 13:1363–1368.

Brecher SM, van Houte J, Hammond BF, 1978. Role of colonization in the virulence of *Actinomyces viscosus* strains T14-Vi and T14Av. Infect Immun 22:603–614.

Brown DA, Fischlschweiger W, Birdsell DC, 1980. Morphological, chemical and antigenic characterization of cell walls of the oral pathogenic strains *Actinomyces viscosus* T14V and T14AV. Arch Oral Biol 25:451–457.

Burckhardt JJ, Guggenheim B, Hefti A, 1977. Are *Actinomyces viscosus* antigens B cell mitogens? J Immunol 118:1460–1465.

Cisar JO, Barsumian EL, Curl SH, Vatter AE, Sandberg EL, Siraganian RP, 1981. Detection and localization of a lectin on *Actinomyces viscosus* T14V by monoclonal antibodies. J Immunol 127:1318–1322.

Cisar JO, Curl SH, Kolenbrander PE, Vatter AE, 1983. Specific absence of type 2 fimbriae on a coaggregation-defective mutant of *Actinomyces viscosus* T14V. Infect Immun 40:759–765.

Cisar JO, Kolenbrander PE, McIntire FC, 1979. Specificity of coaggregation reactions between human oral streptococci and strains of *Actinomyces viscosus* or *Actinomyces naeslundii*. Infect Immun 24:742–752.

Cisar JO, Vatter AE, McIntire FC, 1978. Identification of the virulence associated antigen on the surface fibrils of *Actinomyces viscosus* T14. Infect Immun 19:312–319.

Ellen RP, 1976. Establishment and distribution of *Actinomyces viscosus* and *Actinomyces naeslundii* in the human oral cavity. Infect Immun 14:1119–1124.

Engel D, Clagett J, Page R, Williams B, 1977. Mitogenic activity of *Actinomyces viscosus* I. Effects on murine B and T lymphocytes and partial characterization. J Immunol 118:1466–1471.

Engel D, Schroeder HE, Page RC, 1978. Morphological features and functional properties of human fibroblasts exposed to *Actinomyces viscosus* substances. Infect Immun 19:287–295.

Engel D, van Epps D, Clagett J, 1976. In vivo and in vitro studies of possible pathogenic mechanisms of *Actinomyces viscosus*. Infect Immun 14:548–554.

Hammond BF, Steel CF, Peindl KS, 1976. Antigens and surface components associated with virulence of *Actinomyces viscosus*. J Dent Res 55(special issue A):A19–A25.

Howell A Jr, Jordan HV, 1967. Production of an extracellular levan by *Odontomyces viscosus*. Arch Oral Biol 12:571–573.

Imai S, Kuramitsu H, 1983. Chemical characterization of extracellular polysaccharides produced by *Actinomyces viscosus* T14V and T14Av. Infect Immun 39:1059–1066.

Ivanyi L, Lehner T, 1970. Stimulation of lymphocyte transformation by bacterial antigens in patients with periodontal disease. Arch Oral Biol 15:1089–1096.

Jordan HV, Hammond BF, 1972. Filamentous bacteria isolated from human root surface caries. Arch Oral Biol 17:1333–1342.

Jordan HV, Keyes PH, 1972. Aerobic, gram-positive filamentous bacteria as etiologic agents of experimental periodontal disease in hamsters. Arch Oral Biol 9:401–414.

Jordan HV, Keyes PH, Bellack S, 1972. Periodontal lesions in hamsters and gnotobiotic rats with *Actinomyces* of human origin. J Periodont Res 7:21–28.

Lesher RJ, Gerencser VF, 1977. Levan production by a strain of *Rothia*: activation of complement resulting in cytotoxicity for human gingival cells. J Dent Res 56:1097–1105.

Mangan DF, Lopatin DE, 1981. In vitro stimulation of immunoglobulin production from human peripheral blood lymphocytes by a soluble preparation of *Actinomyces viscosus*. Infect Immun 31:236–244.

Miller CH, Somers PJB, 1978. Degradation of levan by *Actinomyces viscosus*. Infect Immun 22:266–274.

Miller CH, Warner TN, Palenik CJ, Somers PJB, 1975. Levan formation by whole cells of *Actinomyces viscosus ATCC 15987*. J Dent Res 54:906.

Rosan B, Hammond BF, 1974. Extracellular polysaccharides of *Actinomyces viscosus*. Infect Immun 10:304–308.

Socransky SS, Hubersak C, Propas D, 1970. Induction of periodontal destruction in gnotobiotic rats by human oral strain of *Actinomyces naeslundii*. Arch Oral Biol 15:993–995.

Syed SA, Loesche WJ, 1978. Bacteriology of human experimental gingivitis: effect of plaque age. Infect Immun 21:821–829.

Taichman NS, Hammond BF, Tsai C-C, Baehni PC, McArthur WP, 1978. Interactions of inflammatory cells and oral microorganisms VII. In vitro polymorphonuclear responses to viable bacteria and subcellular components of avirulent and virulent strains of *Actinomyces viscosus*. Infect Immun 21:594–604.

van der Hoeven JS, 1974. A slime producing microorganism in dental plaque of rats, selected by glucose feeding. Caries Res 8:193–210.

Wheeler TT, Clark WB, Birdsell DC, 1979. Adherence of *Actinomyces viscosus* T14V and T14AV to hydroxyapatite surfaces in vitro and human teeth in vivo. Infect Immun 25:1066–1074.

Wicken AJ, Broady KW, Evans JD, Knox KW, 1978. New cellular and extracellular amphipathic antigen from *Actinomyces viscosus* NY1. Infect Immun 22:615–616.

Williams RC, Gibbons RJ, 1972. Inhibition of bacterial adherence by secretory immunoglobulin A: a mechanism of antigen disposal. Science 177:697–699.

Chapter 15

Thermophilic Actinomycetes

15.1 INTRODUCTION

Thermophilic actinomycetes grow well and produce numerous spores at the elevated temperatures present in fermenting moist hay. Farmers can become sensitized by repeated contact with dusts containing these spores and filaments. Work in a heavily contaminated hay barn may entail inhalation of 7.5×10^5 actinomycete spores per minute (Lacey and Lacey, 1964). The illness resulting from these exposures takes the form of attacks with respiratory and systemic effects referred to as farmer's lung disease (FLD). Wheezing and shortness of breath are often accompanied by fever, nausea, and weight loss. Over a period of years the disease can progress to loss of respiratory capacity, reflected in fixed pulmonary radiographic changes and fibrosis. The major causative agent of farmer's lung disease is *Faenia rectivirugula (Micropolyspora faeni)*, but other thermophils are known to be involved, including *Thermoactinomyces vulgaris*, *Thermoactinomyces candidus*, *Thermoactinomyces sacchari*, and *Saccharomonospora viridis*.

The fate of the inhaled particulate materials is slow digestion by host alveolar macrophages, because thermophilic actinomycetes cannot invade host tissues. There is, however, uncertainty about whether these spores can germinate in the lung and undergo limited development. The disease is classified as one of the hypersensitivity pneumonitides because the symptoms are similar to an allergic reaction, and because it is not truly an infectious process. Thermophilic actinomycetes are implicated in other forms of hypersensitivity pneumonitis resulting from other occupational and environmental hazards, eg, sugar cane agriculture (bagassosis), mushroom cultivation, or contaminated home humidifiers (Schatz et al, 1977). Removal of the affected person from the inciting environment is often required for improvement in farmer's lung disease.

Antibodies reactive with components of *F rectivirugula* are present in sera from most farmer's lung patients and from cattle with fog fever (Pepys and Jenkins, 1965). The common isolation of this actinomycete from hay and sputa from patients examined by the Pneumoconiosis Unit in Wales and the Marshfield Clinic in Wisconsin are evidence of its wide geographic distribution (Edwards, 1972; Hollingdale, 1974; Marx et al, 1978). Antigens made from *F rectivirugula* isolated in Finland, the United Kingdom, and the USA produce nearly identical immunoelectrophoretograms (Edwards, 1972). *Faenia rectivirugula* grows best at temperatures between 30 °C and 65 °C producing branching filaments 1 μm in diameter and bearing chains of up to ten spores (0.7 to 1.5 μm diameter long) that become easily airborne. Culture filtrates are often used as the antigenic complex for precipitin tests, but washed filaments or spore preparations have also been analyzed for antigens. Surprisingly, spores and spore extracts are deficient in antigens and no reactions are observed in precipitin tests with pooled FLD sera, nor could spores adsorb antibodies from the sera. This finding indicates that limited development of the inhaled spores occurs, that aerosols of moldy hay contain metabolic products of thermophilic actinomycetes, or that spores carry adsorbed "metabolic" antigens on their surfaces into the lungs.

15.2 ANTIGENIC STRUCTURE AND PRECIPITINS

15.2.1 Precipitin Score, C Region Antigens

Before the causative agents of farmer's lung disease were known, antigens were prepared by soaking moldy hay in Coca's solution (Pepys and Jenkins, 1965). Hay made "moldy" by pure culture methods was produced by seeding *F rectivirugula* into sterile, moist hay in a "well-lagged barrel" (Pepys and Jenkins, 1965). Pepys and Jenkins proposed a scheme for grading precipitin reactions according to the appearance of arcs in different regions of the immunoelectrophoretogram (Figure 15.1). Immunoelectrophoresis is less sensitive than immunodiffusion, but more discriminating in detecting antibodies to *F rectivirugula*. Grade 3 precipitins occurred in all three regions, grade 2 in two regions, usually A and B, and in the grade 1 category precipitins were confined to a single region, usually A or B. A total of 89% of patients with clinical evidence of FLD reacted positively in immunoelectrophoresis. In a large series of 205 FLD patients the precipitin responses correlated with the frequency and severity of the attacks: the higher the grade, the more precipitins, and the greater the number of attacks. Precipitins in the C region were infrequent in Grade 2 reactors (less than 20% reacted in the C region) and even less frequent in those patients whose precipitins were confined to a single region (7%). Thus reactions in the C region gave the impression of greater specificity. Trichloroacetic acid-soluble antigens of *F rectivirugula* have the lowest electrophoretic mobility (C region). They are proteinase-resistant, periodate-sensitive and are removed from sera by adsorption with whole *F rectivirugula* bacteria, consistent with their existence as wall-associated glycoproteins.

Inhalation of *F rectivirugula* antigens of A, B, or C region mobility produced fever and pulmonary reactions in sensitized farmers. About 11% of FLD subjects did not produce precipitins to farmer's lung hay or to *F rectivirugula* extracts, and these persons had relatively mild disease. The 15% to 21% positive reactions among asymptomatic exposed farmers poses a more difficult problem of interpretation, showing that precipitins by themselves are not pathogenetic.

15.2.2 Double Dialysis Technique

The double dialysis method of producing thermophilic actinomycete antigens consists of placing a complex medium, peptone broth, sometimes supplemented with

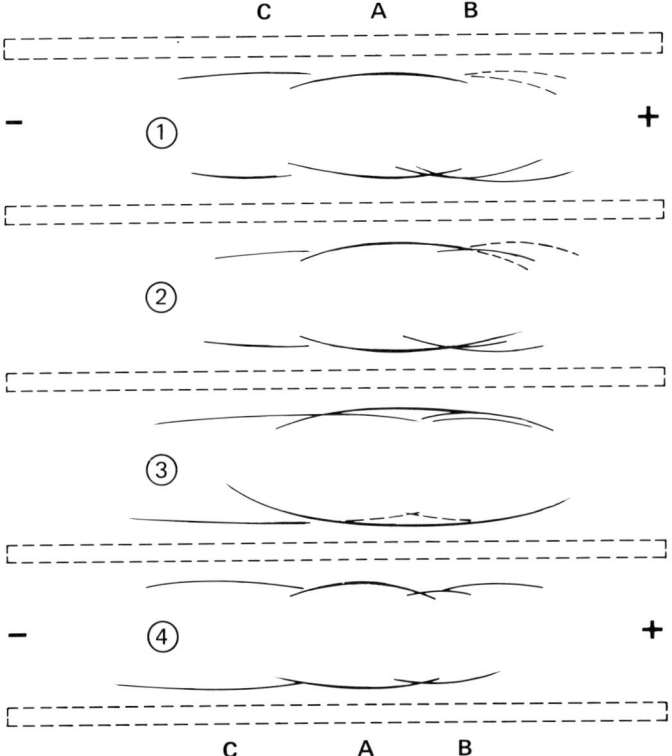

Figure 15.1. Precipitin reactions of an FLD patient with various antigens: 1, extract of moldy hay; 2, mixed actinomycete extract; 3, *Faenia rectivirugula* grown on hay, then extracted; 4, *F. rectivirugula* extract from synthetic medium. The immunoprecipitin arcs have been labeled in 3 regions, A, B, C. Other details are given in Pepys and Jenkins (1965). *Source:* Courtesy of Professor Jack Pepys, London.

yeast extract, in a dialysis sac as the inner phase. The outer phase, consisting of glycine-saline (Edwards, 1972), is then seeded with the actinomycete. The antigens secreted in the outer phase over a 12 day period are recovered, dialyzed, and lyophilized after centrifuging off the organisms (Figure 15.2). This technique was used in several studies until it was realized that antigens in the yeast extract permeated the dialysis membrane and were responsible for the common precipitin reactions observed in several thermophilic antigen complexes, as well as in activating complement.

15.2.3 Glycoprotein "a"

Hollingdale, 1974, reported a *F rectivirugula* culture filtrate antigen forming 16 arcs in immunoelectrophoresis versus pooled FLD sera. Of these factors, four did not occur in supernates of sonicated whole cells. These antigens were sensitive to pronase, with the exception of one that was pronase-resistant and periodate-sensitive. The glycoprotein antigens present in the filaments were separated from proteins by extraction of the whole filamentous growth with cold 90% phenol. Three glycoprotein antigens were partitioned into the aqueous phase, and the 13 pronase-sensitive proteins remained in the phenol phase. The three glycoprotein antigens were also extracted from whole filaments or crude cell walls with cold 5% TCA. The protein

324 THERMOPHILIC ACTINOMYCETES

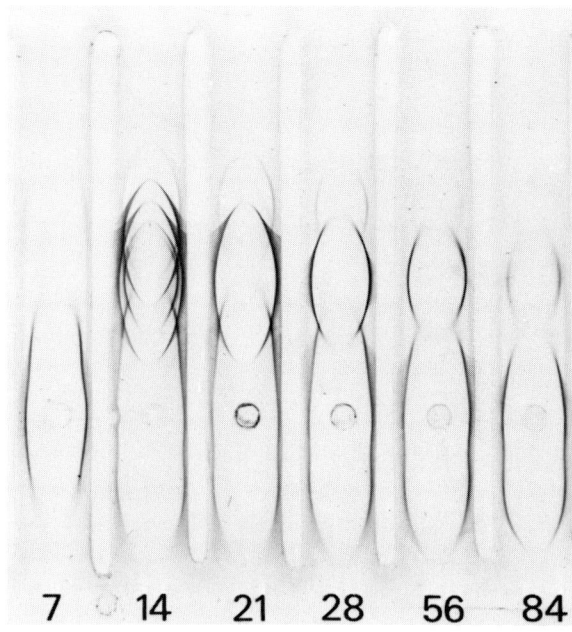

Figure 15.2. Immunoelectrophoresis of *Faenia rectivirugula* antigens produced by the double dialysis technique. The numbers shown are the age of the culture in days when the antigen was sampled (versus pooled farmer's lung sera). The number of immunoprecipitin arcs peaks at day 12. Afterwards, endogenous proteinases digested several protein antigens. *Source:* printed with permission of Dr. John H. Edwards, Medical Research Council Pneumoconiosis Unit, South Glamorgan, Wales, United Kingdom (also see Edwards, 1972).

antigens extracted into the phenol phase reacted with 73 of 76 sera of FLD patients, but also with 62 of 66 non-FLD pulmonary patients' sera, thus limiting the value of this antigenic complex as a diagnostic aid. Glycoproteins in the aqueous phase of the phenol extract or those in the TCA-soluble extract were better able to discriminate between FLD patients (74 positive reactors) and other pulmonary patients (3 reactors). The immunoprecipitin arc due to the *F rectivirugula* "a" glycoprotein was especially reactive in sera of FLD patients.

15.2.4 Glycopeptide "1"

It is not clear why cell wall glycoproteins should provide a selectivity for farmer's lung disease. One possibility is that the protein antigens of the culture filtrate are highly immunogenic when presented on the surface of inhaled or germinated spores, so that even a minute antigenic stimulus triggers precipitins in asymptomatic exposed farmers. Fletcher and Rondle (1973) also showed that glycoproteins of *F rectivirugula* react with precipitins in FLD patients' sera, but no farmers without the disease were screened. Crude cell walls from a sonified filamentous growth were digested with lysozyme. The antigen liberated by lysozyme digestion contained arabinose, galactose, and glucosamine in the neutralized acid hydrolyzate. The protein content of the cell wall derived antigen was very low, only 2% to 6%. Edwards (1972) identified three antigens in the *F rectivirugula* culture filtrate that were considered major because they reacted more frequently or strongly with FLD patients' sera. One of these antigens, the glycopeptide "1" was heat-stable and had a protein : carbohydrate ratio of 1 : 57. This antigen did not migrate in immunoelectrophoresis, and its position corresponds to the C region of the immunoelectrophoretogram of Pepys and Jenkins (1965). Glycopeptide "1" was purified from the culture filtrate by first precipitating the other factors with one volume of ethanol. The alcoholic supernate then was dialyzed, concentrated, and made to 5% with cold TCA. Glycopeptide "1" remained soluble and was chromatographed on Sephadex G200. It was included in the column volume and eluted just before a low molecular weight nonantigenic brown pigment. At this stage, two immunoprecipitin arcs were obtained with this antigen. Further purification was effected by diethylaminoethyl Sephadex chromatography during which the glycopeptide "1" was not adsorbed by the anion exchanger. To summarize, this antigen (i) is present in the 50% ethanol supernatant, (ii) is not precipitable with 5% TCA, (iii) is included by Sephadex G200 and does not bind to DEAE, (iv) gives rise to a single immunoprecipitin arc corresponding to antigen "1" of the original complex. Later it was found that the antibody reactive with glycopeptide "1" could also be adsorbed by intact *F rectivirugula* filaments, indicating that glycopeptide 1 is either a cell surface factor or a secreted glycoprotein.

It is likely that the glycopeptide "1" of Edwards (1972), the "a" glycoprotein of Hollingdale (1974), the C region antigen of Pepys and Jenkins (1965), and the CWL preparation of Fletcher and Rondle (1973), are similar if not the same entity.

The activity of TCA-soluble antigens present in the culture filtrates of several strains of thermophilic actinomycetes were compared in sera of FLD patients and in asymptomatic exposed farmers (Roberts et al, 1976b). The rationale for this approach is that the "precipitin score" of patients (Pepys and Jenkins, 1965) gives the impression that reactions in the C region occur in those who are prone to more frequent and acute attacks. The C region of lowest electrophoretic mobility contains the 5% TCA-soluble antigens that resist digestion by proteinases but are susceptible to periodate oxidation. A significant difference was detected among FLD patients, 68% of whom reacted in immunodiffusion with the TCA-soluble fraction and the 22% of reactors among asymptomatic exposed farmers. Most of the reactive FLD patients respond to the *F rectivirugula* antigen. The results of Roberts et al (1976b) validate the original findings of Pepys and Jenkins but lack the enumeration of precipitins in the three regions that permitted grading of the responses. Perhaps that is why Roberts et al (1976a) recommended the use of an unpurified complex of antigens as a screening device instead of the TCA-soluble fraction.

Although the TCA-soluble glycoproteins have been cited as useful in discriminating between FLD and asymptomatic farmers or persons with other pulmonary diseases, disagreement now exists on this point. Although a significantly higher proportion of FLD patients reacted with the glycoprotein antigens, false-positives and false-negatives were encountered. A more rigorously purified antigen such as Edwards glycopeptide "1" might possibly afford greater specificity.

15.2.5 Two-Dimensional Immunoelectrophoresis, Isoelectric Focusing, and Enzyme Immunoassays

A crossed "rocket" immunoelectrophoresis reference system has been developed for the thermophilic actinomycete antigens that is capable of enumerating precipitins in human sera placed in an intermediate gel (Treuhaft et al, 1979) (Figure 15.3). The goal of such analysis is to distinguish disease-specific antigens from those that merely detect antibodies caused by exposure of otherwise asymptomatic farmers, and thus to extend the pioneering work of Pepys and Jenkins (1965), who proposed the method of grading precipitin responses that accorded with the severity of dis-

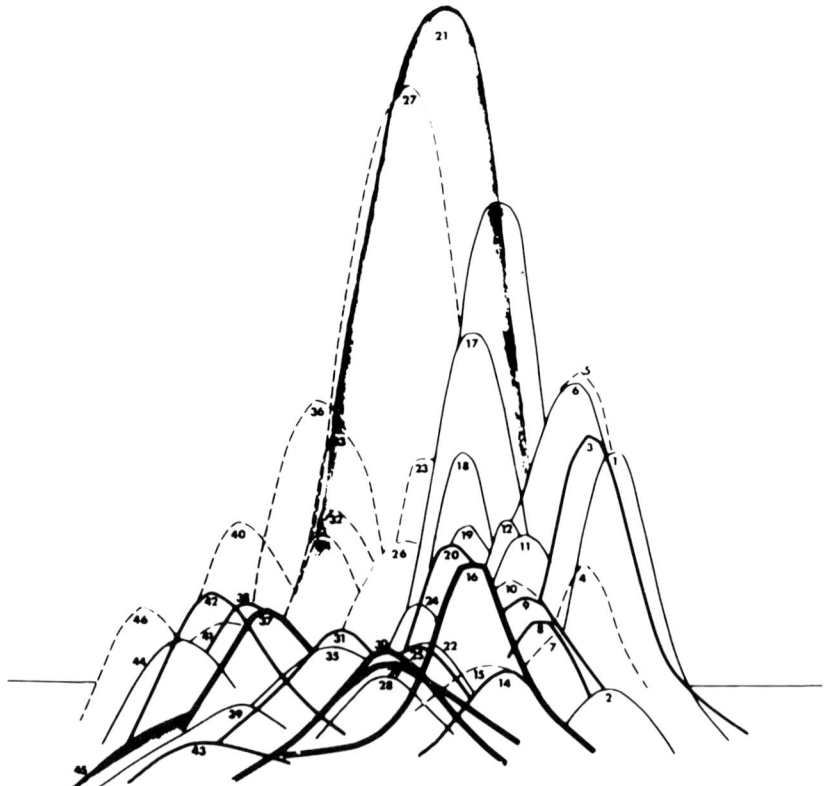

Figure 15.3. Two-dimensional crossed "rocket" immunoelectrophoresis of culture filtrate antigens of *Faenia rectivirugula* grown by a modified double-dialysis technique (Treuhaft et al, 1979). The antiserum was from rabbits immunized with the antigenic complex in complete Freund's adjuvant. There are 46 immunoprecipitates in the reference system; numbers 7, 10, and 12 were present in more than 90% of 13 farmer's lung disease patients but not in asymptomatic farmers. *Source:* reprinted from the American Review of Respiratory Diseases with permission of the publisher, American Lung Association.

ease. Subsequent efforts with the TCA-soluble glycoproteins of *F rectivirugula* or the chymotrypsinlike proteinases have shown that these antigens are potent but not necessarily selective for farmers with clinically recognized disease. Of the maximum of 46 Laurell rockets in the Treuhaft et al reference system, ten appear to have disease specificity. Although no single antigen could discriminate farmer's lung disease from asymptomatic farmers, a combination of factors 7, 10, and 12 detected precipitins in greater than 90% of acute FLD patients. A comparison of these antigens with the glycoproteins discussed above would be interesting.

The optimal conditions for the expression of antigens in the culture filtrates of *F rectivirugula* and *T candidus* were studied in a synthetic medium because yeast extract proteins penetrated the dialysis tubing used in the double dialysis method of Edwards, (1972) (Treuhaft et al, 1981). These extraneous proteins were singled out as responsible for spurious reactions in farmers who responded to all thermophilic species in a test panel. Although the chemically defined medium results in the secretion of fewer proteins, this was considered an acceptable cost for removal of the yeast extract. Monitoring the cultures for the appearance of antigens in the filtrate is important, especially in relation to proteinase secretion, which in *F rectivirugula*

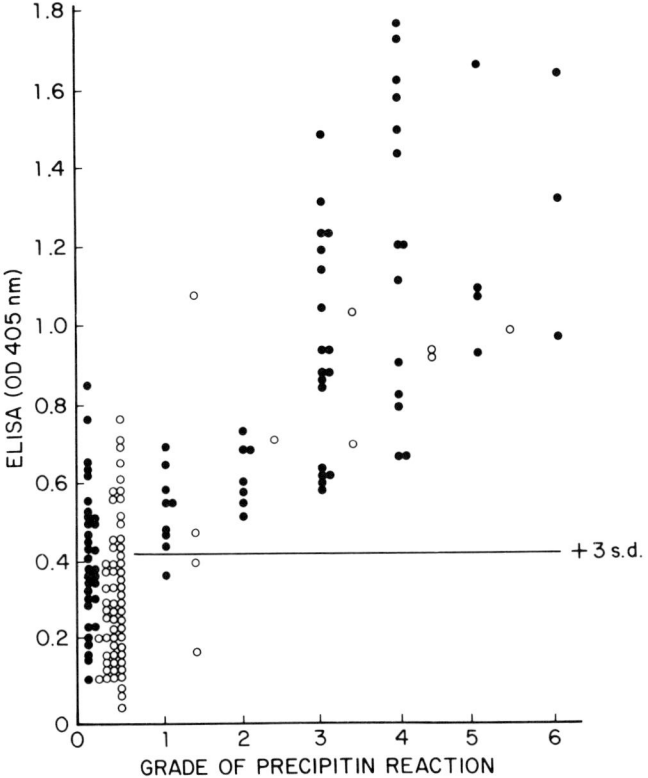

Figure 15.4. Concentration of IgG against antigens of *Faenia rectivirugula* measured by indirect EIA compared to the precipitin score. Antigens were produced by the double-dialysis method. The precipitin score measured the strength of the Ouchterlony double immunodiffusion reaction: 1, weak; 2, medium; 3, strong; 4, very strong. Filled circles are reactions of farmer's lung disease patients; open circles denote those of aymptomatic exposed farmers (Bamdad 1980). *Source:* reprinted from Clinical Allergy with permission from Blackwell Scientific Publications, Ltd.

cultures peaked at 14 days. By that time the precipitating antigens were substantially degraded. Maximal antigen secretion occurred earlier in the growth period, 9 to 12 days; these preparations gave reactions of 25 Laurell rockets with patients' sera. The *T candidus* cultures reached maximal antigen production much earlier, at 3 days, yielding 14 to 16 factors recognized by patients' precipitins. One paradoxical result was that antigenic activity in indirect enzyme immunoassay was directly proportional to the proteinase content. This was interpreted as (i) immunoglobulin binding to proteinase as an enzyme-substrate reaction rather than via the antibody-combining site; or (ii) endogenous proteinase digesting thermophilic proteins to the point that the fragments are unable to react in precipitin tests, but are nevertheless able to react in a primary binding assay; or (iii) that the proteinases themselves are potent antigens of these thermophils, as discussed by Walbaum and Biguet (1973).

Faenia rectivirugula proteins were reserved from the medium after growth for four days at 50 °C in a synthetic broth supplemented with spermidine and lactose (Kurup et al, 1981). The proteins had isoelectric points in the pH 3.5 to 5.5 range. Preparative isoelectric focusing resulted in a pool between pI 3.8 to 4.0 containing two Coomassie-positive proteins and 17% carbohydrate. Four major and four minor antigens were detected in the pool by crossed "rocket" immunoelectrophoresis. This pool reacted in immunodiffusion with all farmer's lung disease patients in this series but not with controls.

Whole dialyzed and lyophilized cultures of *F rectivirugula* have been used to coat microtiter plate walls by passive adsorption for the detection of IgG by indirect EIA (Bamdad, 1980). Although no conclusions can be drawn from this method about which antigens adhere to polystyrene and which may be washed away, this simple approach sufficed to sensitize the plastic to discriminate IgG reactions in one series of FLD patients, from those of asymptomatic exposed farmers, and unexposed controls (Figure 15.4). The concentration of IgG detected by EIA was in agreement with the strength of the immunodiffusion reaction but provided more objective endpoints. Because of the increase in sensitivity, the EIA correlated better with the clinical diagnosis than did immunodiffusion.

To summarize the serologic reactions to unpurified thermophilic actinomycete antigens in precipitin tests: approximately 87% of acute farmer's lung disease patients have precipitins detected in immunodiffusion or immunoelectrophoresis, as do as many as 15% to 21% of asymptomatic exposed farmers (Pepys and Jenkins, 1965). Those patients with other, preexisting lung disease were at the high end of this range (Bamdad, 1980). Precipitins against eight thermophilic actinomycetes were detected by immunodiffusion in 8.9% of 1045 farmers in central Wisconsin (Roberts et al, 1976b). The most reactive antigen was from *F rectivirugula*, followed by *T vulgaris*. The group of positive reactors differed from the group at large in having a history of illness after uncapping silos and having a higher concentration of operators of larger farms with dairy herds. Immune rabbit antisera showed multiple strong precipitins in the reaction with antigens of the same species of actinomycete but weak reactions against antigens from different thermophilic actinomycetes (Kurup et al, 1976). Farmer's lung patients, on the other hand, reacted strongly to various species, reflecting exposure to these different organisms.

15.2.6 Summary of Serology

Up to the present, unpurified culture filtrate antigens of *F rectivirugula*, *T vulgaris*, and other thermophils have been used to provide an important laboratory criterion for establishing the diagnosis. The antigenic characterization discussed above has identified major factors, but none of these has given the sought after clear-cut difference between "disease-specific" versus "exposure-related" antigens. Some promising factors have been identified through two-dimensional crossed "rocket" immunoelectrophoresis and isoelectric focusing. These antigens should be isolated, purified, and evaluated in panels of patients' and control sera, especially with respect to the development of primary binding microplate EIAs.

The presence or absence of precipitins is probably not as selective as the precipitin score as it was originally proposed, and this method has not gained widespread acceptance partly because 30% of the symptomatic group lacked precipitins to the TCA-soluble glycoproteins, and because up to 20% of asymptomatic farmers had precipitins (Roberts and Moore, 1977). However, scrutiny of the original proposal supported by a large series of patients (Pepys and Jenkins, 1965) reveals that precipitins to the glycoproteins (C region antigens) were not considered absolutely specific but were one element in the score pattern.

Typically as newer techniques for mapping antigens become available, they are applied to the culture filtrate antigens of the thermophilic actinomycetes, often without relating the factors detected to those in previous studies. This trend will likely continue until the knowledge of the antigenic structure of *F rectivirugula* and related actinomycetes is consolidated.

15.3 PROTEINASES AND COMPLEMENT ACTIVATION

Proteinases of the thermophilic actinomycetes, *F rectivirugula*, *T vulgaris* (may be synonymous with *T candidus*), and *T sacchari* are detected by zymogram imprints of polyacrylamide gel electrophoresis slabs on casein agar (Roberts et al, 1976a; Roberts et al, 1977; Roberts et al, 1983). More than one zone of proteolysis is usually observed in each species' culture filtrate. *T vulgaris* and *T candidus* proteinases had similar migration indices that differed from the major proteinase of *F rectivirugula*. Earlier, Walbaum and Biguet (1973) showed up to four chymotrypsinlike proteinases localized in immunoprecipitin arcs containing antigens of *F rectivirugula* (Figure 15.5). Five *F rectivirugula* proteins with esterase activity were identified with N-acetyl-D,L-phenylalanine-β-naphthyl ester substrate in immunoprecipitin arcs derived from human and bovine hypersensitivity pneumonitis cases (Bannerman and Nicolet, 1976; Nicolet et al, 1977). Antibodies to one of these antigens, "enzyme 1," were prevalent in most FLD patients and asymptomatic exposed farmers. The reaction to "enzyme 1" was routinely detected only after immunoenzyme staining and was often undetectable in unstained gels, or those stained nonspecifically with protein dyes. Purified "enzyme 1" is a potent immunogen and intratracheal instillation in rabbits induced circulating precipitins.

Diisopropylfluorophosphate inhibited all proteolytic activity consistent with the presence of serine proteinases (Roberts et al, 1977). Soybean trypsin inhibitor in some cases changed the electrophoretic mobility but not the proteolytic activity of the proteinases. α_1-Proteinase inhibitor (α_1-antitrypsin) from human blood inhibits some, but not all, of these actinomycete enzymes.

Two serine-type endopeptidases from supernates of *T candidus* (syn *T vulgaris*)

Figure 15.5. Localization of enzymatic activity in immunoprecipitin arcs of the reaction between rabbit anti-*Faenia rectivirugula* and antigens produced by cultivating the actinomycete on V8 juice agar at 40 °C (Walbaum and Biguet, 1973). *Source:* reprinted from Revue d'Immunologie, Paris, with permission of the Societe Francaise d'Immunologie.

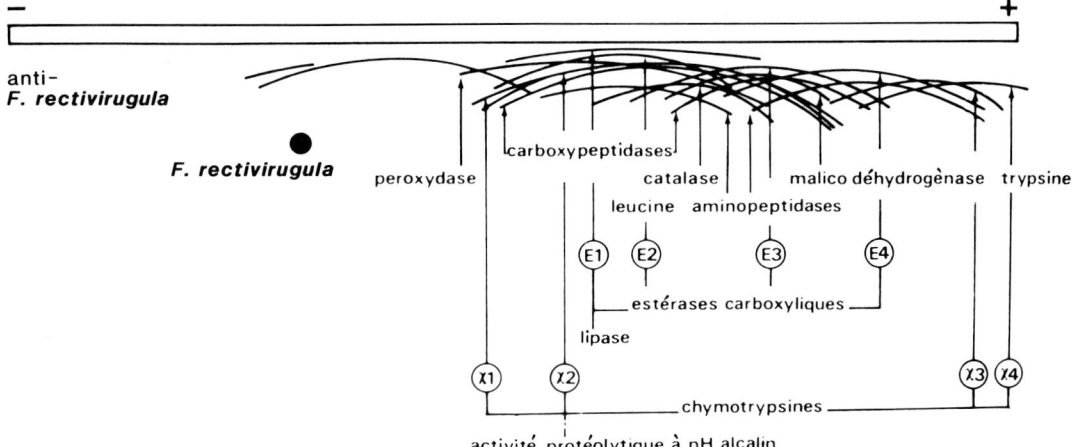

were purified taking into account the differences in their pIs (Roberts et al, 1983). The "P_1" enzyme has a pI of 6 and was purified by adsorption to the cation exchanger CM-Sephadex, and "P_2" with a pI of 4.2 was purified on a DEAE ion exchange column. Further purifications were effected by gel permeation chromatography. The molecular weight of P_1 is 30,000, whereas P_2 occurred as a disulfide-bridged dimer of 52,000 da. These properties differentiate the proteinases from "thermitase", an enzyme from *T vulgaris*, which has a pI of 9.0 (Frommel et al, 1978). The P_1 and P_2 enzymes were not inhibited by plasma proteinase inhibitors, α_2-, and α_1-antichymotrypsin. Inhibition effected by α_2-macroglobulin was noteworthy because this inhibitor is too large to be regularly detected in bronchoalveolar lavage fluids.

The presence of proteinase activity in concentrated extracts of moldy hay dusts from which *T candidus* can be isolated and the presence of IgG against purified proteinases in FLD patients, but not in controls, indicates an association between proteinase and disease. This impression is strengthened by the elastase activity of P_1 and the lack of inhibition by plasma proteinase inhibitors capable of permeating the lung. Such direct damage as may occur is probably amplified by the host IgG response, which could initiate a type III hypersensitivity reaction by deposition of immune complexes in the lung. Roberts et al, (1983) found that the binding of IgG to proteinase did not completely inhibit proteolytic activity.

Moldy hay extract and whole *F rectivirugula* or *T vulgaris* organisms activated complement by the alternative pathway, judging by the electrophoretic demonstration of conversion of C3 proactivator detected with rabbit anti-C3 activator (Edwards et al, 1974). However, more recent studies suggest a different mechanism (Marx et al, 1980). Normal human sera from persons with no farming background were negative for antibodies against *F rectivirugula* by precipitin and complement-fixation tests, and by primary binding assays such as immunofluorescence and radioimmunoassay. Significant reductions in early components of the classical pathway were observed with these sera in the reaction with soluble antigens of *F rectivirugula* grown on synthetic medium. Deposition of C3 and C4 on *F rectivirugula* spores was detected by immunofluorescence. In contrast, serum from an FLD patient reacted strongly, depositing IgG, C1, C3, C4 on the surfaces of spores. Thus *F rectivirugula* appears capable of activating the complement cascade at the level of C4 in the *absence* of detectable antibodies. The conversion of factor B to $\bar{\text{B}}$ was also detected but this was not abolished by Mg_2EDTA suggesting that factor B was converted by C3b resulting from a C_{42} sequence. The possibility that *F rectivirugula* inhibits or inactivates complement components was excluded because chemotactic factors for human polymorphonuclear neutrophils were detected only in those chambers containing both normal serum and *F rectivirugula* extracts. The early changes in lung function resulting from inhalation of spores of thermophilic actinomycetes may result from direct complement activation. This possibility could help explain the symptoms in precipitin-negative farmer's lung disease.

15.4 IMMUNOPATHOLOGY

15.4.1 Humans

Encounters between dusts containing spores of thermophilic actinomycetes and lung tissues cause respiratory illness and damage that results from the immune responses of the susceptible human or animal host. To underscore its immunologic basis, the disease has been classified as one of the hypersensitivity pneumonitides or more precisely, extrinsic allergic alveolitis. The physical symptoms of cough, dyspnea, and fever begin four to eight hours after exposure to the dusts of moldy forage. Histologically, the pattern is one of chronic interstitial and alveolar inflammation followed, after repeated exposures, by a high frequency of noncaseating granulomas (Roberts and Moore, 1977). Infiltrates consist of lymphocytes and activated alveolar macrophages (Karr and Salvaggio, 1980). Pulmonary vasculitis is not consistently found, but has been reported at the time of acute attacks (Karr and Salvaggio, 1980; Ghose et al, 1974).

The relative importance of humoral and cellular mechanisms responsible for lung damage in hypersensitivity pneumonitis caused by the thermophilic actinomycetes is an active area of research. The high incidence of precipitins in farmer's lung disease and the onset of symptoms four to eight hours after exposure are indirect evidence for the operation of Arthus type III immune complex-mediated damage. The histology of lung biopsies from four FLD patients showed two patterns of host responses (Ghose et al, 1974). In two patients, biopsied hours after aerosol challenge or natural exposure to moldy hay, small and medium size blood vessels showed an inflammatory infiltrate of PMN and to a lesser extent, eosinophils and lymphocytes. Extracellular deposits of IgG, IgM, and C3 were also present. The other two patients had chronic FLD and the histology shifted to a mononuclear cell response with interstitial granulomas containing giant cells, macrophages, and lymphocytes. No vasculitis or extracellular immunoglobulin or C3 were detected in the latter two cases. These results and those of Wenzel et al (1971) provide evidence that immune complex deposition is an early event in pathogenesis of the acute phase of farmer's lung.

Circulating immune complexes were detected in 7 of 14 FLD patients by a bovine conglutinin-binding EIA (Terho et al, 1983). These concentrations were found more often in patients than in matched controls (spouses) but the differences between the two groups was not significant. Immune complexes correlated positively with IgG against *F rectivirugula* and *T vulgaris* antigens but negatively against rheumatoid factors. Eight patients with acute FLD had increased rheumatoid factor concentrations which subsided during convalescence. The presence of rheumatoid factor correlated negatively with soluble immune complexes; perhaps the binding of rheumatoid factor occludes the site for conglutinin binding. Rheumatoid factor may be a component of the immunopathological response in FLD, incited by the deposition of immune complexes in the lung.

The similarity of the pulmonary lesions in humans to type IV delayed hypersensitivity has been pointed out in several reviews (Roberts and Moore, 1977; Schatz et al, 1977; Karr and Salvaggio, 1980). Significant differences in the leukocyte migration inhibition evoked by moldy hay extract in farmer's lung patients (mean inhibition, 38.3%) compared with asymptomatic exposed farmers (mean inhibition, 19.5%) gives the impression that leukocyte MIF production correlates with disease and provides some in vitro evidence for the existence of T-cell-mediated immunity (Morell et al, 1982). The incidence of delayed cutaneous hypersensitivity in that study was not surveyed and, in general, positive delayed skin tests are uncommon in farmer's lung disease.

The immunopathology of farmer's lung disease probably also includes an autoimmune component. Braun et al (1983) have found that the presence of detectable antilung immunoglobulin in FLD of greater than 5 years duration correlates significantly with reduced vital and diffusion capacity, and with fibrosis visible on chest radiographs. Recurrent episodes in the presence of antilung Ig may lead to more severe damage. The propensity to produce antilung Ig in some farmers may help explain why there are many asymptomatic exposed farmers and few frank cases of disease.

15.4.2 Animals

Studies in animal models, including adoptive transfer experiments, have yielded information compatible with a composite of immune complex-mediated- and T-cell-mediated-immunopathogenesis in hypersensitivity pneumonitis caused by thermophilic actinomycetes. Rabbits inoculated intratracheally with a nonviable homogenate of *F rectivirugula* spores and mycelia develop pulmonary lesions that resemble hypersensitivity pneumonitis in humans (Harris et al, 1976). The tissue reaction in rabbits is characterized by patchy interstitial mononuclear pneumonitis, but no evidence of immune complex vasculitis was detected. Local and systemic humoral and cellular responses to *F rectivirugula* were also observed. Antibodies of the IgG and IgA class were present in bronchoalveolar wash fluids and in sera. The rabbits reacted in skin tests to *F rectivirugula* antigens with dual Arthus and delayed hypersensitivity. Migration of alveolar macrophages was specifically inhibited by *F rectivirugula*-reactive lymphocytes in the direct MIF assay. Although precipitins were present, the histologic pattern in the lungs was more consistent with a granulomatous type IV hypersensitivity than with immune complex-mediated damage. Soluble *F rectivirugula* antigens elicit MIF responses in alveolar or popliteal lymphocytes in rabbits primed with whole *F rectivirugula* cell homogenate by either the intratracheal or subcutaneous route (Kawai et al, 1973).

Guinea pigs were challenged with aerosols of *F rectivirugula* filtrate antigens after systemic immunization with viable whole *F rectivirugula* in complete Freund's adjuvant (Wilkie et al, 1973). Precipitins appeared in the animals after antigen-priming and aerosol challenge. Elements of both Arthus and delayed pulmonary hypersensitivity were present. The lungs of immunized guinea pigs had interstitial pneumonia marked by (i) vasculitis of the capillaries with parenchymal accumulations of eosinophils and PMN, and (ii) perivascular infiltrates of lymphocytes and macrophages. In this model *F rectivirugula* antigens were not stimulants of IgE because guinea pigs did not react with anaphylactic shock to aerosols of *F rectivirugula* antigens nor did their sera contain specific IgE detectable by passive cutaneous anaphylaxis. Four to eight hours after aerosol challenge, respiratory rates increased and remained elevated for 1 week. The adoptive transfer of hypersensitivity to normal donors with immune plasma or peripheral blood lymphocytes was successful in demonstrating acute vasculitis in plasma recipients after challenge. Otherwise, lymphocyte recipients reacted to aerosols of soluble *F rectivirugula* antigens with a perivascular lymphocytic and macrophage response.

15.4.3 Model for Immunopathogenesis

A mechanism that accounts for the tissue damage in FLD has been proposed (Schorlemmer et al, 1977) (Figure 15.6). The central features of this model are the role of the alveolar macrophage and the direct complement-activating ability of *F rectivirugula* spores and soluble antigens. Direct complement activation results in the generation of C3a, C5a, and $\overline{C567}$ chemotactic factors for PMN and for macrophages, as well as anaphylatoxins that act on mast cells to trigger bronchial contractions. Alveolar macrophages ingest *F rectivirugula* spores and secrete hydrolases that cleave C3 and

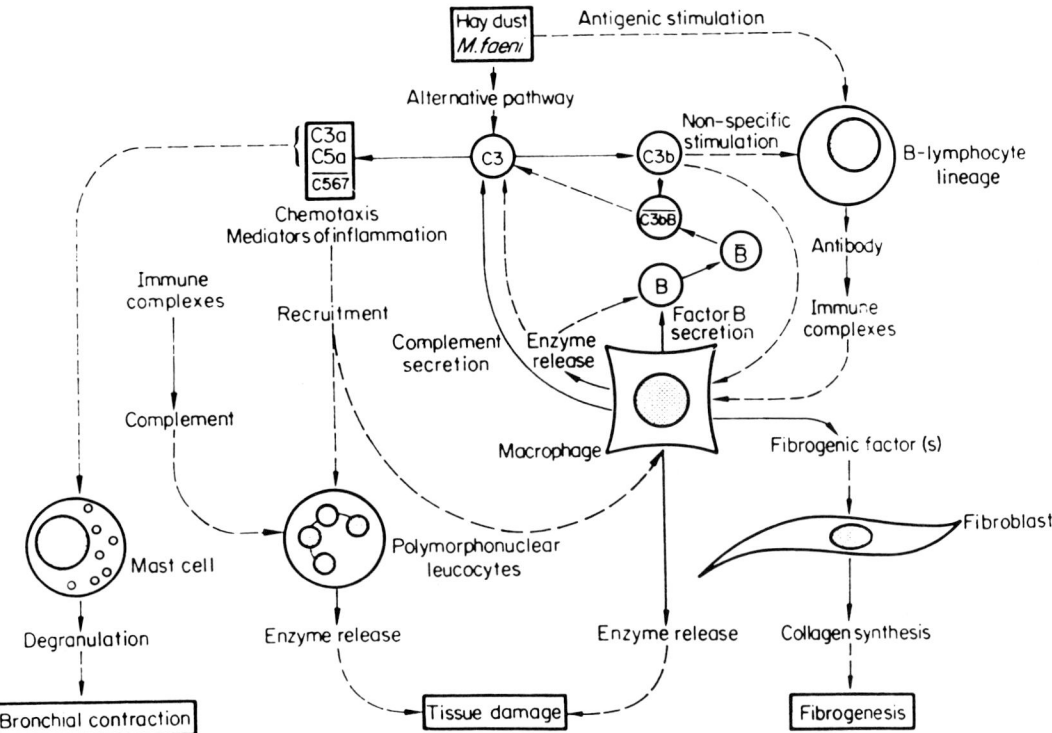

Figure 15.6. Hypothetical model for the immunopathogenesis of farmer's lung disease (Schorlemmer et al, 1977). Upon inhalation, the hay dust activates complement on the alveolar walls. Split products of complement are chemotactic for polymorphonuclear neutrophils. *Faenia rectivirugula* spores and products are phagocytosed by alveolar macrophages resulting in the secretion of lysosomal hydrolases. Accumulated macrophages can form granulomas which persist because *F rectivirugula* cell walls resist degradation and cause chronic inflammation. Stimulated B-cells produce antibodies that can form immune complexes as further inflammatory stimuli. Activated macrophages stimulate the proliferation of fibroblasts and collagen synthesis. *Source:* reprinted from Clinical and Experimental Immunology with permission of Blackwell Scientific Publications, Ltd.

further trigger the amplification phase of complement activation. *Faenia rectivirugula* constituents persist in the macrophages and induce a chronic inflammation. The processing of antigen by macrophages stimulates B-cells to antibody synthesis and the resulting immune complexes perpetuate damage to the lung. A role for T-lymphocytes in this model is implicit: T-cell activation of macrophages and T-cell help in antibody synthesis. Tissue damage, therefore, is caused by lysosomal enzymes of macrophages and PMN and by late complement components. A further role for macrophages in the pathogenesis of FLD is direct stimulation of fibroblasts. The production of proteinases by *F rectivirugula* and by other thermophilic actinomycetes coupled with the occurrence of corresponding antibodies in farmer's lung patients raises the question of what role the direct action of these proteinases on the lung could play in the pathogenesis.

15.5 SUMMARY

Spores of thermophilic actinomycetes, especially those of *F rectivirugula*, are plentiful in moldy hay and repeated inhalation causes farmer's lung disease — a form of extrinsic allergic alveolitis. Precipitins reactive with culture filtrates of *F rectivirugula*,

T vulgaris and other thermophils are found in 80% to 90% of FLD patients. An increased number and intensity of precipitins correlates with the frequency and severity of attacks. Trichloroacetic acid-soluble peptidopolysaccharides of the cell wall are important antigens evoking humoral responses and provoking respiratory reactions in patients with farmer's lung. These antigens are not disease-specific because antibodies to them also occur in asymptomatic exposed farmers. Yet the presence and number of precipitins are an element in the diagnosis of farmer's lung. Antigens that formerly were prepared in complex medium are now produced in one that is chemically defined. Maximal antigen secretion by *F rectivirugula* occurs after nine to 12 days of growth. Afterwards, the number of antigens declines because of the appearance of high levels of proteinase. *Thermoactinomyces candidus* cultures reach a peak of antigen secretion earlier, after 3 days of growth. TCA-soluble glycoproteins of *F rectivirugula* with C region electrophoretic mobility, glycoprotein "a," glycopeptide "1," and cell-wall-lysozyme (CWL) antigens are different designations for similar or identical antigens. Only glycopeptide "1" has been purified by gel permeation and ion exchange chromatography.

A reference pattern of precipitins in crossed "rocket" immunoelectrophoresis has resolved 46 *F rectivirugula* antigens—factors 7, 10 and 12 may be disease-specific. Preparative isoelectric focusing has resulted in a pool of four to eight antigens with pIs in the pH 3.8 to 4.0 range. These antigens have also been reported to be specific for FLD but require further evaluation.

Immunoglobulin G levels measured by indirect EIA against a whole cell antigenic complex of *F rectivirugula* correlate with the number and intensity of precipitins, but monomolecular antigens or pools of well-defined antigens need to be adapted in a primary binding assay.

Proteinases present in *F rectivirugula* and *Thermoactinomyces* species are detected by zymogram imprints of PAGE slabs on casein agar or by localization of enzymatic activity in immunoprecipitin arcs with a chromogenic substrate. One of the *F rectivirugula* proteinases, "enzyme 1" is a potent antigen evoking precipitins in FLD that may also be detected in asymptomatic exposed farmers if the immunoenzyme technique is used. Clearly some method of quantitating immunoglobulin against the proteinases is needed to see if the concentration is proportional to disease activity.

Faenia rectivirugula and *T vulgaris* organisms and their respective extracts activate complement directly by the alternative pathway. Depletion of early complement components of the classical pathway may also occur by an immunoglobulin-independent mechanism. Complement activation is an early event in the pathogenesis of farmer's lung.

The histology of farmer's lung shows a chronic alveolar and interstitial inflammation with lymphocytic, PMN, and macrophage infiltrates. Pulmonary vasculitis indicative of immune complex-mediated damage is not consistently found, but may be present in attacks early in the acute stage of the disease. Extracellular deposits of IgG, and C3 have, in a few instances, been detected in biopsied lung specimens. Chronic FLD is similar histologically to type IV delayed hypersensitivity, although positive delayed skin tests are uncommon.

A current model for the immunopathogenesis of farmer's lung takes into account that direct complement activation may be the initiating event in lung damage. Because spores and filaments of *F rectivirugula* are resistant to digestion they persist, contributing to chronic inflammation. Type III immune complex-mediated damage is suggested by the high incidence of precipitins, especially in bronchoalveolar wash fluid, and the onset of symptoms four to eight hours after exposure.

The histology of lesions early in the acute stage of the disease and detection of extracellular immunoglobulin and C3 further implicate immune complexes. Progression of the disease is accompanied by a maturation of cell-mediated immunity in the lung, judging from the frequent presence of granulomas. The L-MIF test with moldy hay antigens provides proof that in vitro correlates of delayed hypersensitivity are present at a level significantly higher than are found in asymptomatic exposed farmers. Other factors contributing to lung damage may be (i) the direct action of actinomycete proteinases, (ii) fibroblast stimulation by alveolar macrophages and antilung antibodies.

Spores and spore extracts are not fruitful sources of antigens but instead antigens occur in the filtrates of *F rectivirugula* cultures. A question that remains is whether inhaled spores germinate in the lung and undergo limited development, or whether "metabolic" antigens are adsorbed to the spore surfaces and in that form are carried into the lung.

Further studies are needed (i) in animals to resolve contradictions in adoptive transfer of hypersensitivity to farmer's lung sensitins (ii) and in humans to increase confidence that L-MIF reactions to purified thermophilic actinomycete antigens correlate with disease activity. Investigation of these aspects will clarify the contribution of T-cell-mediated immunity to pathogenesis. More quantitative approaches to the measurement of immunoglobulin responses to purified factors may help resolve the current debate about "disease-specific" versus "exposure-related" antigens.

REFERENCES

Bamdad S, 1980. Enzyme-linked immunosorbent assay (ELISA) for IgG antibodies in farmer's lung disease. Clin Allergy 10:161–171.

Bannerman EN, Nicolet J, 1976. Isolation and characterization of an enzyme with esterase activity from *Micropolyspora faeni*. J Appl Env Microbiol 32:138–144.

Braun SR, Flaherty DK, Burrell R, Rankin J, 1983. Importance of anti-lung antibody in farmer's lung disease. Amer J Med 74:535–539.

Edwards JH, 1972. The double dialysis method of producing farmer's lung antigens. J Lab Clin Med 79:683–688.

Edwards JH, Baker JT, Davies BH, 1974. Precipitin test negative farmer's lung—activation of the alternative pathway of complement by mouldy hay dusts. Clin Allergy 4:379–388.

Fletcher SM, Rondle CJM, 1973. *Micropolyspora faeni* and farmer's lung disease. J Hygiene (Cambridge) 71:185–192.

Frommel C, Hausdorf G, Höhne WE, Behnke V, Ruttloff H, 1978. Charakterisierung einer Protease aus *Thermoactinomyces vulgaris* (Thermitase) 2. Einschritt-Feinreinigung und Proteinchemische Charakterisierung. Acta Bio Med Ger 37:1193–1204.

Ghose T, Landrigan P, Kileen R, Dill J, 1974. Immunopathological studies in patients with farmer's lung. Clin Allergy 4:119–129.

Harris JO, Bice D, Salvaggio JE, 1976. Cellular and humoral bronchopulmonary immune response of rabbits immunized with thermophilic actinomycete antigen. Amer Rev Respir Dis 114:29–43.

Hollingdale MR, 1974. Antibody responses in patients with farmer's lung disease to antigens from *Micropolyspora faeni*. J Hygiene (Cambridge) 72:79–89.

Karr RM, Salvaggio JE, 1980. Infiltrative hypersensitivity diseases of the lung. pp. 1336–1371. In Clinical Immunology—Vol II Parker CW, ed. Philadelphia: W. B. Saunders.

Kawai T, Salvaggio J, Harris JO, Arquembourg P, 1973. Alveolar macrophage migration inhibition in animals immunized with thermophilic actinomycete antigen. Clin Exp Immunol 15:123–130.

Kurup VP, Barboriak JJ, Fink JN, Scribner G, 1976. Immunologic cross-reactions among thermophilic actinomycetes associated with hypersensitivity pneumonitis. J Allergy Clin Immunol 57:417–421.

Kurup VP, Ting EY, Fink JN, Calvanico NJ, 1981. Characterization of *Micropolyspora faeni* antigens. Infect Immun 34:508–512.

Lacey J, Lacey ME, 1964. Spore concentrations in the air of farm buildings. Trans Brit Mycol Soc 47:547–552.

Marx JJ, Jr, Emanuel DA, Dovenbarger DV, Reinecke ME, Roberts RC, Treuhaft MW, 1978. Farmer's lung disease among farmers with precipitating antibodies to the thermophilic actinomycetes: a clinical and immunological study. J Allergy Clin Immun 62:185–189.

Marx JJ, Jr, Motszoko C, Roberts RC, 1980. Antibody-independent complement consumption by *Micropolyspora faeni*. Int Arch Allergy Appl Immunol 62:133–141.

Morell F, Jeanneret A, Aiache JM, Molina C, 1982. Leukocyte migration inhibition in farmer's lung. J Allergy Clin Immunol 69:405–409.

Nicolet J, Bannerman EN, de Haller R, Wanner M, 1977. Farmer's lung: immunological response to a group of extracellular enzymes of *Micropolyspora faeni*. Clin Exp Immunol 27:401–406.

Pepys J, Jenkins PA, 1965. Precipitin (F. L. H.) test in farmer's lung. Thorax 20:21–35.

Roberts RC, Moore VL, 1977. Immunopathogenesis of hypersensitivity pneumonitis. Amer Rev Respir Dis 116:1075–1089.

Roberts RC, Nelles LP, Treuhaft MW, Marx JJ Jr, 1983. Isolation and possible relevance of *Thermoactinomyces candidus* proteinases in farmer's lung disease. Infect Immun 40:553–562.

Roberts RC, Wenzel FJ, Emanuel DA, 1976a. Precipitating antibodies in a midwest dairy farming population toward the antigens associated with farmer's lung disease. J Allergy Clin Immunol 57:518–524.

Roberts RC, Zais DP, Emanuel DA, 1976b. The frequency of precipitins to trichloroacetic acid extractable antigens in thermophilic actinomycetes in farmer's lung patients and asymptomatic farmers. Amer Rev Respir Dis 114:23–28.

Roberts RC, Zais DP, Marx JJ Jr, Treuhaft MW, 1977. Comparative electrophoresis of the proteins and proteases in thermophilic actinomycetes. J Lab Clin Med 90:1076–1085.

Schatz MR, Patterson R, Fink J, 1977. Immunopathogenesis of hypersensitivity pneumonitis. J Allergy Clin Immunol 60:27–37.

Schorlemmer HV, Edwards JH, Davies P, Allison AC, 1977. Macrophage responses to mouldy hay dust, *Micropolyspora faeni*, and zymosan, activators of complement by the alternative pathway. Clin Exp Immunol 27:198–207.

Terho EO, Lindström P, Mäntyjärvi R, Tukiainen H, Wager O, 1983. Circulating immune complexes and rheumatoid factors in patients with farmer's lung. Allergy 38:347–352.

Treuhaft MW, Roberts RC, Hackbarth C, Emanuel DA, Marx JJ Jr, 1979. Characterization of precipitin response to *Micropolyspora faeni* in farmer's lung disease by quantitative immunoelectrophoresis. Amer Rev Respir Dis 119:571–578.

Treuhaft MW, Roberts RC, Hackbarth C, Marx JJ Jr, 1981. Characterization of synthetic medium antigens of *Micropolyspora faeni* and *Thermoactinomyces candidus*. J Allergy Clin Immunol 67:375–387.

Walbaum S, Biguet J, 1973. Les extraits antigeniques de *Micropolyspora faeni*. Influence de divers facteurs sur leur qualite; etude d'un extrait total prepare dans les conditions optimales. Rev Immunol (Paris) 36:154.

Wenzel FJ, Emanuel DA, Gray R, 1971. Immunofluorescent studies in patients with farmer's lung. J Allergy Clin Immunol 48:224–229.

Wilkie B, Pauli B, Gygax M, 1973. Hypersensitivity pneumonitis: experimental production in guinea pigs with antigens of *Micropolyspora faeni*. Pathol Microbiol 39:393–411.

Chapter 16

Review of Innate and Humoral Immunity

16.1 INTRODUCTION

The elements of innate immunity are the surveillance provided by normal mucosal barriers, granulocyte fungicidal processes, and, potentially, natural killer lymphocytes. Some aspects of the humoral limb may be regarded as innate immunity, such as the fungicidal properties of normal human serum and the direct activation of complement via the alternative pathway. Less is known about the role of acute phase reactants in immune surveillance. *Aspergillus fumigatus* and *Epidermophyton floccosum* both have been reported to contain C-substance activity, and thus have the capacity to evoke C-reactive protein. The fungi are potent activators of the alternative complement pathway and this theme will receive further attention below.

The humoral limb of the immune system cannot be divorced from the cell-mediated limb because of the immunomodulation effected by T-helper and T-suppressor cells. Bearing that in mind, the participation of antibody- and complement-mediated reactions in the mycoses can be reviewed where there is (i) proof of protection; (ii) no measurable protective effect; (iii) pathogenetic effects.

The information about antigenic stimulants of humoral immunity in the mycoses from the preceding chapters is brought together here in order to underscore the common themes of mannans and secreted protein antigens. Some aspects of immunologic regulatory mechanisms are outlined that help to place the status of mycologic research in perspective. It is apparent that knowledge of mycotic antigens has progressed through the descriptive stage. The roles of these antigens in the immune response are known on the phenomenological level, but insight into the mechanisms of regulation of humoral immunity in the mycoses remains an uncharted area for future research.

Although the clinically important mycotic antigens have been enumerated, much remains to be learned about their structure. Present efforts are directed at purifying them to homogeneity. Monoclonal antibodies and enzyme immunoassays, including the enzyme-linked immunoelectrotransfer blot technique, are important elements of these investigations.

Innate immunity plays a key role in the defense against weak pathogens such as the zygomycetes. *Candida* species are confined in the normal host to an endogenous commensal niche by innate and humoral factors. However, chronic mucocutaneous candidiasis and chronic dermatophyte infections appear to be the result of as yet poorly delineated defects in immunoregulation controlled by thymic functions. Thymic immunity assumes the center stage in preventing and overcoming dissemination of the primary systemic mycotic pathogens.

16.1.1 Serum Fungistasis

Conidiospores of *Rhizopus* sp are inhibited from germinating in normal human serum. Sera from diabetics in ketoacidosis show a loss of fungistasis that is related to the severity of infection in the disease caused by these fungi: rhinocerebral zygomycosis. *Rhizopus* inhibitory factor is present in sera from well-controlled diabetics and is not simply related to glucose concentration because hyperglycemia without ketoacidosis does not ablate the anti-*Rhizopus* factor.

16.1.2 Neutrophil Fungicidal Properties

According to one report, human polymorphonuclear neutrophils fail to kill *Blastomyces dermatitidis* yeast forms after 2 hours in vitro (Brummer and Stevens, 1982) (Table 16.1). Prompt and efficient chemotaxis and aggregation of human PMN with *Blastomyces* yeast forms occurred (Sixby et al, 1979), but only one third of the yeast forms were killed over a wide variety of effector-to-target ratios. A cytophilic inhibitor of PMN chemotaxis was found in sera from five blastomycosis patients.

Polymorphonuclear neutrophils are in evidence early in the primary response to *Coccidioides immitis* arthroconidia in the lung, and then give way to mononuclear cells, except for a brief period when spherules rupture, when the released endospores attract PMN (Stevens, 1980). These events can be better understood in the light of other studies that have shown that human PMN in vitro stimulate the conversion of mycelia to spherules (Baker and Braude, 1956) and that spherulin, or dialyzed mycelial culture filtrates, are chemotaxigenic for PMN (Galgani et al, 1978). No mechanism has been proposed for the stimulatory effect of mycelial-to-spherule transformation apart from a mechanical one in which the mycelial apex is assaulted by PMN and thus assumes a "defensive" position: the spherule. In the case of chemotaxigenesis this is mediated by heat-labile (ie, complement) components of normal human serum.

Both human and cavine PMN are capable of efficient phagocytosis and killing of *Histoplasma capsulatum* yeast forms (Howard, 1973, 1975). Nevertheless, the yeast forms can survive and multiply within macrophages. Purified myeloperoxidase and iodide prevented germination of conidia and yeast forms. Cationic proteins from a granule lysate played a minor role in fungistasis. *H capsulatum* produces catalase, but the amount produced does not correlate with the virulence of the strain for mice (Howard, 1983).

Although PMN from normal controls and patients phagocytosed *Paracoccidioides brasiliensis* yeast forms equally well, those from patients had a 32% lower ability to kill the ingested cells (Goihman-Yahr et al, 1980). Human PMN efficiently phagocytose and kill *Sporothrix schenckii* yeast forms. Killing is mediated by the

Table 16.1. Killing of Fungi by Polymorphonuclear Neutrophils (PMN)

Organism	Form	Efficient	Cidal Pathway	Enhancement by Opsonins	References
Blastomyces dermatitidis	Yeast	No	PMN promote growth	—	Brummer and Stevens, 1982; Sixby et al, 1979
Coccidioides immitis	Mycelia Spherules Endospores	Unknown Unknown Unknown	Chemotaxigenic PMN promote mycelia → spherules	—	Galgiani et al, 1978; Baker and Braude, 1956
Histoplasma capsulatum	Yeast	Yes	MPO[a]-H_2O_2-I^- >[b] cationic proteins	—	Howard, 1975
Paracoccidioides brasiliensis	Yeast	No	Patients' PMN impaired	—	
Sporothrix schenckii	Yeast	Yes	MPO-H_2O_2-I^-	—	Cunningham et al, 1979
Aspergillus fumigatus	Conidia Hyphae	No Yes	MPO-H_2O_2-halide > lysozyme	—	Lehrer and Jan, 1970; Diamond et al, 1978
Rhizopus oryzae	Hyphae	Yes	MPO-H_2O_2-halide > cationic proteins	—	Diamond et al, 1978
Candida albicans	Blastoconidia	Yes	MPO-H_2O_2-Cl^- > cationic proteins lysozyme	Heat-labile suffice	Lehrer et al, 1975
	Pseudohyphae	Yes	Lysozyme	IgG	Diamond et al, 1978
Cryptococcus neoformans	Blastoconidia	No	Unknown	Alternative pathway IgG	Davies et al, 1982 (chemiluminescence); Diamond and Allison, 1976; Miller and Kohl, 1983 (ADCC)[c]

[a] Myeloperoxidase.
[b] "Greater than" indicates the more efficient pathway.
[c] Antibody-dependent cell-mediated cytotoxicity.

MPO-H_2O_2-I^- pathway (Cunningham et al, 1979) which may provide the rational basis for iodide therapy in the cutaneous-lymphatic form of the disease.

Some fungal elements are too large for endophagocytosis, such as the hyphae of *A fumigatus* and *Rhizopus* sp, the pseudohyphae of *Candida albicans,* and large capsule forms of the yeast *Cryptococcus neoformans.*

Damage to hyphae of *Rhizopus oryzae* by human PMN in the absence of serum was detected in electron micrographs and by inhibition of uptake of the RNA precursor ^{14}C-uracil (Diamond et al, 1978). Similar effects were shown against *A fumigatus* caused by the MPO-H_2O_2-halide pathway. *Aspergillus* conidiospores, however, resist killing by PMN (Lehrer and Jan, 1970). Damage to *Candida* pseudohyphae by neutrophil oxidative processes in the absence of serum was verified by Diamond and Krzesicki (1978).

Neutrophils from a patient with chronic granulomatous disease did not damage *Candida* pseudohyphae. Dead pseudohyphae have no affinity for PMN in the absence of serum. *Candida* have a cell surface protein or glycoprotein that is the substrate for PMN attachment and it is released from dead pseudohyphae (Diamond and Krzesicki, 1978).

Opsonization with heat-stable (56° C, 30 minutes) factors enhances the phagocytosis of *C albicans* blastoconidia by PMN. A respiratory burst occurs, H_2O_2 is generated, and candidal proteins are iodinated. The majority of phagocytosed blastoconidia are killed in 1 hour. Patients with myeloperoxidase deficiency have intact nonoxidative fungicidal mechanisms that kill *Candida parapsilosis*, but *C albicans* is more difficult to kill.

Killing of *C neoformans* by human PMN is not efficient, particularly against large capsule forms. Small capsule forms are killed when anti-*C neoformans* immunoglobulin is also present (Diamond and Allison, 1976; Miller and Kohl, 1983). When the encapsulated yeast is too large to be phagocytosed, rosettes of PMN or macrophages have been observed in vitro with pseudopodia extending to contact the plasmalemma (Kalina et al, 1971). Chemiluminescence by human PMN or monocytes requires heat-labile opsonins to be present (Davies et al, 1982). To be more precise, the alternative complement pathway is operative against the cell wall but not the capsule.

In summary, $MPO-H_2O_2-I^-$ is the major fungicidal mechanism, with cationic proteins from granule lysates of secondary importance. Lysozyme has a feeble but definite action on fungi, and a more pronounced enzymatic attack when incorporated with the growing rather than killed yeast forms.

16.1.3 Natural Killer Cells

The nylon wool nonadherent portion of CBA/N splenic lymphocytes significantly inhibit growth of cryptococci by 40% to 85% (Murphy and McDaniel, 1982). Negative selection by treatment of splenic lymphocytes with anti-thy 1, anti-mouse immunoglobulin, or anti-Ia plus complement did not reverse in vitro killing. The only treatment that ablated killing was anti-sialo GM-1, an antibody specific for a glycoprotein found at high density on natural killer (NK) lymphocytes. (Figure 16.1)

The paradoxical resistance of nu/nu mice against *C albicans* and *C neoformans* in the early stages of primary infection is indirect proof of an increment of protection provided by chronically activated macrophages and NK cells which are present in greater abundance than in nu/+ heterozygotes (Cheers and Waller, 1975; Herberman, 1978).

16.2 OVERVIEW OF MECHANISMS REGULATING HUMORAL RESPONSES

The maturation of immunocompetent human T-cells in the thymus corresponds to the segregation of $T4^+$ and $T8^+$ surface markers defined by monoclonal antibodies. Once exported, subsets of these T-cells function as inducer-helpers ($T4^+$) accounting for 55% to 65% of peripheral T-cells, and cytotoxic/suppressors ($T8^+$) making up 20% to 30% of the peripheral population. Only the $T4^+$ population provides inducer-helper function in T–T, T–B, and T–macrophage interactions. More specifically, $T4^+$ cells provide the signals for B-cell proliferation and differentiation to

16.2 OVERVIEW OF MECHANISMS REGULATING HUMORAL RESPONSES 341

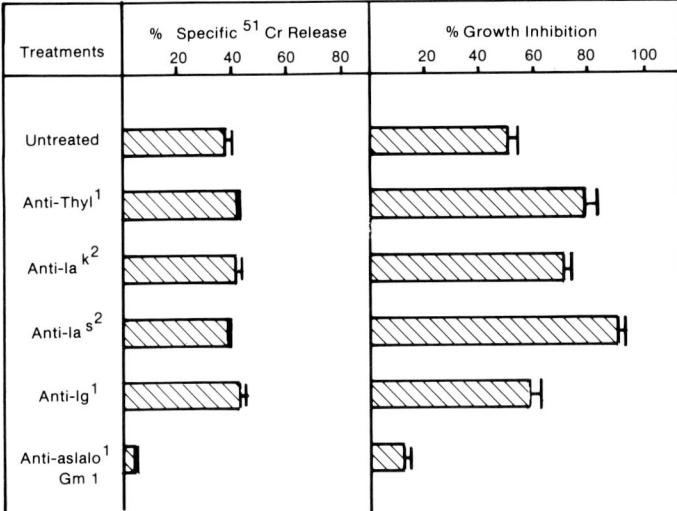

Figure 16.1. Evidence of a role for natural killer (NK) lymphocytes in the innate immune defenses against *Cryptococcus neoformans* (Murphy and McDaniel, 1982). The effector cells are characterized by negative selection procedures (lysis with various antisera and complement). The source of killer lymphocytes was nylon wool nonadherent cells from 7-weeks-old CBA/N mice. The targets for the ^{51}Cr release assay were YAC-1 tumor cells. Growth inhibition of *C neoformans 184*, serotype A, was measured by counting colonies plated on Sabouraud's medium. Bars indicate the mean and standard error. The only significant inhibitor was anti-asialo GM-1, an antiserum reactive with a surface antigen present at reasonably high concentration on NK cells. 1, contained guinea pig complement; 2, contained rabbit complement. *Source*: reprinted from the Journal of Immunology with permission of the authors and of the publisher, Amercian Association of Immunologists, copyright 1982.

immunoglobulin production. The proliferative response to soluble antigen is restricted among T-cells to the $T4^+$ subset (Reinherz and Schlossman, 1981).

The presence of the $T4^+$ subset is also necessary for maximal cytotoxic effects, illustrating their role in T–T interactions. Although both $T4^+$ and $T8^+$ subpopulations proliferate upon concanavalin A stimulation, only the $T8^+$ cells become suppressive. Thus $T4^+$ and $T8^+$ subsets are programmed for separate helper and suppressor tasks beyond their ability to respond to phytomitogens. These two populations are also different in the expression of Ia antigens. Amplification of T-dependent antibody synthesis is controlled by nonimmunoglobulin molecules: the Ia antigens, products of the immune response genes located in the I-A/E region of the major histocompatibility complex (MHC) in the mouse. The Ia antigens are expressed on B-cells and macrophages. Although Ia antigens are absent on resting T-cells, activation causes their synthesis and expression.

The $T4^+$ inducer-helper subset is analogous to the murine lyt $1^+ 2^-$ T-cells, and lyt 2^+ T-cells in the mouse correspond to the human $T8^+$ cytotoxic/suppressor population.

16.2.1 T-Dependent Responses

T-dependent antigens require antigen-specific T-helper cells for antibody induction. Effective collaboration of T-helper and B-cells involves their physical junction brought about by antigen bridging. In this physiologic cooperation T-helper cells recognize carrier determinants and B-cells, of course, recognize the haptenic group or epitope. Other major restrictions influence this interaction: (i) Accessory cells must present the antigen on their surfaces. Typically macrophages are the antigen-presenting cell but other Ia-positive cells, or Langerhans cells in the epidermis, for example, can perform this function. The manner in which accessory cells present the antigen is in association with Ia antigens (Singer and Hodes, 1983). (ii) T-helper cells recognize carrier determinants simultaneously with Ia antigens as shown in experiments where anti-Ia blocks the induction of antibodies (Howie and McBride, 1982).

The binding of T-helper cells to Ia antigens on macrophages and B-cells is a receptor–ligand reaction. Receptor specificity is clonally excluded so that antigen-dependent T-helper cells represent different specificities.

The cluster of antigen-presenting macrophages, T-helper cells, and B-cells enlarges through clonal expansion, perpetuated and amplified by molecules in addition to antigen: monokines and lymphokines, especially the interleukins (Julius, 1982).

The mechanism by which the T-helper cell signal is transmitted is an active area of research. Direct contact through antigen-bridging is the simplest case. The generation of idiotype-bearing T-helper clones interacting with the idiotype network influences the extent and duration of the immune response (Howie and McBride, 1982).

Soluble T-helper factors can spare the requirement for the physical presence of T-cells in driving B-cells to immunoglobulin production in an antigen-specific manner. The ability of T-cells to produce soluble T-helper factors is governed by Ir genes and the ability of B-cells to be triggered by helper factors is similarly restricted. The chemical nature of a soluble T-helper factor has been characterized as expressing antigen-specificity through idiotype determinants that are crossreactive with antibodies. These determinants consist of two chains, one of which is Ia-positive and the other reactive with anti-V_H reagents. These factors therefore, express all the characteristics of antigen-specific T-helper cells (Singer and Hodes, 1983).

Antigen-nonspecific T-helper factors are also powerful reagents in augmenting B-cell activation. The generic term lymphokines covers a multiplicity of molecules, but pertinent to the T–B-cell interaction are interleukins 1 and 2, B-cell growth factor, and T-cell replacing factor. Interleukin 1 is derived from accessory cells (macrophages); it first binds to B-cells and makes them receptive to T-cell derived lymphokines. Interleukin 2 is produced by lyt $1^+2^-3^-$ T-cells (T-helper cells) and aids continued proliferation of antigen-activated T-cell clones. Physiologic, as against allogeneic, induced nonspecific T-helper factors, are antigen-nonspecific and haplotype nonspecific. An example of this is T-cell replacement factor, which has been characterized in pure form from a T-cell hybridoma.

Complexities arise when the selective activation of B-cells is considered. Two subsets of murine B-cells, lyb 5^+ and lyb 5^-, are defined by surface differentiation markers. These subsets also differ in the type of T-helper interaction they require. T-helper cell interaction with lyb 5^+ B-cells, in the response to sheep erythrocytes, does not require T-helper recognition of Ia determinants on the B-cells, but the reaction with lyb 5^- B-cells is Ia-restricted (Julius, 1982; Singer and Hodes, 1983).

Which of these subsets is activated depends on the concentration of antigen and whether the T-helper cells are primed or unprimed. Model in vitro systems contained carrier-primed T-helper cells and hapten-primed B-cells (Singer and Hodes, 1983). Once the T-helper cells were activated by accessory antigen-presenting cells, they could only collaborate with lyb 5^- B-cells of the same Ia type. This restriction provides further evidence that Ir genes regulate only the interaction of T-helper cells with lyb 5^- B-cells and not interactions with lyb 5^+ B cells.

Do antigen-specific B-cells have receptors for Ia determinants that can interact with accessory cells and T-helper cells? This would call for an interactive B-cell phenotype. In the physiologic activation of B-cells, production of soluble T-helper factors would obscure or override the necessity for MHC-restricted B-cell interaction with accessory cells. But some antigens may not trigger production of T-cell replacement factor. In that event self-recognition of accessory cell MHC determinants by both T-helper cells and B-cells could regulate T–B-macrophage interaction (Singer and Hodes, 1983).

16.2.2 T-Independent Responses

Molecules in this category have a common theme of high molecular weight, multiple repeating determinants, and a low rate of catabolism. Microbial polysaccharides are excellent examples of T-independent antigens.

The consequences of activation are likewise similar. Usually a single antigenic dose suffices to induce an immunoglobulin response. In general, the IgM class is induced and there is a limited ability to effect the class switch. IgG_3 production, however, has been noted in the murine response to polysaccharides. The T-independent antibodies produced are restricted in clonality and may be a homogeneous expression of germ-line genes. Supraoptimal amounts of polysaccharide antigens frequently induce immunologic tolerance. Typically few memory cells remain after the primary response subsides.

Some T-independent antigens activate complement by the alternative pathway. One line of thought is that the C3b and C3d fragments proceed to trigger polyclonal B-cell activation via complement receptors on B-lymphocytes.

The concept of T-independence of antibody induction is considered a simplified response to the primordial need to efficiently repel microbial boarders. The sequence of crosslinking of surface immunoglobulin proceeding to patching and capping is probably an oversimplification. It is now well-documented that presentation of the antigen by macrophage-like adherent accessory cells is essential to drive the reaction towards induction instead of to tolerance and that in the event a second signal generated is a macrophage-derived factor. Nonimmunoglobulin receptors on B-cells that may participate in induction are the B-cell differentiation and Ia antigens.

B-Cells exist in distinct subsets, demarcated by lyb 5 surface differentiation antigens in the mouse, and can functionally be separated by their response to T-independent antigens. Responses of lyb 5^- cells to T-independent type 1 antigens such as trinitrophenyl-lipopolysaccharide (TNP-LPS) do not require accessory cells. Ia^+ accessory cells are required for presentation (Singer and Hodes, 1983) of T-independent type 2 antigens such as TNP-levan to lyb 5^+ cells.

16.2.3 T-Suppression

Much emphasis has been placed on the mannan antigens of fungi. Yet little is known about the regulation of synthesis of antibodies against mannans. These molecules are largely polysaccharides, but they also contain a peptido-component, which is less than 10 percent of the dry weight. The working hypothesis is that candidal mannans are T-dependent antigens (see sections 11.3.8 and 16.3.1) but the influence of the polypeptide moiety has not been characterized. It is instructive to consider as a model the regulation of B-cell function to the most well-characterized bacterial polysaccharide, the acidic capsular polysaccharide of *Streptococcus pneumoniae* type III (the soluble specific substance, SSSIII), a glucose-1,4-β-glucuronic acid containing polymer.

Priming mice with a lower than immunogenic dose (5 ng) of SSSIII markedly reduces the capacity of mice to respond in the direct splenic immunoplaque assay to an optimally immunogenic dose (0.5 μg). This low zone tolerance (the historic term is "low dose immune paralysis") persists for months and is mediated by T-suppressor cells. This reaction is antigen-specific and epitope- rather than carrier-specific (ie, there is no suppression of trinitrophenylated SSSIII). In these respects the response to SSSIII differs from the suppressor T-cells indicated by Con A which are not antigen-specific. The tolerance induced is not H-2 related because it is independent of the H-2 haplotype. Suppression can be removed by T-cell depletion maneuvers. There is no evidence that T-cells can recognize SSSIII directly, so the most likely explanation is that the T-suppressor cells recognize the idiotype on the surface of B-cells stimulated by SSSIII (Baker et al, 1979).

Both T-dependent (TNP-bovine γ-globulin) and T-independent (TNP-ficoll) antigens activate the production of anti-idiotype antibodies during the primary immunoplaque-forming B-cell response. This is detected by inclusion of the hapten into the plaque-forming cell assay. The hapten binds to the idiotypic antibody on the B-cell surface and deprives the anti-idiotype antibody of its binding site, resulting in an increase in the number of immunoplaques. Thus in the plaque assay there is sufficient anti-idiotype antibody to show evidence of down-regulation (Siskind et al, 1982).

16.3 FUNGAL POLYSACCHARIDE ANTIGENS

16.3.1 Mannans

Two reports of human B-cell responses to *C albicans* mannans indicate that T-cells or soluble T-helper factors are required for induction of humoral responses. If that proves to be the case then mannans similar to those of *C albicans* and more precisely, the peptidopolysaccharides, may be a distinct category apart from the T-independent bacterial polysaccharides. Blastogenic responses of human B-cells to the "Mangion purified polysaccharide" of *C albicans* required T-helper cells (Piccolella et al, 1981a, 1981b). That antigen contained mannan, glucan, and 3.1% protein. Durandy et al (1983) reported in vitro antibody production by human B-cells stimulated by a more well-characterized *C albicans* mannan. They used mannan prepared by autoclaving in neutral buffer and Fehling precipitation. This product contained 3.4% protein. The molecular weight of this mannan, 8,000 to 15,000 da was considerably lower than would be expected (Nakajima and Ballou, 1974).

No antibodies were stimulated in vitro by mannan alone, but only when complexed with methylated bovine serum albumin (MeBSA). Lymphocytes from a minority of human donors proliferated in the presence of mannan-MeBSA, but no autoradiographic or enzyme immunoassay evidence of anti-mannan immunoglobulin was detected. However, in the majority of normal human donors proliferation led to specific intracellular immunoglobulin production that went on to produce IgM and to a lesser extent IgG. Production of anti-mannan depended on autologous T-cells or T-cells from a family member. Such genetic restrictions suggest the T-cell helper factor is antigen specific. The matter of T-cell dependence of mannan is not settled because the necessity for MeBSA clouds the issue. Perhaps with mannan of higher molecular weight, such as that produced by cetyltrimethylammonium bromide precipitation (Nakajima and Ballou, 1974), the requirement for MeBSA could be spared.

The oligomannoside fragment directing serotype specificity in *C albicans* appears to be a mannohexaose. According to one report, serotype A has a terminal α-1,3 linkage; otherwise, a straight α-1,2 chain is found. Critical β-1,6 linkages connect type A antigenic mannohexaose to the main mannan backbone (Suzuki and Fukazawa, 1982). Alternative structures may be equally likely (see Chapter 11). The nonreducing termini in serotype B mannan are α-1,2-linked and there is a single internal C-1, C-2, C-3 branch point in the immunodominant mannohexaose (Fukazawa et al, 1980).

From the standpoint of antigenic specificity it is interesting that one or two sugar substitutions in mannans are consistent taxonomic markers: thus peptido-L-rhamno-D-mannans occur in *Sporothrix schenckii* (Travassos and Lloyd, 1980) whereas peptido-L-fucomannans mark members of the mucorales (Miyazaki et al, 1980). The latter mannan does not crossreact with baker's yeast mannan or with a galactomannan of *Penicillium chrysogenum*. (3-O-methyl mannose is a marker for the mannan of *C immitis*).

A galactoxylomannan complex is found in *C neoformans* which is separable by affinity for Con A into galactoxylo- and mannan-enriched components (Cherniak et al, 1982; Turner et al, 1984). The glucuronoxylomannan of *C neoformans* is a special case because its accumulation in large amounts as a high molecular weight, viscous, acid capsule (Cherniak et al, 1980) has profound implications for antiphagocytosis and immunologic tolerance. The glucuronoxylomannan has no known peptido moiety.

In baker's yeast peptidomannans, the protein is an important structural component although it comprises less than 10% of the weight. Blocks of mannan are joined to protein by alkali-stable bonds, and oligomannosides are linked by alkali-labile bonds to the same protein. Linear mannan sequences have a majority of α-1,6 linkages and are highly substituted with antigenic side chains. The overall shape is either comblike or treelike. The molecular weight of the aggregate, including the inner core, outer chain, and the base-labile oligomannosides, was estimated for baker's yeast to be 133,000 da (Nakajima and Ballou, 1974).

An important noncarbohydrate constituent is O-phosphonomannan and it is claimed for *C albicans* that during purifications, highly antigenic fractions co- chromatograph with phosphate enrichment (Okubo et al, 1979). Others have suggested in baker's yeast that O-phosphonomannan is the link between glucan and mannan components (Friis and Ottolenghi, 1970).

The A and B serotypes that allow discrimination between *C albicans* isolates were found to reside in mannan. Adsorption of type A antiserum with type B yeast cells leaves an A-specific globulin that can be used as a typing reagent. Antiserum to *Saccharomyces cerevisiae* does not crossreact with type A mannan but does with that of type B. Quantitative precipitin reactions show that *Candida tropicalis* mannan is related to type A. Reciprocal adsorptions between *Candida stellatoidea* and *C albicans* B revealed no differences, but these mannans are not identical because in the quantitative precipitin reaction *C stellatoidea* mannan precipitates less globulin from anti-*C albicans* B than does the homologous mannan.

16.3.2 Galactomannans

Galactomannans are a family of antigenic polysaccharides common to diverse genera of pathogenic fungi that are probably a major contributor to serologic crossreactions (Table 16.2). For example, the major genera of systemic dimorphic pathogens, *Histoplasma, Blastomyces,* and *Paracoccidioides* contain mutually crossreactive galactomannans (Azuma et al, 1974). Antiserum against *C albicans* reacts weakly with the galactomannan from *A fumigatus* and *Trichophyton rubrum* (Suzuki et al, 1967). Galactomannans from black molds of the genera *Phialophora, Wangiella, Fonsecaea* are crossreactive (Suzuki and Takeda, 1977) and anti-*Fonsecaea (Hormodendrum) pedrosoi* serum was crossreactive with galactomannans from *A fumigatus, Exophiala (Cladosporium) werneckii,* and *T rubrum.* Galactomannans exist as the readily soluble outer mural layers that inevitably are present in whole cell lysates used as skin test and serologic reagents. Increased specificity of such reagents demands the rigorous exclusion of the galactomannans, especially for sensitive primary binding assays like enzyme immunoassays. Sometimes removal can be effected by Con A affinity chromatography (Ellsworth et al, 1977; Reiss and Lehmann, 1979) or the offending antigen can be inactivated by periodate oxidation (Brock et al, 1984).

The major structural features of galactomannans are α-1,6-linked linear se-

Table 16.2. Occurrence of Galactomannans among Genera of Pathogenic Fungi[a]

Species	Galactose : Mannose	Affinity for Concanavalin A	Precipitated by Fehling's Solution
Systemics			
Blastomyces dermatitidis	1:2.9	nd[b]	+
Coccidioides immitis	nd	nd	nd
Histoplasma capsulatum	1:1.5 2:5	nd	+
Paracoccidioides brasiliensis	1:2.2	nd	+
Sporothrix schenckii	nd	+	no
Filamentous			
Aspergillus fumigatus	1:1.2	+	nd
Dermatophytes			
I	1:12.3	nd	+
II	1:2.4	nd	no
Other			
Exophiala (Cladosporium) werneckii	1:3.7	+	nd

[a] Refer to text for pertinent citations.
[b] nd, not determined.

quences of mannose residues, substituted, sometimes extensively, with mannose oligosaccharides and terminal nonreducing galactofuranose present either singly or linked to oligogalactosides. The mannosyl oligosaccharides are responsible for two properties that are frequently but not invariably present: reactivity with Con A and precipitation with Fehling solution.

The hypothesis that galactomannans, because of their crossreactivity, may confer crossprotection has not been tested. We are a long way from complete knowledge of the linkage and sequence analysis of these antigens. Existing reports lay the groundwork for definitive studies yet to come. At present it is difficult to judge whether the differences reported for galactomannans of diverse genera are the result of variations in the method of preparation, or are inherent in the molecules themselves. Determinants may exist that direct antigenic specificity, but they may be lost during work-up because of inappropriate conditions. Thus, the presence of peptido components, the ratio of component sugars, and the presence of other noncarbohydrate moieties such as O-acetyl and O-phosphonomannan are first impressions that underscore the need for further work.

Galactomannans of the primary systemic fungal pathogens extracted with 1M NaOH at ambient temperature are precipitated with Fehling solution: *B dermatitidis*, (gal:man = 1:2.9); *H capsulatum*, (gal:man = 1:1.5); and *P brasiliensis* (gal:man = 1:2.2) (Azuma et al, 1974). The galactomannans were studied further by methylation-fragmentation-GLC and 1,6 linkages were found to predominate, with many 2,6-di-O, or 3,6-di-O-substituted mannose branch points.

The crossreactions observed in the extracted galactomannans also provide an explanation for the immunofluorescent cross-staining observed in *H capsulatum* and *B dermatitidis*. Of the four factors identified by adsorption studies, factors 1 and 4 are shared and probably reside in galactomannan. The remaining factors 2 and 3 specific for *H capsulatum* have not been characterized.

Heterogeneity was reported in the galactomannan extracted with dilute alkali from *H capsulatum* (Reiss et al, 1974). One fraction was not bound to anion exchange resin, and a galactomannan-protein complex did bind. The galactomannan and galactomannan-protein had similar serologic activity; in addition the galactomannan-protein was a potent elicitor of migration inhibitory factor in the peritoneal exudate cells of guinea pigs primed with the antigen in adjuvant.

The picture in *C immitis* is complicated by the presence of 3-O-methyl mannose. Whether the 3-O-methyl mannose-containing polymer is part of the galactomannan, has not yet been resolved, but available evidence favors the occurrence of a galactomannan which has not been characterized, and a 3-O-methyl mannose-containing mannoprotein (Wheat et al, 1983). The serologic crossreactions of the *P brasiliensis* galactomannan with galactomannans from other genera of primary systemic fungal pathogens is mentioned above. A 1,6-linked mannan from *Alternaria kikuchiana* having 1,2- and 1,3-mannosyl side chains did not crossreact, giving indirect evidence that galactose is an antigenic determinant. *Alternaria zinniae* galactomannan has a 1,6 linear mannan backbone with galactofuranose linked to C-3 in the main chain, and it crossreacts with anti-heat-killed *P brasiliensis*. This crossreaction indicated the importance of galactofuranose as an antigenic determinant.

Sporothrix schenckii is another dimorphic fungus in which a Fehling-precipitable galactomannan with 6-O and 2,6-di-O-substituted α-mannose units is produced along with the genus-specific peptido-L-rhamno-D-mannan.

Aspergillus fumigatus galactomannan (gal:man = 1:1.2) was not extractable with hot neutral buffer but did become soluble, along with a large glucan component, after exposure to cold alkaline borohydride (Reiss and Lehmann, 1979). The galactomannan was purified by affinity for Con A. The question of C-substance activity in *A fumigatus* (Longbottom and Pepys, 1964) is not satisfactorily resolved because the nature of the *A fumigatus* C-substance is unknown. The Ca^{++}-dependent C-reactive protein binding site for phosphorylcholine has two areas, one specific for the phosphate group, the other interacting with the positively charged quaternary amine. Although galactose is not an effective inhibitor of the reaction of pneumococcal C-polysaccharide with C-reactive protein, galactose can inhibit the binding of C-reactive protein to an agarose column (Gotslich et al, 1982). Ambiguity will continue until the C-substance of *A fumigatus* and galactomannan are compared against C-reactive protein.

The peptidophosphogalactomannan of *Exophiala (Cladosporium werneckii)* is noteworthy because its extraction with hot neutral buffer and selective precipitation with borate-cetyltrimethyl ammonium bromide is a less harsh alternative to the high pH of Fehling solution. In this galactomannan O-acetyl groups contribute to the antigenicity (Lee and Lloyd, 1975).

16.3.3 Mannanemia

The evidence that mannans are the major surface antigens of fungi rests on the following proofs:

1. The agglutination of *C albicans* by Con A and its blockage by methyl-α-mannoside (Cassone et al, 1978).
2. Localization of mannan on the *C albicans* surface by the periodic acid-thiosemicarbazide silver stain (Poulain et al, 1978).
3. Similar ultrastructural localization of the peptidorhamnomannan with Con A on the surface of *S schenckii* (Travassos and Lloyd, 1980).
4. Enzymolysis of cell walls of *H capsulatum* yeast forms with glucanases yields soluble galactomannan-containing antigens (Reiss et al, 1977).
5. Mannan is the only detectable antigen released from purified *C albicans* cell walls with cold dilute alkali (Reiss et al, 1974).

During the proliferation of fungi in tissue, the mannans are sloughed and processed by the phagocytes for eventual excretion in low molecular weight form (de Repentigny et al, 1984). If there is a serious compromise of phagocytosis such as exists in the granulocytopenic host being treated with immunosuppressive agents, opportunities arise for an equilibrium in the plasma between production and excretion. This phenomenon in invasive aspergillosis and candidiasis (Weiner, 1980; Lew et al, 1982; de Repentigny et al, 1985) has permitted the detection of antigenemia in the low ng/ml range as an unequivocal indication of tissue invasion (Meckstroth et al, 1981; Lew et al, 1982; de Repentigny and Reiss, 1984). Mannanemia is a transient phenomenon, because even though the host is immunosuppressed the capability to produce antibodies is not lost, and over a period of days the antigenemia gives way in the face of increased immunoglobulin synthesis (de Repentigny et al, 1984). Antigenemia persisting over a period of weeks suggests an unresolved focus of infection in a host whose immune system is seriously compromised.

Circulating mannan or galactomannan occurs as soluble immune complexes, and high efficiency of detection relies on the observation that these polysaccharides tolerate extremes of pH and heat under conditions that cause irreversible denaturation of immunoglobulins. Obviously, detection of nanogram amounts of antigen requires sensitivity provided by primary binding assays such as radioimmunoassay and enzyme immunoassay.

Antigenemia in cryptococcosis resulting from the glucuronoxylomannan capsule is a more durable indicator of disease activity (Diamond and Bennett, 1974). The capsular material is actively synthesized more abundantly than cell wall polymers (Cherniak et al, 1982), is resistant to digestion because of its chemical composition, and lodges in the renal tubular epithelium (Kozel et al, 1977). Moreover, the glucuronoxylomannan evokes tolerance that suppresses antibodies against the capsule (Murphy and Cozad, 1972; Kozel et al, 1977).

16.4 PROTEIN ANTIGENS

Protein antigens of the fungi are classified as (i) species-specific secreted factors, (ii) secreted proteinases, (iii) mycelial (yeast) extracts, (iv) wall-bound, and (v) membrane-bound. In some instances they are capable of evoking a full spectrum of humoral and cell-mediated responses. Only key features of these proteins will be discussed in order to more fully appreciate comparisons across generic lines. The existing knowledge of the structure versus activity of these antigens has been discussed in the sections devoted to individual agents of disease.

16.4.1 Species-Specific Secreted Factors

The A-antigen of *B dermatitidis* is an acidic protein that has been purified by column chromatography and for which monoclonal antibodies have been prepared (Green et al, 1980, 1982). Humoral responses to the A-factor are species-specific and this antigen represents a promising alternative to blastomycin for tests of cell-mediated functions. The heat-stable TP antigen of coccidioidin evokes an IgM response and serves as an early sign of the primary immune response to *C immitis*. The chemical nature of this antigen has not been established. Antibodies to the heat-labile factor F of coccidioidin, on the other hand, more closely parallel the complement-fixation titer, thus providing a reliable confirmatory test of disease activity. Coccidioidin factors 2 and 11 in Huppert's nomenclature (Huppert et al, 1978) are abundant and specific antigens, but their relation to the TP and F factors is not known.

Histoplasmin contains two proteins, H and M, that are the major antigens of this dimorphic mold. They evoke the full spectrum of humoral and cellular responses. Antibodies to M are first to appear in infection, are stimulated by skin tests in previously sensitized hosts, and endure for months or years after clinical remission. Precipitins to H are less frequently encountered, are more often associated with extrapulmonary dissemination, and subside relatively early in convalescence (Picardi et al, 1976). The M factor is first to elute from anion exchange resins and is pronase resistant, whereas factor H is sensitive to this enzyme, is more acidic, and hence binds more tightly to diethylaminoethyl resins (Pine et al, 1977). The pattern of two distinct immunoprecipitins that migrate towards the anode provides a unique immunoelectrophoretic profile (Pine et al, 1977). *Paracoccidioides brasiliensis* also contains a species-specific protein, factor E, which has been purified by immunoaffinity chromatography (Yarzabal et al, 1976).

Culture filtrate antigens of the allergenic thermophilic actinomycetes have been resolved into immunoelectrophoretic patterns showing a multiplicity of arcs, which several workers have organized into numerical schemes. Glycopeptide 1 of *Faenia rectivirugula* (Edwards, 1972) was purified on the basis of size, charge, and solubility in cold 5% trichloroacetic acid. This antigen is similar or identical to factors given different names by other workers. Later Treuhaft et al (1979) reinvestigated *F rectivirugula* antigens by crossed "rocket" immunoelectrophoresis and emphasized the importance of factors 7, 10, and 12 as being both sensitive for farmer's lung disease and excluding reactions in asymptomatic farmers.

Four serotype-specific secreted factors of *Nocardia asteroides* were characterized by Pier and Fichtner (1971, 1981), but no adsorptions were carried out. A more complete pattern of seven immunotype factors after reciprocal adsorptions of antisera was reported by Kurup and Scribner (1981).

16.4.2 Secreted Proteinases

The demonstration of proteinase activity in vitro and a humoral response in human mycotic infections or in experimentally infected animals is presumptive evidence that the proteinase in question is a factor related to pathogenicity. More direct histologic evidence of proteinase expression in vivo is known in only a few instances. The secretion of proteinases and their complexation with antibodies can contribute to pathology by direct enzymolysis of host tissue and indirect immune complex-mediated damage. With further investigations, the enzymatic activity of some of the species-specific secreted factors will probably come to light. At present, the coincidence of antigens with proteinase activity occurring among different genera of fungi and actinomycetes is good reason for more intensive characterization of their involvement in mycotic disease processes.

Aspergillus fumigatus secretes a proteinase that acts on esterase substrates and thus has a "chymotryptic" action pattern (Tran Van Ky et al, 1966). The proteinase was first detected by localization of enzymatic activity in immunoprecipitin arcs and was purified by affinity for a ligand that is a substrate analog. Secreted proteinase may contribute to pathology in bronchopulmonary and invasive forms of aspergillosis.

The acidic carboxyl proteinase of *C albicans* was purified to near homogeneity by a sequence of chromatographic separations based on charge and ligand-affinity (Rüchel, 1981). Humoral responses to the enzyme have been detected, and it has been localized by immunofluorescence in microabscesses of kidneys from infected animals (Macdonald and Odds, 1983). The action pattern of the enzyme includes attacks on IgA, which can be neutralized only by excess antiproteinase. The characterization of this enzyme is the only clue thus far to the question of how, on a molecular basis, this organism breaks the integrity of the mucosal barrier.

Although frank proteolysis by *C neoformans* is unknown, more sensitive measurements have shown in vitro fibrinolytic activity by whole cryptococci (Müller and Sethi, 1972). However, actual secretion of proteinase by *C neoformans* is not documented.

The growth of dermatophytes within hair shafts and on keratinized layers of the skin are circumstantial evidence for keratinase production by these molds. Direct proof of three inducible proteinases is known for *T mentagrophytes* (Yu et al, 1971). Keratinase I, 48,000 da, is secreted into the medium during growth on horse hair. Two other keratinases loosely associated with the cell wall are extracted by soaking the mycelia in buffer. Cytophilic antibodies against keratinases have been detected in the external sheaths of infected hair follicles (Holden et al, 1981). The inflammatory potential of different dermatophytes was directly related to the level of their proteinase activity by Minocha et al (1972).

Although the zygomycetes are considered weak pathogens, the rapid progression of rhinocerebral zygomycosis in diabetic ketoacidosis argues for the existence of some soluble factors that could account for the growth of the fungus in arteries and cartilage. Reinhardt et al (1986) have shown that only isolates of *Rhizopus oryzae* that produce an alkaline proteinase are capable of producing this disease after intranasal challenge with conidia in alloxan-diabetic rabbits.

Proteinases of the major genera of thermophilic actinomycetes that cause hypersensitivity pneumonitis are detected by zymogram imprints of polyacrylamide gel electrophoresis slabs on casein agar (Roberts et al, 1983). Purified "enzyme 1" of *Faenia rectivirugula* is a potent immunogen (Nicolet et al, 1977).

16.4.3 Extracts of Mycelial and Yeast Forms

The preceding chapters give evidence that a host of cytoplasmic proteins are capable of evoking a multiplicity of antibodies in hyperimmune animals. This fact has led to the description of complex patterns of immunoprecipitin fingerprints, notably in *C albicans* (Axelsen, 1971; Greenfield and Jones, 1981; Syverson et al, 1978) and in *Aspergillus fumigatus* (Kim et al, 1978; Longbottom, 1978). These mapping studies have been extended to more sensitive and high resolution techniques such as the two-dimensional PAGE, combining isofocusing with separation on the basis of molecular weight (Manning and Mitchell, 1980). The enzyme-linked immunoelectrotransfer blot technique with isofocused proteins has been applied in the instance of *Nocardia asteroides* (El-Zaatari et al, 1986). The benefits of mapping techniques such as these are that they provide a frame of reference for standardization of antigens. In this way, single major cytoplasmic proteins have been purified from *C albicans* (Greenfield and Jones, 1981) and *A fumigatus* (Calvanico et al, 1981). Once major antigens are characterized, these may be cut out of the gel and used to immunize mice for fusion experiments to produce monoclonal antibodies.

16.4.4 Wall-Bound

Cell walls of *Aspergillus fumigatus*, when extracted with triton X-100, release five proteins that are serologically active with sera from aspergillosis patients (Hearn and Mackenzie, 1979; 1981). The alkali-soluble, water-soluble extract of *Blastomyces dermatitidis* contains three to four proteins (Cox and Larsh, 1974). This antigenic complex evokes specific blastogenesis and MIF production in lymphocytes from sensitized animals. It is likely that proteins occur in the analogous fraction of *C immitis* walls, but none have yet been characterized. Glycoprotein antigens are extracted from cell walls of the thermophilic actinomycete with cold trichloroacetic acid. The immunoprecipitin arc containing the "a" glycoprotein is cited as especially reactive in sera of patients with farmer's lung disease (Hollingdale, 1974).

Proteins characterized from purified baker's yeast cell walls by two-dimensional PAGE increase the expectation that similar proteins are present in the walls of the pathogenic candidae.

16.4.5 Membrane-Bound

Membrane-bound antigens have not received sufficient attention considering the reported activity of such antigens in stimulating cell-mediated immunity. The hot-water extract of mitochondria-membranes of *C albicans* is superior to other fractions in eliciting delayed cutaneous hypersensitivity (Domer and Moser, 1978). A microsomal fraction of acapsular mutant *C neoformans* is a potent elicitor of lymphocyte blastogenesis in infected guinea pigs (Jones et al, 1981). Protection against challenge with live *H capsulatum* yeast forms is conferred by a ribosomal vaccine. The lineage of cells mediating this protection was characterized by its adoptive transfer with T-cells to irradiated recipient mice (Tewari et al, 1978).

16.5 ENZYME IMMUNOASSAYS

The earliest form of EIA for the mycoses involved the localization of enzymes of *A fumigatus* in immunoprecipitin arcs (Tran Van Ky et al, 1966). This assay did not capitalize on the advantages of speed and sensitivity inherent in a primary binding assay (ie, one that does not depend on secondary phenomena such as lattice formation.) The enzyme immunoassays of the present are based on the passive adsorption of antigens and antibodies on the polystyrene surfaces of microtiter plates (Figures 16.2, 16.3, 16.4). In this configuration either antigen or antibody can be detected, but if the antigen is a mixture of factors, it is not possible to measure the contribution of each to the total activity. A newer type of enzyme immunoassay measures the antibodies directed against each of a multiplicity of antigens: enzyme-linked immunoelectro-transfer blotting (Tsang et al, 1983) (Figures 16.5a, b).

Most commonly an indirect enzyme immunoassay is performed in which antigen is passively adsorbed to the plastic wells, the patient's serum dilutions are added and after washing with buffer, antihuman IgA, IgD, IgE, IgG, or IgM conjugated to horseradish peroxidase or alkaline phosphatase is added (Figure 16.2). After a further incubation with chromogenic substrate, the class-specific reaction is read and the endpoint titer is determined in a recording spectrophotometer.

- *Aspergillus fumigatus.* Filtrates of *A fumigatus* cultures were separated by $(NH_4)_2SO_4$ precipitation into proteins and polysaccharides. Immunoglobulin G levels against the proteins, determined by enzyme immunoassay, in allergic bronchopulmonary aspergillosis correlated positively with the number of precipitins in the immunodiffusion test and with RIA. Most antibodies in allergic bronchopul-

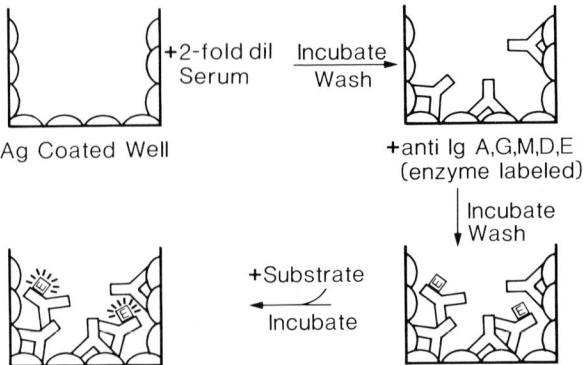

Figure 16.2. General design of an indirect enzyme immunoassay (EIA) performed in polystyrene microtitration plates. Typically, the antigen is passively adsorbed to the plastic surface. Indicator antibodies are labeled with horseradish peroxidase or alkaline phosphatase in most instances. Incubation times are for 30 minutes, usually at 22 °C or 37 °C. The amount of color produced by the action of the chromogenic substrate is measured in a spectrophotometer designed to automatically scan the 96-well plate.

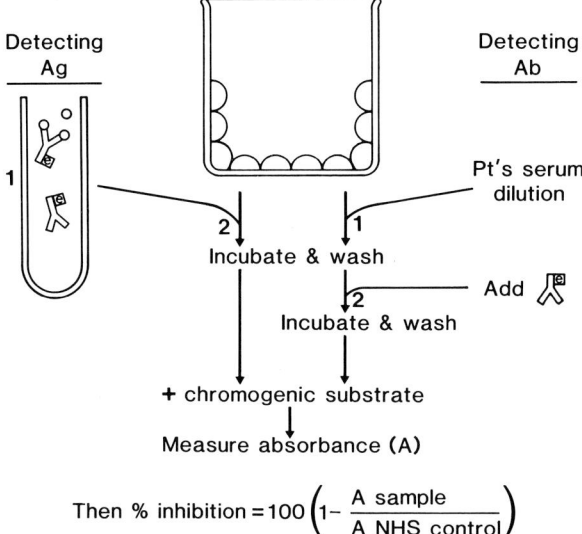

Figure 16.3. Competitive binding enzyme immunoassays, sometimes called EIA-inhibition, can be formatted to measure either antibodies or soluble antigens. If antigens are to be measured, then the original body fluid or solution may have to be treated to irreversibly dissociate any soluble immune complexes that may be present. The common denominator in this type of test is that the result is reported as percent inhibition unless it can be interpolated with respect to a standard curve and reported as μg or ng per ml. The simplest approach is to coat microtitration wells with the reference antigen reagent, as shown. Next a mixture of the body fluid and the enzyme-labeled indicator antibody are incubated briefly (ie, 30 min, 23 °C) and added to the microtitration wells. After incubation and washing with buffer containing a nonionic surfactant, the bound indicator antibody is allowed to react with the chromogenic substrate. The right half of the figure shows how the competitive binding assay can measure antibodies. The EIA-inhibition is well-suited for utilizing monoclonal antibodies, provided that mice respond to the same epitopes as do humans. The symbol with the inscribed "e" is the enzyme-labeled indicator antibody.

monary aspergillosis were directed against the polysaccharide antigen, whereas in aspergilloma, as expected, *A fumigatus* proteins evoked the highest EIA titers because of the multiple precipitins found in that disease (Sepulveda et al, 1979).

The initial enthusiasm for EIA as a substitute for precipitin tests in allergic aspergillosis has been tempered by the findings of Kauffman et al (1983), which include EIA titers in control subjects without aspergillosis and other discrepant results where precipitins were present and EIA tests were negative and vice versa. These findings suggest that refinement of the antigens and increased knowledge of plastic-binding characteristics of the major precipitinogens are goals for future research. Further investigation of EIA for aspergillosis also utilized $(NH_4)_2SO_4$-precipitated antigens (Holmberg et al, 1980). These antigens were characterized by two-dimensional crossed "rocket" immunoelectrophoresis giving 40 antigen–antibody complexes. Immunoglobulin G levels against this preparation were found in invasive aspergillosis, and in some instances the titers increased as the patient's disease grew worse.

- *Blastomyces dermatitidis*. An indirect EIA measurement of IgG against the purified A antigen of *B dermatitidis* achieved higher sensitivity than either immunodiffusion or complement-fixation tests, detecting over 90% of 27 cases. The incidence of crossreactions with histoplasmosis was less than 20%. This crossreactivity should decline further when monoclonal anti-A (Green et al, 1980, 1982) is implemented in this assay.

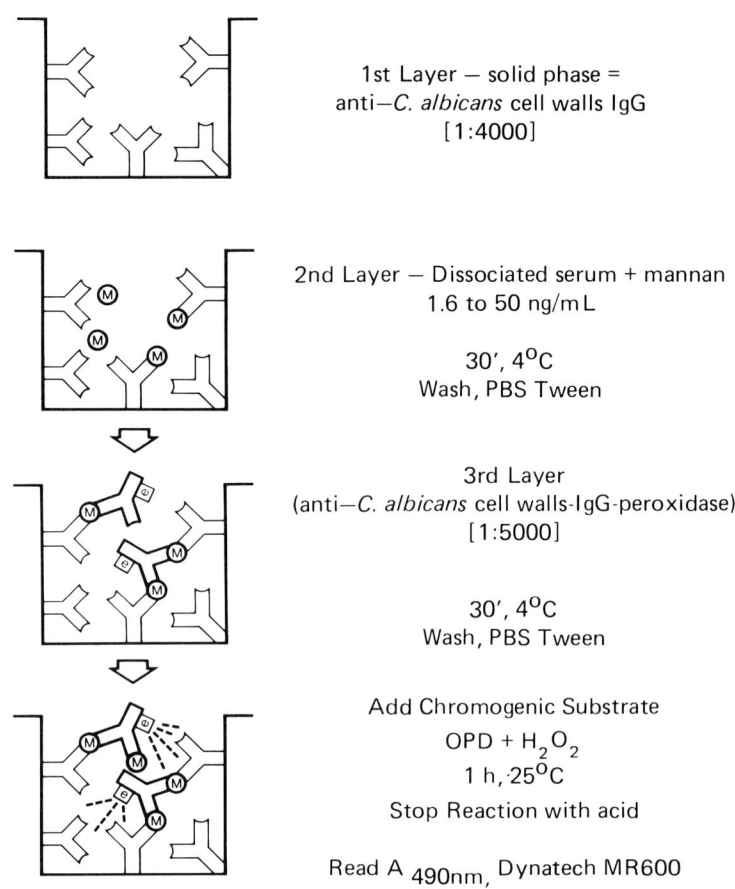

Figure 16.4. Double antibody sandwich enzyme immunoassay to detect circulating antigen. In this instance the antigen being detected is the cell wall polysaccharide, mannan, of *Candida albicans* (de Repentigny and Reiss 1984). Soluble immune complexes must be irreversibly dissociated with heat in the presence of a chelating agent (0.1M Na_2EDTA, pH 7.2) before testing in the sandwich EIA. PBS-Tween, phosphate buffered saline plus a nonionic surfactant; o-PD, ortho-phenylenediamine; w.r.t. with respect to.

- *Candida albicans.* All 640 humans tested had natural IgG against *C albicans* mannan independent of age, sex, or race (Lehmann and Reiss, 1980). Later, it was observed that anti-mannan IgG fluctuates with no apparent relation to clinical condition in leukemia patients with candidiasis (Meckstroth et al, 1981). A more statistical treatment of anti-mannan IgG levels by indirect EIA was carried out (Greenfield et al, 1983). Fifty-four patients colonized with *Candida* were tested and 18.5% had a positive result defined as "multiples of normal activity" (MONA), a cutoff point defining the upper 5% of a standard population as test-positive. Of 33 proven visceral candidiasis patients, 63.6% had a level of anti-mannan IgG rated as positive. The diagnostic sensitivity was lowest (40%) in patients with neoplastic disease.

These results still suffer from the same shortcomings of antibody tests in general for the diagnosis of candidiasis: positive results in patients without invasive disease

A. Example: *Nocardia asteroides* culture filtrate proteins

Isoelectric Focusing in pre-cast polyacrylamide gels "Ampholine PAG plates" pH 4.0 – 6.5 gradient

Electrophoresis Maxima: 15 watts, 1250 volts, 25 milliamps for 3 hr.

Electrotransfer Blot
  Fresh wet gel placed on a nitrocellulose (NC) sheet
  [BA 85 Schleicher and Schuell]
  Sandwiched between filter paper (Whatman 3 mm)
  Loaded into casette of Blot Cell (Bio-Rad)
  Blot buffer, Na Phosphate, 0.025M, pH 7.6
  Electrotransfer 20 V, 30 min; 40 V 1 – 2 hr.

  Remove NC sheet, wash 3X PBS-Tween
  Cut into 5 mm strips and store at 4°C

B. 1. "Blot" strip – 15 ml diluted serum (1:200 H, 1:3000 IR) in a tray with multiple troughs

  ↓ overnight (4° C) or 1 hr (25° C).
     clinical rotator
  Wash 3x, 10 min (PBS/0.2% Tween 20)

2. Add Goat–anti–Human or anti–Rabbit–IgG–peroxidase

  ↓ 1:1000, h, 25° C

  Wash 3x

3. 50 mg Diaminobenzidine – 100 ul H_2O_2/100 ml PBS (stir and filter)

  ↓ Develop color up to 30 min (25° C), dark

  Wash dH_2O

Figure 16.5. Enzyme-linked immunoelectrotransfer blot assay using as an example the conditions for isoelectrically focused *Nocardia asteroides* proteins (El-Zaatari et al, 1986). (A) Isoelectric focusing and electrotransfer to nitrocellulose. (B) Enzyme immunoassay: H, human; IR, immunized rabbit. An illustration of the technique appears as Figure 13.5. The method is also modified for use with SDS-PAGE according to Tsang et al (1983).

and negative results in heavily immunosuppressed patients at greatest risk of the disease. The status of indirect EIA measurement of IgG against *C albicans* unfractionated somatic (= cytoplasmic) or metabolic (= culture filtrate) antigens was reviewed by Richardson and Warnock (1983). These enzyme immunoassays failed to distinguish patients with serious deep infections from subjects with subclinical infection or colonization.

• *Cryptococcus neoformans*. The measurement of IgG against unpurified culture filtrates of *C neoformans* revealed a large overlap of reactions among proven human cases and normal subjects with positive skin tests to cryptococci (Scott et al, 1980).

This lack of selectivity can be minimized by using the galactoxylomannan fraction of the culture filtrate, by adsorbing sera with *C albicans* cell walls, and finally by measuring the IgM class-specific response (Reiss et al, 1984).

There is little interest in replacing the latex agglutination test for cryptococcal polysaccharide in serum or cerebrospinal fluid with an EIA. Even though the double antibody sandwich EIA can detect 6 ng antigen/ml versus 35 ng/ml in the latex agglutination test, there is no proof that the slower, more cumbersome EIA will detect more cases or detect them earlier than the simpler latex test (Scott et al, 1980). The EIA does not appear to be subject to prozones, or when antigen is being measured in serum, to interference from rheumatoid factor. As EIAs for other antigenemia tests become more commonplace, further comparisons in the cryptococcosis antigen detection system may provide a rationale for the EIA.

- Dermatophytes. The lack of clinical usefulness of measuring IgG and IgM against trichophytin glycopeptides was mentioned previously. The reasons for this are that the restricted location of dermatophyte infections makes a systemic stimulus less likely, and, following from that, antibodies against keratinase are cytophilic and restricted to infected hair follicles. The higher incidence of immediate hypersensitivity in chronic dermatophytosis suggests that trichophytin-specific IgE would be a potentially more useful parameter.

- *Histoplasma capsulatum*. Preliminary EIA studies using the M protein antigen of histoplasmin purified by gel permeation and ion-exchange chromatography (Pine et al, 1977) showed an unacceptable overlap between histoplasmosis cases and normal control sera in the measurement of IgG (Brock et al, 1984). Two modifications in the procedure increased the specificity of the reaction. Anti-M, for reference purposes, is produced by immunizing rabbits with immunoprecipitin arcs cut from two-dimensional "rocket" immunoelectrophoretograms. This antiserum is a potentially more specific probe, and it was conjugated to horseradish peroxidase for use in an EIA-inhibition. This format consists of coating microtiter plates with the antigen by passive absorption. Next, the diluted patient's serum is added, incubated, and the unbound proteins are washed away. Then the conjugate is added, and after incubation and washing the color is developed with o-phenylenediamine-H_2O_2. Thus a percent inhibition is selected (in this instance 20%) that discriminates between patients and controls (Figure 5.4). Another improvement consists of treating the antigen for coating microtiter plates by periodate oxidation, thereby inactivating a heat-stable coproduced polysaccharide impurity presumed to be the galactomannan that crossreacts with other genera of systemic fungal pathogens (Brock et al, 1984). This EIA-inhibition discriminates well between patients with "M"-precipitins and controls. Inhibition-type EIAs have the additional benefit of being directly applicable for use with monoclonal antibodies.

- *Nocardia asteroides*. The traditional methods for identifying *N asteroides* cultures are slow and specialized, sometimes requiring weeks to complete. Primary isolation of this organism is frustrated by rapid overgrowth of cultures with other bacteria. Traditional serologic methods such as immunodiffusion in gel are complicated by crossreactions with *Mycobacterium tuberculosis*. These factors have influenced the development of enzyme immunoassays for nocardiosis. It is, first of all, desirable to separate the protein factors that react with the patient's serum. Separation cannot be accomplished in microtiter plates, but is within the scope of the enzyme linked immunoelectrotransfer blot technique. The nocardial proteins obtained from culture supernates and cell extracts are isoelectrically focused in the range pH 4 to 5.5, then electrotransferred to nitrocellulose sheets (Figure 16.5a and b). The sheets are cut into strips and stored in the cold until needed. Nitrocellulose is an ideal medium for conducting enzyme immunoassays because the strips are durable and the electrotransferred proteins do not diffuse. After successive incubations with patient's sera, anti-human IgG conjugated to enzyme, and chromogenic substrate, a reaction pattern emerges.

Certain protein factors have been identified that react with nocardiosis patients' sera but not with sera from tuberculosis patients (El-Zaatari et al, 1986). These factors, 1 (pI 4), 6 (pI 4.43), and 8 (pI 4.68), have been used to prepare monoclonal antibodies for use as immunoadsorbents or as indicator antibodies for EIAs (El-Zaatari et al, 1986).

- *Faenia rectivirugula*. Immunoglobulin G against antigens of this agent of hypersensitivity pneumonitis has been detected by indirect EIA. Whole dialyzed and lyophilized cultures were the antigens used for coating microtiter plates (Bamdad, 1980). Although the antigens were unpurified, good discrimination was made between farmer's lung disease patients and asymptomatic exposed farmers. The ability to quantitate the results is the only obvious advantage over immunodiffusion reactions to *F rectivirugula*. In another series of farmer's lung disease patients (Marx and Gray, 1982), a larger number of farmers reacted in the EIA than had precipitins. This is expected because of the increased sensitivity of enzyme immunoassay. Only a fraction of these showed X-ray changes and symptoms compatible with farmer's lung disease. Serology is only one segment of the diagnosis of this form of hypersensitivity pneumonitis, and it is recognized that many farmers who have precipitins show no clinical symptoms. Thus whether EIA provides a new dimension of clinically useful information in this disease is not clear.

- Enzyme immunoassays for antigenemia. Considerable space was devoted to the developing awareness that mannan polysaccharide circulates in invasive candidiasis and that tests with sensitivity in the low ng/mL range, like enzyme immunoassay or radioimmunoassays, are required to detect such small amounts (Figure 16.4). This section dwells on some of the problems of enzyme immunoassay that limit the useful range of the double antibody sandwich enzyme immunoassay. The limit of sensitivity of the enzyme immunoassay is 1 ng/mL; at that value, the coefficient of variation is uncomfortably large. There are several variables that could be examined to extend the range of this technique.

- Method for dissociating soluble immune complexes. There is general agreement that mannan circulates as soluble immune complexes that must be dissociated before mannan can be reliably detected. This aspect was reviewed by Richardson and Warnock (1983) and de Repentigny and Reiss (1984). The use of heat and extremes of pH, or the presence of a chelating agent such as Na_2EDTA, suggests that recovery is less than complete because of the potential for destruction or trapping of antigen in coagulated protein. An alternative method, using pronase as a dissociating agent, can decrease the yield of antigen even further (Lew et al, 1982), which is not surprising because *Candida* species mannans are glycoproteins. Other methods for concentrating or sequestering mannan from body fluids are necessary. One approach could be to capture the soluble immune complexes in a C1q binding assay.

- Capture and indicator antibodies. Some authors have hinted that the major antigen that circulates is a heat-stable protein and that mannan is a more transient concomitant product (Lew et al, 1982; Araj et al, 1982; Richardson and Warnock, 1983). According to one study, only serum from sublethally infected rabbits could detect antigenemia, and not antisera produced against killed whole cells (Richardson and Warnock, 1983). Araj et al (1982), contended that the antigen they detected was not mannan, although no specific steps were taken to exclude mannan from the homogenate-supernate that was used as the antigenic complex. Antisera prepared against cell walls rigorously washed in sodium dodecyl sulfate to exclude extraneous protein reliably detects antigenemia in rabbits and, subject to the low antigenemia levels observed, in a significant proportion of humans with invasive candidiasis (de Repentigny et al, 1984, 1985).

Thus far, no group has provided evidence that antigens other than mannan (such as cytoplasmic proteins) circulated and can be reliably detected in humans. Until that is done, efforts to improve the reliability of detecting mannanemia are well-placed. Double antibody sandwich assays suffer if the capture and indicator antibodies compete for the same epitopes. Improvements in the configuration of the assay may be helpful, recalling that the original antigenemia enzyme immunoassay was an EIA-inhibition test (Segal et al, 1979). This aspect immediately brings to mind the potential value of monoclonal antibodies. In the "tandem" variety of the sandwich EIA developed by Hybritech, Inc. of Torrey Pines, California, capture and indicator antibodies have different specificities, bind to different epitopes and are thus non-competitive. This assay has two advantages, (i) a potential increase in sensitivity, and (ii) elimination of a separate incubation with the conjugate because a homogeneous assay can be performed with all reactants save the substrate added simultaneously. The progress in developing monoclonal antibodies for this purpose is discussed below.

Amplification of existing enzyme immunoassays can also be explored by using substrate that is fluorescent, eg, 4-methyl-umbelliferyl phosphate for alkaline phosphatase. Otherwise higher substitution of IgG can be achieved with biotin-conjugated antibodies and avidin-coupled enzymes.

There is much scope for combining more than one of these promising modifications to improve the detection of mannanemia in invasive candidiasis and to serve as a model for antigen detection in other mycoses.

16.6 MONOCLONAL ANTIBODIES

Production of monoclonal antibodies by the hybridoma technique is now a predictable approach that will give rise to many antibody secreting clones, especially when whole fungi or actinomycetes are the immunogen. This has already been accomplished for some of the organisms in Table 16.3, and one can fully expect hybridomas to be produced against the remaining genera within a short time. The next stage, characterizing the specificity of the monoclonal antibodies is only now beginning to unfold because of its greater difficulty. More powerful consequences of the monoclonal technology are yet to be realized: (i) The use of monoclonal antibodies as ultrastructural probes for localization of antigens by immunoelectron microscopy; (ii) As replacements for polyclonal antibodies in clinical immunoassays; (iii) For passive transfer of immunity as immunotoxins: homing devices conjugated to antimycotic drugs to achieve higher selective toxicity; and (iv) As aids in imaging the focus of an infection by tracing monoclonal antibodies labeled with radionuclides.

Some of the serendipitous findings of attempts to produce monoclonal antibodies of desired specificity are insights into the complexity of "pure" antigens, and the degree of regulation of antibody synthesis that was previously unrecognized.

Table 16.3. Catalog of Monoclonal Antibodies Against Mycotic and Actinomycotic Agents

	Specificity	Reference
Eumycetes		
Blastomyces dermatitidis	A-antigen (protein) Protein in homogenate-supernatant	Green et al, 1982 Young and Larsh, 1982
Candida albicans	Heat-stable peptidoglucomannan Candidal cytoplasmic proteins	Kerkering and Espinell-Ingroff, 1983 Strockbine et al, 1983
Candida tropicalis	Mannan	Reiss et al, 1984
Coccidioides immitis	nd	Karu et al, 1985
Cryptococcus neoformans	2 *C neoformans*-specific 4 crossreactive with *H capsulatum* 6 crossreactive with *B dermatitidis*	Hall and Blackstock, 1981
Histoplasma capsulatum	Heat-stable polysaccharide of histoplasmin Heat-labile M-protein of histoplasmin	Knowles et al, 1983
Actinomycetes		
Actinomyces viscousus	Fimbriae mediating coaggregation of *S sanguis*	Cisar et al, 1981
Actinomyces naeslundii	Fimbriae	Bragg et al, 1983
Nocardia asteroides	Nocardin proteins, factors 1,6,8	El-Zaatari et al, 1986

Abbreviation: nd, not determined.

The following examples illustrate these points as they relate to an area that is now in its early childhood.

• *Blastomyces dermatitidis*. Monoclonal antibodies were produced against the A-antigen, a species-specific secreted factor (Green et al, 1982). The A-antigen is an acidic protein that is eluted from an anion exchange resin only by a high salt concentration (Green et al, 1980). Antibodies from five of six clones fixed complement and demonstrated complete specificity in the complement-fixation test; no crossreactions with either *C immitis* or *H capsulatum* antigens were observed.

Fusions were made from mice immunized with whole formalin-killed *B dermatitidis* yeast forms (Young and Larsh, 1982). One highly reactive clone producing IgM was propagated because it did not crossreact in EIA with histoplasmin as the solid phase-adsorbed antigen. Small scale immunoaffinity chromatography was conducted with the monoclonal IgM as the solid phase. The affinity-purified antigen was a minor component by weight of the culture supernate and had a different r_f in PAGE than the A-antigen. The monoclonal IgM precipitated the homologous antigen in immunodiffusion, but did not crossreact with the *B dermatitidis* A-antigen or alkali-soluble, water-soluble extract of cell walls. The significance of this antigen in the immune response to *B dermatitidis* is not known.

These antibodies hold great promise for improving serodiagnostic procedures for blastomycosis, namely by overcoming the frequent crossreactions observed because the yeast form antigens used in the conventional complement-fixation test crossreact extensively with the tissue forms of the other primary systemic fungi.

• *Candida albicans*. Considering the popularity of this organism as a model for investigating the host–fungus interaction, it is not surprising that reports of monoclonal antibody production are burgeoning. The thrust appears to be in the area of improvement of immunodiagnostic procedures, although monoclonal antibodies have not yet supplanted conventional rabbit antisera in this regard.

Monoclonal antibodies of the IgG_1 subclass precipitate cytoplasmic protein antigens devoid of mannan (Strockbine et al, 1983). Activity was also assessed by EIA and by autoradiography of PAGE gels. Preliminary characterization was achieved with five IgM-secreting hybridomas induced in BALB/c mice by immunization with a peptidoglucomannan fraction of C albicans cell walls (Kerkering and Espinell-Ingroff, 1983). This antigen, extracted with hot formamide, contained 40% protein by weight and thus does not exactly correspond to the peptidoglucomannan of Reiss et al (1974), which contained 10% protein. The monoclonal antibody remained complexed to peptidoglucomannan after boiling, but could be dissociated by oxidizing the mannan portion with $NaIO_4$.

The double antibody sandwich EIA detects the mannan polysaccharide of C albicans and C tropicalis that circulates in invasive candidiasis of the immunocompromised host. Monoclonal antibodies, as replacements for antisera in this assay, offer the prospect of increased sensitivity, which is needed because clinical levels are in the low ng/mL range. Increased sensitivity could be gained by a higher density of capture antibodies or by using the "tandem" assay, in which capture and indicator antibodies have different specificities and are thus noncompeting.

Three strains of mice were immunized with C tropicalis cell walls and they differed in their ability to respond to mannan in indirect EIA: outbred CFW > C57B1/6 > BALB/c (Reiss et al, 1984). Responses to mannan thus appear to be under some degree of genetic control. The major immunoglobulin class in the mouse antisera was IgM. Thirty-two hybrid clones producing anti-mannan were isolated from the fusion of high responder BALB/c with a nonsecreting plasmacytoma. Of these clones, only one produced IgG. The IgG-producing clone reacted with C albicans A, B, and C tropicalis mannan. This monoclonal antibody and three IgMs were selected for propagation as ascites tumors: one IgM was specific for mannan of C tropicalis and C albicans A, the others reacted with both C albicans serotypes.

The monoclonal antibodies were able to act as both capture and indicator antibodies when mannan was "spiked" into normal human serum, but when sera from cortisone-immunosuppressed rabbits infected with C albicans A or C tropicalis were used as the antigen, detection occurred only when antiserum was used as the capture antibody. The antigen that circulates during infection is different from that which is produced in vitro. One of the monoclonal antibodies was more sensitive as the indicator antibody than antiserum, thus validating the "tandem" hypothesis.

Mannanemia due to C albicans B serotype has proven to be a more difficult problem. Even when 1.75×10^9 colony forming units were detected in the kidney at 4 days after infection, the peak time for antigenemia, neither monoclonal antibodies nor conventional antisera were an efficient detection system. Either the mannan of serotype B is quickly metabolized by the rabbit, "nicked" by host glycosidases, or synthesized differently during infection so that monoclonal antibodies produced against cell walls do not recognize it. In this sense, working with monoclonal antibodies has led to a clearer definition of the problem of antigenemia detection.

- *Histoplasma capsulatum.* The M factor of histoplasmin is a major protein antigen that evokes humoral and cell-mediated immunity in human histoplasmosis. It was purified by gel permeation and ion exchange chromatography before priming mice for fusions (Knowles et al, 1983). The M antigen was coated onto polystyrene microtiter wells in order to screen clone supernates. Reactive IgM clones were produced, but later it was learned that these antibodies were reacting with a heat-stable, periodate-labile, proteinase-stable polysaccharide impurity in the M-antigen. The antigen responsible for inducing these monoclonal antibodies was probably the "C" polysaccharide of Heiner (1958), which is presumed to be a galactomannan based on other experiments reviewed in Chapter 5.

The hypothesis was that the M factor, because of its high antigenicity, would be expressed, and that the response to the polysaccharide would be suppressed because its removal from the whole yeast form would render it nonimmunogenic. This reasoning is not supported by the experiment, which on the contrary showed that the polysaccharide is highly immunogenic for mice.

Because the M factor of histoplasmin was resistant to periodate oxidation, and this simple step inactivated the polysaccharide impurity, mice were immunized with periodate-treated M for additional fusions. Immunoglobulin G-secreting clones were selected that reacted in indirect EIA with both untreated and periodate-treated M preparations in a preliminary screen. Confirmation of the specificity of the monoclonal antibodies was obtained using the EITB assay with gradient SDS-polyacrylamide gels. The monoclonal antibodies as well as antibodies in sera from humans, rabbits, and mice bound to a doublet of proteins with a molecular weight of 70–75,000 da. These are subunits that exist in histoplasmin as a disulfide-linked 150,000 da nonantigenic dimer.

- *Nocardia asteroides*. The antigenic structure of *Nocardia asteroides* cell extracts and culture filtrates has been analyzed by the EITB technique with isofocused polyacrylamide gels. Three proteins focused within the pH range of 4 to 5.5 and were identified as binding antibodies present in sera from nocardiosis but not from tuberculosis patients (El-Zaatari et al, 1986). These proteins were excised from the isofocused gels, and mice were successfully immunized as judged by presumptive microtiter plate EIA and confirmatory immunoblots of the isofocused proteins. Fusions gave rise to clones secreting monoclonal antibodies against these three factors (El-Zaatari et al, 1986). The monoclonal antibody produced against factor 8 bound to several isofocused proteins, indicating either structural homology or the cleavage of a single protein to fragments during growth or autolysis. Monoclonals against the other two factors, 1 and 6, bound only to their respective single antigens. Further characterization of these monoclonals and their adaptation to clinical immunoassays is now possible.

- Microaerophilic actinomycetes. Two types of surface fimbriae occur on *Actinomyces viscosus*. One type mediates adherence to saliva-coated hydroxyapatite and the second is responsible for the lactose-reversible coagglutination with *Streptococcus sanguis*. Nine monoclonal antibodies against a single antigen in the fimbriae responsible for the lectinlike coagglutination were described (Cisar et al, 1981). *Actinomyces naeslundii* has similar fimbriae. A monoclonal antibody against these fimbriae could not by itself precipitate the antigen in agar gel, but a mixture of unlabeled polyclonal antibodies and an ^{125}I-monoclonal were subjected to two dimensional crossed "rocket" immunoelectrophoresis. All of the radioactivity concentrated in one of the two rockets that were observed. This illustrates another way that the binding characteristics of monoclonal antibodies can be studied even when they do not directly form immunoprecipitates (Bragg et al, 1983).

16.7 ROLE OF HUMORAL IMMUNITY IN INFECTION

16.7.1 Polyclonal B-Cell Activation

Good evidence for polyclonal activation exists for the Nocardia water-soluble mitogen in humans and rodents. The homogenate of *Actinomyces viscosus* and purified fractions derived from it also act as a B-cell mitogen in rodents, including athymic nude mice.

Mitogenesis by the NWSM does not require T-helper cells, but their addition increases the quantity of immunoglobulin synthesis fivefold. Although several B-cell mitogens such as pokeweed mitogen, Con A, and phytohemagglutinin activate presuppressor T-cells to become active suppressor-effector cells, NWSM does not (Waldmann and Broder, 1982). Although NWSM does not itself activate suppressor T-cells, the B-cells receptive to it are subject to T-suppressors that have been induced by other phytomitogens. These properties make NWSM valuable for probing abnormalities in the T-suppressor network.

Polyclonal B-cell activation by AVIS is under strict control in humans; T-cell help and depletion of monocytes are required (Mangan and Lopatin, 1981). The blastogenic potential of *A viscosus* probably contributes to the chronic inflammation in periodontal disease.

The widespread use of *C albicans* extracts or whole candidae as an ubiquitous microbial antigen for immunocompetence testing leaves open the question of whether the responses obtained are polyclonal or antigen-specific. The high blastogenic indices that can be obtained with mannan-containing soluble antigens (Piccolella et al, 1981b) suggests the possibility of polyclonal activation. The finding that this reaction is T-dependent and regulated by Ia antigens is, on the other hand, typical of antigen-specific reactions (Piccolella et al, 1981b). The boosting effect of repeat *Candida* skin tests on blastogenic indices also suggests this is an antigen-specific event (Ferguson et al, 1977).

16.7.2 Immune Complex Formation

Mannan of *C albicans* was characterized as the blocking factor in sera of some children with chronic mucocutaneous candidiasis (Fischer et al, 1978; 1982). The detection of soluble immune complexes in coccidioidomycosis increases with disease severity (Cox et al, 1982), and humoral factors in some anergic patients with this disease are responsible for the depressed blastogenic indices (Harvey and Stevens, 1981). The evidence that immune complexes are the blocking factor is, at this point, inferential. Blocking factors have also been reported in histoplasmosis (Cox, 1979), and in paracoccidioidomycosis, where replacement of autologous plasma augmented blastogenic responses to *P brasiliensis* antigens (Musatti et al, 1976). The suggestion that soluble immune complexes are responsible for blocking lymphocyte blastogenesis is strong in the mycoses, warranting further attention to this aspect.

Clinical consequences of immune complexes have been proposed in only a few instances. The immunopathology in bronchopulmonary aspergillosis is considered, in part, to result from the diffusion of antigens into the lung parenchyma and complexation with antibody to give the pulmonary equivalent of an Arthus reaction (Bardana, 1980). Deposition of immune complexes in the kidney is a rare complication of histoplasmosis (Bullock et al, 1979). The Splendore-Hoeppli material surrounding yeast forms of *S schenckii* in tissue and in vitro has been characterized as immune complex deposits (Lurie and Still, 1969).

The high incidence of precipitins against thermophilic actinomycetes in farmer's lung disease and the appearance of symptoms 4 to 8 hours after exposure are indirect evidence for Arthus type III immune complex-mediated damage. Pulmonary vasculitis, which would provide some proof of this mechanism is not consistently found and asymptomatic exposed farmers also carry precipitins (Ghose et al, 1974; Karr and Salvaggio, 1980).

16.7.3 Protective Role for Humoral Immunity

Humoral immunity against the mycotic infections cannot be divorced from T-cell-mediated immunity because of the influence of T-helper and T-suppressor cells on immunoregulation. With that proviso in mind, it is possible to consider instances where humoral immunity has protective value, where there is no measurable protective effect, and where a pathogenetic role for antibodies has been proposed.

It was pointed out that IgG and heat-labile (ie, complement) opsonins enhanced adherence and phagocytosis of *Candida* and *Cryptococcus* cells. Griffin (1981) described circumstances in which macrophages could phagocytose cryptococci via their C3 receptors, but only under the stimulus of immune complexes. There is evidence that C5-deficient mice are more susceptible to *C albicans* infection (Morelli and Rosenberg, 1971), but it is not known whether C5 contributes chemotactic and anaphylatactic functions or whether late complement components of the membrane attack sequence can damage the fungi. As is the case with *Candida*, complement is necessary for efficient phagocytosis of cryptococci (Diamond et al, 1974), and complement depletion (Macher et al, 1978) or genetic deletion of C5 increases susceptibility to cryptococci (Rhodes et al, 1980).

The lack of opportunistic mycoses in boys with Bruton's X-linked agammaglobulinemia, or the lack of reports of fungal complications of deletions in late complement components suggests that antibody-complement plays an adjunctive but not decisive role in host defenses against fungi.

Direct activation of the alternative complement pathway by *Candida* (Ray et al, 1979) and by *C neoformans* (Laxalt and Kozel, 1979) may contribute to innate immunity by providing immune surveillance. Hyperimmune serum protected mice from an intramuscular challenge with *C albicans* (Pearsall et al, 1978). The capacity to make antibodies against T-dependent antigens, "HEX", (hot water extract of mitochondria-membranes) correlates with protection against secondary challenge with *C albicans* in mice (Giger et al, 1978). Thus, T-cell immunologic memory is important to prevent reinfection. Mice survived poorly after cyclophosphamide treatment that selectively depressed B-lymphocytes and reduced antibody production even when delayed cutaneous hypersensitivity was intact (Moser and Domer, 1980).

The humoral response in human cryptococcosis has been partially charted (Diamond et al, 1972; Diamond and Bennett, 1974). The early appearance of antibodies gives way to antigenemia as the blastoconidia disseminate. A favorable response to therapy coincides with gradual replacement of antigenemia by antibody levels. It is not known if the return of antibodies is a "noncausal concomitant" or whether these antibodies contribute to eradication of cryptococci from the deep tissues. In this regard antibody-dependent cell-mediated cytotoxicity has been reported to occur against small capsule cryptococci by monocytes (Diamond and Allison, 1976) and by PMN (Miller and Kohl, 1983). Passive transfer of rabbit anti-*C neoformans* IgG to mice protects them from challenge with live cryptococci, but only when the antiserum and the challenge dose are given simultaneously and by the same intravenous route (Graybill et al, 1981).

16.7.4 No Measurable Protective Effect

Complement-fixing antibodies of the IgG class against *Coccidioides immitis* rise in proportion to disease severity and titers greater than 1/16 create concern about extrapulmonary dissemination (Smith et al, 1956). A decline of these antibodies signals a favorable response to therapy. Clearly, the ability to produce these antibodies is not sufficient to restrict the infection to the lungs. They may, however, add an increment of protection to the immune response. Passive transfer of antibodies in mice before challenge with live *C immitis* has thus far proved ineffective (see section 4.6.2). Passive transfer of immune serum fails to protect mice from challenge with live *B dermatitidis* (Brummer et al, 1982) or *H capsulatum* (Tewari et al, 1978).

Patients with chronic dermatophyte infections were tested for the presence of IgG and IgM by indirect EIA. No differences occurred between chronically infected and non-chronic patients (Kaamen et al, 1981). However, cytophilic antibodies against keratinase have been localized by indirect immunoperoxidase in the external sheath of infected hair follicles (Holden et al., 1981).

Some properties of fungal polysaccharides and their interaction with the immune system do not fit neatly into categories and should stimulate further experimentation. The β-1,3-glucan of *Aphanomyces* induces coagulation of crayfish hemolymph (Söderhäll, 1981) when Ca^{++} exceeds 5mM. The exocellular glucan of *Aureobasidium pullulans*, pullulan, has a polymaltotriose structure. When conjugated to haptens such as the dinitrophenyl group, pullulan is a tolerogen for IgE (Taniguchi et al, 1982), but a good immunogen for IgG or IgM. The hapten pullulan conjugate is a T-independent antigen (type 2).

16.7.5 Pathogenetic Antibodies

In allergic bronchopulmonary aspergillosis rising concentrations of specific IgG, total IgE, and *Aspergillus*-specific IgE correlate positively with flares of the disease and with fleeting pulmonary shadows due to pulmonary eosinophilia (Bardana, 1980; Longbottom, 1983). Direct evidence for immune complex-deposition in the lung has been difficult to obtain because invasive procedures are not warranted in most cases.

Serum IgE levels as high as 82,800 ng/mL have been reported (Rosenberg et al, 1978). The cumulative effects of flares of allergic aspergillosis are well-known: central or saccular bronchiectasis (dilated bronchi ending in blind sacs).

Stimulation of IgE to levels that imply a role for immediate hypersensitivity in the disease process have also been observed in coccidioidomycosis (Cox et al, 1982) and in the tendency of *T rubrum* infections to correlate with atopy (Hunziker and Brun, 1980). Polyclonal gammopathy occurs in paracoccidioidomycosis, and a direct relation exists between inhibition of blastogenesis and the presence of precipitins (Restrepo et al, 1978).

Passive transfer of anti-*Nocardia brasiliensis* either intravenously or by coating live bacteria before injection caused more severe disease in mice (Rico et al, 1982). Splenic B-lymphocytes from immune donor mice were removed by rosetting them with *N brasiliensis* antigen-coated erythrocytes. The specific B-cell-depleted lymphocytes transferred immunity to irradiated recipients, although no antibodies were produced. These findings argue that the humoral response in nocardiosis is detrimental, possibly by blocking recognition of the nocardiae.

Actinomyces viscosus contains a T-dependent B-cell mitogen for humans. This mitogen, when considered along with data suggesting increased blastogenesis of humans with periodontal diseases, implies a causal relationship between polyclonal B-cell activation and chronic inflammation (Ivanyi and Lehner, 1970; Mangan and Lopatin, 1981).

REFERENCES

Araj GF, Hopfer RL, Chestnut S, Fainstein V, Bodey GP, Sr, 1982. Diagnostic value of enzyme-linked immunosorbent assay detection of *Candida albicans* cytoplasmic antigen in sera of cancer patients. J Clin Microbiol 16:46–52.

Axelsen NH, 1971. Antigen-antibody crossed electrophoresis (Laurell) applied to the study of the antigenic structure of *Candida albicans*. Infect Immun 4:525–527.

Azuma I, Kanetsuna F, Tanaka Y, Yamamura Y, Carbonell LM, 1974. Chemical and immunological properties of galactomannans obtained from *Histoplasma duboisii*, *Histoplasma capsulatum*, *Paracoccidioides brasiliensis* and *Blastomyces dermatitidis*. Mycopathol et Mycol Appl 54:111–125.

Baker O, Braude AI, 1956. A study of stimuli leading to production of spherules in coccidioidomycosis. J Lab Clin Med 47:169–182.

Baker PJ, Stashak PW, Amsbaugh DF, Prescott B, 1979. Specificity of suppressor T cells activated during the immune response to type III pneumococcal polysaccharides. pp. 77–93. In Immunomodulation by Bacteria and their Products. Friedman H, Klein TW, Szentivanyi A, eds. New York: Plenum Press.

Bamdad S, 1980. Enzyme-linked immunosorbent assay (ELISA) for IgG antibodies in farmer's lung disease. Clin Allergy 10:161–171.

Bardana EJ Jr, 1980. The clinical spectrum of aspergillosis—part 1: epidemiology, pathogenicity, infection in animals and immunology of *Aspergillus*. CRC Crit Rev Clin Lab Sci 13:21–83.

Borkowsky W, Valentine FT, 1979. The proliferative response of human lymphocytes to antigen is suppressed preferentially by lymphocytes precultured with the same antigen. J Immunol 122:1867–1873.

Bragg SL, Erdos G, Bleiweiss AS, 1983. Detection of a lectin on *Actinomyces naeslundii* N16 with monoclonal antibodies. Abstr Ann Mtg Amer Soc Microbiol, no B13, p. 25.

Brock EG, Reiss E, Pine L, Kaufman L, 1984. Effect of periodate oxidation on the detection of antibodies against the M-antigen of histoplasmin by enzyme immunoassay (EIA)-inhibition. Current Microbiol 10:177–180.

Brummer E, Morozumi PA, Vo PT, Stevens DA, 1982. Protection against pulmonary blastomycosis: adoptive transfer with T lymphocytes, but not serum, from resistant mice. Cell Immunol 73:349–359.

Brummer E, Stevens DA, 1982. Opposite effects of human monocytes, macrophages, and polymorphonuclear leukocytes on replication of *Blastomyces dermatitidis* in vitro. Infect Immun 36:297–303.

Bullock WE, Artz RP, Bhathena D, Tung KS, 1979. Histoplasmosis. Association with circulating immune complexes, eosinophilia, and mesangiopathic glomerulonephritis. Arch Intern Med 139:700–702.

Calvanico NJ, DuPont BL, Huang CJ, Patterson R, Fink JN, Kurup VP, 1981. Antigens of

Aspergillus fumigatus. 1. Purification of a cytoplasmic antigen reactive with sera of patients with *Aspergillus*-related disease. Clin Exp Immunol 45:662–671.

Cassone A, Mattia L, Boldrini L, 1978. Agglutination of blastospores of *Candida albicans* by concanavalin A and its relationship with the distribution of mannan polymers and the ultrastructure of the cell wall. J Gen Microbiol 105:263–273.

Cheers C, Waller RP, 1975. Activated macrophages in congenitally athymic "nude" mice and in lethally irradiated mice. J Immunol 115:844–847.

Cherniak R, Reiss E, Slodki ME, Plattner RD, Blumer SO, 1980. Structure and antigenic activity of the capsular polysaccharide of *Cryptococcus neoformans* serotype A. Mol Immunol 17:1025–1032.

Cherniak R, Reiss E, Turner SH, 1982. A galactoxylomannan antigen of *Cryptococcus neoformans* serotype A. Carbohydr Res 103:239–250.

Cisar JO, Barsumian EL, Curl SH, Vatter AE, Sandberg AL, Siraganian RP, 1981. Detection and localization of a lectin on *Actinomyces viscosus* T14V by monoclonal antibodies. J Immunol 127:1318–1322.

Cox RA, 1979. Immunologic studies of patients with histoplasmosis. Amer Rev Respir Dis 120:143–149.

Cox RA, Baker BS, Stevens DA, 1982. Specificity of immunoglobulin E in coccidioidomycosis and correlation with disease involvement. Infect Immun 37:609–616.

Cox RA, Larsh HW, 1974. Yeast and mycelial phase antigens of *Blastomyces dermatitidis*: comparison using disc gel electrophoresis. Infect Immun 10:48–53.

Cox RA, Pope RM, Stevens DA, 1982. Immune complexes in coccidioidomycosis — correlation with disease involvement. Amer Rev Respir Dis 126:439–443.

Cunningham KM, Bulmer GS, Rhoades ER, 1979. Phagocytosis and intracellular fate of *Sporothrix schenckii*. J Infect Dis 140:815–817.

Davies SF, Clifford DP, Hoidal JR, Repine JE, 1982. Opsonic requirements for the uptake of *Cryptococcus neoformans* by human polymorphonuclear leukocytes and monocytes. J Infect Dis 145:870–874.

de Repentigny L, Kuykendall RJ, Chandler FW, Broderson JR, Reiss E, 1984. Comparison of serum mannan, arabinitol and mannose in experimental disseminated candidiasis. J Clin Microbiol 19:804–812.

de Repintigny L, Marr LD, Keller JN, et al, 1985. Comparison of enzyme immunoassay and gas-liquid chromatography for the rapid diagnosis of invasive candidiasis in cancer patients. J Clin Microbiol 21:972–979.

de Repentigny L, Reiss E, 1984. Current trends in immunodiagnosis of candidiasis and aspergillosis. Rev Infect Dis 6:301–312.

Diamond RD, Allison AC, 1976. Nature of the effector cells responsible for antibody-dependent cell-mediated killing of *Cryptococcus neoformans*. Infect Immun 14:716–720.

Diamond RD, Bennett JE, 1974. Prognostic factors in cryptococcal meningitis. Ann Intern Med 80:176–181.

Diamond RD, Krzesicki R, 1978. Mechanisms of attachment of neutrophils to *Candida albicans* pseudohyphae in the absence of serum, and of subsequent damage to pseudohyphae by microbicidal processes of neutrophils in vitro. J Clin Invest 61:360–369.

Diamond RD, Krzesicki R, Epstein B, Yao W, 1978. Damage to hyphal forms of fungi by human leukocytes in vitro: a possible host defense mechanism in aspergillosis and mucormycosis. Amer J Pathol 91:313–323.

Diamond RD, May JE, Kane MA, Frank MM, Bennett JE, 1974. The role of the classical and alternate complement pathways in host defenses against *Cryptococcus neoformans* infection. J Immunol 112:2260–2270.

Diamond RD, Root RK, Bennett JE, 1972. Factors influencing killing of *Cryptococcus neoformans* by human leukocytes in vitro. J Infect Dis 125:367–376.

Domer JE, Moser SA, 1978. Experimental murine candidiasis — cell mediated immunity after cutaneous challenge. Infect Immun 20:88–98.

Durandy A, Fischer A, Griscelli C, 1983. Specific in vitro antimannan-rich antigen (sic) of

Candida albicans antibody production by sensitized human blood lymphocytes. J Clin Investig 71:1602–1613.

Edwards JH, 1972. The isolation of antigens associated with farmer's lung. Clin Exp Immunol 11:341–355.

Ellsworth JH, Reiss E, Bradley RL, Chmel H, Armstrong D, 1977. Comparative serological and cutaneous reactivity of candidal cytoplasmic proteins and mannan separated by affinity for concanavalin A. J Clin Microbiol 5:91–99.

El-Zaatari FA, Reiss E, Yakrus MA, Bragg SL, Kaufman L, 1986. Monoclonal antibodies against isoelectrically focused *Nocardia asteroides* proteins characterized by the enzyme-linked immunoelectro-transfer blot method. Diagn Immunol: in press.

Ferguson AC, Kershnar HE, Collin WK, Stiehm ER, 1977. Correlation of cutaneous hypersensitivity with lymphocytic response to *Candida albicans*. Amer J Clin Pathol 68:499–504.

Fischer A, Ballet J-J, Griscelli C, 1978. Specific inhibition of in vitro *Candida* induced lymphocyte proliferation by polysaccharide antigen present in the serum of patients with chronic mucocutaneous candidiasis. J Clin Invest 62:1005–1013.

Fischer A, Pichat L, Audinot M, Griscelli C, 1982. Defective handling of mannan by monocytes in patients with chronic mucocutaneous candidiasis resulting in a specific cellular unresponsiveness. Clin Exp Immunol 47:653–660.

Friis J, Ottolenghi P, 1970. The genetically determined binding of Alcian blue by a minor fraction of yeast cell walls. Compt Rend Trav Lab Carlsberg 37:327–341.

Fukazawa Y, Nishikawa A, Suzuki M, Shinoda T, 1980. Immunochemical basis of serologic specificity of the yeast: immunochemical determinants of several antigenic factors of yeasts. Zentralblatt f. Bakteriologie Suppl 8:127–136.

Galgiani JN, Isenberg RA, Stevens DA, 1978. Chemotaxigenic activity of extracts from the mycelial and spherule phases of *Coccidioides immitis* for human polymorphonuclear leukocytes. Infect Immun 21:862–865.

Ghose T, Landrigan P, Kileen R, Dill J, 1974. Immunopathological studies in patients with farmer's lung. Clin Allergy 4:119.

Giger D, Domer J, Moser S, McQuitty J, 1978. Experimental murine candidiasis: pathological and immune responses in T lymphocyte depleted mice. Infect Immun 21:729–737.

Goihman-Yahr M, Essenfeld-Yahr E, de Albornoz MC, Yarzabal L, de Gomez MH, San Martin B, Ocanto A, Gil F, Convit J, 1980. Defect of in vitro digestive ability of polymorphonuclear leukocytes in paracoccidioidomycosis. Infect Immun 28:557–566.

Gotslich EC, Liu T-Y, Oliveira E, 1982. Binding of C-reactive protein to C-carbohydrate and PC-substituted protein. Annals N Y Acad Sci 389:163–171.

Graybill JR, Hague M, Drutz DJ, 1981. Passive immunization in murine cryptococcosis. Sabouraudia 19:237–244.

Green JH, Harrell WK, Aloisio C, 1982. The preparation of monoclonal antibodies for use as control reagents in the serological diagnosis of blastomycosis. Abst Ann Mtg Amer Soc Microbiol, no F67, p. 337.

Green JH, Harrell WK, Johnson JE, Benson R, 1980. Isolation of an antigen from *Blastomyces dermatitidis* that is specific for the diagnosis of blastomycosis. Curr Microbiol 4:293–296.

Greenfield RA, Jones JM, 1981. Purification and characterization of a major cytoplasmic antigen of *Candida albicans*. Infect Immun 34:469–477.

Greenfield RA, Stephens JL, Bussey JM, Jones JM, 1983. Quantitation of antibody to *Candida* mannan by enzyme linked immunosorbent assay. J Lab Clin Med 101:758–771.

Griffin FM Jr, 1981. Roles of macrophage Fc and C3b receptors in phagocytosis of immunologically coated *Cryptococcus neoformans*. Proc Nat Acad Sci USA 78:3853–3857.

Hall NK, Blackstock R, 1981. Production of specific antibody to *Cryptococcus neoformans* by hybridomas in vitro. Sabouraudia 19:157–160.

Harvey RP, Stevens DA, 1981. In vitro assays of cellular immunity in progressive coccidioidomycosis: evaluation of suppression with parasitic-phase antigen. Amer Rev Respir Dis 123:665–669.

Hearn VM, Mackenzie DWR, 1979. The preparation and chemical composition of fractions from *Aspergillus fumigatus* wall and protoplasts possessing antigenic activity. J Gen Microbiol 112:35–44.

Hearn VM, Mackenzie DWR, 1981. Analysis of wall antigens of *Aspergillus fumigatus* by two dimensional immunoelectrophoresis. J Med Microbiol 14:119–129.

Heiner DC, 1958. Diagnosis of histoplasmosis using precipitin reactions in agar gel. Pediatrics 22:616–629.

Herberman RB, 1978. Natural cell mediated cytotoxicity in nude mice. pp 135–166. In The Nude Mouse in Experimental and Clinical Research, Vol 1. Fogh J, Giovanella BC, eds. New York: Academic Press.

Holden CA, Hay RJ, MacDonald DM, 1981. The antigenicity of *Trichophyton rubrum*: in situ studies by an immunoperoxidase technique in light and electron microscopy. Acta Derm Venereol (Stockholm) 61:207–211.

Hollingdale MR, 1974. Antibody responses in patients with farmer's lung disease to antigens from *Micropolyspora faeni*. J Hygiene (Cambridge) 72:79–89.

Holmberg K, Berdischewsky M, Young LS, 1980. Serological diagnosis of invasive aspergillosis. J Infect Dis 141:656–664.

Howard DH, 1973. Fate of *Histoplasma capsulatum* in guinea pig polymorphonuclear leukocytes. Infect Immun 8:412–419.

Howard DH, 1975. The role of phagocytic mechanisms in defense against *Histoplasma capsulatum*. pp. 50–59. In Mycoses, Proc Third Intern Conf Mycoses, Sci Pub No 304. Washington, DC: Pan American Health Organization.

Howard DH, 1983. Studies on the catalase of *Histoplasma capsulatum*. Infect Immun 39:1161–1166.

Howie S, McBride WH, 1982. Cellular interactions in thymus dependent antibody responses. Immunol Today 3:273–278.

Hunziker N, Brun R, 1980. Lack of delayed reaction in presence of cell-mediated immunity in trichophytin hypersensitivity. Arch Dermatol 116:1266–1268.

Huppert M, Spratt NS, Vukovich KR, Sun SH, Rice EH, 1978. Antigenic analysis of coccidioidin and spherulin determined by two dimensional immunoelectrophoresis. Infect Immun 20:541–551.

Ivanyi L, Lehner T, 1970. Stimulation of lymphocyte transformation by bacterial antigens in patients with periodontal disease. Arch Oral Biol 15:1089–1096.

Jones AE, Reiss E, Spira T, 1981. A microsomal fraction of *Cryptococcus neoformans* elicits in vitro blastogenesis of lymphocytes from infected guinea pigs. Mycopathol 75:129–138.

Julius MH, 1982. Cellular interactions involved in T-dependent B-cell activation. Immunol Today 3:295–299.

Kaaman T, von Stedingk LV, von Stedingk M, Wasserman J, 1981. ELISA-determined serological reactivity against purified trichophytin in dermatophytosis. Acta Derm Venereol (Stockholm) 61:313–317.

Kalina M, Kletter Y, Shahar A, Aronson M, 1971. Acid phosphatase release from intact phagocytic cells surrounding a large-sized parasite. Proc Soc Exptl Biol Med 136:407–410.

Karr RM, Salvaggio JE, 1980. Infiltrative hypersensitivity disease of the lung. pp. 1336–1371. In Clinical Immunology Vol II. Parker CW, ed. Philadelphia: W. B. Saunders, Co.

Karu AE, Gennevois DJP, Hoffman JW, Kraeger SJ, Levine HB, 1985. Research and diagnostic applications of monoclonal antibodies to *Coccidioides immitis*. Abst Ninth Intern. Cong. Intern. Soc. Human and Animal Mycol. (unpublished) Abst. No. S19–3.

Kauffman HF, Beaumont F, Meurs H, van der Heide S, de Vries K, 1983. Comparison of antibody measurements against *Aspergillus fumigatus* by means of double-diffusion and enzyme-linked immunosorbent assay (ELISA). J Allergy Clin Immunol 72:255–261.

Kerkering TM, Espinell-Ingroff A, 1983. Production of monoclonal antibodies (MA) against a cell wall peptido-glucomannan (PGM) of *Candida albicans*. Abstr Ann Mtg Amer Soc Microbiol, no F41:389.

Kim SJ, Chaparas SD, Brown TM, Anderson MC, 1978. Characterization of antigens from *Aspergillus fumigatus* II. Fractionation and electrophoretic, immunologic and biologic activity. Amer Rev Respir Dis 118:553–560.

Knowles JB, Reiss E, Brock EG, Bragg SL, Pine L, Aloisio CH, 1983. Characterization of monoclonal antibodies to histoplasmin factors. Abstr Ann Mtg Amer Soc Microbiol, p. 389.

Kozel TR, Gulley WF, Cazin J, Jr, 1977. Immune response to *Cryptococcus neoformans* soluble polysaccharide: immunological unresponsiveness. Infect Immun 18:701–707.

Kurup VP, Scribner GH, 1981. Antigenic relationship among *Nocardia asteroides* immunotypes. Microbios 31:25–30.

Laxalt KA, Kozel TR, 1979. Chemotaxigenesis and activation of the alternative complement pathway by encapsulated and nonencapsulated *Cryptococcus neoformans*. Infect Immun 26:435–440.

Lee W-L, Lloyd KO, 1975. Immunological studies on a yeast peptidogalactomannan. Nature of antigenic determinants reacting with rabbit antisera and those involved in delayed hypersensitivity in guinea pigs. Arch Biochem Biophys 171:624–630.

Lehmann PF, Reiss E, 1980. Detection of *Candida albicans* mannan by immunodiffusion, counterimmunoelectrophoresis and enzyme-linked immunoassay. Mycopathol 70:83–88.

Lehrer RI, Jan RG, 1970. Interaction of *Aspergillus fumigatus* spores with human leukocytes and serum. Infect Immun 1:345–350.

Lehrer RI, Ladra KM, Hake RB, 1975. Nonoxidative fungicidal mechanisms of mammalian granulocytes. Demonstration of components with candidacidal activity in human, rabbit and guinea pig leukocytes. Infect Immun 11:1226–1234.

Lew MA, Siber GR, Donahue DM, Maiorca F, 1982. Enhanced detection with a enzyme-linked immunosorbent assay of *Candida* mannan in antibody-containing serum after heat extraction. J Infect Dis 145:45–56.

Longbottom JL, Pepys J, 1964. Pulmonary aspergillosis: Diagnostic and immunological significance of antigens and C-substance in *Aspergillus fumigatus*. J Pathol Bacteriol 88:141–151.

Longbottom JL, 1978. Immunological aspects of infection and allergy due to *Aspergillus* species. Mykosen Suppl 1:201–217.

Longbottom JL, 1983. Antigens/allergens of *Aspergillus fumigatus*. Identification of antigenic components reacting with both IgG and IgE antibodies of patients with allergic bronchopulmonary aspergillosis. Clin Exp Immunol 53:354–362.

Lurie HI, Still WJS, 1969. The capsule of *Sporotrichum schenckii* and the evolution of the asteroid body. A light and electron microscopic study. Sabouraudia 7:64–70.

Macdonald F, Odds FC, 1983. Virulence for mice of a proteinase-secreting strain of *Candida albicans* and a proteinase-deficient mutant. J Gen Microbiol 129:431–438.

Macher AM, Bennett JE, Gadek JE, Frank MM, 1978. Complement depletion in cryptococcal sepsis. J Immunol 120:1686–1690.

Mangan DF, Lopatin DE, 1981. In vitro stimulation of immunoglobulin production from human peripheral blood lymphocytes by a soluble preparation of *Actinomyces viscosus*. Infect Immun 31:236–244.

Manning M, Mitchell TG, 1980. Analysis of cytoplasmic antigens of the yeast and mycelial phases of *Candida albicans* by two-dimensional electrophoresis. Infect Immun 30:484–495.

Marx JJ Jr, Gray RL, 1982. Comparison of the enzyme-linked immunosorbent assay and double immunodiffusion test for the detection and quantitation of antibodies in farmer's lung disease. J Allergy Clin Immunol 70:109–113.

Meckstroth KL, Reiss E, Keller JW, Kaufman L, 1981. Detection of antibodies and antigenemia in leukemic patients with candidiasis by enzyme-linked immunosorbent assay. J Infect Dis 144:24–32.

Miyazaki T, Yadomae T, Yamada H, Hayashi O, Suzuki I, Ohshima Y, 1980. Immunochemical examination of the polysaccharides of mucorales. pp. 81–94. In Fungal Polysaccharides. Sandford PA, Matsuda K, eds. Washington, DC: American Chemical Society.

Miller GP, Kohl S, 1983. Antibody-dependent leukocyte killing of *Cryptococcus neoformans*. J Immunol 131:1455–1459.

Minocha Y, Pastricha JS, Mohapatra LN, Kandhari KC, 1972. Proteolytic activity of dermatophytes and its role in the pathogenesis of skin lesions. Sabouraudia 10:79–85.

Morelli R, Rosenberg LT, 1971. Role of complement during experimental *Candida* infection in mice. Infect Immun 3:521–523.

Moser SA, Domer JE, 1980. Effects of cyclophosphamide on murine candidiasis. Infect Immun 27:376–386.

Müller HE, Sethi KK, 1972. Proteolytic activity of *Cryptococcus neoformans* against human plasma proteins. Med Microbiol Immunol 158:129–134.

Murphy JW, Cozad GC, 1972. Immunological unresponsiveness induced by cryptococcal capsular polysaccharide assayed by the hemolytic plaque technique. Infect Immun 5:896–901.

Murphy JW, McDaniel DO, 1982. In vitro reactivity of natural killer (NK) cells against *Cryptococcus neoformans*. J Immunol 128:1577–1583.

Musatti CG, Rezkallah MT, Mendes E, Mendes F, 1976. In vivo and in vitro cell mediated immunity in patients with paracoccidioidomycosis. Cell Immunol 24:365–378.

Nakajima T, Ballou CE, 1974. Characterization of the carbohydrate fragments obtained from *Saccharomyces cerevisiae* mannan by alkaline degradation. J Biol Chem 249:7679–7684.

Nicolet J, Bannerman EN, de Haller R, Wanner M, 1977. Farmer's lung: immunological response to a group of extracellular enzymes of *Micropolyspora faeni*. Clin Exp Immunol 27:401–406.

Okubo Y, Honma Y, Suzuki S, 1979. Relationship between phosphate content and serological activities of the mannans of *Candida albicans* strains NIH A-207, NIH B-792 and J-1012. J Bacteriol 137:677–680.

Pearsall NN, Adams BL, Bunni R, 1978. Immunologic responses to *Candida albicans*. III. Effects of passive transfer of lymphoid cells or serum on murine candidiasis. J Immunol 120:1176–1180.

Picardi JL, Kauffman CA, Schwarz J, Phair JP, 1976. Detection of precipitating antibodies to *Histoplasma capsulatum* by counterimmunoelectrophoresis. Amer Rev Respir Dis 114:171–176.

Piccolella E, Lombardi G, Morelli R, 1981a. Generation of suppressor cells in the response of human lymphocytes to a polysaccharide from *Candida albicans*. J Immunol 126:2151–2155.

Piccolella E, Lombardi G, Morelli R, 1981b. Mitogenic response of human peripheral blood lymphocytes to a purified *C albicans* polysaccharide fraction: lack of helper activities is responsible for the in vitro unresponsiveness to a second antigenic challenge. J Immunol 128:2156–2160.

Pier AC, Fichtner RE, 1971. Serologic typing of *Nocardia asteroides* by immunodiffusion. Amer Rev Respir Dis 103:698–707.

Pier AC, Fichtner RE, 1981. Distribution of serotypes of *Nocardia asteroides* from animal, human and environmental sources. J Clin Microbiol 13:548–553.

Pine L, Gross H, Bradley-Malcolm G, George JR, Gray SB, Moss CW, 1977. Procedures for the production and separation of H and M antigens in histoplasmin: chemical and serological properties of the isolated products. Mycopathol 61:131–141.

Poulain D, Tronchin G, Dubremetz JF, Biguet J, 1978. Ultrastructure of the cell wall of *Candida albicans* blastospores: study of its constituitive layers by use of a cytochemical technique revealing polysaccharides. Annal Microbiol (Inst Pasteur) 129:141–153.

Ray TL, Hanson A, Ray LF, Wuepper KD, 1979. Purification of a mannan from *Candida albicans* which activates serum complement. J Invest Derm 73:269–274.

Reinhardt DJ, Hon PJ, Abdelal AT, 1986. Purification and characterization of an alkaline protease from *Rhizopus oryzae*. In preparation.

Reinherz EL, Schlossman SF, 1981. The characterization and function of human immunoregulatory T-lymphocyte subsets. Immunol Today 2:69–75.

Reiss E, Cherniak R, Eby R, Kaufman L, 1984. Enzyme immunoassay detection of IgM to galactoxylomannan of *Cryptococcus neoformans*. Diagn Immunol 2:109–115.

Reiss E, de Repentigny L, Kuykendall RJ, Carter AW, Aloisio CH, 1984. Applications of monoclonal antibodies (MAbs) against *Candida tropicalis* mannan. Abstr Ann Mtg Amer Soc Microbiol no F40:299.

Reiss E, Lehmann PR, 1979. Galactomannan antigenemia in invasive aspergillosis. Infect Immun 25:357–365.

Reiss E, Miller SE, Kaplan W, Kaufman L, 1977. Antigenic, chemical and structural properties of cell walls of *Histoplasma capsulatum* yeast-form chemotypes 1 and 2 after serial enzymatic hydrolysis. Infect Immun 16:690–700.

Reiss E, Mitchell WO, Stone SH, Hasenclever HF, 1974. Cellular immune activity of a galactomannan-protein complex from mycelia of *Histoplasma capsulatum*. Infect Immun 10:802–809.

Reiss E, Stone SH, Hasenclever HF, 1974. Serological and cellular immune activity of peptidoglucomannan fractions of *Candida albicans* cell walls. Infect Immun 9:881–890.

Repine JE, Clawson CC, Rasp FL Jr, Sarosi GA, Hoidal JR, 1978. Defective neutrophil locomotion in human blastomycosis: evidence for a serum inhibitor. Amer Rev Respir Dis 118:325–334.

Restrepo A, Restrepo M, de Restrepo F, Aristizábal LH, Moncada LJ, Vélaz H, 1978. Immune responses in paracoccidioidomycosis. A controlled study in 16 patients before and after treatment. Sabouraudia 16:151–163.

Rhodes JC, Wicker LS, Urba WJ, 1980. Genetic control of susceptibility to *Cryptococcus neoformans* in mice. Infect Immun 29:494–499.

Richardson MD, Warnock DW, 1983. Enzyme-linked immunosorbent assay and its application to the serological diagnosis of fungal infection. Sabouraudia 21:1–14.

Rico G, Ochoa R, Oliva A, Gonzalez-Mendoza A, Walker SM, Ortiz-Ortiz L, 1982. Enhanced resistance to *Nocardia brasiliensis* infection in mice depleted of antigen specific B cells. J Immunol 129:1688–1693.

Roberts RC, Nelles LP, Treuhaft MW, Marx JJ Jr, 1983. Isolation and possible relevance of *Thermoactinomyces candidus* proteinases in farmer's lung disease. Infect Immun 40:553–562.

Rosenberg M, Patterson R, Roberts M, Wang J, 1978. The assessment of immunologic and clinical changes occurring during corticosteroid therapy for allergic bronchopulmonary aspergillosis. Amer J Med 64:599–606.

Rüchel R, 1981. Properties of a purified proteinase from the yeast *Candida albicans*. Biochim Biophys Acta 659:99–113.

Scott EN, Felton FG, Muchmore HG, 1980. Development of an enzyme immunoassay for cryptococcal antibody. Mycopathol 70:55–59.

Scott EN, Muchmore HG, Felton FG, 1980. Comparison of enzyme immunoassay and latex agglutination methods for detection of *Cryptococcus neoformans* antigen. Amer J Clin Pathol 73:790–794.

Segal E, Berg RA, Pizzo PA, Bennett JE, 1979. Detection of *Candida* antigen in sera of patients with candidiasis by an enzyme linked immunosorbent assay inhibition technique. J Clin Microbiol 10:116–118.

Sepulveda R, Longbottom JL, Pepys J, 1979. Enzyme linked immunosorbent assay (ELISA) for IgG and IgE antibodies to protein and polysaccharide antigens of *Aspergillus fumigatus*. Clin Allergy 9:359–371.

Singer A, Hodes RJ, 1983. Mechanisms of T cell–B cell interaction. Ann Rev Immunol 1:211–241.

Siskind GW, Hayama T, Shepherd GM, Schrater AF, Weksler ME, Thorbecke GJ, Goidl EA, 1982. Autoantiidiotype production following antigen injection and immune regulation. Ann N Y Acad Sci 392:345–349.

Sixby JW, Fields BT, Sun CN, Clark RA, Nolan CM, 1979. Interactions between human granulocytes and *Blastomyces dermatitidis*. Infect Immun 23:41–44.

Smith CE, Saito MT, Simons SS, 1956. Patterns of 39,500 serologic tests in coccidioidomycosis. J Amer Med Assn 160:546–552.

Söderhäll K, 1981. Fungal cell wall β-1,3-glucans induce clotting and phenoloxidase attachment to foreign surfaces of crayfish hemocyte lysate. Dev Comp Immunol 5:565–573.

Stevens DA, 1980. Immunology of coccidioidomycosis. pp. 87–95. In Coccidioidomycosis—a Text. Stevens DA, ed. New York: Plenum Medical Book Co.

Strockbine N, Largen MT, Buckley HR, 1983. Monoclonal antibody production of cytoplasmic proteins from *Candida albicans*. Abst Ann Mtg Amer Soc Microbiol no F42:389.

Suzuki M, Fukazawa Y, 1982. Immunochemical characterization of *Candida albicans* cell wall antigens: specific determinant of *Candida albicans* serotype A mannan. Microbiol Immunol 26:387–402.

Suzuki S, Suzuki M, Yokota K, Sunayama H, Sakaguchi O, 1967. On the immunochemical and biochemical studies of fungi. XI. Cross reactions of the polysaccharides of *Aspergillus fumigatus, Candida albicans, Saccharomyces cerevisiae* and *Trichophyton rubrum* against *Candida albicans* and *Saccharomyces cerevisiae* antisera. Jpn J Microbiol 11:269–273.

Suzuki S, Takeda N, 1977. Immunochemical studies on the galactomannans isolated from mycelia and culture broths of three *Hormodendrum* strains. Infect Immun 17:483–490.

Syverson RE, Buckley HR, Gibian JR, 1978. Increasing the predictive value positive of the precipitin test for the diagnosis of deep-seated candidiasis. Amer J Clin Pathol 70:826–831.

Taniguchi N, Usui M, Okuda K, Matuhasi T, 1982. Hapten-pullulan conjugate-induced CMI suppression: demonstration of a common pathway of suppressor cells involving idiotypic interactions. J Immunol 129:1816–1822.

Tewari RP, Sharma DK, Mathur A, 1978. Significance of thymus-derived lymphocytes in immunity elicited by immunization with ribosomes or live yeast cells of *Histoplasma capsulatum*. J Infect Dis 138:605–613.

Tran Van Ky P, Uriel J, Rose F, 1966. Caracterisation de type d'activities enzymatiques dans des extraits antigeniques d'*Aspergillus fumigatus* apres electrophorese et immunoelectrophorese en agarose. Ann Inst Pasteur 111:161–170.

Travassos LR, Lloyd KO, 1980. *Sporothrix schenckii* and related species of *Ceratocystis*. Microbiol Rev 44:683–721.

Treuhaft MW, Roberts RC, Hackbarth C, Emanuel DA, Marx JJ Jr, 1979. Characterization of precipitin response to *Micropolyspora faeni* in farmer's lung disease by quantitative immunoelectrophoresis. Amer Rev Respir Dis 119:571–578.

Tsang VCW, Peralta JM, Simons AR, 1983. Enzyme-linked immunoelectrotransfer blot techniques (EITB) for studying the specificities of antigens and antibodies separated by gel electrophoresis. Meth Enzymol 92:377–391.

Turner SH, Cherniak R, Reiss E, 1984. Fractionation and characterization of galactoxylomannan from *Cryptococcus neoformans*. Carbohydr Res 125:343–349.

Waldmann TA, Broder S, 1982. Polyclonal B-cell activators in the study of the regulation of immunoglobulin synthesis in the human system. Adv Immunol 32:1–65.

Weiner MH, 1980. Antigenemia detected by radioimmunoassay in systemic aspergillosis. Ann Intern Med 92:793–796.

Wheat RW, Woodruff WW III, Haltiwanger RS, 1983. Occurrence of antigenic (species-specific?) partially 3-*O*-methylated heteromannans in cell wall and soluble cellular (nonwall) components of *Coccidioides immitis* mycelia. Infect Immun 41:728–734.

Yarzabal LA, Andrieu S, Bout D, Naquira F, 1976. Isolation of specific antigen with alkaline phosphatase activity from soluble extracts of *Paracoccidioides brasiliensis*. Sabouraudia 14:275–280.

Young KD, Larsh HW, 1982. Production and characterization of a hybridoma-derived antibody to *Blastomyces dermatitidis*. J Clin Microbiol 15:204–207.

Yu RJ, Harmon SR, Grappel SF, Blank F, 1971. Two cell-bound keratinases of *Trichophyton mentagrophytes*. J Invest Dermatol 56:27–32.

Chapter **17**

Review of Cell-Mediated Immunity

17.1 INTRODUCTION

A complete survey of the lymphokines and T-cell effector mechanisms is outside the text's scope, but several of these have been invoked in the text to interpret the immune responses to fungi. Our interpretations are filtered through the models of T-cell mediated immunity that continue to evolve into a complex hierarchy of interactions. An overview of T-cell-mediated immunity can serve as a benchmark to place this specialized work within the context of the dynamic of cellular immunology.

Topics of particular interest to those involved with the mycotic pathogens are: (i) T-cell functions that are effective or critical in the successful host defense; (ii) a basis for understanding anergy at the level of T-cell-mediated suppression; and (iii) factors that affect the maturation of the delayed type hypersensitivity response and granuloma formation.

17.1.1 T-Cell–B-Cell-Macrophage Collaboration

T-helper cells have receptors for the carrier portion of the antigen. When a B-lymphocyte binds at the immunodominant epitope, a bridge is formed by antigen between the two cell types. T-cells recognize antigen when it is presented on the surface of a macrophage (or other Ia-positive antigen presenting cell) in association with an I-region determinant, demonstrating the phenomenon of associative recognition (Kimball, 1983). Thus, T–B-macrophage collaboration is required for the generation of a second signal necessary to evoke immunoglobulin synthesis by the B-cell, in the response to a T-dependent antigen (Figure 17.1). The second signal is a soluble T-helper factor. Macrophages and dendritic cells also provide two kinds of stimuli in the antibody response: (i) the binding and presentation of antigen; (ii) the secretion of low 15,000 da non-antigen-specific polypeptide, interleukin 1 (Mizel, 1982). This factor stimulates maturation of T-helper cells so they can respond

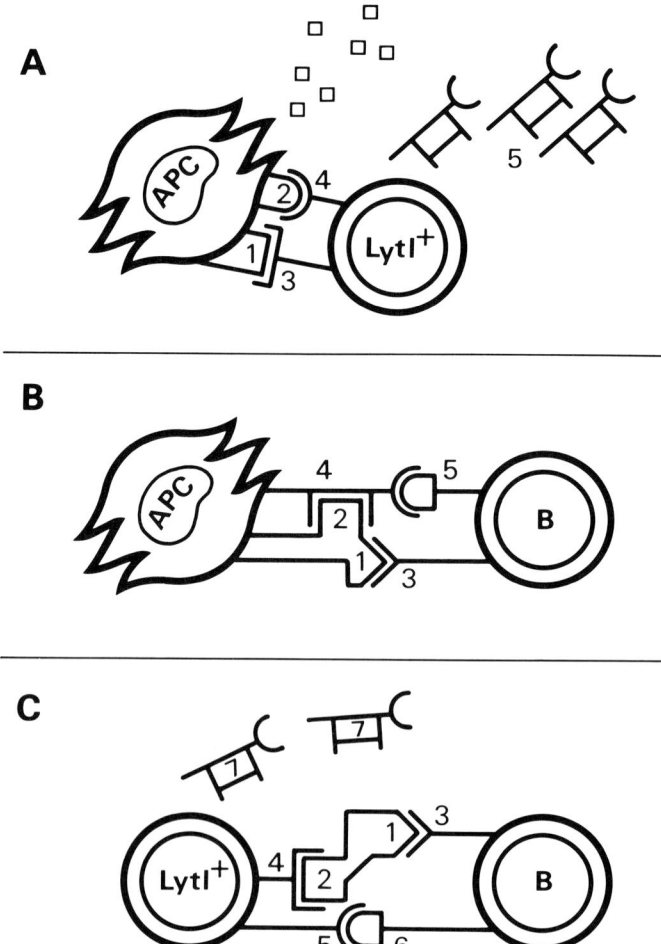

Figure 17.1. Three aspects of the murine T-cell–B-cell–accessory cell cooperation in the antibody response to a T-dependent antigen. *Panel A*. The macrophage, dendritic cell, or other Ia$^+$ antigen presenting cell (APC) has processed and re-expressed a microbial antigen (1) in association with an Ia molecule (2) shown here in the form of dual recognition. The antigen and Ia molecule may in fact be recognized as a complex antigenic determinant. The lyt 1$^+$ T-helper cell has engaged the carrier portion of the antigen with the T-cell receptor (3) and the Ia molecule with a complementary receptor (4). An antigen-specific, soluble T-helper factor (5) is being released into the fluid phase. The small rectangles are the monokine, interleukin 1. *Panel B* applies to B-cells which do not require direct contact with T-helper cells. Such B-cells can receive help that is not restricted by the I region gene products and carry the lyb 5$^+$ surface marker. The epitope (1) of the microbial antigen and the carrier portion (2) have bound to surface immunoglobulin (3) on the B-cell and to the antigen-specific region on the T-helper factor (4), respectively. The helper factor Ia-receptor engages the Ia molecule on the B-cell (5) completing the circuit to deliver the second signal for B-cell activation. *Panel C*. T-cell–B-cell contact is required for some B-cells illustrating the *antigen bridge* model in which the epitope of the microbial antigen (1) binds to surface immunoglobulin (3) while the carrier portion (2) is engaged by the T-cell receptor (4). Associative recognition and H-2 restriction is satisified when the complementary T-cell receptor (5) binds to the Ia molecule (6). Antigen-specific, soluble T-helper factor (7) is generated in the response.

to antigen. They in turn secrete interleukin 2, or T-cell growth factor, a secondary proliferative signal that amplifies immune responses (Palacios, 1982). One means of suppression that has been suggested is the adsorption of interleukin 2 by T-suppressor cells. Further readings in the cellular cooperation in the immune response are in Golub (1981) and Kimball (1983).

17.2 LYMPHOKINES

Lymphokines are nonimmunoglobulin factors produced by activated lymphocytes with biologic effects on cells involved in immune reactions. Most of them are glycoproteins with great potency at low concentrations.

- Macrophage migration inhibitory factor. Produced by T-helper cells, MIF increases the adhesiveness of macrophages, impairs their random mobility, and detains them where T-cell activation is occurring. Macrophages, under the influence of MIF, are activated metabolically and immunologically. Molecular weight measurements indicate heterogeneity of active species in the 12,000 to 70,000 da range. The macrophage receptor for MIF is blocked by L-fucose or L-rhamnose. Macrophage activation factor is immunochemically similar to MIF but its function is measured by the ability to stimulate microbicidal and tumoricidal functions (Oppenheim, 1981).

- Macrophage fusion factor. This factor is found in supernates of lymphocytes cultured with *Mycobacterium bovis* BCG and may promote giant cell formation in granulomas (Clough, 1981).

- Chemotactic factors. Lymphocyte-derived chemotactic factors specifically attract each type of granulocyte, polymorphonuclear neutrophils (PMN), eosinophils, basophils, as well as macrophages. These factors direct movement along a gradient, for example, across the membrane of a Boyden chamber (Oppenheim, 1981).

- Leukocyte inhibitory factor. This factor is distinct from MIF and selectively inhibits the random migration of PMN. The production of LIF is an in vitro correlate of delayed-type hypersensitivity. LIF has a molecular weight of 68,000 and proven esterase activity inhibitable by proteinase inhibitors such as phenylmethyl sulfonyl fluoride and diisopropyl fluorophosphate (Oppenheim, 1981).

- Lymphotoxins. These are glycoproteins produced in vitro by lectin-activated T- and B-lymphocytes and human B-cell lines. This lymphokine is the effector molecule of cytotoxic T-lymphocyte-mediated cytolysis of tumor cells. Evidence for their presence in human inflammatory processes is weak, but lymphotoxins may contribute to necrotic ulceration present in active cell-mediated inflammation (Oppenheim, 1981; Granger et al, 1980). A potent lytic macromolecular lymphotoxin complex of 200,000 da dissociates during purification into α_H, α_L, β and γ subunits in order of decreasing molecular weight and lytic power. A combination of gel permeation, lectin affinity, hydrophobic interaction chromatography, and preparative isoelectric focusing resulted in 60,000-fold purification of the α_L subunit (Klostergaard et al, 1983).

- Interleukin 2. Supernates of lymphocyte cultures stimulated by mitogens or in the mixed lymphocyte reaction contain a mitogenic factor that supports continuous growth and proliferation of T-cells. The production of interleukin 2 (IL-2) by T-helper cells is specifically induced by the macrophage product IL-1. The interaction of Ia molecules (HLA-DR in humans) in association with foreign antigens on antigen-presenting cells with the T-cell receptor results in the expression of receptors for IL-2 (Palacios, 1982). If IL-2 is not present, cloned T-cells will cease to proliferate, revert to small lymphocytes and die in vitro.

Human T-cells that produce IL-2 have been cloned and propagated (Palacios, 1982). Interleukin 2 has been purified by gel permeation, ion exchange, and isoelectric focusing. The molecular weight of human IL-2 is 15,500 da and the isoelectric point is 6.5.

The production of IL-2 may be a natural part of maintenance of homeostasis in the immune system. One proposal suggests that a mechanism for the action of T-suppressor cells is by the adsorption of IL-2. Hydrocortisone treatment prevents IL-2 producer T-cells from responding to IL-1 thus shutting down IL-2 secretion (Palacios, 1982). The fungal immunosuppressant cyclosporin A acts at different points in the T-cell activation process by preventing (i) help of T-4^+ cells to macrophages for IL-1 synthesis, (ii) IL-2 producer T-4^+ cells from responding to IL-1, and (iii) T-cell responsiveness to IL-2. The three effects occur by blockade of the activation receptor on T-cells.

17.3 T-CELL–T-CELL INTERACTIONS

There are several examples of T-cell–T-cell interactions in the immune response. In T-cell-mediated cytotoxicity, T-helper-inducer cells (lyt 1^+) recognize the antigen on the target cell in association with I-region determinants. A factor is secreted that helps T-cytotoxic cells (lyt 2^+ 3^+) to mature and recognize the target cells by their K/D determinants. This example is known as the classical "3 cell" experiment (Golub, 1981). Cooperation between T-cells in the delayed hypersensitivity reaction is also probable although the effector cell in this reaction is lyt 1^+. The secretion of IL-2 by T-helper cells aids the maturation and proliferation of other T-cells (section 17.2). Modulation of the immune response by T-suppressor factors is a different type of T–T interaction, "down-regulation" (Golub, 1981), that may involve idiotype–anti-idiotype reactions (sections 16.2.3, 17.5).

17.4 T-CELL EFFECTOR MECHANISMS

The state of knowledge of the mechanisms responsible for successful host defenses against yeasts and molds are observations that cannot at present be integrated into a general scheme. Furthermore, the mycoses resist being lumped in a single category, which should come as no surprise to students of other classes of microbial pathogens and parasites. Cell-mediated responses are on the whole considered beneficial, whereas the significance of antibody for protection is unproven in the major systemic mycoses. The ability to marshal and activate monocytes into a granulomatous response correlates with protection, subject to the clarification that damage can occur if the target organism is not destroyed, leading to calcifications and fibrosis. Granulomatous responses are considered protective, and the induction of MIF, leukocyte inhibitory factor, and leukocyte chemotactic factors are important preconditions for such responses.

There is no evidence yet for T-cell-mediated cytotoxicity against the fungi, an important proven protective response against intracellular pathogens such as viruses. Fungi have rigid cell walls that confer resistance against osmotic shock, making direct T-cell-mediated cytotoxicity less likely. The findings that mice with high levels of NK cells can inhibit cryptococci should prompt reinvestigation of other mechanisms of antibody-mediated and direct T-cell-mediated cytotoxicity against fungi.

17.5 GENERATION OF T-SUPPRESSOR CELLS AND SOLUBLE SUPPRESSOR SUBSTANCES

Approximately 30% to 40% of the human peripheral blood T-lymphocytes bind to monoclonal OKT-8 (T8)(Goldstein et al, 1982). This T-cell set has the potential to become either cytotoxic or suppressor T-cells. Studies of human T-cells show they bear receptors for μ, γ, Fc-α fragments. Generally, Fc-γ receptors are assumed to reside on T-suppressor cells, whereas T-helper cells have Fc-μ receptors. In the mouse, alloantisera have defined a corresponding set of lymphocytes, lyt2^+, 3^+ that further assort into Qa1^+ suppressor-effector cells and Qa1^- cytotoxic T-cells (Clough, 1981).

Suppressor cells prevent inappropriate responses (afferent suppression) and terminate appropriate responses (efferent suppression). Suppression functions not only against antibody production, but regulates mitogen and antigen-induced T-cell proliferation, cell-mediated cytotoxicity, and delayed hypersensitivity. Two categories of suppression are known: antigen-specific and general suppression, according to the means of induction and nature of the soluble factor that is produced.

17.5.1 Antigen-Specific Suppression

T-suppressor cells are induced by either anti-idiotype antibodies or when an antigen is presented in such a way that macrophage processing is evaded favoring suppression over immunity. Although lyt 1^+ cells are usually destined to become T-helper cells there is a subset of these cells that is lyt 1^+, I-J$^+$, and Qa-1^+. They function not as helper cells but as antigen-specific T-suppressor-inducer cells and recruit suppressor cells from the large pool of resting lyt $1,2,3^+$ cells (Figure 17.2). Ts1 cells are activated, bearing the appropriate antigenic (idiotype) specificity. They are the first order suppressor cells. The function of these cells is to secrete a factor, TsF, that induces the effectors of suppression—second order suppressor cells, Ts2. The soluble suppressor factor, TsF, consists of 2 chains. One chain contains the I-J determinant, and the other chain has idiotype specificity. Since the antigen-binding specificity has no region with homology for the heavy chain constant region (C_H), it is not antibody. The actual effectors of suppression, Ts2, like the soluble factors they produce, do not bind antigen but instead bear anti-idiotype determinants. Ultimate down-regulation of the immune response mediated by second order T-suppressor cells acts upon T-helper cells by direct binding to the complementary idiotype (Kimball, 1983).

A carrier-specific T-suppressor factor against keyhole limpet hemocyanin has been characterized from cloned T-cell hybridoma cell lines. It consists of two polypeptides, one carrying I-J determinants and the other having V_H and idiotype structure (Cathcart, 1981).

Examining the mechanism of cell-mediated suppression of delayed hypersensitivity is of interest because of the propensity of fungi to evoke this type of reaction in skin tests and in the lung. Many of the experiments performed have been conducted with contact sensitins which do not provide an exact parallel to the fungal antigens used to evoke delayed hypersensitivity. Three cases are instructive of the soluble mediators of suppression (Cathcart, 1981): (i) supernates of lymphocyte cultures obtained from mice primed with high sensitin (picryl chloride) doses prevented the adoptive transfer of sensitization with cells to normal recipients; (ii) a suppressor factor specific for the hapten in association with K or D products on the T-helper cell; (iii) p-azobenzenearsonate conjugated to syngeneic spleen cells was used to induce hapten-specific tolerance. A soluble factor was identified that passively transferred tolerance to the hapten in footpad challenge tests. The suppressor factor was bound to an anti-idiotype column and also contained I-J determinants. The suppressor factor acts by inducing a second suppressor population that displays the anti-idio-

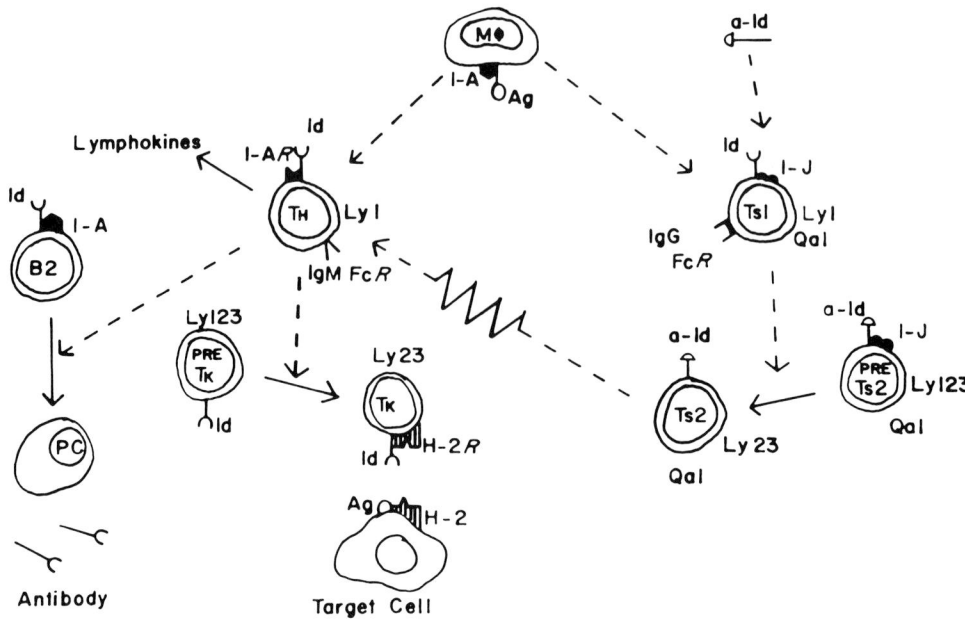

Figure 17.2. T-Cell immunoregulatory circuits in the mouse responding to a T-dependent antigen shown above as the "target cell". Macrophages, or other Ia$^+$ accessory cells, process and express the antigen on their surfaces in association with Ia molecules. The T-helper cell receptor for Ia, IaR, and the receptor for the carrier portion of the antigen are engaged, resulting in a positive stimulus releasing lymphokines and T-helper factors. Lyt 1$^+$ and lyt 1,2,3$^+$ interactions result in the maturation of cytotoxic T_K cells. These cells are genetically restricted to react with the antigen and the H-2 K/D gene products. B-cells (B2) stimulated by the antigen also interact with T-helper cells (lyt 1$^+$) and this reaction is restricted by I-region molecules. Stimulated B-cells that receive a second helper signal proliferate and differentiate into antibody-forming plasma cells (PC). Down-regulation of the immune response is triggered by one of three types of stimuli acting upon lyt 1$^+$, Qa 1$^+$ suppressor-inducer cells (Ts 1). The stimuli are the antigen, the antibody, or anti-idiotype immunoglobulin (anti-Id). Once activated, Ts 1 cells react with anti-Id on pre-Ts 2 cells. This is the V_H restriction caused by the structural homology of the T-cell receptor to the variable portion of the immunoglobulin heavy chain. The pre-Ts 2 cell can differentiate into a fully functional Ts 2 cell that secretes soluble suppressor factors, or in the adult mouse, pre-Ts 2 and Ts 2 can coexist; the pre-Ts 2 influencing Ts 2 by an anti-Id–anti-anti-Id interaction (diagram from Clough, 1981). *Source*: reprinted from Suppressor Cells and Their Factors with permission from CRC Press Inc., Copyright 1981.

type. Thus, antigen induces Ts with idiotype determinants to produce a suppressor factor with idiotype and I-region determinants. This in turn evokes a population of Ts with anti-idiotype determinants—a feedback regulatory mechanism.

17.5.2 Modulation of Suppression by Immune Complexes

The association of antigen with antibody sometimes inhibits immune responses. B-cells have complement receptors and T-cells have Fc receptors so the potential for interaction with immune complexes exists, especially because Fc-γ receptors formed in antigen excess lead to enhanced T-helper functions whereas those found in antibody excess trigger suppression (Clough, 1981).

17.5.3 General Suppression

Concanavalin A activates suppressor T-cells to produce a factor that mediates non-specific suppression of in vitro antibody formation in response to various antigens:

for example, suppression of splenic immunoplaque forming cell responses to sheep erythrocytes. The factor, soluble immune response suppressor substance (SIRS), acts on the inductive phase of the antibody response, inhibiting IgM and IgG against T-dependent and T-independent antigens. SIRS is not genetically restricted by the major histocompatibility complex. It does not suppress the generation of cytotoxic lymphocytes or the proliferative responses to lectins and allogeneic cells (Cathcart, 1981).

Histamine-induced suppressor factor (HSF) suppresses antigen-induced proliferation and MIF production by sensitized lymphocytes. It is produced by T-cells with H2 receptors and its synthesis is blocked by H2 receptor antagonists (Cathcart, 1981).

Lymphotoxin is found in supernates of in vitro cultured immune rodent and human lymphocytes exposed to the specific antigen. The source of the factor is considered to be T-cells. This type of suppression is the most drastic since the potential responding cell is lysed.

17.5.4 Suppression by Anti-Idiotype Antibodies

Jerne's network theory predicted the existence of anti-idiotype antibodies that would contribute to the homeostatic control of the immune system by specifically down-regulating antibody production when it reached a high level (Jerne, 1974). This has been demonstrated not only in the instance of humoral immunity but in the suppression of delayed-type hypersensitivity. One of the classic examples showing how anti-idiotype antibodies can suppress antibody synthesis is the case of anti-phosphorylcholine immunoglobulin in mice injected with the C-polysaccharide from pneumococcal cell walls.

This antigen induces a response of restricted clonality. Frequently the idiotype is the same as that from the mouse myeloma protein TEPC-15. Antibodies against TEPC-15 can suppress antiphosphorylcholine immunoplaque forming responses in mice and the injection of anti-TEPC blocks the induction of this idiotype in mice injected with pneumococcal cell walls (Kimball, 1983). Homeostatic control is exercised when the production of the idiotype reaches a high level, thus triggering the anti-idiotype that can shut off the production of antibodies either by direct receptor blockade or indirectly through the activation of T-suppressor cells.

17.5.5 Contrasuppression

The hierarchy of immune responses appears to also include the phenomenon of contrasuppression. This type of modulation is differentiated from T-cell help in an operational way. T-contrasuppressor cells (Tcs) in mice are lyt 1^+, I-J$^+$ cells whose induction depends on the activation of a lyt 2^+ T-suppressor population. The contrasuppressor effector cell acts on T-helper cells, making them resistant to suppressor T-cell signals. The physiologic role of contrasuppression is (i) to activate an immune response after suppression has occurred, (ii) to localize an immune response, while maintaining systemic suppression, and (iii) to establish a suppression-resistant state such as that found in hyperimmune animals (Green et al, 1983a). We can postulate that the anergy that accompanies the acute disseminated state of systemic mycoses is mediated by T-suppression and that in the recovery from anergy the contrasuppressor circuit should become operational.

17.6 CELLULAR BASIS OF DELAYED HYPERSENSITIVITY

Delayed hypersensitivity occurs when sensitized T-cells (lyt 1^+ in the mouse) secrete lymphokines that attract monocytes and PMN causing them to migrate into the extravascular space. An early component of this reaction has been described in which sensitized T-cells release an antigen-specific factor, that is not antibody, that binds to tissue mast cells (Askenase and van Loveren, 1983). Upon encountering antigen, as in a skin test, the mast cells release serotonin, but do not undergo degranulation or histamine release. Serotonin causes local endothelial cells to contract leaving gaps that facilitate the migration of monocytes and PMN. This is the delayed phase of the reaction which peaks at 24 hours. A second T-cell population passes through the endothelial gaps and reacts with antigen presented on the surface of Ia-positive Langerhans cells (Figure 17.3).

Figure 17.3. Proposed cascade of steps, numbers 1–12, involved in eliciting delayed-type hypersensitivity. The antigen stimulates production of antigen-binding T-cell factor (TCF, 1) which is carried in the blood and sensitizes tissue mast cells (2,3). When challenged with the antigen, sensitized mast cells react and serotonin is released (4) but there is no histamine release or degranulation (5,6). Serotonin reacts with its receptors on endothelial cells of postcapillary venules (7). This reaction causes gaps between epithelial cells (8) through which other antigen-specific T-cells can pass into the reaction site (9). Once on the scene these T-cells interact with antigen in association with Ia antigens on Langerhans cells (10). This step accounts for the H-2 restriction in delayed-type hypersensitivity. Lymphokines are produced (11) and these attract leukocytes to complete the maturation of the cellular infiltrate (12) (Askenase and van Loveren, 1983). *Source*: reprinted from Immunology Today with permission of the authors and from Elsevier Publications (Cambridge).

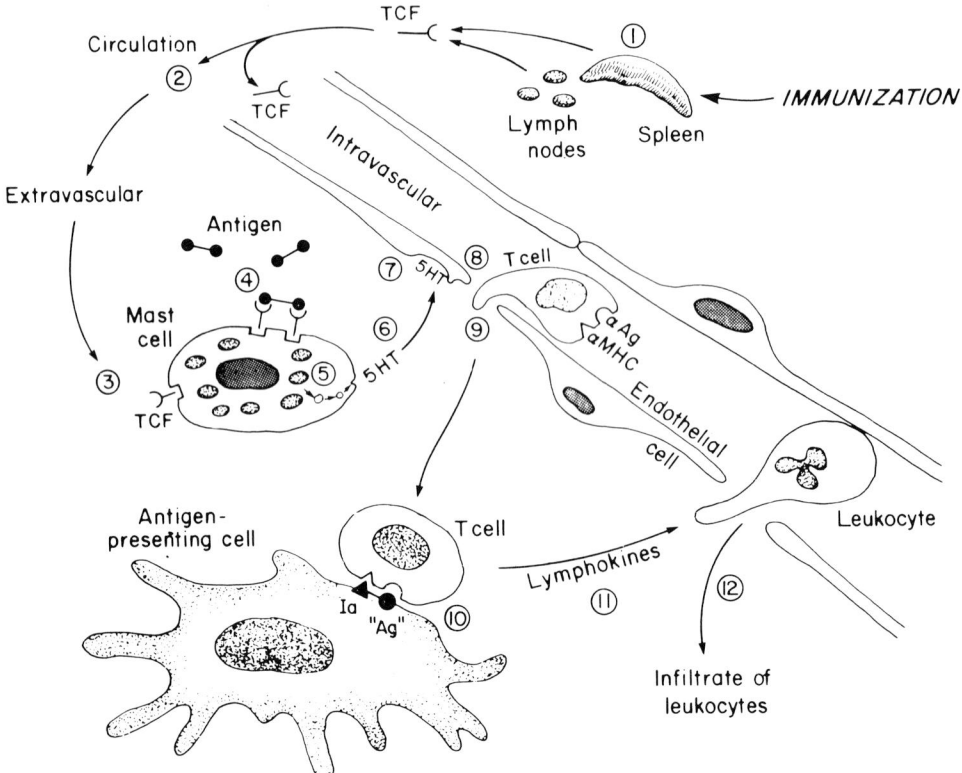

17.6.1 Organization of the Hypersensitivity Granuloma and the Associated Immunopathology

At the center, surrounding the infecting organisms is the "epithelioid" cell region containing large, round or oval macrophages with much interdigitation between cells, moderate endoplasmic reticulum, and an increase in lysosomes. Macrophages at the center of granulomas have the most active hydrolase content. In the next layer are macrophages with a high metabolic potential to phagocytose and kill microorganisms. T-Lymphocytes at the periphery activate blood-borne monocytes and maintain the cellular flux (McGee et al, 1980). The production by T-cells of lymphotoxin is implicated in the necrosis of granulomas.

The nature of the cellular infiltrate in granulomas is exemplified by the cutaneous lesions of leprosy (van Voorhis et al, 1982). In tuberculoid leprosy, macrophages exist as giant cells or epithelioid cells and only remnants of *Mycobacterium leprae* are found in the macrophage phagosomes. The predominant lymphoid cell types are T4(+) Leu 3a(+) T-helper cells clustered in well-organized cuffs surrounding the macrophages in the granulomas. In contrast, lesions of lepromatous leprosy contain large numbers of acid-fast bacilli within macrophages in large, dispersed, and poorly organized infiltrates. The T-cells in the lesions were exclusively T8(+), Leu 2a(+) suppressor cells.

The character of the T-cell infiltrate in granulomas portends either help or suppression of cell-mediated immunity. Extending these observations to the granulomas of histoplasmosis might prove profitable, especially if the granulomas from patients whose disease is under control or remission were compared with those of patients suffering from chronic pulmonary histoplasmosis. Such studies could provide benchmarks for possible ways to manipulate the immune system so as to influence the level of T-helper versus T-suppressor cells.

17.6.2 Neutral Proteinases and Lysosomal Hydrolases

Damage to the offending microbe and, coincidentally, against the host tissues is the product of proteinases and lysosomal hydrolases. β-Cathepsin, plasminogen activator, collagenase, and elastase are proteinases that are active at neutral pH and are released only during inflammation and endocytosis of antigenic material (McGee et al, 1980; Gordon, 1980). Plasminogen activator is a trypsinlike serine proteinase which is studied by its action on ^{125}I-fibrin. Lysosomal hydrolases are active at acidic pH and degrade a variety of proteins, carbohydrates, and lipids (eg, lysozyme, acid phosphatase, β-glucuronidase, hyaluronidase).

17.6.3 Fibrosis

Chronic granulomatous inflammation, of the kind associated with the persistence of tissue forms of fungi in pulmonary lesions, may resolve and the lesion may disappear. Otherwise, fibrosis with permanent damage of the involved organ may occur. Collagen is the major constituent of fibrous tissue. But collagenase is produced by activated macrophages in granulomas, so that collagen deposition is the result of its increased synthesis, brought on by factors secreted by macrophages and lymphocytes that stimulate the proliferation of fibroblasts (Dunn, 1980). Even the target microbe itself has the potential to stimulate fibrosis. In many immunologic granulomas, active collagen-secreting fibroblasts are a regular feature (Turk, 1980). Stimulation of fibrosis may also be mediated by a lymphokine, fibroblast activation factor, produced by T-cells (Wahl et al, 1983).

17.7 ANERGY OF INFECTION

Cutaneous anergy is commonplace in the active phase of mycotic infections. This anergy is best considered in the context of *either* disseminated or localized infections because disseminated infections often result in parasitism of the reticuloendothelial system and of the paracortical regions of lymph nodes (histoplasmosis, coccidioidomycosis, and paracoccidioidomycosis). This granulomatous pathology causes displacement of T-lymphocytes and disturbs lymphocyte traffic (Bullock, 1979).

17.7.1 Primary Systemic Mycoses

In the major systemic fungal infections, blastomycosis (Witorsch and Utz, 1968), coccidioidomycosis (Catanzaro et al, 1975; Cox and Vivas, 1977), histoplasmosis (Artz et al, 1980), and paracoccidioidomycosis (Musatti et al, 1976) extrapulmonary dissemination results in a higher incidence of anergy. As the extent of dissemination increases, a more profound anergy occurs, extending to antigen-specific MIF unresponsiveness (Catanzaro et al, 1975), broad anergy to common recall antigens and to phytomitogens (Cox and Vivas, 1977; Catanzaro et al, 1975; Artz et al, 1980; Restrepo et al, 1978), failure to respond to contact sensitins (Rea et al, 1979; Mok and Greer, 1977; Musatti et al, 1976) lower mixed lymphocyte reactions (Cox and Vivas, 1977) and reduced number of peripheral T-cells (Artz et al, 1980; Musatti et al, 1976).

Blastogenic responses to specific fungal antigens are uncoupled from delayed cutaneous hypersensitivity in disseminated infection so that positive blastogenesis can be detected in cutaneously anergic patients in coccidioidomycosis (Catanzaro et al, 1975) and upon favorable response to therapy, positive blastogenic indices return before delayed cutaneous hypersensitivity in coccidioidomycosis (Cox and Vivas, 1977; Catanzaro, 1977) and paracoccidioidomycosis (Musatti et al, 1976). Typically, blastogenic indices slowly return during convalescence to values intermediate between those of healthy skin test positive controls and persons with active infections.

If persons receiving immunosuppressive therapy or those with primary diseases that compromise their immune systems are excluded, the anergy of infection in the mycoses is a reversible, specific defect that begins at a time when the antigenic load is near its peak. Three possible mechanisms to explain the anergy of infection are: (i) disturbance of lymphocyte traffic, mentioned above; (ii) the production of humoral substances that suppress lymphocyte blastogenesis; (iii) a shift in immunoregulation towards production of T-suppressor cells. Soluble suppressor substances could be the product of T-suppressor cells or may be soluble immune complexes. The evidence for plasma factors that inhibit blastogenesis of lymphocytes in the primary systemic mycoses was reviewed in the discussion of humoral immunity to coccidioidomycosis (Harvey and Stevens, 1981), histoplasmosis (Cox, 1979) and paracoccidioidomycosis (Musatti et al, 1976). Antigen-specific T-suppressor cells have been characterized in human histoplasmosis (Couch et al, 1978; Stobo et al, 1976).

Soluble immune complexes can inhibit antigen-specific lymphocyte proliferation because they are good inducers of anti-idiotype antibodies. Cells bearing idiotypic determinants include regulatory T-cells (Casali et al, 1979). The IgG$_2$ subclass of antibody suppresses specific immune responses (Eichmann and Simon, 1980). When mice are immunized with the IgG$_2$ subclass of antibody against *Streptococcus* group A carbohydrate, idiotype-specific suppressor cells are induced that prevent the formation of T-helper-effector and memory cells for the group A carbohydrate.

Whole killed cryptococci stimulate blastogenesis in normal persons, but this reactivity is not usually expressed as delayed cutaneous hypersensitivity when cryptococcin is used (Graybill and Alford, 1974). These skin tests are positive, however, in a majority of mycology laboratory workers (Schimpff and Bennett, 1975). This disparity could be the result of testing with an antigenic complex that lacks appropriate specificity. The apparent dissociation of recognition and effector functions towards *C neoformans* antigens in normal persons deserves further investigation. Cutaneous anergy occurs in cryptococcal meningitis even when there is no underlying primary immunosuppressive condition (Graybill and Alford, 1974). Where no generalized anergy was present, half of active cryptococcosis patients tested had positive delayed skin tests with the 1/100 dilution of cryptococcin (Atkinson and Bennett, 1967) and this figure increased to 59% in treated patients whose disease was in remission.

17.7.2 Chronic Dermatophyte and *Candida* Species Infections

The anergy of infection in the mycoses localized in the skin, chronic dermatophyte and candidal infections, is a special case. The immunological unresponsiveness of patients with these infections has been the subject of several studies, most of them at the phenomenological level. Chronic *Trichophyton rubrum* infections are associated with anergy in skin tests for delayed hypersensitivity and in the in vitro correlates, blastogenesis and MIF production. Several explanations that have been offered to account for this failure include—the presence of a preexisting immunodeficit, or immunosuppressive therapy (Hay, 1979; Brahmi, 1980); serum suppressor substances (Green and Balish, 1979); *T rubrum* as an inducer of IgE and consequent suppression of delayed cutaneous hypersensitivity by the immediate reaction (Hanifin et al, 1974; Jones et al, 1974); the superficial layers of the skin as a site not favorable for systemic antigenic stimulation (Kaaman, 1979); evasion of immune surveillance by *T rubrum* via its hypoproduction of keratinase or other proteinases; and a low ratio of T-helper to T-suppressor cells (Petrini and Kaaman, 1981).

The most common immunological finding in chronic mucocutaneous candidiasis, independent of the clinicogenetic classification of this disease, is cutaneous anergy to *C albicans*. Anergy is a reversible, specific defect that is probably secondary to the infection, because with a favorable response to therapy and a decreased antigenic load, conversion to positive delayed cutaneous hypersensitivity returns. Anergy frequently extends to an inability of peripheral blood leukocytes to produce MIF when stimulated with *Candida* antigens. In healthy adults delayed cutaneous hypersensitivity to *C albicans* correlates more closely with the blastogenic index, than with the leukocyte migration index (Bice et al, 1974; Ferguson et al, 1977). In some chronic mucocutaneous candidiasis patients, however, (groups I–III) significant antigen-induced blastogenesis occurs when delayed cutaneous hypersensitivity and MIF reactions are flat. One mechanism of suppression of blastogenic responses is the effect of autologous serum containing mannan or mannan-antibody complexes (Fischer et al, 1982). Those authors suggest that defective metabolism of mannan by host macrophages is a key event in the development of chronic mucocutaneous candidiasis.

Others have pointed out that in genetic nonresponder strains of mice, the nonresponders lack a necessary macrophage function which normally allows antigen presentation to precursors of T-helper cells, but instead permits a preferential activation of T-suppressor cells (Sinclair, 1980).

Women who experience recurrent vaginitis were studied for immunocompetence (Witkin et al, 1983). Suppressor T-cells and soluble antigen-specific suppressor factors from them could inhibit the blastogenesis of lymphocytes of normal women to C albicans.

17.8 PROTECTIVE FUNCTIONS OF CELL-MEDIATED IMMUNITY

17.8.1 Experimental Mycoses

Protection against *B dermatitidis* is adoptively transferred with T-cells but not with serum (Brummer et al, 1982). Immune T-cells from mice act primarily through normal or activated macrophages. The endospores of *C immitis* are rapidly phagocytosed by macrophages but are not killed in the absence of T-cells (Beaman et al, 1981). Endospores and arthroconidia of *C immitis* inhibit the fusion of phagosomes and lysosomes of macrophages in mice, a blockage that is overcome by the presence of lyt 1^+, 2^- lymphocytes from immune donors. These lymphocytes activate macrophages and stimulate phago-lysosome fusion (Beaman et al, 1981). Athymic nu/nu mice are very susceptible to *H capsulatum* and can be protected by adoptive transfer with T-lymphocytes from immune donors (Williams et al, 1981). Ribosomal vaccines derived from yeast forms confer protection in adoptive transfer experiments against *H capsulatum* via T-lymphocytes (Tewari et al, 1978). Resistance to *S schenckii*, like the primary systemic fungi seems to require intact thymic function because no resistance is encountered in athymic nu/nu mice (Shiriashi et al, 1979).

Successful adoptive transfer of immunity with lymphocytes from immune donors protected mice from a systemic challenge with live *C albicans*. Neonatally thymectomized, X-irradiated, and bone marrow reconstituted mice are equally susceptible to cutaneous challenge in the primary infection, but upon a secondary systemic challenge these "B mice" are more susceptible. Increased susceptibility correlates with lack of T-lymphocytes and the inability to respond to a candidal immunostimulant "HEX" (candidal membrane-bound proteins) reflecting poor ability to produce antibodies to T-dependent antigens.

Innate resistance to *C albicans* is important in the early stages of primary infection. The high level of natural killer lymphocytes and chronically activated macrophages of athymic nu/nu mice probably explain their reduced renal pathology in the early days after infection, compared with euthymic littermates (Cheers and Waller, 1975; Herberman, 1978). The early resistance of nu/nu mice gives way to wasting and death (Miyake et al, 1977). Immune T-lymphocytes from previously infected mice successfully transfer immunity against a lethal challenge of *Candida* in normal mice (Miyake et al, 1977) measured as reduced mortality and reduced kidney counts.

Other evidence of a role for T-cell-mediated immunity to *Candida* has been obtained. The rapidity of epidermal scaling of *C albicans* in skin lesions of guinea pigs is a function of cell mediated immunity and can be adoptively transferred with peritoneal exudate cells from immune donors (Sohnle, 1976a; 1976b). This mechanism may function at the level of an as yet uncharacterized keratinization-stimulating lymphokine.

Cryptococcal infection in athymic nu/nu mice at first progress more slowly than in euthymic heterozygotes but by the time that the heterozygotes developed delayed type hypersensitivity, their infection began to plateau, whereas the infection in nu/nu mice disseminated in spite of the presence of IgM agglutinins. Adoptive transfer of T-lymphocytes from immune donors prolonged the survival of recipients after challenge with live cryptococci.

Cooperation between humoral and T-cell-mediated immunity may help to explain the reason why defects in cellular immunity predispose individuals to cryptococcal infection. Murine macrophages take up soluble immune complexes containing cryptococcal polysaccharide (Griffin, 1981) that block the membrane Fc receptors and inhibit phagocytosis of opsonized cryptococci. Macrophage C3b receptors can be activated by a lymphokine, which in turn is stimulated by immune complexes. This lymphokine enables macrophages that have ingested immune complexes to phagocytose C3b-coated cryptococci, thus providing a potentially critical link between the humoral and cell mediated limbs of the immune response to *C neoformans* (see Table 17.1).

- Actinomycetes. Athymic nu/nu mice are hypersusceptible to *N asteroides* infection, but immunity can be transferred to them with *N asteroides* primed T-cells (Deem et al, 1982). T-lymphocytes from immune mice killed *N asteroides* directly in the absence of macrophages (Deem et al, 1983). The role of T-cell mediated immunity against intracellular pathogens is thought to proceed by activation of macrophages, but the discovery of this direct T-cell mediated cytotoxicity offers a potent alternative mechanism especially useful against pathogens like virulent strains of *N asteroides* that escape after ingestion by macrophages.

The B-lymphocyte mitogenic properties of *A viscosus* extracellular heteroglycan and of a 20,000 × g supernate of homogenized bacteria are T-independent in rodents and the assessment of T-cell mediated immunity in experimental periodontal disease of rodents is not yet developed. Murine models of T-cell function in hypersensitivity pneumonitis resulting from inhalation of spores of thermophilic actinomycetes are likewise an area of potential interest. Rabbits have been used as a model to demonstrate the cell-mediated nature of immunopathology in hypersensitivity pneumonitis caused by thermophilic actinomycetes. Inhalation of particulate *Faenia rectivirugula* spores results in granulomatous pulmonary inflammation. The cell-mediated aspect of the reaction is demonstrated by the presence of MIF in bronchoalveolar lavage fluids and adoptive transfer of the lesions to normal recipients with lymphocytes (Fink and Moore, 1980).

17.8.2 Human Mycoses

- *Blastomyces dermatitidis*. Experience with a large series of blastomycosis patients was reviewed to determine the association of immunodeficits with this disease (Recht et al, 1982). Only six patients were identified as significantly immunocompromised. Three had hematologic malignancies but these were not among the categories that specifically impair T-cell responses (chronic myelogenous leukemia, chronic lymphocytic leukemia, multiple myeloma). The other three patients received systemic corticosteroid therapy for asthma or chronic interstitial lung disease. In two of these patients blastomycotic pulmonary infiltrates were, in retrospect, considered to antedate steroid therapy. The course of these infections was chronic, not fulminant as often occurs in coccidioidomycosis or blastomycosis in the im-

Table 17.1. Assessments of the Athymic State on the Host Response to Experimental Mycotic Infection in Rodents

Organism	Athymic nu/nu Rodents	Other Effects of Thymic Immunity
Blastomyces dermatitidis	nd	Macrophage activation (McDaniel and Cozad, 1983)
Coccidioides immitis	Susceptible[a] (Beaman et al, 1977)	Activated macrophages kill endospores (Beaman et al, 1981)
Histoplasma capsulatum	Susceptible (Williams et al, 1981)	Adoptive transfer of immunity with T cells (Tewari et al, 1978) Lymphokine stimulation of macrophages (Wu-Hsieh et al, 1984)
Sporothrix schenckii	Susceptible[a] (Shiriashi et al, 1979)	
Dermatophytes	Chronic infection does not clear, Rnu rats (Green et al, 1983b)	
Rhizopus oryzae	More extensive lesions (Corbel and Eades, 1977)	
Candida albicans	Less initial pathology (Cutler, 1976) Thigh lesions resolve faster Tabeta et al, 1984 T-X-BM mice have no anamnestic response[b] (Giger et al, 1978)	Adoptive transfer of immunity with T cells (Miyake et al, 1977)
Cryptococcus neoformans	Initially slower multiplication in reticuloendothelial system. At day +14 nu/+ begin to clear infection, nu/nu disseminate despite IgM agglutinins (Cauley and Murphy, 1979)	Adoptive transfer of immunity with T cells (Graybill and Mitchell, 1979)

[a] More susceptible than euthymic heterozygotes.
[b] Thymectomized-X-irradiated and bone marrow reconstituted mice.
nd, not determined.

munocompromised host. Obvious alterations of cell-mediated immunity have not been identified as risk factors in human blastomycosis. Alternatively, the reservoir of persons with primary blastomycosis is small relative to the other primary systemic mycoses so that the statistical chances for endogenous reactivation are low.

Cutaneous hypersensitivity and in vitro lymphocyte blastogenic and MIF responses to common microbial antigens and to *B dermatitidis* were normal in two patients with relapsing pulmonary blastomycosis (Sohnle et al, 1980). The defect in handling *B dermatitidis* may therefore lie in the inability of macrophages to become activated by lymphokines. Evaluation of this type of function in an animal model would be a helpful step. Defective neutrophil chemotaxis in blastomycosis (Repine et al, 1978) may also contribute to an inadequate host defense.

• *Coccidioides immitis.* Several host factors have been implicated to account for the observed 1% to 5% of persons who experience moderate to severe infection after exposure to *C immitis.* Among these factors are: Black or Filipino ancestry (10 times and 175 times greater frequency of dissemination, respectively) (Deresinski et al, 1979); association with human leukocyte antigen, HLA-A9 (section 4.7); the third trimester of pregnancy, except in women with a history of arrested coccidioidomycosis (the latter category is more immune) (Drutz and Huppert, 1983); diabetes mellitus (Sarosi et al, 1970) and patients with Hodgkin's disease or other lymphoma. Age has also been implicated as a risk factor because children under five years old and adults over 50 years old have a greater incidence of dissemination (Drutz and Catanzaro, 1978). Endogenous reactivation of quiescent lesions containing *C immitis* can occur in association with the use of systemic corticosteroids or other immunosuppressive agents (Drutz and Catanzaro, 1978; Deresinski and Stevens, 1974).

The acquisition of delayed cutaneous hypersensitivity to coccidioidin as a result of residence in the endemic zone also seems to protect against dissemination. Clinical improvements obtained with transfer factor support a role for T-cell-mediated immunity in overcoming infection. T-cell-mediated immunity appears to be a common thread running through the diverse factors that contribute to host defense against *C immitis*. This view is given further credence by the adoptive transfer experiments in mice (section 4.6.2) and by the anergy that results from dissemination in humans (section 4.4). Much remains to be done in order to delineate the effector functions in humans that are critical for destroying the pulmonary focus of infection and preventing dissemination.

- *Histoplasma capsulatum*. A hypothetical composite of immune responses that have been implicated in the host response during histoplasmosis is shown in Figure 17.4. Among the "experiments of nature" that shed light on which limbs of the immune response contribute to protection in histoplasmosis are the large number of case reports in which serious histoplasmosis was associated with Hodgkin's disease, prompting Schwarz (1981) to conclude that this form of lymphoproliferative disease is one of the most frequent underlying diseases involved in disseminated histoplasmosis in the immunocompromised host. Even without exogenous immunosuppressive therapy, disseminated histoplasmosis has occurred in patients with advanced Hodgkin's disease (Kauffman et al, 1978). Hodgkin's lymphoma profoundly depresses delayed cutaneous hypersensitivity including reactions to common microbial antigens, phytomitogens, and contact sensitins (Wing and Remington, 1980). These defects are manifest even in the early stages of the disease. Hodgkin's disease induces a disproportionate excess of suppressor T-cells (Wells and Ries, 1980). The malignant cell type in Hodgkin's lymphoma is presumed to be of macrophage-monocyte lineage.

There is insufficient evidence to know if *chronic pulmonary* histoplasmosis results from endogenous reinfection in the face of a defect in cellular immunity or is related more to a structural defect in the lung such as emphysema.

If, as postulated, cell-mediated immunity is paramount in restricting the growth of *H capsulatum* yeast forms in the lung then the calcification granuloma responses in histoplasmosis represent a failure to eradicate and absorb remnants of the yeast forms. In the case of the deposition of eggs in the liver in schistosomiasis the granulomas that ensue are an exaggerated immune response that contributes to immunopathology (Colley, 1981). We are at a primitive lack of understanding of (i) what cellular constituents of *H capsulatum* are responsible for lysis resistance (section 5.2), (ii) whether the yeast forms secrete factors or induce lymphocytes to secrete factors that promote collagen synthesis by fibroblasts and calcification (Turk, 1980), (iii) what immunomodulation occurs as a result of the long-term exposure to soluble antigens that seep out of encircled, dormant but probably viable yeast forms. Pulmo-

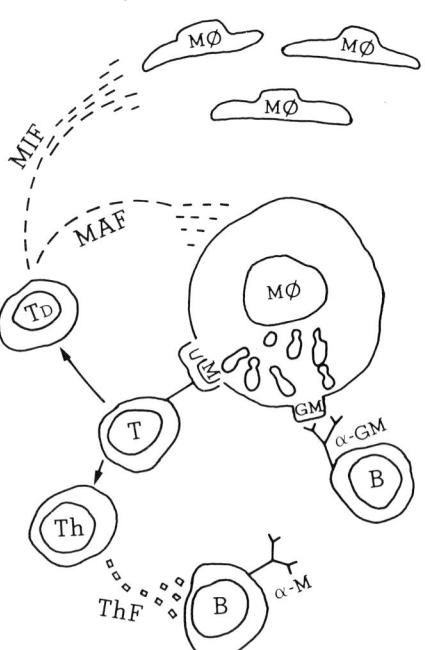

Figure 17.4. Cellular interactions in the immune response to *Histoplasma capsulatum* in the lung. Yeast forms are avidly phagocytosed by macrophages but only those macrophages that are activated by T-cells are effective in killing yeast forms. The antigenic stimulus for activating T-cells may be M-protein, a major antigen of histoplasmin. Products of activated T-cells are various including macrophage migration inhibitory factor (MIF), macrophage activating factor (MAF), and T-helper cell factors that drive B-cells to produce anti-M IgG. Cell surface galactomannan (GM) is antigenic and, as a polysaccharide, its stimulation of B-cells may not require T-cell help. Antigalactomannan immunoglobulins could function as opsonins. The migration of macrophages into the site of yeast form replication is the first step in the organization of the hypersensitivity granuloma that is the hallmark of acute pulmonary histoplasmosis.

nary nodules—histoplasmomas—can often enlarge with time to give concentric rings of calcification and fibrosis (Diamond, 1981). Yet in another granulomatous disease, schistosomiasis, there is diminished reactivity in the face of continued egg production, a phenomenon termed "spontaneous modulation" (Colley, 1981) mediated by lyt 2^+ 3^+ T-cells. Clearly, further investigation is needed to determine the effect of persistent *H capsulatum* yeast forms in granulomas on the immune responses of the host.

Another site of cell-mediated immunopathology thought to follow subclinical primary histoplasmosis is granulomatous uveitis—presumed ocular histoplasmosis syndrome (POHS). In his book, Schwarz, 1981, reviews the evidence pro and con for the involvement of *H capsulatum* in this illness. The difficulties in linking *H capsulatum* to this disease include the subclinical nature of the primary infection, the long lag (years) between exposure and occurrence of uveitis, and the scarcity of reports in which yeast forms are present in eye lesions. Nevertheless, there is a high incidence of positive histoplasmin skin tests and complement fixing antibodies to *H capsulatum* in POHS patients. These patients have higher blastogenic responses to histoplasmin than do matched controls and the intensity of blastogenesis varies as a direct function of ocular pathology. Increased frequencies of HLA-B7 and HLA-DR-w2 alleles have been observed in patients with POHS.

- *Paracoccidioides brasiliensis*. Paracoccidioidomycosis occurs more often in males (Lacaz, 1955–56). An association was found between a histocompatibility antigen (HLA-A9) and progressive pulmonary paracoccidioidomycosis in a series of 41 patients (de Restrepo et al, 1983). The immunologic ramifications of these findings are unknown.

- *Sporothrix schenckii*. There is no known basis to suspect that an immune deficit is responsible for susceptibility to cutaneous-lymphatic sporotrichosis, but pulmonary or other extracutaneous sporotrichosis signifies a failure of normal resistance barriers. Diamond (1981) mentions that alcoholism, and systemic corticosteroid therapy for sarcoidosis or lymphoproliferative malignancy have been involved as concurrent predisposing factors for extracutaneous sporotrichosis. The far-reaching effects of steroids on the immune system obscure interpretation of specific defects, and thus monitoring effector functions of cell mediated immunity will be necessary in sporotrichosis patients. Agglutinins against *S schenckii* yeast forms fluctuate over months or years during osseous or articular sporotrichosis depending on the response to therapy, with lower antibody levels corresponding to periods of remission (section 7.2.1). There are no parallel studies in these patients at the level of T-lymphocyte functions.

- *Candida* species. A hierarchy of innate, humoral, and T-cell mediated functions make up the normal immune response against *Candida* species (Figure 17.5). Conversion of *C albicans* from a benign, commensal existence on the mucosae to an infectious agent often is an early warning of a compromise of the cell mediated limb of the immune system. This has been noted in children with the DiGeorge syndrome, congenital thymic aplasia, and in Nezelof's syndrome, cellular immunodeficiency with increased Ig concentration, especially IgE. Acquired immune deficiency syndrome (AIDS) is an example of an acquired deficit of T-cell immunity in which mucosal candidiasis is a frequent infectious complication (Buckley, 1983). Late onset mucocutaneous candidiasis in conjunction with malignant thymoma and reduced T_H cells is also known (Ruiz-Argüelles et al, 1983).

Candida albicans or other *Candida* species are not killed by PMN of children with chronic granulomatous disease (section 11.7) and these yeastlike fungi are among the pathogens that cause infections in these children.

Theories advanced to account for the immune defects that lead to chronic mucocutaneous candidiasis (CMC) include: suppression of lymphocyte blastogenesis by a blocking factor in the plasma, and inhibition of MIF production by histamine-sensitive lymphocytes (section 11.6.2). An extension of these theories (section 11.3.8) postulates that purified human T-lymphocytes do not undergo blastogenesis in response to a *C albicans* metabolic extract in the absence of monocytes. Although

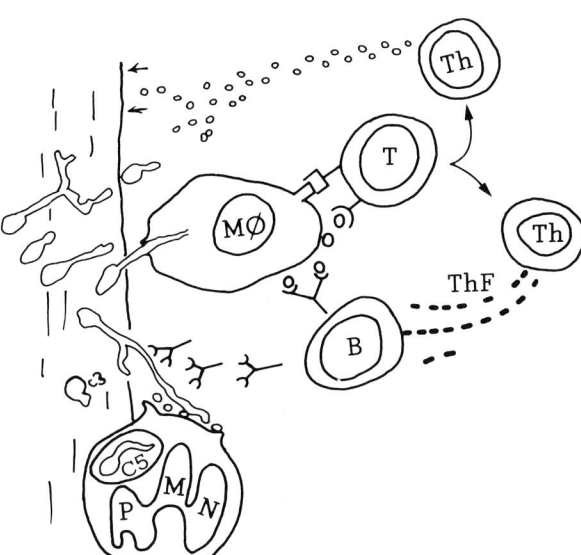

Figure 17.5. Hypothetical composite of immune responses against *Candida albicans* in the skin. Innate immune functions provide the first line of defense. Yeast forms and mycelia activate complement directly and opsonized candidae are subject to killing by oxidative and nonoxidative processes of polymorphonuclear neutrophils (PMN) and macrophages (MΦ). Endophagocytosis is the fate of yeast forms, but mycelia are adhered to and the discharge of phagocytic granules containing lysosomal hydrolases causes damage. Heat-stable opsonins occur as IgG against cell surface mannan. Their production in humans requires processing of mannan by macrophages and T-cell help. A lymphokine that stimulates keratinization may affect the ability of the host to slough infected epidermal tissue.

Candida mannan polysaccharide is taken up normally by monocytes of children with CMC, the antigen is released very slowly compared with its release from normal controls (Durandy et al, 1983). As a result, these monocytes are seriously impaired in the ability to collaborate with T-cells in the proliferative response to *C albicans* mannan, a major cell wall antigen.

Thus the available evidence at present implies that defects in antigen processing and presentation by macrophages, coupled with the persistence of antigens or immune complexes in the plasma, block the blastogenic and MIF responses and cut off the expression of delayed cutaneous hypersensitivity that normally prevents *Candida* from growing in the stratum corneum of the epidermis. Moreover the ability to rapidly achieve scaling of the infected skin is also a function of cell mediated immunity and can be adoptively transferred with cells from immune donor animals (section 11.6.2) which is evidence for a keratinization-stimulating lymphokine.

Favorable clinical responses to transfer factor and fetal thymus grafts in CMC (section 11.6.2) are further illustrations that the defects in this disease-group occur in the cell-mediated limb of the immune response.

• *Cryptococcus neoformans.* Even though our knowledge is fragmentary, a model of the known effector functions that contribute to the defense against cryptococci can serve a heuristic purpose (Figure 17.6). "Despite the presumed association of cryptococcosis with defects in cell mediated immunity, the majority of patients have no obvious predisposing factors." (Diamond, 1981). Even so, Hodgkin's lymphoma or other lymphoproliferative malignancy, in the absence of corticosteroids or other immunosuppressive therapy, is a risk factor (Diamond and Bennett, 1974). The frequent occurrence of positive delayed skin test reactions in mycology labora-

Figure 17.6. Hypothetical composite of immune responses against *Cryptococcus neoformans* in the lung. The responses can be discussed in terms of innate, humoral, and thymic functions. Innate immunity occurs via the direct activation of complement by the cell wall, and by natural killer (NK) lymphocytes. Oxidative microbicidal processes (MPO, myeloperoxidase-H_2O_2-iodide) of polymorphonuclear neutrophils (PMNs) and alveolar macrophages can damage cryptococci. Immunoglobulin M agglutinins are produced against the capsular glucuronoxylomannan, but their production is regulated by T-suppressor cells, and perhaps by direct tolerance of B-lymphocytes. Anticapsular IgM opsonizes cryptococci which are more readily phagocytosed or, in the case of large capsule forms, are subject to cytotoxic effects by rosetting macrophages. Immune complexes (IC) stimulate T-lymphocytes to produce a lymphokine that enables macrophages to phagocytose cryptococci via C3 receptors.

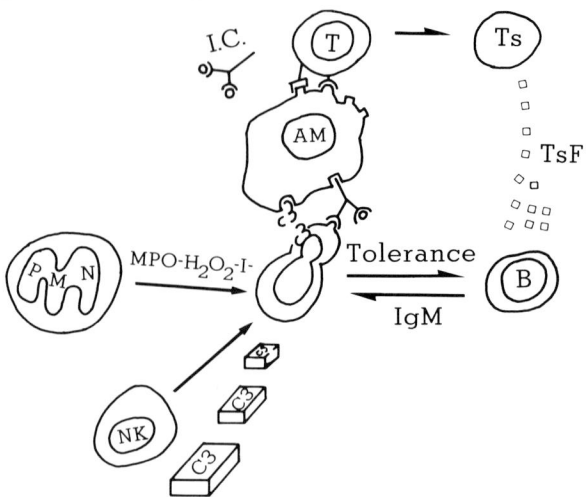

tory workers, in the absence of any disease, suggests a high degree of acquired immunity. Unlike other systemic mycoses, high levels of circulating antibodies are a good prognostic sign, implying a protective role for antibody which is expected in the defense against an organism with an acidic polysaccharide capsule, if bacterial agents of meningitis are any guide (Robbins, 1978). The understanding of critical effector functions in cryptococcosis is stalled by insufficient knowledge of the *C neoformans*—host interaction.

New insights into immunity to *C neoformans* have been derived from animal models. Guinea pigs infected with encapsulated cryptococci have lymphocytes that proliferate in vitro against protein antigens of an acapsular mutant (section 12.5.2) showing that protein antigens are potent elicitors and perhaps stimulants of cell-mediated immunity in cryptococcosis. The barrier to effective destruction of cryptococci is the tolerogenic acidic polysaccharide capsule. Conjugation of the major glucuronoxylomannan to protein amplifies the humoral response probably by making it T-dependent (section 12.6). This is important, because if production of opsonins against the capsular polysaccharide could be stimulated, one could expect a favorable response in infected animals, but this possibility has not been tested, nor have such polysaccharide-protein conjugates been tested as a vaccine.

Nonphagocytic, antibody-dependent cytotoxicity against cryptococci has been demonstrated and is potentially an important mechanism against encapsulated forms too large for endophagocytosis (section 12.6.2). In vitro killing of cryptococci by natural killer (NK) lymphocytes in mice and improved survival of mice with intact NK functions demonstrates an alternative mechanism that by definition does not require a primed immune system and has potential importance in maintaining surveillance against cryptococci (section 12.6.2).

A linkage between humoral and cell-mediated immunity has been proposed (Griffin, 1981) to explain how susceptibility to cryptococci is related to cell mediated immunity (section 12.4). This linkage is at the level of production of a lymphokine regulating the ability of macrophages to phagocytose cryptococci by their C3b receptors. The stimulus for lymphokine production is the presence of immune complexes.

- *Aspergillus fumigatus*. Studies of the immunopathology of allergic bronchopulmonary aspergillosis have not progressed to the point where it is possible to estimate the contribution of T-cell-mediated immunity to pathogenesis. Reports of antigen specific in vitro T-cell responses in these patients are scarce. The absence of type IV cutaneous responses in infected, immunocompetent individuals and the low frequency of granulomatous changes in allergic aspergillosis suggest that the classical delayed-type hypersensitivity in the lung that occurs in the systemic mycoses does not apply to this essentially noninvasive process. The immune responses in pulmonary allergic diseases are placed in a deep perspective in Pepys' (1981) review.

- Dermatophytes. The anthropophilic dermatophytes have adapted so that they provoke a chronic rather than acute inflammatory response (section 9.5). Diminished thymic function exemplified in the DiGeorge syndrome increases susceptibility to chronic dermatophytosis, but the mechanism of cell-mediated resistance is still obscure. The restricted niche of dermatophytes in the stratum corneum suggests that a lymphokine stimulating keratinization may be involved in the sloughing of infected tissue, but this is conjectural (section 11.6.2). The association of chronic dermatophytosis with atopy is notable because IgE reactions may interfere with and suppress delayed-type hypersensitivity (section 9.2.3). Serum substances capable of suppressing blastogenesis of lymphocytes in chronic dermatophytosis, and low

T-helper/T-suppressor ratios have been observed (section 9.6.6). These phenomenological findings have yet to be organized into a coherent scheme to help understand what regulatory mechanisms are faulty in chronic dermatophytosis.

If no obvious T-effector functions or mechanisms can be ascribed as critical then other evidence for cell mediated immunity might turn on the host's ability to demonstrate immunity against reinfection. Grappel (1981), has reviewed this aspect. Some acquired immunity in humans to zoophilic species is expressed in the need, in a secondary challenge, for a larger inoculum and for a shorter time to resolve the smaller lesions that follow. There is no conclusive evidence in humans of acquired immunity against infections of the smooth skin with anthropophilic *T rubrum*. Acquired immunity in guinea pigs to cutaneous re-infection with dermatophytes is strongest at the site of previous infection, with only a low order of general immunity (Grappel, 1981).

- Zygomycetes. Innate immunity in the form of normal anatomical barriers, neutrophil cidal mechanisms and the *Rhizopus* inhibitory factor of serum appear sufficient for successful host defense against these weak pathogens, and it has not been necessary to invoke a critical role for T-cell-mediated immunity.
- Actinomycetes. The evidence in experimental murine infections strongly recommends a crucial role for T-cell mediated immunity and a detrimental role for antibody responses. Delineation of immune responses in humans is, in contrast, a neglected subject. It is pertinent that 50% to 75% of patients with serious nocardiosis have a recognized primary disease (section 13.1). Further progress awaits the application of knowledge gained in the murine model to evaluate the role of direct T-cell-mediated cytotoxicity in humans (section 13.6).

Hypersensitivity pneumonitis caused by thermophilic actinomycetes is a noninvasive process in which the spores lodge in the alveoli and soluble proteinases diffuse into the lung parenchyma. The presence of precipitins and onset of symptoms 4 to 8 hours after exposure argue for immune complex deposition as the major factor inciting lung damage in humans, but direct evidence of complement components, immunoglobulin, and antigen in the lung tissues and vascular beds has not been consistently demonstrated in any large series of patients (section 15.4). Histologic evidence of noncaseating granulomas is notable in chronic farmer's lung. The resistance to lysis of spores of *Faenia rectivirugula* and other thermophils contributes to chronic inflammation (Schorlemmer et al, 1977) and may be the factor that evokes a granulomatous cell-mediated response.

The response of human peripheral blood lymphocytes to the mitogenic activity of *A viscosus* depends on T-cell help (section 14.4.1). Moreover, blastogenic indices of these lymphocytes in severe periodontitis are higher than in normal persons. These findings imply that the major pathway of immunopathology is humoral, but that it is regulated by T-helper cell activity.

17.9 ANTIGENIC STIMULANTS OF CELL-MEDIATED IMMUNITY

At this point in our state of knowledge, the structure versus activity relationships of mycotic antigens are just being explicated. The general availability of column chromatography and polyacrylamide techniques of high resolving power are having a decisive role. Another major impetus for the development of antigenically specific probes is that manipulation of lymphocyte sets has reached a stage that important insights into immunoregulatory mechanisms in the mycoses are now subject to practical experiments. The degree of T-dependence of major antigens can only be addressed by using probes that are produced according to objective criteria for purity. Furthermore, we must know how the biologic function of the antigen, ie, its enzymatic function, can further our understanding of its role in pathogenesis.

We need to discover what antigens are able to mediate protective responses and which, during the course of chronic infections, contribute to immunopathology. There is much scope for interdisciplinary studies. Table 17.2 includes those antigens that have the greatest promise of fulfilling these needs. We should strive towards the goal of making all of these preparations available with agreed upon standards for purity and activity.

Table 17.2. Mycotic Antigens Proposed as Biological Standards

Organism	Antigen Identity	Source, Cellular or Exocellular	Functions	Text Reference Section
Blastomyces dermatitidis	B-ASWS	Mural macromolecular complex	Delayed cutaneous hypersensitivity, MIF, blastogenesis	3.3
	Factor A	Secreted acidic protein	Species-specific	3.5.2
Coccidioides immitis	Coccidioidin	Autolyzate of mycelial culture, complex of proteins	Delayed cutaneous hypersensitivity, blastogenesis	4.4
	Coccidioidin factors 2,11	Proteins	Monomolecular, specific, dominant antigens	4.41
	Spherulin	Autolyzate of spherules, protein complex	Delayed cutaneous hypersensitivity, lymphocyte blastogenesis	4.4.3
	C-ASWS	Mural macromolecular complex of mycelia = "c", of spherules = "s"	MIF, lymphocyte blastogenesis	4.4.2
	Spherule vaccine	Formalin killed spherules	Protection	4.6.1
Histoplasma capsulatum	Cell walls	Yeast form walls	Granulomagenic	5.2
	Histoplasmin	Autolyzate of mycelial cultures, a protein complex	Delayed cutaneous hypersensitivity, MIF, lymphocyte blastogenesis	5.3
	H and M factors	Histoplasmin proteins	Species-specific, pluripotent	5.3
	Ribosomes	Microsomal fraction of yeast forms	Protection	5.7.1
Paracoccidioides brasiliensis	Paracoccidioidin	Autolyzed supernate of mycelial or yeast-form cells = protein complex	Delayed cutaneous hypersensitivity but prone to crossreactions with other fungi	6.3.2
	Factor E	Secreted alkaline phosphatase	Species-specific	6.3.1

(continued)

Table 17.2. Mycotic Antigens Proposed as Biological Standards *(continued)*

Organism	Antigen Identity	Source, Cellular or Exocellular	Functions	Text Reference Section
Sporothrix schenckii	Peptido-L-rhamno-D-mannan	Cell wall	Species-specific precipitinogen	7.2.1
	Sporotrichin	Autolyzate of yeast form cultures	Delayed cutaneous hypersensitivity	7.4.1
Aspergillus fumigatus	Aspergillin	Secreted protein complex	Immediate-, Arthus type, and delayed-type hypersensitivity	8.3
	Chymotryptic proteinases	Secreted enzyme	Presumptive aggressin, precipitinogen	8.3.2
	Major cytoplasmic antigens	Purified mycelial extract proteins	Precipitinogens, species-specific	8.3.2
Dermatophytes	Trichophytin glycopeptides	Mural and soluble polysaccharide-protein complex	Immediate-, delayed-type hypersensitivity	9.3
	Keratinases	Mural and soluble enzymes	Aggressins	9.5
Zygomycetes	None available			
Candida albicans	Mannan	Cell wall extract (CTAB procedure)	Circulates in infection, precipitinogen	11.3.2
	Peptido-glucomannan complex	Cell wall extract (cold alkali procedure)	Immunogenic	11.3.3
	Acidic carboxyl proteinase	Exocellular enzyme	Aggressin	11.5.2
	Major cytoplasmic protein	Homogenate-supernate	Precipitinogen	11.5.3
Cryptococcus neoformans	Glucuronoxylomannan	Capsule	Aggressins circulate in infection	12.3.1
	Mannoprotein from cell wall	Released into medium	Evokes IgM	12.3.2
	Cryptococcin	Urea extract	Delayed cutaneous hypersensitivity	12.5.1
	Microsomal fraction 105,000 g supernatant	Supernatant of homogenized acapsular mutant	MIF, lymphocyte blastogenesis	12.5.2
	B-HEX	A butanol and hot aqueous buffer extract of mitochondria-membranes	Lymphocyte blastogenesis	12.5.2
Actinomycetes *Nocardia asteroides*	*Nocardia*-active polypeptides	Dilute acid-solubilized cell extract	Delayed cutaneous hypersensitivity, lymphocyte blastogenesis	13.4.1

Table 17.2. *(continued)*

Organism	Antigen Identity	Source, Cellular or Exocellular	Functions	Text Reference Section
	Factors 1,6	Proteins released during growth and/or autolysis	Potentially species-specific	13.4.3
Actinomyces viscosus	AVIS	Supernatant of homogenized bacteria	T-dependent B-cell mitogens in humans	14.4
	Fimbrial antigens 1 and 2	Structural proteins	Mediate lactose-reversible coaggregation with *Streptococcus sanguis* (type 2); adherence to saliva-coated hydroxyapatite (type 1)	14.3
Faenia rectivirugula	Glycopeptide 1	Culture supernatant (double dialysis)	Precipitinogen	15.2.4
	Factors 7, 10, 12	Culture supernatant	Precipitinogen	15.2.5
	Proteinases	Secreted enzymes	Presumptive aggressins	15.3

Abbreviations: B-ASWS, alkali-soluble, water-soluble fraction of *Blastomyces dermatitidis*; C-ASWS, alkali-soluble, water-soluble fraction of *Coccidioides immitis*; MIF, migration inhibitory factor.

REFERENCES

Artz RP, Jacobson RR, Bullock WE, 1980. Decreased suppressor cell activity in disseminated granulomatous infections. Clin Exp Immunol 41:343–352.

Askenase PW, van Loveren H, 1983. Delayed-type hypersensitivity: activation of mast cells by antigen-specific T-cell factors initiates the cascade of cellular interactions. Immunol Today 4:259–264.

Atkinson AJ, Bennett JE, 1968. Experience with a new skin test antigen prepared from *Cryptococcus neoformans*. Amer Rev Respir Dis 97:637–643.

Beaman L, Benjamini E, Pappagianis D, 1981. Role of lymphocytes in macrophage-induced killing of *Coccidioides immitis* in vitro. Infect Immun 34:347–353.

Beaman L, Pappagianis D, Benjamin E, 1977. Significance of T cells in resistance to experimental murine coccidioidomycosis. Infect Immun 17:580–585.

Bice DE, Lopez M, Rothchild H, Salvaggio J, 1974. Comparison of *Candida* delayed hypersensitivity skin test size with lymphocyte transformation, migration inhibitory factor production and antibody titer. Inter Arch Allergy 49:54–62.

Brahmi Z, Liautaud B, Marill F, 1980. Depressed cell mediated immunity in chronic dermatophytic infections. Annales d'Immunol 131:143–153.

Brummer E, Morozumi PA, Vo PT, Stevens DA, 1982. Protection against pulmonary blastomycosis: adoptive transfer with T lymphocytes, but not serum, from resistant mice. Cell Immunol 73:349–359.

Buckley RH, 1983. Immunodeficiency. J Allergy Clin Immunol 72:627–641.

Bullock WE, 1979. Mechanisms of anergy in infectious diseases. pp. 269–294. In Immunological Aspects of Infectious Diseases. Dick G, ed. Baltimore: University Park Press.

Casali P, Perrin LH, Lambert PH, 1979. Immune complexes and tissue injury. pp. 295–342. In Immunological Aspects of Infectious Diseases. Dick G, ed. Baltimore: University Park Press.

Catanzaro A, 1977. Development of immunologic and clinical staging for immunotherapy. pp. 325–334. In Coccidioidomycosis, Proc Third Intern Symp. Ajello L, ed. New York: Stratton.

Catanzaro A, Spitler LE, Moser KM, 1975. Cellular immune responses in coccidioidomycosis. Cell Immunol 15:360–371.

Cathcart MK, 1981. Suppressor factors. pp. 73–102. In Suppressor Cells and their Factors. Krakauer RS, Clough JD, eds. Boca Raton: CRC Press.

Cauley LK, Murphy JW, 1979. Response of congenitally athymic nude and phenotypically normal mice to *Cryptococcus neoformans* infection. Infect Immun 23:644–651.

Cheers C, Waller RP, 1975. Activated macrophages in congenitally athymic "nude" mice and in lethally irradiated mice. J Immunol 115:844–847.

Clough JD, 1981. Immune effectors. pp. 22–41. In Suppressor Cells and their Factors. Krakauer RS, Clough JD, eds. Boca Raton: CRC Press.

Colley DG, 1981. Immune responses and immunoregulation in experimental and clinical schistosomiasis, pp. 1–83. In Parasitic Diseases—Vol 1—The Immunology. Mansfield JM, ed. New York: Marcel Dekker.

Corbel MJ, Eades SM, 1977. Experimental mucormycosis in congenitally athymic (nude) mice. Mycopathol 62:117–120.

Couch JR, Abdou NI, Sagawa A, 1978. *Histoplasma* meningitis with hyperactive suppressor T cells in cerebrospinal fluid. Neurology 28:119–123.

Cox RA, Vivas JR, 1977. The spectrum of in vivo and in vitro cell-mediated immune responses in coccidioidomycosis. Cell Immunol 31:130–141.

Cox RA, 1979. Immunologic studies of patients with histoplasmosis. Amer Rev Respir Dis 120:143–149.

Cutler JE, 1976. Acute systemic candidiasis in normal and congenitally thymus-deficient (nude) mice. J Reticuloendothel Soc 19:121–124.

Deem RL, Beaman BL, Gershwin ME, 1982. Adoptive transfer of immunity to *Nocardia asteroides* in nude mice. Infect Immun 38:914–920.

Deem RL, Doughty FA, Beaman BL, 1983. Immunologically specific direct T-lymphocyte mediated killing of *Nocardia asteroides*. J Immunol 130:2401–2406.

Deresinski SC, Pappagianis D, Stevens DA, 1979. Association of ABO blood group and outcome of coccidioidal infection. Sabouraudia 17:261–264.

Deresinski SC, Stevens DA, 1974. Coccidioidomycosis in compromised hosts. Medicine 54:377–395.

de Restrepo FM, Restrepo M, Restrepo A, 1983. Blood groups and HLA antigens in paracoccidioidomycosis. Sabouraudia 21:35–39.

Diamond RD, 1981. Immunology of invasive fungal infections. Comprehensive Immunol 8:585–633.

Diamond RD, Bennett JE, 1974. Prognostic factors in cryptococcal meningitis. Ann Intern Med 80:176–181.

Drutz DJ, Catanzaro A, 1978. State of the art, coccidioidomycosis, part II. Amer Rev Respir Dis 117:727–771.

Drutz DJ, Huppert M, 1983. Coccidioidomycosis: factors affecting the host-parasite interaction. J Infect Dis 147:372–390.

Dunn MA, 1980. Fibrosis in granulomas. pp. 133–152. In Basic and Clinical Aspects of Granulomatous Diseases. Boros DL, Yoshida T, eds. New York: Elsevier North Holland.

Durandy A, Fischer A, Griscelli C, 1983. Specific in vitro anti-mannan antibody production by human blood lymphocytes. Birth Defects 19:41–45.

Eichmann K, Simon MM, 1980. The generation of effector functions as a result of communication between subclasses of T-lymphocytes. pp. 405–415. In Regulatory T Lymphocytes. Pernis B, Vogel HJ, eds. New York: Academic Press.

Ferguson AC, Kershnar HE, Collin WK, Stiehm ER, 1977. Correlation of cutaneous hypersensitivity with lymphocytic response to *Candida albicans*. Amer J Clin Pathol 68:499–504.

Fink JN, Moore VL, 1980. Experimental hypersensitivity pulmonary granulomas. pp. 174–179. In Basic and Clinical Aspects of Granulomatous Diseases. Boros DL, Yoshida T, eds. New York: Elsevier North Holland.

Fischer A, Pichat L, Audinot M, Griscelli C, 1982. Defective handling of mannan by monocytes in patients with chronic mucocutaneous candidiasis resulting in a specific cellular unresponsiveness. Clin Exp Immunol 47:653–660.

Giger D, Domer J, Moser S, McQuitty J, 1978. Experimental murine candidiasis: pathological and immune responses in T-lymphocyte depleted mice. Infect Immun 21:729–737.

Goldstein G, Lifter J, Mittler R, 1982. Monoclonal antibodies to human lymphocyte surface antigens. pp. 71–89. In Monoclonal Antibodies and T Cell Products. Katz DH, ed. Boca Raton, FL: CRC Press.

Golub ES, 1981. The Cellular Basis of the Immune Response, 2nd edition. Sunderland, MA: Sinauer Associates, Inc.

Gordon S, 1980. Macrophage neutral proteinases and lysosomal hydrolases—role in tissue destruction. pp. 119–132. In Basic and Clinical Aspects of Granulomatous Diseases. Boros DL, Yoshida T, eds. New York: Elsevier North Holland.

Granger GA, Hiserodt JC, Yamamoto RS, Ross MW, 1980. L T molecules form a subunit system of cell toxins. pp. 279–283. In Biochemical Characterization of Lymphokines. de Weck AL, Kristensen F, Landy M, eds. New York: Academic Press.

Grappel SF, 1981. Immunology of surface fungi–dermatophytes. pp. 495–524. In Immunology of Human Infection, Part 1—Bacteria, Mycoplasmae, Chlamydiae and Fungi. Nahmias AJ, O'Reilly RJ, eds. New York: Plenum Medical Book Co.

Graybill JR, Alford RH, 1974. Cell-mediated immunity in cryptococcosis. Cell Immunol 14:12–21.

Graybill JR, Mitchell L, 1979. Host defense in cryptococcosis III. In vivo alteration of immunity. Mycopathol 69:171–178.

Green DR, Flood PM, Gershon RK, 1983a. Immunoregulatory T-cell pathways. Ann Rev Immunol 1:439–463.

Green F III, Balish E, 1979. Suppression of in vitro lymphocyte transformation during an experimental dermatophyte infection. Infect Immun 26:554–562.

Green F III, Weber JK, Balish E, 1983b. The thymus dependency of acquired resistance to *Trichophyton mentagrophytes* dermatophytosis in rats. J Invest Derm 81:31–38.

Griffin FM Jr, 1981. Roles of macrophage Fc and C3b receptors in phagocytosis of immunologically coated *Cryptococcus neoformans*. Proc Nat Acad Sci (USA). 78:3853–3857.

Hanifin JM, Ray LF, Lobitz WC Jr, 1974. Immunological reactivity in dermatophytosis. Brit J Dermatol 90:1–8.

Harvey RP, Stevens DA, 1981. In vitro assays of cellular immunity in progressive coccidioidomycosis: evaluation of suppression with parasitic-phase antigen. Amer Rev Respir Dis 123:665–669.

Hay RJ, 1979. Failure of treatment in chronic dermatophyte infections. Postgrad Med J 55:608–610.

Herberman RB, 1978. Natural cell mediated cytotoxicity in nude mice. pp. 135–166. In The Nude Mouse in Experimental and Clinical Research, Vol 1. Fogh J, Giovanella BC, eds. New York: Academic Press.

Jerne NK, 1974. Towards a network theory of the immune system. Annales d'Immun Inst Pasteur 125C:373–389.

Jones HE, Reinhardt JH, Rinaldi MG, 1974. Immunologic susceptibility to chronic dermatophytosis. Arch Dermatol 110:213–220.

Kaaman T, 1978. The clinical significance of cutaneous reactions to trichophytin in dermatophytosis. Acta Derm Venereol 58:139–143.

Kalina M, Kletter Y, Shahar A, Aronson M, 1971. Acid phosphatase release from intact phagocytic cells surrounding a large-sized parasite. Proc Soc Exptl Biol Med 136:407–410.

Kauffman CA, Israel KS, Smith JW, White AC, Schwarz J, Brooks GF, 1978. Histoplasmosis in immunosuppressed patients. Amer J Med 64:923–932.

Kimball JW, 1983. Introduction to Immunology. New York: Macmillan Publishing Co., Inc.

Klostergaard J, Devlin JJ, Orr SL, Yamamoto RS, Granger GA, 1983. Studies of the role of α-heavy and complex lymphotoxins in human cell-mediated cytotoxic reactions in vitro. pp. 527–533. In Interleukins, Lymphokines, and Cytokines. Oppenheim JJ, Cohen S, eds. New York: Academic Press.

Lacas, C da Silva. 1955–1956. South American blastomycosis. An Fac Med (São Paolo) 29:1–120.

McDaniel LS, Cozad GC, 1983. Immunomodulation by *Blastomyces dermatitidis*: functional activity of murine peritoneal macrophages. Infect Immun 40:733–740.

McGee MP, Myrvik QN, Thompson BY, Eaton LJ, Ockers J, 1980. Macrophage heterogeneity in BCG-induced granulomas. pp. 51–66. In Basic and Clinical Aspects of Granulomatous Diseases. Boros DL, Yoshida T, eds. New York: Elsevier North Holland.

Miyake T, Takeya K, Nomoto K, Muraoka J, 1977. Cellular elements in the resistance to *Candida* infection in mice. I. Contribution of T lymphocytes and phagocytes at various stages of infection. Microbiol Immunol 21:703–725.

Mizel SB, 1982. Interleukin 1 and T-cell activation. Immunol Rev 63:51–72.

Mok PN, Greer DL, 1977. Cell-mediated immune responses in patients with paracoccidioidomycosis. Clin Exp Immunol 28:89–98.

Musatti CG, Rezkallah MT, Mendes E, Mendes F, 1976. In vivo and in vitro cell mediated immunity in patients with paracoccidioidomycosis. Cell Immunol 24:365–378.

Oppenheim JJ, 1981. Lymphokines. pp. 259–282. In Cellular Functions in Immunity and Inflammation. Oppenheim JJ, Rosenstreich DL, Potter M, eds. New York: Elsevier North Holland.

Palacios R, 1982. Mechanism of T-cell activation: role and functional relationship of HLA-DR antigens and interleukins. Immunol Rev 63:73–110.

Pepys J, 1981. Fungi in pulmonary allergic diseases. pp. 561–584. In Immunology of Human Infection, Part I: Bacteria, Mycoplasmae, Chlamydiae and Fungi. Nahmias AJ, O'Reilly RJ, eds. New York: Plenum Medical Book Co.

Petrini C, Kaaman T, 1981. T-lymphocyte subpopulations in patients with chronic dermatophytosis. Int Arch Allergy Appl Immunol 66:105–109

Rea TH, Johnson R, Einstein H, Levan NE, 1979. Dinitrochlorobenzene responsivity: difference between patients with severe pulmonary coccidioidomycosis and patients with disseminated coccidioidomycosis. J Infect Dis 139:353–356.

Recht LD, Davies SF, Eckman MR, Sarosi GA, 1982. Blastomycosis in immunosuppressed patients. Amer Rev Respir Dis 125:359–362.

Restrepo A, Restrepo M, Restrepo F de, Aristizábal LH, Moncada LH, Vélez H, 1978. Immune responses in paracoccidioidomycosis—a controlled study of 16 patients before and after treatment. Sabouraudia 16:151–163.

Robbins JR, 1978. Vaccines for the prevention of encapsulated bacterial diseases: current status, problems and prospects for the future. Immunochemistry 15:839–854.

Ruiz-Argüelles A, Jett JR, Ritts RE Jr, 1983. Impaired generation of helper T cells in a patient with chronic mucocutaneous candidiasis and malignant thymoma. J Clin Lab Immunol 10:165–169.

Sarosi GA, Parker JD, Doto II, Tosh FE, 1970. Chronic pulmonary coccidioidomycosis—a National Communicable Disease Center cooperative mycoses study. New Engl J Med 283:325–329.

Schimpff SC, Bennett JE, 1975. Abnormalities in cell mediated immunity in patients with *Cryptococcus neoformans* infection. J Allergy Clin Immunol 55:430–441.

Schorlemmer HU, Edwards JH, Davies P, Allison AC, 1977. Macrophage responses to mouldy hay dust, *Micropolyspora faeni*, and zymosan, activators of complement by the alternative pathway. Clin Exp Immunol 27:198–207.

Schwarz J, 1981. Histoplasmosis. 472 pp. New York: Praeger.

Shiriashi R, Nakagaki K, Arai T, 1979. Experimental sporotrichosis in congenitally athymic (nude) mice. J Reticuloendothel Soc 26:333–336.

Sinclair NRStC 1980. Regulation by antibody feedback and other nonactive site control. pp. 211–223. In Strategies of Immune Regulation. Sercarz EE, Cunningham AJ, eds. New York: Academic Press.

Sohnle PG, Frank MM, Kirkpatrick CH, 1976a. Deposition of complement components in the cutaneous lesions of chronic mucocutaneous candidiasis. Clin Immunol Immunopathol 5:340–350.

Sohnle PG, Frank MM, Kirkpatrick CH, 1976b. Mechanisms involved in elimination of organisms from experimental cutaneous *Candida albicans* infections in guinea pigs. J Immunol 117:523–530.

Sohnle PG, Varkey B, Rose H, 1980. Lymphocyte function in relapsed pulmonary blastomycosis. J Allergy 65:376–380.

Stobo JD, Paul S, van Scoy RE, Hermans PE, 1976. Suppressor thymus-derived lymphocytes in fungal infection. J Clin Invest 57:319–328.

Tabeta H, Mikami Y, Abe F, Ommura Y, Arai T, 1984. Studies on defense mechanisms against *Candida albicans* infection in congenitally athymic nude (nu/nu) mice. Mycopathol 84:107–113.

Tewari RP, Sharma DK, Mathur A, 1978. Significance of thymus-derived lymphocytes in immunity elicited by immunization with ribosomes or live yeast cells of *Histoplasma capsultum*. J Infect Dis 138:605–613.

Turk JL, 1980. Delayed Hypersensitivity, 3rd Edition, rev. Amsterdam: Elsevier-North Holland.

van Voorhis WC, Kaplan G, Sarno EN, Horowitz MA, Steinman RM, Levis WR, Nogueira N, Hair LS, Gattass CR, Arrick BA, Cohn ZA, 1982. The cutaneous infiltrates of leprosy. Cellular characteristics and the predominant T-cell phenotypes. New Engl J Med 307:1593–1597.

Wahl SM, Gately CL, Helsel WH, 1983. Fibroblast growth factor production by human peripheral blood and leukemic cell line lymphocytes. pp. 335–340. In Interleukins, Lymphokines and Cytokines. Oppenheim JJ, Cohen S, Landy M, eds. New York: Academic Press.

Wells JV, Ries CA, 1980. Hematologic Diseases. pp. 473–512. In Basic and Clinical Immunology, 3rd Edition. Fudenberg JJ, Stites DP, Caldwell JL, Wells JV, eds. Los Altos CA: Lange Medical Publications.

Williams DM, Graybill JR, Drutz DJ, 1981. Adoptive transfer of immunity to *Histoplasma capsulatum* in athymic nude mice. Sabouraudia 19:39–48.

Wing EJ, Remington JS, 1980. Delayed hypersensitivity and lymphocyte function. pp. 129–143. In Basic and Clinical Immunology, 3rd Edition. Fudenberg JJ, Stites DP, Caldwell JL, Wells JV, eds. Los Altos CA: Lange Medical Publications.

Witkin SS, Yu IR, Ledger WJ, 1983. Inhibition of *Candida albicans*-induced lymphocyte proliferation by lymphocytes and sera from women with recurrent vaginitis. Amer J Obstet Gynecol 147:809–811.

Witorsch P, Utz JP, 1968. North American blastomycosis. A study of 40 patients. Medicine (Baltimore) 47:169–200.

Wu-Hsieh B, Zlotnik A, Howard DH, 1984. T-cell hybridoma-produced lymphokine that activates macrophages to suppress intracellular growth of *Histoplasma capsulatum*. Infect Immun 43:380–385.

Epilogue

The innate, humoral, and cell-mediated limbs of the immune response all contribute increments of protection against the mycoses. Indeed, fungi are well-represented among those microbes that activate complement directly. The inborn and acquired immunodeficits that are the accidental experiments of nature show that defects in thymus-dependent functions predispose to mycotic infections. This view is also supported by the results of experimental infections in athymic nude mice and in other murine models.

There is obvious danger in generalizing about so diverse a group as the pathogenic fungi and actinomycetes. The following comments and speculations about each major etiologic agent are intended to raise questions that remain about the immune response especially as they relate to clinically important antigens.

The A-protein of *Blastomyces dermatitidis* probably has the desired specificity to make it a suitable skin test antigen, and possibly a vaccine. Clinical trials with the *Coccidioides immitis* spherule vaccine are still in progress, and its efficacy is not yet known. The relation between the 3-O-methyl mannose-containing antigen and Huppert's factor 2 remain to be demonstrated. The latter antigen, it will be recalled, is the heat-, trypsin-, and alkali-stable protein present in coccidioidin that is responsible for the classical tube precipitin test. Initial findings showing that lysozyme has some ability to digest spherules of *C immitis* would benefit from corroboration. During disseminated coccidioidomycosis anergy occurs that is both broad and also *C immitis*-specific. Is it the result of T-suppressor cell activity or because of a disturbance in lymphocyte traffic? Soluble immune complexes and elevated IgE concentrations were detected in coccidioidomycosis, but their clinical significance is not known.

Similar questions about the basis for the anergy of disseminated infection apply in histoplasmosis and the other systemic mycoses. Hyperreactivity, the opposite of anergy, results in granuloma formation with its attendant immunopathology. Factors in the fungal cell wall such as chitin may contribute to the survival of yeast forms within granulomas for months or years. The slow leakage of soluble antigens around granulomas can also result in pathology. The protein factors of histoplasmin may have potential to act as vaccines to limit the production of granulomas by stimulating fungicidal processes of macrophages via the activation of T-lymphocytes.

The presumed ocular histoplasmosis syndrome has gained credence as a late immunopathologic consequence of *Histoplasma capsulatum* infection, but too few positive correlations have been made in humans between retinal damage and in vitro cell-mediated immunity to *H capsulatum* antigens, so that it is premature to draw firm conclusions. There has not been a systematic attempt to differentiate the immunopathologic effects of delayed hypersensitivity in the lung from protective aspects of cell-mediated immunity with respect to *H capsulatum* or to the other primary systemic fungi.

Unresolved difficulties also remain in understanding how labile or unknown epitopes in the otherwise crossreactive galactomannans could confer species-specificity for *H capsulatum, B dermatitidis, C immitis* and *P brasiliensis*.

Paracoccidioides brasiliensis infections are also marked by severe and broad anergy during dissemination in the face of high concentrations of complement-fixing antibodies. Later, after remission, cell-mediated parameters return to positive values below that of healthy exposed persons. The residual yeast forms may remain dormant for months or years. Further investigations of the damage to T-cells by *P brasiliensis* may benefit from the use of cell wall derived glucan, galactomannan and the secreted "E" protein as probes.

Less is known about *Sporothrix schenckii* and about the in vitro correlates of cell-mediated immunity during sporotrichosis than is the case for other primary systemic fungi. Thymic immunity is important in host resistance to *S schenckii* as shown in experiments with athymic nude mice. Knowing whether thymic immunity limits functions by providing help for responses to important T-dependent antigens would be useful in view of findings that antibody production correlated positively with protection in infected hamsters.

The peptido-L-rhamno-D-mannan of *S schenckii* is a well-characterized surface antigen that is species-specific. Apart from its documented and important serologic activity, nothing is yet known about its ability to evoke cell-mediated immunity. In a separate matter, although iodide therapy is efficacious in the cutaneous-lymphatic form of the disease, the link is tenuous between it and iodide-dependent oxidative fungicidal processes of polymorphonuclear neutrophils.

Infection with *Aspergillus fumigatus* and related species is a multifaceted problem because of the presence of three distinct clinical entities: aspergilloma, allergic bronchopulmonary aspergillosis and invasive aspergillosis of the immunocompromised host. The antigens evoking IgE in allergic aspergillosis are not well characterized and there is no explanation for the proposed polyclonal IgE activation. The contribution of T-cell-mediated immunity to the host defenses against aspergilli has not been assessed. Although a chymotrypsinlike proteinase of *A fumigatus* has been described, its role in pathogenesis is unknown.

Thymic immunity was shown to be important in accelerating the healing of lesions after cutaneous infection with *Trichophyton mentagrophytes* in euthymic rats compared with the chronic infection that develops in athymic littermates. Intense inflammation may be sufficient to eradicate the fungus from the keratinized epidermis, but a role for lymphokines that stimulate keratinization could be explored. More proof is needed to support the idea that because zoophilic dermatophytes produce increased concentrations of proteinase this antigenic stimulus contributes to brisker and more successful immune response.

In contrast, anthropophilic dermatophytes, especially *Trichophyton rubrum*, produce lower proteinase or keratinase concentrations and may evade detection thus promoting chronicity. Further analysis is needed of the antigenic role of trichophytin glycopeptides and keratinases in the dynamic of delayed cutaneous hypersensitivity in the skin and surrounding the infected hair follicles. Other mechanisms of down-regulating the immune response in dermatophytosis have come to light. Mediation by serum suppressor substances, low T-helper:T-suppressor cells and feedback suppression by IgE of delayed cutaneous reactions require further investigation.

After surveying the literature of the other mycotic pathogens it is not surprising that in the zygomycetes there is a recurrent theme of cell surface mannans and secreted proteinases. Studies of peptido-L-fuco-D-mannans should be extended to the thermotolerant *Rhizopus* species isolated from rhinocerebral zygomycosis. Structural features of this antigen should be delineated to show how it could contribute to genus specificity. Information is needed about the occurrence of normal levels of natural antibodies and the capacity of peptido-L-fuco-D-mannans to evoke in vitro correlates of cell-mediated immunity to these ubiquitous weak pathogens. Experiments in athymic nu/nu mice have shown that T-cell functions provide an increment of protection against zygomycetes. The effector functions involved should be characterized. Then it will be easier to discover what perturbations of immunity in poorly regulated diabetics enable these normally benign organisms to invade blood vessels, cartilage, and bone to produce a rapid and, if untreated, fatal disease. Diabetic animals have depressed cell-mediated immune functions. The plasma factors in diabetic ketoacidosis that permit the growth of zygomycetes require further definition.

Zygomycetes, because they are normally such weak pathogens provide an excellent opportunity to study opportunism at the level of innate and natural immunity.

Candida albicans and related *Candida* species remain the major pathogenic fungi used as models to study the antigenic structure and its interactions with the immune system. This emphasis has brought us to the threshold of understanding immunoregulatory mechanisms in the mycoses. Antigenic preparations containing mannan polysaccharide have been used to evoke stimulation and suppression of in vitro blastogenesis of human lymphocytes. T-cells were required to help B-cells respond to mannan, and continued cultivation of human lymphocytes with mannan resulted in an abrupt shift from stimulation to suppression that was antigen-specific and mediated by T-suppressor cells. This interaction also required monocytes as antigen-presenting cells. Such a model is ideally suited for further delineation of the activation and down-regulation of blastogenic responses and antibody production.

Preliminary experiments to perturb this system with mannan-containing immune complexes should be continued especially as immune complexes have been detected in other mycoses (eg, coccidioidomycosis, histoplasmosis) and their significance is unknown.

It would clarify matters if in future experiments objective criteria for purity of the antigenic mannan were developed in terms of the overall degree of polymerization and the composition of the peptido component. Models of the antigenic structure of *C albicans* mannan are still equivocal. Therefore, further linkage-sequence analysis of serotype A epitopes and of the elusive "factor 13b" B-specific epitopes are justifiable. The analytical technology for accomplishing these tasks relies on methylation-fragmentation gas-liquid chromatography–mass spectrometry and carbon 13 nuclear magnetic resonance spectrometry of haptenic oligomannosides produced by acetolysis or β-elimination. Implementation of immunochemical lasers—monoclonal antibodies—will be indispensable as probes in this work.

Studies designed to determine covalent linkages, if they exist, between glucan and mannan components of the *C albicans* cell wall would help to bring order to an unresolved aspect of the literature. Phosphate-enriched fragments of *Saccharomyces cerevisiae* glucomannan were precipitated with Alcian blue some years ago. Similarly, glycopeptide bonds between glucan and chitin have been inferred in a variety of fungal saprophytes. One anticipates with great expectation a revival of interest in extending knowledge about the cell wall of mycotic pathogens.

Proof is accumulating that the acidic carboxyl proteinase of *C albicans* is a promoter of IgA digestion, an activator of coagulation factors IX, and X and of angiotensin-mediated vasoconstriction. It would be timely to integrate this antigen-aggressin into in vitro assays of lymphocyte functions that have proved promising for studying immunoregulation evoked by other *Candida* antigens (mannan).

Research on the molecular and cellular level with *Cryptococcus neoformans* has progressed and the fruitful lines of investigation have become discernible. On the molecular side, strides were made in the compositional analysis of the viscous, acidic, capsular polysaccharide glucuronoxylomannan, the coproduced galactoxylopolymer, and a mannoprotein. Yet the epitopes that encode serotype specificity and the role of *O*-acetyl groups are unknown. Further fine structural analysis of these glycosidic antigens is needed to control for strain-to-strain structural variations within the four serotypes. Such an effort will increase confidence in the proposed minimum repeat unit structures of the glucuronoxylomannan. Corroboration of structural analysis by ^{13}C-NMR spectrometry is also needed. Knowledge of the three-dimensional shape of the capsular polysaccharide in solution would help us understand its antiphagocytic properties.

T-Cells are capable of adoptively transferring suppression of *C neoformans*-specific cutaneous hypersensitivity and in vitro growth of cryptococci. Otherwise, athymic nu/nu mice survived poorly against challenge with live cryptococci despite the production of IgM agglutinins. Independent of these studies, protein antigens of *C neoformans* were described that evoke delayed cutaneous hypersensitivity and its in vitro correlates.

Clearly these strands of research should converge to elucidate the protein antigens responsible for evoking stimulation and suppression. If suppressor T-cells are also evoked by the major viscous, acidic glucuronoxylomannan, the regulation of this phenomenon should be explored to find out if the mechanism is similar to immunologic tolerance resulting from priming mice with type III pneumococcal polysaccharide.

No preliminary analysis of the *C neoformans* cell wall has been made. Apart from satisfying the need for knowledge about this important yeast, the wall has been implicated as the structure responsible for direct complement activation. Curious thickening and laminations occur in the *C neoformans* wall after prolonged cultivation, but the implications of this event in pathogenicity are not understood.

About half of the patients with cryptococcal meningitis have no discernible primary immunologic defect. It would be of interest to know how immunologic tolerance in the context of persistent antigenemia and a bland tissue response could be broken and how the hosts' cell-mediated responses could be stimulated to promote eradication of cryptococci from the central nervous system. Passive transfer of monoclonal anti-glucuronoxylomannan could opsonize excess antigen. Protein antigens devoid of the capsular material could provide immunostimulation. Experiments in guinea pigs or mice would be suitable to test these possibilities.

The medically important actinomycetes are bacteria that have, however, historically attracted the attention of investigators whose primary interest is yeasts and molds. These actinomycetes have distinct ecologic niches: nocardiae in the soil, microaerophils in the gingival mucosae of warmblooded animals, and thermophils in the elevated temperatures of fermenting moist hay.

Practical methods for removal of crossreactive arabinogalactan and arabinomannan from nocardial antigens have not yet been applied. This has hindered achievement of specific primary binding assays such as enzyme immunoassays that can serologically differentiate mycobacterial infections from nocardiosis. This situation may be further complicated because it is likely that one or more protein antigens are held in common with mycobacteria. In that event a library of monoclonal antibodies will be needed to purify specific *Nocardia asteroides* antigens, including those responsible for serotypes.

The "*Nocardia*-active polypeptides" produced by dilute acid extraction of *N asteroides* have intermittently been investigated and show promise as specific antigens for demonstrating delayed cutaneous hypersensitivity and lymphocyte blastogenesis. Comparison of these antigens is warranted with proteins purified by other means from *N asteroides* culture supernatant fluids.

The superoxide dismutase of *N asteroides* is a major protein on the cell surface that is also released into the culture supernatant. It is important to establish whether it contributes to the ability of nocardiae to avoid intracellular killing by phagocytes. Moreover, an antigenic role for superoxide dismutase is probable and should be defined.

Protective immunity against *N asteroides* can be transferred to nu/nu mice with sensitized T-cells. The T-cell functions important in conferring protection are T-cytotoxic cells but a role is also involved for T-cells mediating delayed hypersensitivity phenomena. The relative importance of these two effector functions could be quantitated in adoptive transfer experiments by first fractionating the T-cells of sensitized donors by fluorescence-activated cell sorting. Similarly, the role of T-helper cells in nocardiosis is important to develop because some data convincingly show that humoral immunity is *detrimental* in nocardiosis, especially in actinomycotic mycetoma. It appears that opsonization of nocardiae may actually mask them so they are not recognized directly by T-cells. They can then multiply in the tumefactions of mycetomas and intracellularly within macrophages of the reticuloendothelial system. The possibility should be further explored that opsonization by immunoglobulins promotes the conversion of nocardiae to L-forms and thereby leads to their evasion of immune recognition.

Characterization of the B-cell mitogen isolated from nonpathogenic nocardiae and *Nocardia brasiliensis* is worthwhile extending to find out if this mitogen occurs in the *N asteroides*. The ambiguous occurrence of mitogenic activity in two distinct subcellular compartments should be clarified.

Microaerophilic actinomycetes including *Actinomyces viscosus* and related species increase in proportion to other oral bacteria when oral hygiene is poor and as gingivitis becomes apparent. Their contribution to periodontal disease has, as its probable cause, inflammation induced by bacterial products including mitogens. Disease production may be augmented by a viscous levan synthesized by *A viscosus* in the presence of sucrose. It may function by direct complement activation, contribution to the cementing substance of dental plaque, or as an antiphagocytic molecule. Other polysaccharides such as the exocellular *N*-acetylglucosamine polymer,

6-deoxytalose polymer, and amphipathic lipopolysaccharide make a veritable garden of polysaccharides of *A viscosus,* the contributions of which to disease production are unknown.

Fimbriae of *A viscosus* mediate adherence to teeth and coaggregation with *Streptococcus sanguis,* thus having major implications for initiating pathogenesis. It is likely that the host is capable of responding to fimbrial proteins and the dimensions of humoral and cellular responses could be measured especially to find out if antibodies to fimbriae inhibit adherence and coaggregation.

The T-dependent, B-cell mitogen of *A viscosus* requires further characterization in view of the activity observed in cell walls, in an extracellular heteroglycan, and in the supernatant of the disrupted actinomycete. The biologic activities of the *A viscosus* mitogens are not fully delineated because although the mitogens are capable of driving the B-cells of normal individuals to polyclonal immunoglobulin synthesis, the blastogenic indices of periodontal disease patients are much higher. Sorting out of the subsets of responding lymphocytes is needed to differentiate normal responses to the mitogen from the hyperreactivity found in periodontal disease.

Much groundwork has been laid for identifying the culture supernatant antigens of *Faenia rectivirugula* and *Thermoactinomyces vulgaris* that evoke precipitins in farmer's lung disease patients. Unifying existing knowledge in order to produce standard preparations for workers in different areas of the world is a timely endeavor. A rational approach includes growing the thermophils on chemically defined medium and monitoring the production of proteinase to prevent degradation of key factors. The proteinases are themselves antigens and complement activators which should be further purified and characterized. A numerical taxonomy of the antigens based on two dimensional crossed "rocket" immunoelectrophoresis has been proposed and provides a useful guide in monitoring purifications.

No definite proof has been found relating proteinases found in extracts of moldy hay dusts to the immunopathology of farmer's lung. A paradox of this disease is the variable presence of precipitins in asymptomatic exposed farmers at one extreme and the uncommon occurrence of persons with frank symptoms but no precipitins. Primary binding enzyme immunoassays with purified antigens may help us to understand the humoral response on a more quantitative and sensitive scale.

There is potentially more than one mechanism of immunopathology in farmer's lung disease. Early in the acute stage immune complex deposition in the lung appears to occur, whereas later in the disease process granulomas and type IV delayed hypersensitivity are more frequent. These statements are based on a few, small series of patients and supporting lymphocyte function tests are scarce. Further clinical studies are needed and should make provisions for testing the proteinases as cutaneous or in vitro probes of lymphocyte functions. The search for antilung antibodies and their significance in patients with hypersensitivity pneumonitis should also continue to be pursued.

The obstacles to achieving a more complete knowledge of mycotic antigens are gradually being overcome. Major antigens of clinical significance have been identified. Their production in chemically defined media is reaching an acceptable level. Mild procedures for the isolation of antigens with maximal preservation of epitopes have been proposed in several instances. The spectrum of biological activities is an active area of research. Criteria for purity of antigens and the establishment of biologic standards remain to be accomplished. Beyond these issues lie the definition of biophysical properties of the antigens: their molecular weights, subunit structures, isoelectric points and shapes in solution. Only fragmentary data are available. The complete amino acid sequences of mycotic protein antigens eludes us. The minimum repeat units of the polysaccharide antigens in terms of sequence, configuration of linkage and glycosidic bond arrangements remain largely uncharted. An increasing impact of monoclonal antibodies as specific probes in defining epitopes of mycotic antigens is beginning to be felt. Neoantigens made by conjugating haptenic sequences to new carriers offer the future prospect of new classes of immunostimulatory molecules.

Index

A

A-antigen, *B dermatitidis*, 349
  monoclonal antibodies to, 47
  purification, 46–47
  serologic activity of, 46–47
  as suitable skin test antigen, 401
ABPA. *See* Allergic bronchopulmonary aspergillosis
Absidia cylindrospora (Zygomycetes), peptidofucomannan in cell wall, 179
Absidia ramosa-corymbifera, as zygomycotic cause of animal disease, 186
Acetolysis
  of *C albicans* mannan, 202–204
  fragments, for fingerprinting yeast mannans, 201
N-Acetylglucosamine polymer, *A viscosus*, as polyclonal B-cell activator, 312, 313–314
Acidic carboxyl proteinase
  C albicans, as promoter of IgA digestion, 404
  expression of, during *Candida* infection, 222
  mural enzyme of *C albicans*, 29
Acid phosphatase
  mural enzyme of *C albicans*, 29
  presence of, in *C neoformans* cell wall, 266
  protein antigen, *C albicans*, 220
Acquired immune deficiency syndrome, association of, with *Candida* infection, 389
Acquired immunity, mice, *C neoformans*, 270
Acquired resistance, dermatophytes, thymic dependency of, 168
Actinomyces viscosus
  N-Acetylglucosamine polymer, in, 313–314
  6-Deoxy-L-talose polymer, possible pathogenic factor, 313
  monoclonal antibodies to, 361
  pathogenic nature of T-dependent B-cell mitogen, 365
  as producer of levan, 312
Actinomycetes, 281–336
  immunodeficits associated with, 392
  microaerophilic, 311–319

  as agents of periodontal disease, 311
  B-cell mitogen of, 315–318
  polysaccharides in, 312–314
  thermophilic, 321–336
    antigenic structure of, 322–329
    complement activating activity of, 329–330
    infection with, 321–322
    precipitins, 322–329
    proteinases of, 329–330
  transfer of immunity in, 385
Actinomycotic mycetoma, localized extrapulmonary nocardiosis, 282
Agglutinin factor 13b, *C parapsilosis*, 198
Agglutinins
  nocardiae, 287
  sporotrichosis, 123
Aggressin
  capsule of *C neoformans*, 252–256
  products of *B dermatitidis* yeast forms, 49–50
  proteinase of *C albicans* as, 224
  role of cryptococcal capsule as, 265
Ajellomyces capsulata, ascomycetous perfect form, *H capsulatum*, 77
Alcian blue, and binding of glucomannan complex, 207
Alkaline proteinase, as presumed virulence factor in zygomycosis, 185
Alkali-soluble-water-soluble fraction (ASWS)
  in *B dermatitidis* (yeast) cell wall, 43–44
  extraction of antigen from, 44–45
  fractionation of, 44
  specificity of, *B dermatitidis*, 44
Allergic bronchopulmonary aspergillosis, 129, 143–144
Alloxan, diabetes in rabbits and zygomycosis, 184–187
Alternaria zinniae, and galactomannan crossreactions, 347
Alternative complement pathway activation, by *Candida*, *C neoformans*, 363
Alveolar macrophage, role of, in Farmer's lung disease, 332–333

409

Amphotericin B
  and chronic mucocutaneous candidiasis therapy, 235
  treatment for disseminated coccidioidomycosis, 64
Anergy
  antigen-specific, in histoplasmosis, 93
  C immitis, 63–65
  chronic dermatophytosis, 167
  cutaneous
    C neoformans, 266–268
    in chronic mucocutaneous candidiasis, 231
    mechanism of, in dermatophytosis, 169
    P brasiliensis, 109
  of infection, 381–384
    chronic dermatophyte and Candida infections, 383–384
    mechanisms of, 382
    primary systemic mycoses, 381–383
Antibiotics, block wall synthesis, 31
Antibodies
  anti-idiotype, suppression by, 379
  capture and indicator, 357–358
  circulating, detection of, in histoplasmosis, 85–86
  complement-fixing
    H capsulatum and crossreactions, 98
    to histoplasmin factors, 87–88
  H capsulatum, detection of, with enzyme immunoassay, 88–89
  histoplasmosis, detection of, with radioimmunoassay, 88–89
  pathogenic, 364–365
  tests to detect in aspergillosis, 146–149
Antibody production, down-regulation of, 379
Antigen. See also Alkali-soluble-water-soluble fraction (B-ASWS); Antigens
  A-antigen, B dermatitidis, 349
  C albicans, major cytoplasmic proteins, 225–226
  C neoformans, polysaccharide, 252
  capsular polysaccharides, C neoformans, 268
  cell wall, C neoformans, 268
  coccidioidin factors 2 and 11, 349
  cryptococcin, 266–268
  detection of, circulating in candidiasis, 214
  galactomannan in P brasiliensis cell wall, 105–106
  galactomannans, 346–348
  galactoxylomannan complex, C neoformans, 268
  glycopeptide 1, F rectivirugula, 324–326, 350
  glycoprotein a, 323–324
  histoplasmin H and M proteins, 349
  mannan, C albicans, 198–209
  P brasiliensis, E factor, 106–107
  P brasiliensis, paracoccidioidin, 108
  peptidofucomannans, 182
  peptidomannans as major mural, 6–7
  polysaccharide, detection of, for A fumigatus, 136
  postmitochondrial supernatant, C neoformans, 268
  potential, O-phosphonoglucomannan, 207
  protein
    acidic carboxyl proteinase, C albicans, 220–225
    acid phosphatase, C albicans, 220
    C albicans, 219–226
  soluble cytoplasmic substances, C neoformans, 268
  specificity, role of mannan in, 345
  specific suppression, 377–378
  TP antigen, coccidioidin, 349
  trichophytin, 170–171
Antigenemia, 349. See also Mannanemia
  cryptococcal, 252, 269
  detection of, in active coccidioidomycosis, 68
  double antibody sandwich enzyme immunoassays for, 136
  enzyme immunoassay for, 357
  in invasive aspergillosis, 133–137
  modified Farr-type radioimmunoassay for, 135–136
Antigenic complex, mural, H capsulatum yeast form, 79
Antigenic stimulants, humoral immunity, mycoses, 337–372
Antigens
  A fumigatus, 130–143
  B dermatitidis. See also Alkali-soluble-water soluble fraction; Factor A.
    cytoplasmic soluble, 45–46
    specific, methods to produce, 48
  C immitis, 54–61
  crossreactivity of, among Aspergillus species, 149
  double dialysis method of producing, actinomycetes, 322–323
  histoplasmosis, H- and M-factors, 82–83
  keratinases in dermatophytes, 166–167
  membrane-bound, C albicans, 226
  method to determine, actinomycetes, 326–328
  mycotic
    proposed as biological standards, 393–395
    as stimulants of cell-mediated immunity, 392–395
  nocardiae, serologic analysis of, 285–291
  P brasiliensis, 106–108
  polysaccharide, mannans, 344–346
  processed by macrophages, 93
  protein, 349–352
    A fumigatus, 137–143
    extracts of mycelial and yeast forms, 351
    immunoenzyme analysis in A fumigatus, 138–139
    membrane bound, 352
    species-specific secreted factors, 349–350
    wall bound, 351
  purification of, in A fumigatus, 147
  S schenckii, 121
  T-dependent, mannans, 343–344
  thermophilic actinomycetes, 322–329
  T-independent, 343
Anti-Nocardia brasiliensis, pathogenic nature of, 364–365
Apical growth, in fungi, 5
Arabinogalactan, nocardiae cell wall, 283–285
D-Arabino-D-galactans, mycobacteria and nocardiae, 284
Arabinomannan, nocardiae cell wall, 283–285
Arteries, infection of, with zygomycete hyphae, 177

Arthrobacter GJM, exo-α-mannan hydrolase in, 208–209
Arthroconidia, *C immitis*, cell wall composition of, 56–58
Arthroderma, sexual ascogenous form of dermatophytes, 157
Arthus reaction, evoked by galactomannan, *A fumigatus*, 132
Arthus type III immune complex-mediated damage, in FLD, 331
Aspergillin, 138
Aspergilloma, 129
Aspergillosis
  detection of antibodies in, 146–149
  evidence of immune complexes in, 146
  extent of, 129–130
  invasive
    antigenemia in, 133–137
    A fumigatus antigen-like extract in, 135
    pathogenic mechanisms of, 130
  tests to detect antibodies in, 146–149
Aspergillus awamori, nigeran content in, 13
Aspergillus flavus, as causative agent of aspergillosis, 129
Aspergillus fumigatus, 129–156
  antigenic activity of cytoplasmic protein of, 141–142
  antigens of, 130–143
  binding assays, 146–149
  cell-mediated immunity to, 149–150
  characterization of antigens of, 142
  chymotryptic action pattern of proteinase, 350
  conidiophore of, 129–130
  crossreactivity of antigens of, 149
  enzyme immunoassay of, 352–353
  galactomannan in, 347–348
  glucan-chitin component of cell wall of, 131
  high reactivity of glycoproteins of, 147–148
  humoral immunity and immunopathology, 144–149
  IgE, IgG responses to, 144–145
  immunodeficits associated with, 391
  infection by, 129–130
  as major causative agent of aspergillosis, 129
  mycelial extracts, 139–142
  polysaccharide antigens of, 130–137
  precipitins, 146–149
  protein antigens of, 137–143
  proteinase isolated from culture filtrate of, 142–143
  protein in cell wall of, 26
  purified antigens of, 142–143, 147
  secretion of antigens during growth, autolysis, 138
Aspergillus nidulans, as causative agent of aspergillosis, 129
Aspergillus niger
  as causative agent of aspergillosis, 129
  glucan in, 13
Aspergillus species, cell wall composition of, 131
Asteroid bodies, and *S schenckii*, 124
ASWS. *See* Alkali-soluble water-soluble fraction
Athymic nude mice
  and *C neoformans* infection, 271
  and experimental nocardiosis, 297–300

Athymic state, and host response to mycotic infection, 386
Atopy, and chronic dermatophytosis, 168
Autohydrolysis, cell wall, 22
Autoimmune component, in FLD, 331–332
Autoimmune disease, and chronic dermatophytosis, 168
AVIS
  B-cell mitogen, *Actinomyces*, 315–318
  effect of, on fibroblast cultures, 317
  polyclonal B-cell activation by, 362

B

BALB/c mice, and *C neoformans* infections, 272
B-cell mice, and *Candida* infection, 228
B-cell mitogen, *Actinomyces*, 315–318
B-cells, role of, in T-dependent humoral response, 342
Binding assays, *A fumigatus*, 146–149
Blastogenesis
  to *A viscosus* antigens, 317
  C albicans, regulation of, 210–211
  C immitis, 66–67
  C immitis, as sensitive indicator of cell-mediated immunity, 66
  C neoformans, 267–268
  in chronic mucocutaneous candidiasis, 233–234
  correlation of, with periodontitis, 317–318
  elicited by *A fumigatus* mycelial extracts, 139
  lymphocyte
    elicited by coccidioidin, 73
    fluctuation, chronic mucocutaneous candidiasis, 237
  paracoccidioidomycosis, 108
  to *P brasiliensis* antigens, 109–110
Blastogenic responses
  A fumigatus, 149–150
  in blastomycosis, 48
  S schenckii, 122
Blastomyces dermatitidis, 41–51
  ability of, to replicate in vivo, 47
  antigens, 43–45
  cell-mediated immunity, 48–50
  cell wall composition of, 43
  cell wall extracts of, 42–45
  complement fixation and immunodiffusion tests, 46–47
  and effectiveness of PMNs, 338
  enzyme immunoassay of, 353
  humoral response, 46–47
  immunodeficits associated with, 385–386
  infection by, 41
  monoclonal antibodies against, 359
  phagocyte interactions, 47
  protein in cell wall of, 26
Blastomycin
  crossreaction of, with histoplasmin, 91
  and skin-test reaction, 41–42
  as supernatants of old mold form *B dermatitidis*, 41–42
  unpurified, as antigenic complex, 42
Blastomycosis
  anergy of infection in, 383
  diagnosis of, 46–47
  experimental murine, 48–50

Blastomycosis [cont.]
  murine
    lung as target organ in, 49
    protection of subcutaneous priming infection, 49
  serum, detection of PMN chemotaxis inhibitor in, 47
Blocking factor
  in chronic mucocutaneous candidiasis, 233
  in coccidioidomycosis, 67
  and immune complex formation, 362
Bronchial asthma, contribution of *A fumigatus* to, 129
Bruton's X-linked agammaglobulinemia, role of, 363

C

C-reactive protein, role of, 22
C-substance
  activity, 337
  A fumigatus polysaccharide antigen, 137
  in dermatophytes, 163–164
C5-deficient mice, and *C albicans* challenge, 230
Calcifications, lung, in histoplasmosis, 77
Calcofluor white M2R, as specific probe for chitin in yeast, 18
Candida albicans, 191–250
  agglutinated by Con A, 6
  amino acid synthetic medium for, 192
  antigen, acidic carboxyl proteinase, 220–225
  blastoconidia, reaction of, to probe, 7
  cell wall composition of three forms of, 28
  cell wall of, 193–196
  enzyme immunoassay of, 354
  glucan in cell wall of, 112–113, 218–219
  hyper-IgE-recurrent infection syndrome, 234
  immune responses against, in skin, model, 389
  immunodeficits associated with, 389–390
  infection, immune responses to, 226–236
  in inherited defects of innate phagocytic immunity, 238
  innate resistance to, 384
  isolates, serotypes of, 197–200
  mannanemia, 212–218
  mannan immunochemistry and immunology, 196–218
  mannan structure in cell wall of, 7–9
  membrane-bound antigens of, 226
  as model to study antigenic structure, 403
  monoclonal antibodies to, 359
  oxidative mechanism for killing, 236
  phagocytic candidacidal mechanisms, 236–239
  and PMN effectiveness, 339–340
  proteinase action pattern, 350
  regulation of blastogenesis in, 210–211
  ultrastructure of cell wall of, 195–196
Candidacidal mechanisms, phagocytic, 236–239
Candida guilliermondii, 191
Candida krusei, 191
Candida parapsilosis, 191
Candida pseudotropicalis, 191
Candida species, conversion of, to pathogen, 192
Candida tropicalis, 191
Candidiasis
  chronic mucocutaneous, 230–236
  blastogenesis in, 233–234
  clinical features of, 231
  genetic association of, 230
  immune defects leading to, 389–390
  immunologic defects in, 211
  lesions of, 234–235
  classification of types of, 231–232
  diagnosis of, 212–218
  incidence of, in leukemia patients, 213
  murine, 226–230
  nature of, 212
Capsule, *C neoformans*
  antiphagocytic nature of, 255–256
  characteristics of, 252
  polysaccharide content of, 252
  serological heterogeneity in, 261–264
β-Cathepsin, proteinase, 380
Cell-mediated cytotoxicity, and *C neoformans*, 270–271
Cell-mediated immunity
  B dermatitidis, 48–50
  C immitis, 63–69
  nocardiosis, 298
  review of, 373–399
Cell-mediated responses, to *P brasiliensis*, 108–111
Cell shape, role of glucan in, 11
Cellular responses, in vitro. *See also* Blastogenesis; E rosettes; Gammopathy; Leukocyte inhibitory factor
  antigens evoking, *C immitis*, 68–69
Cell wall
  amino acids in, 6
  B dermatitidis (yeast), alkali-soluble-water soluble fraction in, 43
  B dermatitidis (yeast), biologic activity of, 45
  C albicans, 193–196
    glucan in, 218–219
  carbohydrates in, 6
  chitin in, 5–6, 14–20
  components, *H capsulatum*, fluctuation of, during morphogenesis, 79
  components of, in *Aspergillus* species, 131
  criteria for purity, 22–26
  dermatophytes, 158
  enzymes associated with, 27, 29
  extracts of, *B dermatitidis*, 42–45
  as foreign antigenic complex, 5–40
  fractions, method of preparation and determining purity, 24–26
  functions of, 20–22
  fundamental differences between bacterial and fungal, 6
  glucans in, 5, 11–14
  glycanases in, 6
  H capsulatum
    antigenic nature of, in yeast form, 81–82
    characterization of, 97
    chemotypes of, 79–80
    yeast form, 78–82
  hypothetical model for organization in *C albicans*, 32
  as locus for enzymatic activity, 27
  as locus for periplasmic enzymes, 21
  mannan in, 5, 6–11
  molecular organization of, 5–40
  nocardiae, 282–283

mitogenic activity in, 301–303
P brasiliensis, 103–106
peptidomannans in, 6–11
porous nature of, 21
preparation of, 22–26
protective nature of, 20–21
protein in, 26–29
role of, in adherence, 21
as source of antigens, 22
Sporothrix schenckii, 116–120
of zygomycetes, 177–183
Central nervous system, as sites of *C neoformans* infection, 251
Ceratocystis stenoceras, coexists with *S schenckii*, 120
Cetyltrimethylammonium bromide complex, for mannan preparation, 199
Chemotactic factors, 375
Chemotypes
H capsulatum, 79–80
yeast form, differing cell wall composition in, 19–20
Chitin
in *C albicans* cell wall, 194–195
in cell wall, 14–20
as component in yeast, 17–19
content of, in zoopathogenic fungi, 19–20
effect of nutritional conditions on content of, 19
high content of, in *H capsulatum* yeast form cell walls, 80
in *H capsulatum* yeast form cell wall, 78
increase of, in cell wall through form transitions, 29
as major component of cell walls of filamentous fungi, 14
methods of detection of, in fungi, 15–16
possible effect of, on PMN or macrophage killing, 20
presence of, in molds, 16–17
probes for, 18
role of, in yeast morphogenesis, 18–19
synthesis of, in yeast, 18
Chitinase, from *S griseus, S marcescens*, 15
Chronic dermatophytosis. *See* Dermatophytosis, chronic
Chronic granulomatous disease, and defective killing of *C albicans*, 238
Chronic mucocutaneous candidiasis. *See* Candidiasis, chronic mucocutaneous
Chymotrypsinlike proteinase, possible mural protein of *A fumigatus*, 29
CIE, use of, to detect antibodies in aspergillosis, 146–149
Cimetidine, effect of, on anergy to *Candida*, 233
Coccidioides immitis, 53–76
antigenic structure, 54–61
antigens evoking cellular responses, 68–69
antigen-specific responses, 66–69
arthroconidia, cell wall composition of, 56–58
arthroconidial wall fractions, chemical composition, 58
carbohydrate composition of extracts of, 59
cell-mediated immunity, 63–69
characterization of antigens of, 54–55
as dimorphic fungus, 53
distribution of, 53
and galactomannan crossreactions, 347
humoral responses to, 61–63
immunodeficits associated with, 386–387
infection by, 53
mycelial form, preparation of coccidioidin from, 54
non-antigen-specific responses, 65–66
protein in cell wall of, 26
and role of PMNs, 338
spherules, freeze-fracture preparation of, 56
wall antigens of, 55–60
Coccidioidin, 54–55
antigens, 349
crossreaction of, with histoplasmin, 91
delayed cutaneous hypersensitivity, lymphocyte blastogenesis, 58
as elicitor of lymphocyte blastogenesis, 73
erythema nodosum as reaction to, 65
licensed for skin tests, 54
methods of preparation, 54
reaction of patients to, 64–65
Coccidioidomycosis
active, detection of antigenemia in, 68
anergy of infection in, 381
blocking factor in, 67
and diminished blastogenesis to *C immitis* antigens, 66
disseminated, and increase in cutaneous anergy, 66
disseminated, reduced sensitization by contact sensitins, 72
E-rosette response in, 65–66
experimental murine, 70–72
human, immune status as function of disease progress, 72
hyper IgE production in, 63
immunotherapy for, with transfer factor, 69–70
mixed lymphocyte reaction in, 65
murine
protective cellular mechanisms, 71–72
T-lymphocyte activity in, 71–72
polyclonal gammopathy in, 62
response to PHA as index of immunosuppression in, 65
sensitivity of patients to DNCB, 65
soluble immune complexes in, 67–68
symptoms of, 53
value of, in vitro tests for, 69
Coin lesions, 91
Cold alkali extraction, for mannan preparation, 199
Collagenase, proteinase, 380
Competitive binding enzyme immunoassay (EIA inhibition), 353
Complement, direct activation of, 22
Complement activation, yeast mannan, 211–212
Complement fixation
B dermatitidis, 46–47
nocardiae, 287
Complement-fixing antibodies
C immitis, 61–62
to histoplasmin factors, 87–88
Concanavalin A
and activation of suppressor T-cells, 378
agglutinates *C albicans*, 6, 193

Concanavalin A [cont.]
  binds to S schenckii cell wall rhamnomannan, 118
  reactive sites, and C-substance activity, dermatophytes, 164
Congenital thymic aplasia, association of Candida infection with, 389
Conidiophore, A fumigatus, 129–130
Contrasuppression, 379
Controlled yeast lysate (CYL antigen), to measure delayed cutaneous hypersensitivity, H capsulatum, 91
Corticosteroid therapy, and chronic dermatophytosis, 168
Cortisone, immunosuppression induced by, 216–217
Counterimmunoelectrophoresis, to measure anti-H and anti-M, 87–88
Crossreactions
  C albicans and S cerevisiae serotypes, 206
  involving galactomannans, 347–348
  Nocardia and Mycobacterium species, 288–289
Crossreactivity, C albicans and C neoformans, 264
Cryptococcal granuloma, 251
Cryptococcal meningitis, 251
  immunologic tolerance in, 404–405
Cryptococcosis, cell-mediated immune functions in, 273
Cryptococcus neoformans, 251–280
  asexual stages of, 251
  capsular and cell wall heteroglycans of, 256–264
  capsule as aggressin, 252–256
  cryptococcin, 266–268
  as encapsulated yeast, 251
  enzyme immunoassay of, 355–356
  host defenses against, 268–274
  immune responses against, in lung, composite, 390
  immunodeficits associated with, 390
  infection by, 251
  and PMN effectiveness, 340
  protein antigens of, 265–267
  research directions, 404
  secreted mannans of, 10
  serological heterogeneity of capsule, 261–264
  varying tissue reaction of host to, 252
  virulence of, in rodents, 253
Cutaneous anergy
  in blastomycosis, 48
  nature of, in coccidioidomycosis patients, 64. See also Anergy, cutaneous
Cutaneous hypersensitivity
  A fumigatus antigens, ABPA, 143–144
  and blastomycin, 41–42
  C immitis, 63–65
  chronic dermatophytosis, 167–168
  evoked by paracoccidioidin, 108
Cyclophosphamide, suppression of immunity to Candida induced by, 228–229
CYL. See Controlled yeast lysate
Cytoplasmic protein, A fumigatus, antigenic activity of, 141–142

D
Delayed cutaneous hypersensitivity
  C immitis, 72
  C neoformans, 267
  chronic mucocutaneous candidiasis, 232
  course of, in histoplasmosis, 93–94
  in dermatophytosis, 171
  evoked by alkaline extracts, C immitis, 59
  evoked by spherulin, C immitis, 60
  H capsulatum, 90–91
  return of, after recovery, coccidioidomycosis, 67
  nocardiae, 290–291
  to nocardial cytoplasmic extract, 296
  P brasiliensis, 109
  to S schenckii, 120–122
Delayed hypersensitivity
  cellular basis of, 380–381
  hypersensitivity granuloma, 380
  mechanism of cell-mediated suppression of, 377–378
Dental plaque, constituents of, 311. See also Actinomycetes
6-Deoxy-L-talose polymer, possible pathogenic factor in A viscosus, 313
Dermatophytes, 157–175
  cell wall of, 158–159
  crossreactive and species-specific factors, 164
  C-substance in, 163–164
  enzyme immunoassay of, 356
  galactomannan antigens of, 159–165
  glucans in, 165–166
  glycogenlike glucans in, 14
  glycopeptides in, 159–165
  humoral responses to, 164–165
  immunodeficits associated with, 391–392
  infection by, 157
  keratinase production by, 350–351
  properties of glycoproteins of, 161
  properties of polysaccharides of, 161
  proteinase production by, 166–167
  sexual ascogenous forms of, 157
Dermatophytosis, 157–158
  chronic, 167–171
    immunodeficits in, 168
    T rubrum as agent of, 170
  cutaneous anergy and T-helper : T-suppressor lymphocytes, 171
  IgE-mediated suppression in, 169
  IgG and IgM levels in, 165
  mechanism of cutaneous anergy in, 169
  soluble suppressor substances in, 168–169
Diabetics, susceptibility of, to zygomycetes, 184
Dimorphic fungus. See Blastomyces dermatitidis; Candida albicans; Coccidioides immitis; Histoplasma capsulatum; Paracoccidioides brasiliensis; Sporothrix schenckii
Dinitrochlorobenzene
  depression of sensitization to, paracoccidioidomycosis, 109
  sensitivity coccidioidomycosis patients to, 65
DiGeorge syndrome
  association of Candida infection with, 389
  association of, with dermatophyte infection, 391
Double antibody sandwich enzyme immunoassay, general method of, 354

Double dialysis method, for producing antigens, actinomycetes, 322–323

E

EIA inhibition, 353
Elastase, proteinase, 380
Endocarditis, and *C albicans*, 210
Enzyme 1, proteinase of thermophilic actinomycetes, *F rectivirugula*, 351
Enzyme immunoassays, 352–358. *See also* specific names
  and actinomycete antigens, 326–328
  to detect antibodies in histoplasmosis patients, 88–89
Enzyme-linked immunoelectrotransfer blot assay
  general method of, 355
  to identify *Nocardia* antigens, 293–294
Epidermophyton floccosum, cell wall composition of, 158
E-rosette response, in coccidioidomycosis, 66
E-rosettes, in paracoccidioidomycosis patients, 110
Erythema nodosum, as reaction to coccidioidin, 65
Exoantigen test, and association of H- and M-factors with histoplasmosis, 82
Exo-α-mannan hydrolase, in *Arthrobacter* GJM, 208–209
Exophiala (*Cladosporium*) *werneckii*
  galactomannan in, 348
  peptidophosphogalactomannan of, 163
  phosphodiester-linked mannans in, 10
Exopolysaccharides, *C neoformans*, properties of, 260
Extrinsic allergic alveolitis, FLD, 331

F

Factor A, *B dermatitidis* antigen, 45
Factor E, *P brasiliensis* antigen, 106–107, 349
Faenia rectivirugula. *See also* Farmer's lung disease
  as causative agent of Farmer's lung disease, 321
  cell extract antigens of, 323
  enzyme immunoassay of, 357
  localization of enzymatic activity, 329
Farmer's lung disease
  animal models of, 332
  Arthus type III immune complex-mediated damage in, 331
  autoimmune component of, 331–332
  circulating immune complexes in, 331
  granulomas in, 331
  immunopathology of, 330–333, 406
  and infection by thermophilic actinomycetes, 321–322
  leukocyte migration inhibition in, 331
  lung biopsies in, 331
  model for immunopathogenesis in, 332–333
  physical symptoms of, 331
  precipitins in, 331
  symptoms of, 321
Farr-type radioimmunoassay, for detection of mannan, *C albicans*, 216
Favus, 157

Fehling precipitation, for preparing mannan, 199
Fibrosis, and chronic granulomatous inflammation, 381
Filamentous fungi, 129–189
Filobasidiella neoformans var *neoformans*, perfect state, *C neoformans*, 251
Fimbriae, actinomycetes, association of, with virulence, 314–315
FLD. *See* Farmer's lung disease
Focal pneumonitis, lesion of pulmonary nocardiosis, 281
Fungi, glycopeptide bonds between chitin and β-1,3-glucan in, 26
Fungi, pathogenic
  chitin content in cell wall by species, 15
  filamentous, 129–189
  primary systemic dimorphic, 41–128
  yeastlike, 191–280
Fungi, zoopathogenic, chitin content of, 19–20
Fungistasis, serum, 338

G

Galactomannan
  A fumigatus, 131–133
    radioimmunoassay detection of, 136
    structural features of, 134
  antigenic polysaccharides, 346–348
  antigens, in dermatophytes, 159–165
  characterization of, *H capsulatum* mycelial forms, 81
  comparison of, in fungi, 106
  and crossreactions, 347
  dermatophytes
    purification of, 160–161
    repeat unit structure of, 163
  in *H capsulatum* yeast form cell wall, 78–79
  method of preparation, 106
  methods to separate from glucan, *A fumigatus*, 133
  methylation analysis of, 106, 107
  occurrence of, among pathogenic fungal genera, 346
  in *P brasiliensis* cell wall, 105–106
  role of, 31
  structural features of, 346–347
Galactose
  as constituent of mannans, 11
  as crossreactive determinant, 11
  in mycelial form cell wall of *P brasiliensis*, 104
Galactoxylomannan, *C neoformans*, structural features of, 260–261
Gammopathy, in paracoccidioidomycosis patients, 110
Gingivitis, involvement of actinomycetes in, 311
α-1,3-Glucan
  B dermatitidis, in cell walls of mycelial form, 43
  in *H capsulatum* yeast form cell wall, 78
  role of, in *P brasiliensis* cell wall, 105
  synthesis of, in dimorphic systemic fungi, 13–14
  synthesis of, in *P brasiliensis* cell wall, 103–104
β-1,3-Glucan
  in *H capsulatum* yeast form cell wall, 78

β-1,3-Glucan [*cont.*]
  possible presence of, in *A fumigatus*, 137
  synthesis of, in dimorphic systemic fungi, 12–14
  synthesis of, in *P brasiliensis* cell wall, 104
Glucan. *See also* α-1,3-Glucan; β-1,3-Glucan
  association of, with chitin in mycelial forms, 11–12
  B dermatitidis, exposure of, to glycosidases, 42–43
  C albicans
    functions of, 219
    role of, in cell wall, 194
    role of, in infection, 218–219
  in dermatophytes, 14, 165–166
  purification of, 160–161
    extraction method, for preparation of, 12
  microfibrils, 12
  presence in cell walls of dimorphic systemic fungi, 13–14
  role of, in fungal cell wall, 11–14
  S schenckii cell wall, 116
β-1,3-Glucanases, as mural enzymes, 27
Glucan-chitin, in *A fumigatus* cell wall, 131
Glucuronic acid
  as determinant of glucuronoxylomannan serotypes A,C, 262
  as structural component of *M rouxii* cell wall, 178–179
Glucuronoxylomannan, *C neoformans*
  composition of, 258
  heterogeneity among serotypes of, 262
  as major capsular polysaccharide, 256
  repeat unit structures, serotypes, 258–259
  structure of, 257–261
Glycans
  fluctuation of, in yeast, mycelial *H capsulatum* cell wall, 81
  in *H capsulatum* yeast form cell wall, 78–79
Glycoenzymes (mural enzymes), 27, 29
Glycolipid, nocardiae, 296
Glycopeptide 1
  F rectivirugula, 350
  F rectivirugula antigen, 324–326
  purification of, 325
Glycopeptides, in dermatophytes, 159–165
Glycoprotein, *C laurentii*, 261
Glycoprotein a, *F rectivirugula* antigen, 323–324
Glycosidic components, *S schenckii* cell wall, 116
Granulocytes, killing by, in paracoccidioidomycosis, 110–111
Granuloma
  association of, with hyperreactivity, 401
  and *B dermatitidis* infection, 41
  presence of, in FLD, 331
  role of glucan in formation of, 218–219
Granulomatous disease, association of *H capsulatum* with, 388–389
Granulomatous uveitis, association of, with *H capsulatum*, 388

H

H-and M-antigens
  and crossreactivity with *B dermatitidis* antiserum, 85
  and synthesis of *H capsulatum* cell wall, 85
H-factor, associated with active histoplasmosis, 82–83
H-protein
  histoplasmin antigen, 349
  possible mural enzyme of *H capsulatum*, 29
Hansenula capsulata, O-phosphonomannan in, 205
Hansenula holstii, pentasaccharide monophosphate of, 207
Hansenula wingei, phosphodiester-linked mannans in, 10
Heiner's C, galactomannan, 85
Hemagglutination, use of, to detect antibodies in aspergillosis, 146–149
Hen egg white lysozyme, digestion of chitin by, 20
Heteroglycans
  capsular and cell wall of *Tremella* species, 256–264
  Cryptococcus capsular and cell wall, 256–264
HEX antigen, membrane-bound of *Candida*, 226
  of *Cryptococcus*, 267–268
Histamine-induced suppressor factor, 379
H-2 histamine receptor antagonists, and *C albicans* infection, 233
Histoplasma capsulatum, 77–102
  association of, with granulomatous disease, 387–388
  association of, with granulomatous uveitis, 388
  cell wall barrier to macrophages, 78
  chemotypes of cell walls of, 79–80
  distribution of, 77
  and effectiveness of PMNs, 338
  enzyme immunoassay of, 356
  and galactomannan crossreactions, 347
  glucans in cell walls of, 13
  humoral responses to, 85–89
  immune response to, cellular interactions, 388
  immunodeficits associated with, 387–388
  immunomodulation of, 93–97
  interaction of, with macrophages, 91–93
  monoclonal antibodies to, 360–361
  murine infection by, 95–96
  as pathogen of reticuloendothelial system, 91
  resistance of, to lysis, 78
  yeast form
    and crossreactions with other mycoses, 88
    and false-positive reactions, 88
Histoplasmomas, lesions of histoplasmosis, 91
Histoplasmin, 82–85
  antigenic factors H and M, immunoelectrophoretic patterns, 84
  characterization of, 97–98
  crossreaction of, with coccidioidin, 91
  crossreaction with blastomycin, 91
  H- and M-antigens, purification of, 83–85
  mycelial, deficient in M-factor, positive delayed cutaneous response, 87
  skin tests with, 87
    validity of, 98

Histoplasmosis
   acute pulmonary
      and suppression of blastogenic response, 93
      immune responses in, 98
   age-related decline in immunocompetence in, 90
   anergy of infection in, 381
   antigen-specific anergy in, 93
   areas of high prevalence, 90
   characteristics of, 77–78
   chronic pulmonary, characteristics of, 94
   course of delayed cutaneous hypersensitivity in, 93–94
   detection of circulating antibodies, 85–86
   as disease of reticuloendothelial system, 80
   disseminated, subnormal T-suppressor activity in, 94
   H- and M-factors in, 82–83
   immunosuppression in, 98
   murine, suppressor cell activity in, 96
   murine model of immunosuppression for, 98
   presumed ocular syndrome, 95
   as pulmonary disease, 77, 80
   treated, potential for endogenous reactivation, 94–95
HLA. See Human leukocyte antigen
Hodgkin's lymphoma
   association of, with histoplasmosis, 387
   as risk factor for C neoformans infection, 390
Hot alkali extraction, for mannan preparation, 199
Human leukocyte antigen (HLA)
   relation of, to coccidioidomycosis, 387
   relation of, to paracoccidioidomycosis, 388
   relation of, to POHS, 95
Human leukocyte lysozyme, digestion of chitin by, 20
Human mycoses, immunodeficits associated with, 385–392
Humoral immunity
   A fumigatus, 144–149
   genetic control of, mice, 265
   histoplasmosis, stimulated by H and M proteins, 83
   nocardiosis, 298
   polyclonal B-cell activation, 362
   protective role for, 363–364
   review of elements of, 337–372
   role of, in infection, 362–365
Humoral response
   B dermatitidis, 46–47
   to C albicans, 226–236
   H capsulatum, 85–89
   mycoses, mechanisms regulating, 340–344
   T-dependent responses, 341–343
      role of B-cells in, 342
      role of lymphokines in, 342
      role of mannans in, 344–346
      role of T-helper cells in, 342
   T-independent, 343
   T-suppression, 343–344
Hyper-IgE-recurrent infection syndrome, C albicans, 234
Hyperkeratosis, in chronic mucocutaneous candidiasis, 231
Hyperreactivity, and granuloma formation, 401
Hypersensitivity granuloma, 380
Hypersensitivity pneumonitis, associated with actinomycetes, 321
Hypha, structure of, at growing tip, N crassa, 16

I

IgG. See Immunoglobulin G
IgM. See Immunoglobulin M
Immune complexes
   circulating, in FLD, 331
   evidence for, in aspergillosis, 146
   modulation of suppression by, 378
   soluble, method for dissociating, 357
Immune complex formation, in mycotic infection, 362
Immune responses, to C albicans, 226–236
Immunity, adoptive transfer of, 384
Immunity, cell-mediated
   antigenic stimulants of, 392–395
   delayed hypersensitivity, 380–381
   generation of T-suppressor cells, substances, 376–379
   lymphokines, 375–376
   protective functions of, 384–392
   review of, 373–399
   T-cell–B-cell macrophage collaboration, 373–374
   T-cell–T-cell interactions, 376
   T-cell effector mechanisms, 376
Immunity, humoral, review of elements of, 337–372
Immunity, innate, review of elements of, 337–341
Immunoassays, C albicans, 212–218. See also specific names
Immunochemistry, C albicans mannan, 196–218
Immunocompromised patients, detection of antibodies, A fumigatus, 148–149
Immunodeficiency, and occurrence of candidiasis, 231–233
Immunodiagnosis. See individual diseases
Immunodiffusion
   for diagnosis of candidiasis, 213
   use of, to detect antibodies in aspergillosis, 146–149
Immunodiffusion studies, Nocardia and Mycobacterium species, 287–289
Immunodiffusion test
   B dermatitidis, 46–47
   and complement-fixing antibodies, C immitis, 61–62
   to determine H- and M-factor activity, 83
   S schenckii, 123–124
Immunoenzyme analysis, A fumigatus protein antigens, 138–139
Immunogen, peptidoglucomannan, C albicans, 216
Immunoglobulin E
   antigen-specific, in disseminated coccidioidomycosis, 62–63
   pathogenic nature of, aspergillosis, coccidioidomycosis, 364

Immunoglobulin E [*cont.*]
  response evoked by peptidorhamnomannan, *S schenckii*, 120
  in response to *A fumigatus*, 144–145
Immunoglobulin G
  accounts for heat-stable opsonic activity, 264
  antibodies, as major opsonins of cryptococci, 269
  in dermatophytosis, 165
  mediated suppression, in dermatophytosis, 169
  pathogenic nature of, in aspergillosis, 364
  in response to *A fumigatus*, 145
  in response to mycelial extract, *A fumigatus*, 141
Immunoglobulin M
  in dermatophytosis, 165
  and early diagnosis, *C neoformans*, 263
  evidence of humoral response to *C immitis*, 72
Immunologic memory, and lethal systemic challenge with *C albicans*, 228
Immunology, *C albicans* mannan, 196–218
Immunopathology, *A fumigatus*, 144–149
Immunoperoxidase staining, and presence of *Candida* antigens, 216
Immunoprecipitin arc, in aspergillosis, 135
Immunoprecipitin fingerprints, 351
Immunostimulants, nocardial, 300–303
Immunosuppression, in paracoccidioidomycosis, 109. See also Immunodeficiency
Immunotherapy, paracoccidioidomycosis, 111
Indirect enzyme immunoassay, general design of, 352
Indirect fluorescent antibody, to detect antibodies in aspergillosis, 146–147
Infection, mycotic
  immune complex formation in, 362–363
  polyclonal B-cell activation in, 362
  role of humoral immunity in, 362–365. See also specific names of organisms, diseases
Innate immunity, review of elements of, 337–341
Interleukin 1 and 2, role of, in T-dependent humoral response, 342
Interleukin 2 (T-cell growth factor), 274, 375–376
In vitro cellular responses, *P brasiliensis*, 109–111
In vitro tests, value of, in coccidioidomycosis, 69. See also other diseases
Isoelectric focusing, and actinomycete antigens, 326–328

K

Keratinase I, II, possible mural enzyme of *T mentagrophytes*, 29
Keratinase
  in dermatophytes, 166
  production of, by dermatophytes, 350–351
Kerions, 157
Kloeckera brevis, O-phosphonomannan in, 205

L

Latex agglutination test
  to detect heat-stable antigen, *C immitis*, 62
  for diagnosis of candidiasis, 213
  S schenckii, 123

Lesions
  of chronic mucocutaneous candidiasis, 234–235
  pulmonary nocardiosis, 281
  sporotrichosis, 115
Leukocyte inhibitory factor, 375
  correlation of, with cutaneous reaction, *P brasiliensis*, 110
Leukocyte migration inhibition, in FLD, 331
Levan, induction of, in oral actinomycetes, 311, 312
L-forms, nocardiae, 298–300
β-Linkages, in yeast mannan, 200, 202
Lipid content, cell wall, critical in purification, 24
Lung
  antibodies against, in Farmer's lung, 331–332
  involvement of, in histoplasmosis, 77–78
  as one site of *C neoformans* infection, 251
  as site of *P brasiliensis* infection, 103
  as target organ in murine blastomycosis, 49
  as target organ of nocardiae, 281
Lung parenchyma, invaded by *A fumigatus*, 130
Lymphocytes. See also Blastogenesis; E-rosette
  activity of, in chronic mucocutaneous candidiasis, 233
  blastogenesis of, in *C neoformans* infection, 267–268
  effect of cyclophosphamide on, in candidiasis, 228–229
  effectiveness of, against *C neoformans*, 270–271
  and estimation of *S schenckii* antibody response, 125–126
  and lymphokine production, 375–376
  peripheral, response of, to Mangion purified polysaccharide, 210
  response of, in blastomycosis patients, 48
  T-helper: T-suppressor in cutaneous anergy, dermatophytosis, 171
Lymphokines, 375–376. See also Delayed hypersensitivity
  and activation of C3b receptors, macrophages, 265
  chemotactic factors, 375
  interleukin 2, 375
  leukocyte inhibitory factor, 375
  lymphotoxins, 375
  macrophage fusion factor, 375
  macrophage migration inhibitory factor, 375
  role of, in T-dependent humoral response, 342
Lymphotoxin, as suppressor, 379
Lymphotoxins, 375
Lysate antigens, development of, *C neoformans*, 267

M

Macrophage fusion factor, 375
Macrophage migration inhibitory factor, 375
Macrophages
  alveolar, parasitism of, by *H capsulatum*, 97
  and *A fumigatus* hyphae, 150
  enzymes produced by, 92
  interaction of, with nocardiae, 296–297
  interaction of *H capsulatum* with, 91–93
  phagocytosing, physiology of, 92–93

possible effect of chitin on killing by, 20
role of, in T-dependent humoral response, 342
survival of *H capsulatum* yeast forms in, 80
T-cell–B-cell collaboration, 373–374
Mangion purified polysaccharide, *C albicans*, 210
α-1,3-Mannan backbone, in
 glucuronoxylomannan, *C neoformans*, 257, 259
Mannan
 antigenic, need for criteria for purity, 403
 antigenic structure of, 198–209
 capsular, 9–10
 capsular, serotypes of, 9–10
 C albicans
  acetolysis of, 202–204
  cell wall, 193–194
  immunochemistry and immunology, 196–218
  immunomodulator, 209–212
  methylation-fragmentation of, 200–202
  preparation of, 199–200
  serotypes, fractionation of, 208
  serotypes in, 197–198
 and complement activation, 211–212
 as component of fungal cell wall, 5, 6–11
 content of, in cell wall, 30
 elevated in candidiasis patients' serum, 218
 extracted, properties of, 199
 general structure of, 196–197
 phosphodiester-linked, structural role of, 10
 role of, in serotype specificity, *C albicans*, 345
 S cerevisiae, Alcian blue treatment of, 208
 seroactivity of, according to method of preparation, 8–9
 serum, detection of, for diagnosis of candidiasis, 215
 structure of, in *H capsulatum*, 80–82
 subfractions, *C albicans*, chemical composition of, 207
 surface location of, in *C albicans*, 194
 yeast, structure of, 196
Mannanemia, 348–349
 C albicans, 212–218
 C albicans serotypes, monoclonal antibodies to detect, 360
 and candidiasis diagnosis, 216–218
Mannans
 circulation of, in chronic mucocutaneous candidiasis, 211
 dermatophyte, structural analyses of, 161–163
 galactose as constituent of, 11
 mural, model structure of, 7–9
 as polysaccharide antigens, 344–346
 possible T-cell dependence of, 345
 as principal surface antigens, 31
 purification of, 11
 regulation of synthesis of antibodies against, 343–344
 saprophytic yeast, immunodominant side chains in, 9
 secreted, 10
  galactoxylomannan of *C neoformans*, 10
  peptidophosphogalactomannan of *P charlesii*, 10

M-antigen
 detection of, with enzyme immunoassay, 89–90
 histoplasmin, monoclonal antibodies to, 361
 induction of monoclonal antibodies to, 85
Meningitis
 C neoformans diagnosis of, 252
 as result of *C neoformans* infection, 251
Methylation analysis, galactomannan, 106, 107
3-*O*-Methylated heteromannans, in *C immitis* cell wall, 59
M-factor
 H capsulatum, enzyme-linked immunoelectrotransfer blot assay, 86
 mycelial antigen, histoplasmosis, 82–83
 as mural enzyme, 27
Mice, experimental blastomycosis in, 48–50
Microaerophilic actinomycetes, 311–319. *See also* Actinomycetes, microaerophilic
 contribution of, to periodontal disease, 405
Microsporum audouinii, and ringworm, 157
Migration inhibitory factor (MIF)
 B-ASWS stimulation of, 44
 C immitis, 66–67
 in CMC, 231–233
Mitogen, B-lymphocyte, nocardiae cell wall, 301–303
Mixed-lymphocyte reaction, in coccidioidomycosis, 65
Molds, presence of chitin in, 16–17
Monoclonal antibodies, 358–361
 against mycotic, actinomycotic agents, 359
 to *B dermatitidis* A-factor, 47
 induction of, to M-antigen, 85
Monocytes
 myeloperoxidase-deficient, and *C albicans* infection, 236–239
 T-cell activation of, in histoplasmosis, 98. *See also* Delayed hypersensitivity
M protein
 histoplasmin antigen, 349
 possible mural enzyme of *H capsulatum*, 29
Mucoran, in zygomycetes cell wall, 178–179
Mucoric acid, extracted from *M rouxii*, 178
Mucormycosis, clinical and laboratory parameters of, 180–181
Mucor rouxii
 cell wall of, 177–178
 hyphal wall composition of, 182
 synthesis of chitin microfibrils in, 15–16
Mural enzymes. *See* Cell wall, enzymes
Muramic acid, in nocardiae cell wall, 285
Muramyl dipeptide, and relation to adjuvant activity, nocardia cell wall, 300
Mycelial extracts, *A fumigatus*, 139–142
 localization of antigenic activity in, 140–141
Mycetomas, nocardial, nature of, 299–300
Mycobacteria
 arabinogalactomannan in cell wall of, 284
 priming with, enhances blastogenesis of splenocytes, *C albicans*, 230
Mycobacterium species, polysaccharides in cell wall of, 283
Mycobacterium tuberculosis, reference precipitins, 290

Mycoses, primary systemic, anergy of infection, 381–383
Mycotic infection, athymic state and host response to, 386
Myeloperoxidase deficiency, and defective killing of C albicans, 238

N

Nannizzia, sexual ascogenous form of dermatophytes, 157
Natural killer cells, and protection against mycotic infection, 340
Neurospora crassa
  location of chitin in cell walls of, 16–17
  septum from 5-day culture of, 17
  X-ray diffraction patterns of chitin in, 15
Neutral proteinases, in delayed hypersensitivity, 380–381
Neutrophil cidal mechanisms, and zygomycetes, 185
Neutrophils, myeloperoxidase-deficient, and C albicans infection, 236–239
Nezelof's syndrome, association of Candida infection with, 389
Nigeran
  glucan in A niger cell wall, 13
  present in cell wall of A awamori, 13
Nocardia-active polypeptides, 291–293, 405
Nocardia asteroides, 281
  cell wall composition of, 286
  enzyme immunoassay of, 356
  monoclonal antibodies to, 361
  transfer of protective immunity against, 405
Nocardia brasiliensis, 281
Nocardia caviae, 281
Nocardiae, 281–309
  antigenic complexes in, 285–291
  arabinogalactan in cell wall of, 284
  arabinomannan in cell wall of, 284
  delayed cutaneous hypersensitivity, 290–291
  infection by, 281
  interaction of, with macrophages, 296–297
  L-forms, 298–300
  mural covalent skeleton, 282–283
Nocardia extracts. See Nocardia-active polypeptides; Ribosomal proteins; Superoxide dismutase
Nocardial extracts, fractionation of, 291–296. See also Glycolipid
Nocardial immunostimulants, 300–303
Nocardial water-soluble mitogen, 302
Nocardia water-soluble mitogen, polyclonal activation by, 362
Nocardiosis
  diagnosis of, 282
  experimental murine, 297–300
  humoral response is detrimental, 298
  predisposing factors for, 282
  pulmonary, unit lesion of, 281
Nocardomycolic acids, nocardiae cell wall, 283
Nu/nu mice, and infection with Candida, 228

O

Ocular histoplasmosis syndrome, possible association with H capsulatum, 402
Oerskovia xanthineolytica, exo-α-mannan hydrolase in, 209
Oligosaccharides, isolated from C albicans serotypes A and B mannan, 203–204
Opsonins
  C neoformans, 264–265
  protective role of, in infection, 363
Oxidative pathway
  and neutrophil fungicidal properties, 338–339
  provides optimal resistance to C albicans, 238

P

Paracoccidioides brasiliensis, 103–113
  cell-mediated responses to, 108–111
  cell wall of, 103–106
  correlation of cutaneous reaction, LIF, 110
  as dimorphic fungus, 103
  E-factor antigen in, 106–107
  galactomannan in cell wall of, 105–106
  glucans in cell walls of, 13–14
  immunodeficits associated with, 388
  infection by, 103
  mycelial
    content of cell wall of, 104
    protein in cell wall, 104
  in vitro cellular responses to, 109–111
  and PMN effectiveness, 338
  role of α-1,3-glucan in cell wall of, 105
  secreted antigens of, 106–108
  yeast, structure of cell wall of, 105
Paracoccidioidin
  blastogenic indices to, 109–110
  P brasiliensis antigen, 108
Paracoccidioidomycosis
  anergy of infection in, 381
  characteristics of, 103
  E-rosettes in, 110
  gammopathy in patients with, 110
  immunosuppression in, 109
  immunotherapy for, 11
  killing by granulocytes in, 110–111
  mucocutaneous-lymphangitic involvement in, 103
  organ involvement in, as function of age, 104
  precipitins and complement-fixing antibodies, 107
  pulmonary involvement in, 103
  reticuloendothelial involvement in, 103
Pathogenic antibodies, 364–365
Penicillium charlesii, secreted mannans of, 10
Peptido-L-fuco-D-mannan
  structural features, composition of, 183
  in zygomycete cell wall, 179
Peptidofucomannans, as major antigen in zygomycetes, 182
Peptidoglucomannan, C albicans, as effective immunogen, 216

Peptidoglycan
  A viscosus, activator of alternative complement pathway, 312
  cell wall nocardiae, locus of mitogenic activity, 301
  in nocardiae cell wall, 285
Peptidomannans
  in fungal cell wall, 6–11
  as principal cell wall glycan producing antigenic stimulus, 10
  probes for, 6–7
  role of, in antigenic specificity, 345
  structure of, 10–11
Peptidophosphogalactomannan, of *Exophiala (Cladosporium) werneckii*, 163
Peptidorhamnomannan
  complexes, cell wall, *S schenckii*, 116–120
  precipitin and IgE responses evoked by, 120
  S schenckii cell wall, fractionation of, 118–120
Peptido-L-rhamno-D-mannan
  in cell wall, *S schenckii*, 116–120
  repeat unit structure, *S schenckii*, 117
Periodontal disease
  actinomycetes as agents of, 311
  severity of correlates with degree of blastogenesis, 317
Phagocytes. *See also* Polymorphonuclear neutrophils
  chemotaxis and adherence of, to cryptococci, opsonin mediated, 264
  interactions, *B dermatitidis*, 47–48
Phagocytosis, affected by capsule size of *C neoformans*, 255–256
O-Phosphonomannan, *C albicans*, 204–208
O-Phosphonoglucomannan, potential antigenicity of, 207
Phytohemagglutinin, response to, as index of immunosuppression, coccidioidomycosis, 65
Pichia polymorpha, glucan-mannan complexes in cell walls of, 13
Plasminogen activator, proteinase, 380
PMN. *See* Polymorphonuclear neutrophils
Pneumonia, from *C immitis*, 53
POHS. *See* Presumed ocular histoplasmosis syndrome
Poly-β-1,4-N-Acetylglucosamine, as component of chitin, 14
Polyclonal B-cell activation, 362
Polyclonal gammopathy, in coccidioidomycosis patients, 62
Polymorphonuclear neutrophils (PMN). *See also* Innate immunity
  ability of, to kill *P brasiliensis* yeast, 110
  and *C neoformans* killing, 271
  cidal activity of, against *B dermatitidis* (yeast), 47
  exudate, and *B dermatitidis* infection, 41
  fungicidal properties of, 338–340
  and in vitro killing of *S schenckii*, 124
  possible effect of chitin on killing by, 20
Polyoxin D, as inhibitor of chitin synthesis in yeast, 18
Polysaccharide antigen, *A fumigatus*, C-substance, 137
Polysaccharides, antigenic
  A fumigatus, 132
  galactomannans, 346–348
Polysaccharides, microbial, as T-independent antigens, 343
Polysaccharides, nocardial, function of, 294–295
Polysaccharides, nocardiae cell wall, 283–285
Polyuronides, in *M rouxii* cell walls, 179
Precipitin analysis, for detection of antibodies to *Candida*, 213
Precipitins
  A fumigatus, resolution of, 139, 146–149
  in acute histoplasmosis, 85–86
  detectable in active histoplasmosis, 83
  mycobacterial reference, 290
  response evoked by peptidorhamnomannan, *S schenckii*, 120
  sporotrichosis, 123
  tests. *See also* other organisms, diseases
    C albicans mannan, 198
    histoplasmosis, 98
    thermophilic actinomycetes, 322–329
    weak in allergic aspergillosis, 138
Presumed ocular histoplasmosis syndrome (POHS), 95, 388
  and blastogenic response, *H capsulatum*, 95
Primary systemic dimorphic fungi, 41–128
Protein antigens, 349–352
  C albicans, 219–216
  C neoformans, 265–268
Protein
  in *C albicans* cell wall, 193
  as component of cell wall, 30
  presence of, in cell wall, 26–29
  presence of, in *P brasiliensis* mycelial cell wall, 104
Proteinase
  acidic carboxyl, *C albicans* antigen, 220–225
  alkaline, zygomycetes, *R oryzae*, 351
  chymotrypsinlike, in *A fumigatus*, 142–143
  dermatophytic, 166–167
  enzyme 1, thermophilic actinomycetes, *F rectivirugula*, 351
  as mural enzyme, 27
  secreted, 350–351
  of thermophilic actinomycetes, 329–330
  treatment, to disrupt cell walls, 23–24
Proteins
  cytoplasmic, *C albicans*, 2-D PAGE of, 223
  major cytoplasmic, *C albicans*, 225–226
Pulmonary lesions, nature of, in FLD, 331
Pulmonary nodules, residual fibrocaseous, in histoplasmosis, 77

Q

Quantitative precipitin reactions, for study of yeast mannan, 197–198

R

Radioimmunoassay
  to detect antibodies in aspergillosis, 139, 141, 145

to detect antibodies in histoplasmosis, 88–89
to quantitate antibody titer, *C immitis*, 62
Repeat unit, mannan, determination of, with methylation-fragmentation, 200
Reticuloendothelial system
 histoplasmosis as disease of, 80, 91
 involvement of, in *Paracoccidioides* infection, 103
Rheumatoid factor
 as component of response in FLD, 331
 interferes with complement fixation in histoplasmosis, 88
Rhinocerebral zygomycosis, 183–184
Rhizopus (Zygomycetes), as cause of rhinocerebral infection, 177
Rhizopus inhibitory factor, 184–185
Rhizopus oryzae
 alkaline proteinase produced by, 351
 and PMN effectiveness, 339
Rhodotorula glutinis, chitin content in, 19
Ribosomal proteins, nocardiae, 295–296
Ribosomal vaccine, *H capsulatum*, mice, 96
Ringworm, 157
Rosettes, lymphocyte, specific for *S schenckii*, 126

S

Saccharomyces cerevisiae
 glucan in cell walls of, 12
 location of chitin in cell wall of, 17–18
 mannan structure in cell wall, 7–8
 repeat unit structure of mannoprotein, 8
 ultrastructural arrangement of glucans in, 12
Schistosomiasis, and histoplasmosis, 387
Schizophyllum commune, glucan-chitin linkage in, 11
Schizosaccharomyces pombe, surface location of galactomannan in, 6–7
Secreted proteinases, 350–351
Septa, *N crassa*, location of chitin in, 17
Serological heterogeneity, *C neoformans* capsule, 261–264
Serotypes
 C albicans, hypothetical models of, 202
 C neoformans, 261–264
 capsular mannan, 9–10
 determined by peptidomannans, 6
 N asteroides, 289–290
Serotype specificity, *C albicans* mannan, 197–198, 204, 345
Serratia marcescens, chitinase from, 15
Serum fungistasis, 338
Silver proteinate-depositing electron-dense granules, *C albicans* cell wall, 195
Skin cancer, and chronic dermatophytosis, 168
Skin test antigens, *C neoformans*, 266
Skin tests
 C immitis, CDN and SPH antigens used for, 63
 with histoplasmin, 87
Slime polysaccharides, in *A viscosus*, 313–314
Soluble immune complexes
 in coccidioidomycosis, 67–68
 in histoplasmosis, 94
Soluble immune response suppressor substance, 379
Sonication, for fungal cell wall disruption, 23, 56

Species-specificity, role of galactomannans in, 402
Species-specific secreted factors, protein antigens, 349–352. *See also* individual diseases, organisms
Spherule, *C immitis*, structure, 56–57
Spherule vaccine
 coccidioidomycosis, 73
 in experimental murine coccidioidomycosis, 70–71
 in humans, 71
Spherulin
 C immitis antigen, 60–61
 effects of storage temperature and dialysis on activity of, 60
 lack of crossreactivity of, in skin tests, 64–65
 sensitivity of, as indicator, delayed cutaneous hypersensitivity, 68
Splendore-Hoeppli phenomenon, 124
Splenic enlargement, in murine histoplasmosis, 95
Sporothrix schenckii, 115–128
 and asteroid bodies, 124–125
 binds Con A, 6
 cell wall of, 116–120
 coexists with *C stenoceras*, 120
 cutaneous and blastogenic responses to, 122
 delayed cutaneous hypersensitivity to, 120–122
 as dimorphic fungus, 115
 estimation of antibody response to, cell level, 125–126
 experimental infections with, 125–126
 and Fehling-precipitable galactomannan, 347
 immunodeficits associated with, 388–389
 infection by, 115
 interaction of, with phagocytes, 124–125
 latex agglutination tests, 123
 and PMN effectiveness, 338–339
 positive complement-fixation tests, 123
 skin test surveys for, 122
Sporotrichin
 delayed cutaneous hypersensitivity to, 120
 and skin test surveys, 122
Sporotrichosis
 humoral response to, 115, 123–124
 immunodiagnosis of, 123–124
 sera, human, precipitin test, 120
 sex-related difference in, 122
Streptomyces griseus, chitinase from, 15
Streptomyces tatsumaensis, glucan digestion with chitinase from, 11–12
Supernatants, blastomycin, *B dermatitidis*, 41–42
Superoxide dismutase, secreted by *N asteroides*, 293, 405
Suppression
 antigen-specific, 377–378
 anti-idiotype antibodies, 379
 functions, 376–379
 modulation of, by immune complexes, 378
Suppressor cell activity, in murine histoplasmosis, 96
Suppressor factor, from T-lymphocytes of *C neoformans*-infected mice, 272

T

T–B-cell interaction, 342
TCA-soluble antigens, to discriminate FLD from asymptomatic exposure, 326

T-cell deficit, in coccidioidomycosis, 69, 73
T-cell effector mechanisms, 376
 paracoccidioidomycosis, 108–109
T-cell-mediated immunity, stimulated by H and M proteins, histoplasmosis, 83
T-cell regulation, possible failure of, in coccidioidomycosis, 62–63
T-cell replacing factor, 342
T-cell responses to coccidioidin, and transfer factor, 70
T-cell–T-cell interactions, 376
T-cells
 and *N asteroides*, 298
 role of, in humoral response to mycoses, 340–344
 role of, in macrophage ability to kill *H capsulatum* (yeast), 91–92
T-contrasuppressor cells, 379
T-dependent B-cell mitogen, *A viscosus*,
 further characterization of, 406
 pathogenic nature of, 365
T-dependent responses, 341–343
Teichozyme Y, 209
T-helper cells, role of, in T-dependent response, 342
Thermophilic actinomycetes, 321–336. *See also* Actinomycetes, thermophilic
Thymic defects, and chronic mucocutaneous candidiasis, 234
Thymic immunity, role of, in mycotic infection, 402
Thymic involution, in murine histoplasmosis, 95
Thymic responses, to *C albicans*, 226–236
T-independent humoral response, 343
Tinea capitis, 157
T-lymphocyte function
 and host defense against *C immitis*, 73
 human, and *C neoformans*, 273–274
 mice, and *C neoformans*, 271–273
T-lymphocytes
 and monocytes for response to *C albicans* extract, 211
 suppressor, in histoplasmal meningitis, 94
Tolerogen, major viscous polysaccharide, *C neoformans*, 253–255
Torulopsis (Candida) glabrata, 191
TP antigen, coccidioidin, 349
Transfer factor
 and conversion of anergy to hypersensitivity, 73
 as immunotherapy for coccidioidomycosis, 69–70
 and immunotherapy of chronic mucocutaneous candidiasis, 235–236
 use of, in paracoccidioidomycosis, 111
Tremella, heteroglycans of capsule and cell wall, 256–264
Trichophytin
 antigenic potency of, 170–171
 glycopeptides, *T mentagrophytes*, 159–160
 preparation of, 162
Trichophyton mentagrophytes
 cell wall composition of, 158
 galactosylated mannans of, 10
Trichophyton rubrum
 glycogenlike glucans in, 14
 tolerogenic potential of, 170
Trichophyton schoenleinii, and favus, 157
Trichophyton tonsurans, and ringworm, 157
Trichophyton violaceum, and ringworm, 157
Trichophyton species, IgE-mediated suppression and, 169
Trypsin, for digestion of cell walls, 23
Ts1 cells. *See* T-suppressor cells
TsF factor. *See* T-suppressor cells
T-suppression, of humoral responses, 343–344
T-suppressor activity, subnormal in disseminated histoplasmosis, 94
T-suppressor cells, 377
Tube precipitin, primary immune response to *C immitis*, 61
Two-dimensional immunoelectrophoresis
 and actinomycete antigens, 326–328
 for detection of antibodies, *Candida*, 213

V

Vaccine, ribosomal, *H capsulatum* in mice, 96
Vaccine, spherule, in experimental murine coccidioidomycosis, 70–71
Valley fever X-ray pattern, in coccidioidomycosis, 64

W

Wheat germ agglutinin, used to identify chitin in *S cerevisiae*, 18
β-1,4-Xylanase, mural enzyme of *C albidus*, 29

Y

Yeastlike fungi, 191–280
Yeasts
 chitin as component of, 17–19
 differing cell wall composition of chemotypes, 19–20
 O-phosphonomannans in, 204–208

Z

Zygomycetes, 177–189
 cell-surface mannans and secreted proteinases in, 403
 cell wall of, 177–183
 distribution of, 177
 immunodeficits associated with, 392
 infection by, 177, 186
 major antigen in, 182
 neutrophil cidal mechanisms in, 185
 proteinase production by, 351
 susceptibility of diabetics to, 184
Zygomycosis
 agents that cause, 177
 alkaline proteinase as presumed virulence factor in, 185
 experimental, 185–187
 host response in, 186–187
 human pulmonary, 182
 rhinocerebral, 183–184
Zymolyase, 209
Zymosan, and reduced in vitro killing of *C albicans* by leukocytes, 230

RAYMOND H. FOGLER LIBRARY
DATE DUE

ARE SUBJECT TO
TWO WEEKS